D1187915

Advances in Clinical Nutrition

Advances in Clinical Nutrition

Proceedings of the 2nd International Symposium
held in Bermuda, 16–20th May 1982

Edited by
Ivan D. A. Johnston

Department of Surgery, University of Newcastle-upon-Tyne, England

MTP PRESS LIMITED · LANCASTER · BOSTON · THE HAGUE
International Medical Publishers

Falcon House
Lancaster, England

British Library Cataloguing in Publication Data

Advances in clinical nutrition
   1. Nutrition—Congresses
   I. Johnston, I.D.A.
   613.2   TX345

   ISBN 0-85200-496-6

Published in the USA by

MTP Press
A division of Kluwer Boston Inc
190 Old Derby Street
Hingham, MA 02043, USA

Library of Congress Cataloging in Publication Data

Main entry under title:

Advances in clinical nutrition

   Includes bibliographies and index
   1. Diet therapy—Congresses.   I. Johnston, Ivan
      David Alexander.
   [DNLM: 1. Nutrition—Congresses. QU 145 A2437 1982]
   RM216.A37 1982   615.8'54   82-20831
   ISBN 0-85200-496-6

Printed in Great Britain
by McCorquodale (Scotland) Ltd.

Contents

	Page
Preface	ix
List of Contributors	xi
The Sir Hans Krebs Memorial Lecture Tissue-specific partitioning of substrates in the circulation D. H. Williamson	xiii
Introduction Current developments in nutritional care I. D. A. Johnston	xxv

SECTION 1 BRANCHED-CHAIN AMINO ACIDS

1. Branched-chain amino acids: metabolic roles and clinical applications D. C. Madsen	3
2. Intracellular and plasma branched-chain amino acid interrelationships P. Fürst, A. Alvestrand, J. Bergström, J. Askanazi, D. Elwyn and J. Kinney	25
3. Branched-chain amino acid supplementation in the injured rat model H. R. Freund, Z. Gimmon and J. E. Fischer	37
4. Review of branched-chain amino acid supplementation in trauma F. B. Cerra	51
5. The importance of study design to the demonstration of efficacy with branched-chain amino acid enriched solutions L. L. Moldawer, M. M. Echenique, B. R. Bistrian, J. L. Duncan, R. F. Martin, E. G. St. Lezin and G. L. Blackburn	65

6. Branched-chain amino acids in surgically stressed patients   77
   J. Takala, J. Klossner, J. Irjala and S. Hannula

7. Selective nutritional support for hepatic regeneration: 83
   experimental and clinical experiences with branched-chain amino
   acids
   H. Joyeux, B. Saint-Aubert, C. Astre, P. C. Andriguetto, P. Vic,
   C. Humeau and Cl. Solassol

8. The effect of branched-chain amino acids on body protein   103
   breakdown and synthesis in patients with chronic liver disease
   J. D. Holdsworth, M. B. Clague, P. D. Wright and I. D. A.
   Johnston

SECTION 2 NUTRITIONAL SUPPORT IN RENAL FAILURE
9. Nutritional therapy for patients with acute renal failure   113
   J. D. Kopple and E. I. Feinstein

10. Can low protein diet retard the progression of chronic renal   123
    failure?
    J. Bergström, M. Ahlberg and A. Alvestrand

11. Experience with prolonged intradialysis hyperalimentation in   131
    catabolic chronic dialysis patients
    D. V. Powers and A. J. Piraino

SECTION 3 SPECIAL CARE ENTERAL FEEDINGS
12. Enteral feeding in liver failure   149
    J. E. Wade, M. Echenique and G. L. Blackburn

13. Nutritional support in the thermally injured patient   163
    D. E. Beesinger, K. Gallagher and S. Manning

14. Breast milk in neonatal care   171
    H. C. Børresen

SECTION 4 CURRENT PERSPECTIVES IN THE USE OF LIPID EMULSION
15. Intravenous fat emulsions: a neonatologist's point of view   179
    D. H. Adamkin

16. Clinical applications of 20% fat emulsions   189
    J. D. Hirsch and D. L. Bradley

17. Changes in binding of bilirubin due to intravenous lipid nutrition 193
    P. F. Whitington and G. J. Burckart

CONTENTS

18. Clinical experience with fat-containing TPN solutions   207
    A. Sitges-Serra, E. Jaurrieta, R. Pallares, L. Lorente and
    A. Sitges-Creus

19. The stability of fat emulsions for intravenous administration   213
    S. S. Davis

20. The stability and comparative clearance of TPN mixtures   241
    with lipid
    G. Hardy, R. Cotter and R. Dawe

21. Problems and prospects in lipid metabolism during parenteral
    nutrition   261
    T. P. Stein

SECTION 5 ENERGY REQUIREMENTS IN STRESSED PATIENTS
22. Alterations in fuel metabolism in stress   275
    P. R. Black and D. W. Wilmore

23. The carbohydrate content of parenteral nutrition   283
    J. M. Kinney

24. Fat utilization in critically ill patients   293
    Y. A. Carpentier

25. Liver malfunction associated with parenteral nutrition   303
    R. K. Ausman, E. J. Quebbeman and C. L. Altmann

SECTION 6 SOME NEW CONCEPTS IN CLINICAL NUTRITION
26. Role of carnitine supplementation in clinical nutrition   315
    P. R. Borum

27. Experience with alternative fuels   325
    R. H. Birkhahn

28. Topical nutrition – a preliminary report   339
    A. W. Goode, I. A. I. Wilson, C. J. C. Kirk and P. J. Scott

29. Nutritional and immunological parameters for identification of 351
    the high risk patient
    R. Dionigi, L. Dominioni, P. Dionigi, S. Nazari and
    D. H. Colombo

SECTION 7 HOME PARENTERAL NUTRITION
30. Intestinal failure and its treatment by home parenteral nutrition 369
    M. Irving

31. Implementation of a home nutritional support programme   383
    J. P. Grant and M. S. Curtas

32. Techniques of longterm venous access   389
    I. W. Goldfarb

33. Home parenteral nutrition   401
    E. Steiger

34. Quality of life for the home-care patient   411
    K. Ladefoged and S. Jarnum

35. Experience with ambulatory total parenteral nutrition in Crohn's 419
    disease
    J. M. Müller, H. W. Keller, J. Schindler and H. Pichlmaier

36. Preliminary report on collaborative study for home parenteral   433
    nutrition patients
    G. L. Blackburn, C. Di Scala, M. M. Miller, C. Champagne,
    M. Lahey, A. Bothe and B. R. Bistrian

SECTION 8 CONCLUSION
37. Future directions   451
    J. M. Kinney

38. Thoughts for the future of clinical nutrition   457
    A. W. Goode

39. Concluding remarks to the symposium on clinical nutrition   465
    G. F. Cahill

    Index   471

Preface

The advent of any new and effective therapy is soon followed by large numbers of publications in which the indications and benefits are explored critically. It is not unexpected, therefore, that within five years of the first Bermuda Symposium on advances in parenteral nutrition that a second Symposium was considered appropriate to review progress and explore new areas of investigation, as well as enlarging the scope of the meeting to include enteral nutrition.

The rate of progress can be judged by the number of subjects which were not discussed at the first Symposium. For example, home parenteral nutrition, computer assisted assessment and prescribing, Studies of body protein synthesis and breakdown and the role of branched-chain amino acids are all new subjects for this Symposium which were not covered at all in the first meeting.

Much progress has also been made to our understanding of the biochemical complications of parenteral nutrition and the problems related to long term access to the circulation. Nutritional care has become safer and more effective.

There is an increasing awareness of the difficulties in making a true nutritional assessment in selecting patients for total parenteral nutrition and more attention has also been focussed on different approaches to enteral support in the management of undernourished patients.

There is also continuing debate on the cost effectiveness of this expensive method of treatment and critics look in vain for evidence of efficacy based on controlled trials in specific groups of patients.

Clinical nutrition is a new and rapidly growing clinical discipline and the opportunity is being seized to examine critically the value of total or partial parenteral nutrition with or without specific enteral regimens in different clinical situations. Clinical investigators in the nutrition field are also adding significantly to our knowledge of body metabolism under conditions of stress and physiological knowledge of the response to injury is being increased by careful clinical studies.

I. D. A. Johnston

List of Contributors

DAVID H. ADAMKIN
Department of Pediatrics, University of Louisville, School of Medicine, c/o Norton Children's Hospital, PO Box 35070, Louisville, Kentucky 40232, USA

ROBERT K. AUSMAN
Clinical Research, Travenol Laboratories Inc, One Baxter Parkway, Deerfield, Illinois 60015, USA

DAVID E. BEESINGER
Department of Surgery, The University of Texas, Health Science Center at Houston, Medical School Main Building 4156, 6431 Fannin, Texas Medical Center, Houston, Texas 77030, USA

RONALD H. BIRKHAHN
Medical College of Ohio at Toledo, Arlington and South Detroit Avenues, Toledo, Ohio 02115, USA

BRUCE BISTRIAN
Harvard Medical School, New England Deaconess Hospital, 194 Pilgrim Road, Boston, Massachusetts 02215, USA

PRESTON R. BLACK
Laboratory for Surgical Metabolism, Brigham and Women's Hospital, 75 Francis Street, Boston, Massachusetts 02215, USA

GEORGE L. BLACKBURN
Harvard Medical School, New England Deaconess Hospital, 194 Pilgrim Road, Boston, Massachusetts, 02215, USA

HANS CHR. BORRESEN
Department of Clinical Chemistry Rikshospitalet, Oslo 1, Norway

PEGGY R. BORUM
Biochemistry Department, Vanderbilt University, Nashville, Tennessee, 37232, USA

GEORGE F. CAHILL
Howard Hughes Medical Institute, 398 Brookline Avenue, Suite Eight, Boston, Massachusetts 02215, USA

YVON CARPENTIER
Department of Surgery, Hospital St. Pierre, 332 rue Haute, B − 1000 Brussels, Belgium

FRANK CERRA
University of Minnesota − Twin Cities, Department of Surgery, Medical School, Phillips − Washington Building, 516 Delaware Street, SE, Minneapolis, Minnesota 55455, USA

STANLEY S. DAVIS
Department of Pharmacy University of Nottingham, University Park, Nottingham NG72RD, UK

R. DIONIGI
Istituto di Patologie Chirurgica, University of Pavia, Policlinico S. Matteo, 27100 Pavia, Italy

HERBERT R. FREUND
Department of Surgery, and Nutritional Support Unit, Hadassah University Hospital and Hebrew University − Hassadah Medical School, Jerusalem, Israel

PETER FÜRST
Institute for Biological Chemistry and Nutrition, University of Hohenheim, Garbenstrasse 30, 7000 Stuttgart, West Germany

I. WILLIAM GOLDFARB
Nutrional and Metabolic Support Service The Western Pennsylvania Hospital, Pittsburgh, Pennsylvania, 15224 USA

ANTHONY W. GOODE
Surgical Unit, London Hospital, London E1, UK

JOHN GRANT
Duke University Medical School, Box 3105,
Durham, North Carolina, 27710, USA

GILBERT HARDY
Clinical Research, Travenol International
Services Inc., Chaussée de la Hulpe 130,
1050 Brussels, Belgium

JOHN D. HIRSCH
Jewish Hospital of St. Louis,
216 South Kingshighway Blvd, St Louis,
Missouri 63110, USA

MILES IRVING
University of Manchester, Hope Hospital,
Eccles Old Road, Salford, M6 8HD, UK

IVAN D. A. JOHNSTON
Department of Surgery, University of
Newcastle upon Tyne, UK

HENRI JOYEAUX
Laboratoire de Nutrition et Cancerologie,
Expérimentale, Centre Lamarque, B P 5054,
Montpelier, France

JOHN KINNEY
College of Physicians and Surgeons of
Columbia University, 630 West 168th Street,
Box 44, New York, 10032, USA

JOEL KOPPLE
Division of Nephrology and Hypertension,
Los Angeles County Harbor, University of
California at Los Angeles Medical Center,
1000 West Carson Street, Torrance,
California 90509 USA

KARIN LADEFOGED
Medical Department A, Rigshospitalet,
2100 Copenhagen, Denmark

DAVID C. MADSEN
Nutrition and Flow Control Division,
Travenol Laboratories Inc., One Baxter
Parkway, Deerfield, Illinois 60015, USA

JOACHIM MÜLLER
Chirurgische Universitäts Poliklinik, Köln,
Joseph Stelzmannstrasse 9, 5000 Köln 41,
West Germany

ANTHONY J. PIRAINO
Dialysis Incorporated, 1230 Burmont Road,
Drexel Hill, Pennsylvania 19026, USA

DONALD POWERS
Dialysis Incorporated, 1230 Burmont Road,
Drexel Hill, Pennsylvania 19026, USA

ANTONIO SITGES-SERRA
Nutrition Unit, Departmento de Cirugia,
Hospital 'Principes de Espana',
L'Hospitalet, Barcelona, Spain

EZRA STEIGER
Cleveland Clinic, 9500 Euclid Avenue,
Cleveland, Ohio 44106, USA

T. PETER STEIN
Surgical Research Laboratory, Graduate
Hospital, University of Pennsylvania,
19th and Lombard Streets, Philadelphia,
Pennsylvania 19146, USA

JUKKA TAKALA
Department of Anesthesiology, Turku
University Central Hospital, SF − 20520
Turku 52, Finland

HUGH N. TUCKER
Nutrition and Flow Control Division,
Travenol Laboratories Inc., One Baxter
Parkway, Deerfield, Illinois 60015, USA

PETER F. WHITINGTON
The Pediatric Research Laboratory,
951 Court Avenue − 545 Dobbs, Memphis,
Tennessee 38163, USA

DERMOT H. WILLIAMSON
Metabolic Research Laboratory, Nuffield
Department of Clinical Medicine, Radcliffe
Infirmary, Oxford, UK

The Sir Hans Krebs Memorial Lecture

Tissue-specific partitioning of substrates in the circulation

D. H. WILLIAMSON

The aim of parenteral nutrition is to provide all the tissues of the body with the optimum mixture of substrates to allow them to carry out their normal functions. Intravenous feeding does not always achieve this because the underlying pathological condition of the patient cannot be corrected or because certain tissues fail to respond to the substrate supplied. It is in the latter case that the study of the factors which regulate substrate utilization by tissues may be of value and it is in this area that the physiological biochemist can make an important contribution.

All tissues of the body receive the substrates (carbohydrates, lipids and amino acids) present in the circulation and a key question is what controls the amount and type of substrate extracted by a particular tissue. The intention of this contribution is to briefly discuss the basic principles involved in the control of substrate utilization by tissues and, in particular, to emphasize the physiological importance of the partitioning of substrates to specific tissues.

REGULATION OF SUBSTRATE UTILIZATION

For the purposes of this contribution it is intended to limit the discussion to those substrates present in the circulation that take part in major energy-yielding or biosynthetic processes. The major factors involved in the regulation of the utilization of a particular substrate by a tissue include the concentration of the substrate in the circulation, the blood flow to the tissue, the permeability of the tissue and its intracellular compartments to the substrate, the presence of the enzyme required for the initiation of the metabolism of the substrate and the kinetic and regulatory properties of the enzymes involved in the further metabolism of the substrate (Table 1). These factors will now be briefly reviewed.

Table 1   Factors affecting substrate utilization by tissues

Substrate availability (concentration) in the blood stream
Blood flow
Substrate permeability (plasma and mitochondrial membranes)
Intracellular concentration of substrate
Presence of necessary enzymes (initiating enzyme)
Regulation of metabolic pathway

Substrate availability

The utilization of most substrates is directly proportional to their concentration in the circulation which in turn represents a balance between production (either from endogenous or dietary sources) and utilization. Changes in concentration are important not only for the utilization of substrate, but they may also act as signals to regulate the production and utilization of other substrates. This interaction between substrates in the circulation may be direct (e.g. ketone bodies inhibit the utilization of glucose by muscle[1]) or indirect, by alteration of hormone concentrations (e.g. the increase in plasma insulin produced by an elevation of blood glucose decreases the release of non-esterified fatty acids (NEFA) from adipose tissue and consequently the production of ketone bodies by the liver[2]).

The total amount of oxidizable substrate is remarkably constant in various physiological states, but the proportions of glucose, NEFA, triacylglycerols, ketone bodies and amino acids may vary considerably. An extreme example is provided by the diurnal variations of glucose and ketone bodies in a child with hepatic glycogen synthetase deficiency[3] (Figure 1). This inborn error of metabolism illustrates the importance of hepatic glycogen in 'buffering' the blood glucose in response to short-term fasting.

Blood flow

The role of blood flow in the control of tissue metabolism is often underestimated. A specific example is the depression of blood flow to the splanchnic area during prolonged exercise and the reciprocal increase to the peripheral tissues which provides more substrate, especially NEFA, for working muscle. More recently it has been demonstrated that changes in blood flow are important in the increase of metabolism and concomitant heat production by brown adipose tissue in response to catecholamines[4]. Redistribution of the cardiac output can play a part in the partitioning of substrates to specific tissues (see below).

Substrate permeability

Substrates enter cells by simple diffusion or by specific transport systems, whose activity may be altered by hormones. Thus the transport of glucose

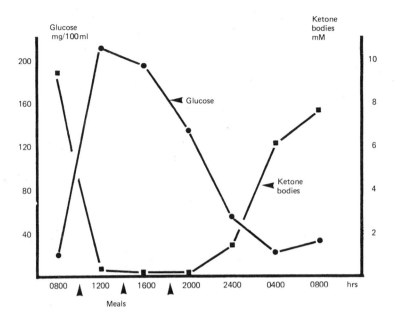

Figure 1 Diurnal blood metabolite levels in a child with a hepatic glycogen synthetase deficiency[3]

into muscle and adipose tissue is stimulated by insulin and the pyruvate transporter on the inner mitochondrial membrane may be activated by glucagon[5]. There appear to be inter-tissue differences in the method of entry of a particular substrate, for example, the rate of diffusion of ketone bodies into skeletal muscle does not limit their utilization[6], whereas in brain there is evidence for a transport system for these substrates[7]. Triacylglycerols (chylomicrons, VLDL) occupy a special position because they are not transported into cells and the initial enzyme, lipoprotein lipase, concerned in their metabolism is located on the capillary endothelium.

Initiation of metabolism
Once the substrate has entered the cell the next potential site of regulation is the concentration (activity) of the first or initiating enzyme which allows the substrate to enter a metabolic pathway. Although the majority of substrates eventually yield acetyl-CoA which enters the tricarboxylic acid cycle for oxidation, each substrate has its own initiating enzyme(s) (Table 2) and these may be located in different cellular compartments (Figure 2). The activity of these initiating enzymes may be regulated by acute changes in the concentration of hormones or effector molecules (e.g. inhibition of muscle hexokinase by glucose-6-phosphate limits further utilization of glucose) or by

longterm changes in the amount of enzyme (e.g. decrease in hepatic gluco-kinase concentration in starvation or insulin-deficiency limits extraction of glucose).

Table 2 Initiating enzymes for substrates in the circulation

Substrate	Initiating enzyme	Product	Intracellular site
A. *Carbohydrate*			
Glucose	Hexokinase, glucokinase	Glucose-6-phosphate	Cytosol
Lactate	Lactate dehydrogenase	Pyruvate	Cytosol
Pyruvate	Pyruvate dehydrogenase	Acetyl-CoA	Mitochondrial inner membrane
	Pyruvate carboxylase	Oxaloacetate	Mitochondrial matrix
Glycerol	Glycerol kinase	Glycerol-3-phosphate	Cytosol
B. *Lipid*			
Triacylglycerols	Lipoprotein lipase	Fatty acids and glycerol	Extracellular
Non-esterified fatty acids	Fatty acyl-CoA synthetase	Fatty acyl CoA	(1) Microsomes (2) Mitochondrial outer membrane
Hydroxybutyrate	Hydroxybutyrate dehydrogenase	Acetoacetate	Mitochondrial inner membrane
Acetoacetate	(1) 3-Oxoacid CoA transferase	Acetoacetyl-CoA	(1) Mitochondrial matrix
	(2) Acetoacetyl-CoA synthetase	Acetoacetyl-CoA	(2) Cytosol
Acetate	Acetyl-CoA synthetase	Acetyl-CoA	(1) Cytosol (2) Mitochondrial matrix
C. *Amino acids*			
Alanine	Alanine amino transferase	Pyruvate	(1) Cytosol (2) Mitochondrial matrix
Glutamine	Glutaminase	Glutamate	Mitochondrial matrix
Branched-chain amino acids (leucine, isoleucine and valine)	Corresponding amino transferases	Corresponding ketoacids	(1) Cytosol (2) Mitochondrial matrix

The initiating enzymes are grouped according to the main metabolic fuels. Apart from lipo-protein lipase they are all intracellular and may be found in different locations within the cell.

Although, if substrate is freely available, the activity of the initiating enzyme will determine the flux of substrate into the pathway, other regulatory sites exist within the metabolic pathway. These other regulatory enzymes are often situated at branch points in the pathway and are low-activity enzymes whose reactions are far from equilibrium. It is not the intention to discuss metabolic regulation at this level and the reader should refer to a standard text[8].

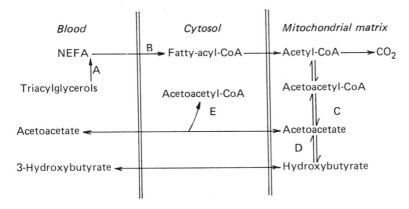

Figure 2   Sites of initiating enzymes for lipid substrates. (A) Lipoprotein lipase; (B) fatty acyl-CoA synthetase; (C) 3-oxoacid-CoA transferase; (D) hydroxybutyrate dehydrogenase; (E) acetoacetyl-CoA synthetase

TISSUE-SPECIFIC PARTITIONING OF SUBSTRATES

The fact that each substrate has its own initiating enzyme and that the concentration of these initiating enzymes varies from tissue to tissue is the basis for the partitioning or direction of a particular substrate to a particular tissue. The concept of tissue-specific direction of substrates is not meant to imply that a blood-borne substrate is solely removed by a single tissue of the body but rather that a particular tissue(s) predominates in its removal. The importance of tissue-specific direction of substrates in terms of whole body substrate supply depends to some extent on the size of the tissue and the blood flow to it. If tissue-specific direction is demonstrated in a particular situation it is important to decide the reasons for the partitioning of the substrate to the tissue. It may be to supply a respiratory fuel or a biosynthetic substrate or even a signal to regulate the metabolism of other substrates.

The inter-tissue differences in amounts of the initiating enzymes may be constitutive and therefore not alter with change in physiological state or they may be adaptive and therefore changeable. A key question is what are the signals that bring about changes in the concentration (activity) of initiating enzymes in particular tissues in response to an alteration of physiological state?

One of the first pieces of evidence for tissue-specific direction of substrate came from studies of lipoprotein lipase activity in adipose tissue in response to alteration in nutritional state[9]. In the well-fed rat the activity of lipoprotein lipase in adipose tissue is high and chylomicrons derived from dietary fat or VLDL produced by the liver are taken up for storage. On starvation the lipoprotein lipase activity falls dramatically in adipose tissue and triacylglycerols available in the blood are now directed to other tissues such as muscle. Other examples concern the partitioning of ketone bodies to brains of developing rats[10] and gluconeogenic amino acids, alanine and serine to liver[11].

PARTITIONING OF SUBSTRATES IN LACTATION

In lactation there is a need to direct substrates to the mammary gland for the synthesis of milk products (Table 3). It is of considerable interest to discover how this is achieved and some of the information is now available.

At the physiological level lactation is characterized by increased cardiac output, hypertrophy of liver, heart, intestine and mammary gland and an increase in dietary intake (300% at peak lactation)[12]. Despite this dramatic hyperphagia the lactating rat does not increase its adipose tissue mass[13]. As might be expected there are profound alterations to the metabolism of the tissues of the lactating rat to allow the partitioning of substrates to the mammary gland[12].

Table 3 Substrate requirements of the lactating mammary gland

Substrate	Product
Glucose	Lipid and lactose
Triacylglycerols	Lipid
NEFA	Lipid
Ketone bodies	Lipid
Amino acids	Milk proteins

Triacylglycerols

The concentration of lipoprotein lipase in adipose tissue of the rat is high during pregnancy, but falls to low levels just before parturition and remains low throughout lactation. Conversely, the activity of lipoprotein lipase is low in mammary gland throughout pregnancy but increases to high levels in this tissue with the onset of lactation (Figure 3)[14]. The net results of these reciprocal changes in lipoprotein lipase activity in adipose tissue and mammary gland is that during lactation a considerable proportion of the available triacylglycerols is diverted to the mammary gland for the formation of milk fat. These changes in the activity of lipoprotein lipase are rapidly reversed on removal of the pups or suppression of prolactin secretion[15].

Glucose

At peak lactation the lactating mammary gland of the rat utilizes approximately 30 mmol of glucose per day which is about equal to the glucose turnover of a non-lactating animal[12]. Part of this increased glucose require-

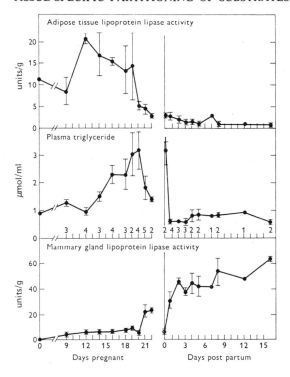

Figure 3 Partitioning of plasma triglyceride during pregnancy and lactation in the rat

ment is met by the increase in dietary intake, but inhibition of glucose utilization in adipose tissue for fat synthesis (lipogenesis) appears to be an additional mechanism of providing more glucose for the gland. Of the glucose taken up by the gland a small proportion is oxidized (10%), about 15% is converted to lactose and some 60% to milk lipids. Measurements of lipogenesis *in vivo* with ^3H_2O illustrate the partitioning of glucose to the gland at the expense of adipose tissue (Table 4)[12]. This is one reason for the absence of an increase in adipose tissue mass during lactation despite the hyperphagic state. As in the case of lipoprotein lipase, the changes in rates of lipogenesis are rapidly reversed on removal of pups or suppression of prolactin secretion with bromocryptine[16].

Table 4 Rates of lipogenesis *in vivo* in adipose tissue and mammary gland

State of rats	Adipose tissue	Mammary gland
Virgin	9.6	7.8
Lactating (10−14 days)	2.2	118

Units are μmol ^3H_2O incorporated into saponified lipid per g fresh tissue per hour. Values are from Reference 19

Ketone bodies

The enzymes concerned in ketone body utilization (3-hydroxybutyrate dehydrogenase, 3-oxoacid-CoA transferase and acetoacetyl-CoA thiolase) are present at high activity in lactating mammary gland of the rat and this tissue is potentially the most important site of ketone body metabolism[17]. Ketone bodies are extracted by the gland[18] and the acetyl-CoA formed could act as a lipogenic precursor. However, starvation results in inhibition of milk fat synthesis[19] and therefore there must be another reason for the partitioning of ketone bodies to mammary gland. As in muscle (heart and skeletal) ketone bodies inhibit glucose utilization and pyruvate oxidation by the gland and it has been concluded that the major role of ketone bodies in the mammary gland is to act as regulatory signals to depress glucose utilization when carbohydrate is restricted[10]. A novel finding is that insulin can reverse the inhibition of glucose utilization by ketone bodies in mammary tissue[20]. There is no evidence at present that prolactin affects ketone body metabolism.

Leucine

In non-lactating rats the major sites of leucine catabolism are skeletal muscle and adipose tissue. The hyperphagia of lactation means that more leucine is available and there is evidence from the tissue activity of branched-chain aminotransferase (the initiating enzyme)[21] and whole-body leucine turnover that the mammary gland becomes an important new site of leucine catabolism[22]. Leucine is an effective lipogenic precursor in mammary gland[22] and its conversion to lipid in this tissue can be increased by inhibition of protein synthesis with cycloheximide[23]. Prolactin deficiency decreases the extraction of most amino acids including leucine by the rat mammary gland[24].

Signals for the partitioning of substrates to mammary gland

As indicated above prolactin deficiency induced by either removal of the suckling stimulus or direct inhibition of prolactin secretion with bromocryptine rapidly decreases (within 24 h) the partitioning of substrates to the lactating mammary gland; administration of ovine prolactin prevents this change. It is therefore tempting to conclude that a raised plasma prolactin is the sole signal for the re-direction of substrates in lactation. However, recent evidence has indicated that despite the hyperphagia, lactation is characterized by a low plasma insulin compared to non-lactating rats[19,25]. In addition, in short-term experiments the plasma insulin increases when prolactin secretion is inhibited[16]. The reasons for these reciprocal changes in concentration of insulin and prolactin are not at present known, but it is possible that the ratio of the two hormones may be important in the short-term regulation of the partitioning of substrates in the lactating rat.

It has been postulated that the mammary gland acts as a 'sink' not only for substrates but also for insulin, and that this may explain the low plasma insulin during lactation[12]. Clearly, hypoinsulinaemia would be advantageous in this situation because it would depress glucose uptake by adipose tissue and stimulate lipolysis; the NEFA released being transported to the liver for conversion to VLDL and subsequent export to the lactating mammary gland.

The possible importance of alterations in blood flow to the gland, especially in relation to dietary intake, has not been investigated.

SUBSTRATE PARTITIONING IN DISEASE STATES

At first sight it might appear that lactation is a special situation which has little relevance to disease states requiring parenteral nutrition. However, if one considers the lactating mammary gland as a substrate 'sink' which is of no benefit to the mother, there is a clear analogy with the rapidly growing tumour in a cancer patient. The difference is that lactation is a finely controlled state in which the increased energy requirement of the gland is balanced by the increased energy intake, whereas in the cancer patient the demands of the tumour result in cachexia and eventual death.

The main substrate requirements of a rapidly growing tumour are glucose for energy production via glycolysis and amino acids for protein synthesis. Recent *in vivo* studies in rats bearing tumours indicate that there is a degree of substrate partitioning to the tumour, particularly if it is a rapidly growing one. Protein metabolism is altered in tumour-bearing rats in such a way that adequate supplies of amino acids are directed to the tumour for growth even in the presence of decreased dietary intake[26]. Plasma insulin decreases in rats with large tumours[27], and this may provide more glucose for the tumour because the lower plasma insulin would be expected to impair glucose utilization by peripheral tissues such as muscle and adipose tissue and to promote glucose efflux from the liver. This relative hypoinsulinaemia may also explain the decrease in lipoprotein lipase activity and lipogenesis in adipose tissue of tumour-bearing mice[28]. However, some caution in the interpretation of these results is necessary because the changes may in part be due to the decreased dietary intake in tumour-bearing animals. A similar stricture applies to the recent observation that adipose tissue lipoprotein lipase activity and lipogenesis are depressed in rats with Gram-negative sepsis[29].

Clearly much requires to be discovered about the role of substrate partitioning in normal and disease states, but the study of a number of different animal models should provide information which is of value for future progress in parenteral nutrition.

The author is a member of the External Staff of the Medical Research Council, U.K.

References

1 Randle, P. J., Garland, P. B., Hales, C. N., Newsholme, E. A., Denton, R. M. and Pogson, C. I. (1966). 1. Protein hormones. Interactions of metabolism and the physiological role of insulin. *Recent Prog. Horm. Res.*, **22**, 1–48

2 Williamson, D. H. and Whitelaw, E. (1978). Physiological aspects of the regulation of keto-genesis. *Biochem. Soc. Symp.*, **43**, 137–161

3 Aynsley-Green, A., Williamson, D. H. and Gitzelmann, R. (1977). Hepatic glycogen synthetase deficiency. Definition of syndrome from metabolic and enzyme studies on 9-year-old girl. *Arch. Dis. Child.*, **52**, 573–579

4 Foster, D. O. and Frydman, M. L. (1978). Measurements of blood flow with microspheres point to brown adipose tissue as the dominant site of the calorigenesis induced by noradren-aline. *Can. J. Physiol. Pharmacol.*, **56**, 110–122

5 Denton, R. M. and Halestrap, A. P. (1979). Regulation of pyruvate metabolism in mammalian tissues. *Essays in Biochemistry*, **15**, 37–77

6 Owen, O. E., Markus, H., Sarshik, S. and Mozzoli, M. (1973). Relationship between plasma and muscle concentration of ketone bodies and free fatty acids in fed, starved and alloxan diabetic states. *Biochem. J.*, **134**, 499–506.

7 Moore, T. J., Lione, A. P., Sugden, M. C. and Regen, D. M. (1976). β-Hydroxybutyrate transport in rat brain: developmental and dietary modulations. *Am. J. Physiol.*, **230**, 619–630

8 Newsholme, E. A. and Start, C. (1973). *Regulation in Metabolism.* (London and New York: Wiley)

9 Robinson, D. S. (1970). The function of the plasma triglycerides in fatty acid transport. In: M. Florkin and E. H. Stotz (eds.). *Lipid Metabolism, Comprehensive Biochemistry*, Vol. 18, pp. 51–116 (Amsterdam: Elsevier)

10 Robinson, A. M. and Williamson, D. H. (1980). Physiological roles of ketone bodies as substrates and signals in mammalian tissues. *Physiol. Rev.*, **60**, 143–187

11 Williamson, D. H. (1973). Tissue-specific direction of blood metabolites. *Symp. Soc. Exp. Biol.*, **27**, 283–298

12 Williamson, D. H. (1980). Integration of metabolism in tissues of the lactating rat. *FEBS Lett.*, **117** Suppl., K93–K105

13 Steingrimsdottir, L., Greenwood, M. R. C. and Brasel, J. A. (1980). Effect of pregnancy, lactation and a high fat diet on adipose tissue in Osborne-Mendel rats. *J. Nutr.*, **110**, 600–609

14 Hamosh, M., Clary, T. R., Chernick, S. S. and Scow, R. O. (1970). Lipoprotein lipase activity of adipose tissue and mammary tissue and plasma triglyceride in pregnant and lactat-ing rats. *Biochim. Biophys. Acta*, **210**, 473–482

15 Zinder, O., Hamosh, M., Fleck, T. R. C. and Scow, R. O. (1974). Effect of prolactin on lipo-protein lipase in mammary gland and adipose tissue of rats. *Am. J. Physiol.*, **226**, 744–748

16 Agius, L., Robinson, A. M., Girard, J. R. and Williamson, D. H. (1979). Alterations in the rate of lipogenesis *in vivo* in maternal liver and adipose tissue on premature weaning of lactating rats. A possible regulatory role of prolactin. *Biochem. J.*, **180**, 689–692

17 Page, M. A. and Williamson, D. H. (1972). Lactating mammary gland of the rat: a potential major site of ketone body utilization. *Biochem. J.*, **128**, 459–460

18 Hawkins, R. A. and Williamson, D. H. (1972). Measurements of substrate uptake by mammary gland of the rat. *Biochem. J.*, **129**, 1171–1173

19 Robinson, A. M., Girard, J. R. and Williamson, D. H. (1978). Evidence for a role of insulin in the regulation of lipogenesis in lactating rat mammary gland. Measurements of lipogenesis *in vivo* and plasma hormone concentrations in response to starvation and refeeding. *Biochem. J.*, **176**, 343–346

20 Robinson, A. M. and Williamson, D. H. (1977). The control of glucose metabolism in isolated acini of the lactating mammary gland of the rat. The ability of glycerol to mimic some of the effects of insulin. *Biochem. J.*, **168**, 465–474

21 Ichihara, A., Noda, C. and Ogawa, K. (1973). Control of leucine metabolism with special reference to branched-chain amino acid isoenzymes. *Adv. Enzym. Reg.*, **11**, 155–166

22 Viña, J. R. and Williamson, D. H. (1981). Effects of lactation on L-leucine metabolism in the rat. Studies *in vivo* and *in vitro*. *Biochem. J.*, **194**, 941–947

23 Roberts, A. F. C., Viña, J. R., Munday, M. R., Farrell, R. and Williamson, D. H. (1982). Effects of inhibition of protein synthesis by cycloheximide on lipogenesis in mammary gland and liver of lactating rats. *Biochem. J.*, **204**, 417–423

24 Viña, J., Puertes, I. R., Saez, G. T. and Viña, J. R. (1981). Role of prolactin in amino acid uptake by the lactating mammary gland of the rat. *FEBS Lett.*, **126**, 250–252

25 Flint, D. J., Sinnett-Smith, P. A., Clegg, R. A. and Vernon, R. G. (1979). Role of insulin receptors in the changing metabolism of adipose tissue during pregnancy and lactation in the rat. *Biochem. J.*, **182**, 421–427
26 Kawamura, I., Moldawer, L. L., Keenan, R. A., Batist, G., Botte, A., Bistrian, B. R. and Blackburn, G. L. (1982). Altered amino acid kinetics in rats with progressive tumor growth. *Cancer Res.*, **42**, 824–829
27 Singh, J., Grigor, M. R. and Thompson, M. P. (1981). Glucose tolerance and hormonal changes in rats bearing a transplantable sarcoma. *Int. J. Biochem.*, **13**, 1095–1100
28 Thompson, M. P., Koons, J. E., Tan, E. T. H. and Grigor, M. R. (1981). Modified lipoprotein lipase activities, rates of lipogenesis and lipolysis as factors leading to lipid depletion in C57BL mice bearing the preputial gland tumor (ESR-586). *Cancer Res.*, **41**, 3228–3232
29 Lanza-Jacoby, S., Lansey, S. C., Cleary, M. P. and Rosato, F. E. (1982). Alterations in lipogenic enzymes and lipoprotein lipase activity during Gram-negative sepsis in the rat. *Arch. Surg.*, **117**, 144–147

Introduction

Current developments in nutritional care

I. D. A. JOHNSTON

INTRODUCTION

The widespread introduction of an effective therapy such as parenteral nutrition was soon followed by the identification of a number of problems which required solution to enable further progress to be made. The selection of appropriate patients, the identification of different requirements in various clinical conditions and the precise and accurate provision of requirements at minimum risk to the patient and at least.cost are examples of current activity to improve all aspects of our understanding of the metabolic needs of patients during nutritional support.

UNDERNUTRITION IN HOSPITAL

Selection of patients

The extent and severity of the problem of undernutrition in hospital is well documented but there is still doubt about which patients to feed intravenously before surgery, and critics point out that there is little evidence that aggressive nutritional support before or after major surgery in underweight patients will hasten recovery and reduce the incidence of complications.

Difficulties of assessment

The difficulties of assessing malnutrition were emphasized by Forse and Shizgal[1] in Montreal in 1980 in a study of over 200 patients. Body weight, height, plasma proteins, skinfold thickness and arm circumference, creatinine height ratios, and hand grip strength were all measured as part of nutritional assessment and compared with measurements of body composition. The best correlation was between weight/height and body cell mass and the poorest was between creatinine/height and body cell mass. Regression analysis indicated that the nutritional parameters only predicted malnutrition as measured by compositional studies in 68% and failed to predict it in 11% of patients.

It is accepted that undernourished patients are at increased risk of surgical morbidity and mortality, and certain patients such as those with upper gastrointestinal malignancy, are recognized as being particularly likely to develop complications. However, difficulty arises in identifying those individuals who are likely to obtain most benefit from nutritional support.

With a view to producing biochemical evidence which could be used for the basis of such a selection the metabolic response to starvation was examined in 80 undernourished surgical patients prior to operation[2]. The hormonal and substrate profiles were examined in those patients who were 15% or more below their ideal weight and could be regarded as undernourished. The patients were divided into two groups depending on whether their serum ketone level (acetoacetate + β-hydroxybutyrate) after 8 hr fasting was below the upper normal limit of 0.2 mmol/l (normoketonaemic group) or greater than it (hyperketonaemic group). About two-thirds of the patients were normoketonaemic and the remainder hyperketonaemic. The two groups were comparable in terms of weight, body fat and muscle, age and diagnosis and their daily dietary intakes were similar.

The normoketonaemic group excreted significantly more nitrogen and urea, suggesting that they were continuing to break down body protein rather than utilize stored fat. The normoketonaemic group lost less body fat and more muscle as measured by triceps skinfold thickness and arm muscle circumference, and suffered a greater fall in their serum albumin and transferrin levels.

Ten of the 12 deaths occurred in the normoketonaemic group. Analysis of the factors involved in mortality suggests no common pattern and it seems possible that the excess mortality in the normoketonaemic group could be related to continuing protein depletion at a cellular level resulting from their failure to become adequately starvation adapted during the period prior to admission.

It would appear that undernourished patients who do not display a postoperative hyperketonaemia have failed to become starvation adapted during their period of weight loss. Such patients risk a continuing severe protein depletion. Normal ketone levels may thus be used as predictors of such an outcome and thus encourage the introduction of full nutritional support preoperatively or in the immediate postoperative period. These observations require to be submitted to a prospectively controlled clinical trial.

Substances have been found in the plasma of patients with cancer which depress cell-mediated immune responses. A study was undertaken to examine plasma suppressive activity in undernourished surgical patients and in patients with malignant disease[3]. The TEEM test was used to measure the lymphocyte depressant activity of plasma samples of 25 patients with malignant disease and 25 with benign disease before operation. The results were expressed as microlitres of plasma required to produce 50% inhibition of lymphocyte response. A high suppressive activity is thus indicated by a small plasma volume. Plasma suppressive activity was raised significantly in undernourished patients with benign disease. The plasma suppressive activity was related to nutritional status based on anthropometric and biochemical measurements in patients with benign disease ($r = 0.75$), $p<0.001$; see Table 1) whereas plasma suppressive activity was raised in all patients with malignant disease irrespective of nutritional status. The extent to which the high suppressive activity can be reversed in the undernourished patients with benign disease has not yet been established.

There has been much written on the results of skin tests to allergens in undernourished patients as a prognostic guide. Many factors other than malnutrition can alter skin reactivity to allergens. The presence of cancer, zinc deficiency, age and non-specific stress all reduce skin test responsiveness while treatment with the H_2 receptor antagonists stimulates lymphocyte activity as measured in various ways. The value of skin testing is thus limited and cannot be used as an index of undernutrition.

Table 1   Plasma suppressive activity in malnourished surgical patients

	Benign		Malignant	
	Well-nourished	Malnourished	Well-nourished	Malnourished
n	17	8	9	16
Mean plasma vol. (μl) producing 50% inhibition	22.99	7.60	1.10	1.14
± SEM	± 3.15	± 0.63	± 2.85	± 1.03
Students t-test		$p<0.001$	$p = NS$	

PROTEIN TURNOVER STUDIES IN SURGICAL PATIENTS

The metabolic demands of the body after major injury are known to be increased mainly on evidence of increased oxygen consumption and nitrogen excretion.

The net nitrogen balance gives no information about protein synthesis or breakdown, either of which can be changed significantly without being reflected in an altered nitrogen balance.

Whole body protein metabolism can be determined by measuring turnover of some component of protein, an amino acid or nitrogen tagged with an appropriate radio isotope label and extrapolating the results to body protein.

A technique was developed using a tracer dose of L[1−^{14}C] leucine to measure synthesis and breakdown of body protein in a number of clinical situations to enlarge our understanding of protein and energy requirements.

Measurements were made on two occasions only on each patient to comply with the requirements of the radio isotope panel controlling clinical research.

An infusion of [^{14}C]bicarbonate was given initially followed by a constant rate infusion of L[l–^{14}C] 1 d.p.m/h for 2.5 h during steady state conditions.

Venous samples were taken during the infusion with chemical separation of the various ^{14}C components and the plasma proteins precipitated.

Plasma bicarbonate was evolved using lactic acid and when related to the amount of bicarbonate infused the rate of oxidation of amino acids could be calculated without the use of a canopy to collect exhaled gases.

The rate of entry and exit of leucine into the free pool from protein was calculated and values for body protein metabolism were then derived assuming body protein contains 8% leucine.

This model has proved to be a reliable method of measuring changes in protein metabolism in man in spite of the limitation imposed by only two measurements. Other methods using [^{15}N]glycine have some advantages but it will require the use of stable isotopes to enlarge the range of measurements and enhance our understanding of protein turnover even further.

It was first observed that changes in body protein balance could be brought about by modifying the rate of protein synthesis with no apparent change in protein breakdown.

Protein breakdown increased significantly following injury irrespective of whether or not the patients nutritional needs were met.

The increase in breakdown was related to the severity of the injury and following major injury covered by total parenteral nutrition there was still a significant gap between breakdown and synthesis. These observations explain the increased negative nitrogen balance recorded with increasing severity of injury. Breakdown of protein is related not only to the severity of injury but also to nutritional status and age, whereas synthesis responds to the energy and nitrogen available.

A study was carried out in four patients in the week prior to colorectal surgery for malignant disease. The rates of protein synthesis and breakdown were measured before and after 1 week of supplemental parenteral nutrition. The oral intake was modest but constant throughout the study and consisted of 120 mg N/kg and 15 kcal/kg daily. The intravenous supplement consisted of 144 mg N/kg and 30 kcal/kg daily. The results are shown in Table 2. The improved nitrogen retention associated with supplemental parenteral nutrition is derived both from the stimulation of protein synthesis and a reduction in protein breakdown. The increase in synthesis occurs as a response to the higher nitrogen intake while the reduction in breakdown may be due in part to the improved energy to nitrogen ratio leading to a reduction in gluconeogenesis from protein.

This evidence of improved protein economy during intravenous feeding before surgery is important and indicates clearly a benefit to the patient in metabolic terms which are measurable. It is still uncertain how soon protein synthesis improves and breakdown diminishes after the introduction of intra-

Table 2 Effect of supplemental parenteral nutrition on protein synthesis and breakdown in patients with colorectal cancer prior to surgery

	Synthesis ($g\,kg^{-1}d^{-1}$)				Breakdown ($g\,kg^{-1}d^{-1}$)			
Patient	1	2	3	4	1	2	3	4
Oral feeding	2.09	1.66	2.89	1.16	1.78	1.26	2.49	0.54
After 1 week of parenteral feeding	2.52	1.96	3.08	1.76	1.53	0.56	1.76	0.32
Mean change		+ 0.38				− 0.52		
Paired 't'-test		$p<0.03$				$p<0.04$		

venous feeding and the effect of preoperative support on subsequent post-operative metabolic changes in patients still requires further study.

Direct measurement of body protein metabolism using L[l–^{14}C]leucine during total parenteral nutrition following cholecystectomy demonstrated no significant difference between isocaloric isonitrogenous regimens containing glucose, alone ($n = 5$) or with a fat emulsion and glucose[4] ($n = 5$), as the energy substrates. It would appear that, providing the obligatory requirement for glucose is met (about 150 g/day), fat and carbohydrate calories are interchangeable with regards to fuelling protein metabolism in the early period following trauma of moderate severity.

As a basis for assessing protein metabolism in cancer patients, whole body protein turnover, synthesis and breakdown were measured preoperatively using a constant rate infusion of L[–^{14}C]leucine in patients with differing stages of colorectal carcinoma. The levels of protein synthesis and breakdown increased with the extent of disease as measured by the percentage incorporation of the labelled amino acid into plasma protein[5] and the subsequent modified Dukes' classification[6]. Eleven apyrexial patients were divided into two groups; six of whom had normal appetites while five were anorectic. Protein synthesis increased with advancement of disease in both groups, as did protein breakdown. Protein synthesis and breakdown were lower in the anorectic group, suggesting some degree of starvation adaptation. All patients were in positive balance, despite anthropometric data to support loss of host body protein. This suggests translocation of protein stores from muscle to areas of more rapid protein synthesis such as tumour. Remodelling of body protein is an important facet of metabolism in cancer patients.

These studies in our laboratory have shed new light on the metabolic response to major injury, the effect of total parenteral nutrition on protein metabolism and the relationship between protein turnover and the presence of a large tumour mass.

It is now possible to study ways of modifying protein metabolism by hormone manipulation or different substrate and amino acid intakes. It should also be possible to gain further understanding of the changing patterns of response in patients with malignant disease undergoing aggressive chemotherapy or radiotherapy.

ADVANCES IN PRESCRIBING NUTRITIONAL SUPPORT

The task of nutritional assessment, monitoring and the prescribing of daily requirements for a variety of patients has become increasingly complex. Multiple calculations are required particularly if the patient is having both enteral and parenteral support. There have been a number of reports on the use of computer programs to assist in day to day management of critically ill patients.

A programmable pocket calculator can be used to estimate nutritional and electrolyte requirements[7] and computer programs to match intravenous[8] or enteral requirements[9] with pharmacy availability are available. The program should be able to handle intravenous, enteral and oral intake presented to it in any order; it should contain details of the constituents of a reasonable number of proprietary and nonproprietary preparations; it should be more rapid than existing methods; results should be clear and accessible without the need for a printed copy.

A program was written for the Texas TI 59 (Texas Instruments, Lubbock, TX), a battery operated pocket calculator with the capability of reading pre-programmed magnetic cards into its memory[10].

The program was originally written for use in Great Britain and later revised for use in the United States. The tendency for hospitals in the United States to prepare intravenous solutions in their own pharmacy allowed expanded use of the program.

The actual computations carried out by the program are relatively simple. The data storage areas contain details of the constituents of each item in terms of grams of fat, protein and carbohydrate per litre of normal solution or per food exchange. The program recalls the appropriate figures, adjusts them for strength, and by simple multiplication, calculates the amount of each constituent in the quantity of each item the patient received.

In order to assess the value of the system in clinical use, 29 daily intake charts for 14 patients were reviewed. The patients were obtaining their nutrition by a variety of routes and were reported as apparently receiving their full prescribed requirements. The actual intakes of energy and nitrogen were computed by the calculator system from the input charts, which had been filled in by nursing personnel. These figures were compared with the prescribed intakes, which were taken from the medical orders.

When the major part of nutrition was received parenterally, two-thirds of the daily intakes were inadequate in energy or nitrogen, or both. If the major part of nutrition was received orally and enterally, almost half of the daily intakes were deficient. This implied a shortfall of approximately 610 kcal and

7.2 g nitrogen each day. The prescribed amounts of nitrogen were less frequently achieved than those of energy.

The program and calculator system has proved easy to use, even by those unfamiliar with dietetics or programmable calculators. This, and the rapidity with which intake can be calculated, has meant that an up-to-date intake record for any patient is almost instantly available to the nutritional support service. Any shortfall is quickly identified and can be rapidly corrected, in contrast to manual intake monitoring systems which may fall 24 h or more behind.

The system has also highlighted the considerable discrepancies which can exist between the nutritional prescription and the actual intake. These errors usually remain unnoticed for a variety of reasons. Considerable time savings can probably be achieved with this system.

It is reasonable to expect that the dietetic workload for a nutritional support service responsible for 20 patients can be reduced by 10 h/day. Alternatively, the system is simple enough to allow intake assessment to be performed by less highly trained personnel.

In use, the programmable pocket calculator has proved a valuable asset in the management of patients under the care of the nutritional support service. The forthcoming generation of pocket computers with even larger data capacity, alphanumeric displays and permanent memory offer still further advantages to aid in the day to day care of nutrition for hospital in-patients.

Computer-assisted assessment and prescribing, however, requires local enthusiasm to take available programs and alter them to meet local needs.

The success of this approach has been reported widely in district hospitals and for special categories of patients[11-14]. There are also considerable economic benefits to be obtained from computer controlled nutritional support. The daily costs are easily identified and since the requirements are known accurately it is a simple matter to select the least expensive daily prescription, provided all the nutritional requirements have been met.

References

1 Forse, R. A. and Shizgal, H. M. (1980). The assessment of malnutrition. *Surgery*, **88**, 18
2 Rica, A. J. and Wright, P. D. (1979). Ketosis and nitrogen excretion in undernourished surgical patients. *J. Parent. Ent. Nutr.*, **3**, 350-354
3 Johnston I. D. A., Wright, P. D., Lennard, J. W. J., Clague, M. B., Carmichael, M. J., Francis, D. M. A. and Williams, R. N. P. (1981) Malnutrition and Cancer. *Clinical Oucology*, **7**, 83-91
4 Rogaly, E. Clague, M. B., Carmichael, M. J., Wright, P. D. and Johnston, I. D. A. (1982). Comparison of body protein metabolism during total parenteral nutrition using glucose or glucose and fat as the energy source. *Clin. Nutr.*, **1**, 81-89
5 Clague, M. B., Carmichael, M. J., Keir, M. J., Rogaly, E., Wright, P. D. and Johnston, I. D. A. (1982). Increased incorporation of an infused labelled amino acid into plasma proteins as a means of assessing the severity of injury or activity of disease in surgical patients. *Ann. Surg.*, **196**, 53
6 Holyoke, E. D., Lokich, J. and Wright, H. (1975). Adjuvant therapy of adenocarcinoma of the colon following clinically curative resection. *Proceedings of the Gastrointestinal Tumour*

Study Group, Division of Cancer Treatment, Bethesda, Md., National Institutes of Health, National Cancer Institute

7 Rich, A. J. and Wright, P. D. (1980). A pocket calculator program for intravenous requirements. *Br. J. Surg.*, **67**, 313

8 Wright, P. D., Shearing, G., Rich, A. J. and Johnston, I. D. A. (1978). The role of a computer in the management of clinical parenteral nutrition. *J. Parent. Enteral Nutr.*, **2**, 652

9 Geller, R. J., Blackburn, S. A., Glendon, D. H., Henneman, W. H. and Steffee, W. (1979). Computer optimization of enteral hyperalimentation. *J. Parent. Enteral Nutr.*, **3**, 79

10 Rich, A. J. (1981). A programmable calculator system for the estimation of nutritional intake of hospital patients. *Am. J. Clin. Nutr.*, **34**, 2276

11 Mohamed el Lozy, M. D. (1978). Programable calculators in the field assessment of nutritional status. *Am. J. Clin. Nutr.*, **31**, 1718

12 Goggin, M. J. (1979). The use of a small programmable calculator in intravenous feeding. *Med. Inform.*, **4**, 115

13 James, R. M., Roberts, J. M., Harvey, P. W., Bellis, J. D. and Cooper, R. I. (1978). A computerized scheme for the preparation of parenteral nutrition regimes. *Med. Inform.*, **3**, 77

14 Schlaepfer, L. V. and Shmerling, D. H. (1979). The use of a programmable pocket calculator in clinical dietetics. *Res. Exp. Med. (Berl.)*, **174**, 267

Section 1
Branched-Chain Amino Acids

1
Branched-chain amino acids: metabolic roles and clinical applications

D. C. MADSEN

INTRODUCTION

The three branched-chain amino acids (BCAA), leucine, isoleucine and valine, are 'essential' amino acids, and together comprise about 40% of the minimum daily requirement for essential amino acids in man. Plasma levels of these amino acids are more drastically affected than the other AA following changes in caloric or protein intake[1-4]. Starvation for even 24 h will increase the plasma concentration of all three BCAA in humans[1,5,6] and in rats[3,7], while most other amino acid levels decline. Starvation beyond 1 week shows a fall in plasma BCAA to basal levels. Protein deprivation (in days, or long-term as in kwashiorkor) lowers the BCAA to below basal levels[8,9].

Following a protein meal, peripheral blood levels of BCAA increase to a much greater extent than for most other amino acids, due to the relatively minimal hepatic extraction of the BCAA from portal blood[4]. Most of the arterial BCAA load is removed by muscle, in contrast to marked hepatic uptake of other amino acids.

The BCAA are important not only as substrates for protein synthesis but are believed to be peripheral (i.e., skeletal muscle) calorie substrates and regulators of protein metabolism[2,10-14]. The prominent metabolic roles of the BCAA receiving emphasis in the current literature are outlined in Table 1.1 Leucine in particular seems to have a prominent role[2,15-19].

Table 1.1 Proposed metabolic roles of branched-chain amino acids

Peripheral calorie source
'Anabolic' − i.e. promoting protein synthesis
  Decreasing catabolism
  Increasing synthesis
Normalizing metabolism of brain neurotransmitters

BRANCHED-CHAIN AMINO ACIDS: PERIPHERAL METABOLISM

Catabolism of most amino acids for calories occurs predominantly in the liver. The BCAA, however, are also catabolized in extrahepatic (peripheral) tissues. The organ distribution of the catabolic enzymes for BCAA is unique. Figure 1.1 indicates that the two initial enzymes in BCAA catabolism are, sequentially, the transaminase, resulting in branched-chain keto-acid formation, and the dehydrogenase(s). The BCAA transaminases are distributed predominantly in skeletal muscle, with lesser amounts in liver[20–24]. In contrast, the BCAA dehydrogenases (the irreversible step in BCAA catabolism) are in greater concentrations in hepatic and other tissues than in muscle[3,20–25] (see Table 1.2). Thus, the transaminase is rate-limiting in liver, kidney, and other tissues; while in skeletal muscle the dehydrogenase is the rate-limiting step.

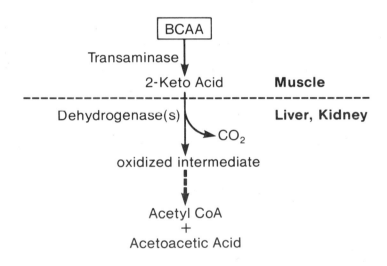

Figure 1.1 Catabolism of branched-chain amino acids

The activities of these two enzymes respond to a large variety of biological factors: dietary manipulations[7,22,23,26–29] (e.g., low protein; fasting) (Figure 1.2), circadian rhythms[30], differences between anatomical muscle groups[13,24,31–33], species differences[21,34], and hormonal factors[3,7,34–37]. However, the direction and extent of change in enzyme activity differs between the two enzymes (e.g., Figure 1.2). Many intracellular factors influence the enzyme assay procedures, which can in turn influence interpretation of results[16,21,33,35,38,38a]. The increased oxidation of BCAA seen in starvation[1,7,12,22,23,33] is due to complete oxidation via the citrate cycle[7,39]. A greater caloric contribution from BCAA also occurs in diabetes[3,4,22,33,37,40],

Table 1.2 Tissue distribution of branched-chain catabolic enzymes

	% of total body activity	
	Transaminase	Dehydrogenase
Rat		
Muscle	80	10
Liver	15	75
Man		
Muscle	80	60
Liver	5	30

(Adapted from Refs. 20,21,23)

in exercise[41-44], and in the stress of injury, sepsis or liver disease. In these states of pathophysiology, BCAA take on greater importance as a 'peripheral calorie source'.

Elucidation of the broader features of BCAA catabolism was presented by Odessey, *et al.*[45], who proposed that, in fasting, a 'BCAA–alanine cycle' supplies calories to peripheral tissues. In this scheme, BCAA (of splanchnic origin) were suggested to be the source of nitrogen for intramuscular synthesis of alanine (and/or glutamine), which was then released to blood for hepatic (or renal, in the case of glutamine) gluconeogenesis (Figure 1.3). This pathway has been amply confirmed by others[11,12,46-49] and serves, in effect, as a 'nitrogen shuttle' among organs.

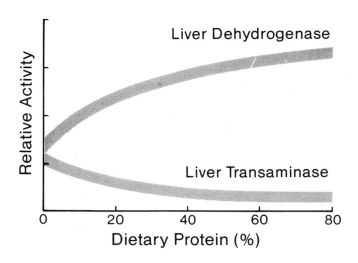

Figure 1.2 Response of hepatic enzymes of branched-chain amino acid catabolism to level of dietary protein (Combined from several sources.)

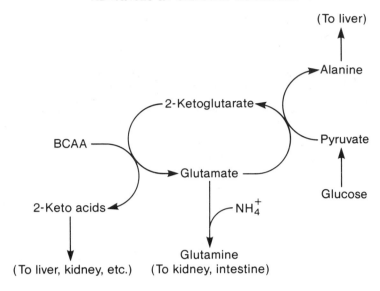

Figure 1.3 Probable pathway of nitrogen flow in muscle

Further elaboration of these pathways led to the concept of 'interorgan participation'[7,50] for the catabolism of BCAA. Transamination in skeletal muscle supplies branched-chain keto acids (BCKA) which are released for transport to other organs, where the appropriate enzymes enable irreversible oxidation to proceed at optimal rates (Figure 1.3; Table 1.2). In support of this concept are several studies, indicating muscle to be the quantitatively primary source of BCKA found in blood[51–54]. Livesey and Lund[54] demonstrated this pathway in a perfused rat model; the liver removed from the blood a quantity of BCKA equal to that released by the hindlimb. Although skeletal muscle is probably the site for generation of BCKA, the origin of the BCAA substrate can vary. The normal source of BCAA for muscle is the blood draining the splanchnic bed, since the liver extracts little of the BCAA input[4,52]. An exogenous supply of BCAA can increase alanine production (assuming adequate glucose is provided as a source of the pyruvate-carbon skeleton), which is thus an indicator of increased BCAA oxidation[46,47,51,52]. Odessey, et al.[52] observed that exogenous BCAA seemed to decrease phenyl-alanine release (breakdown of protein) in rat diaphragm incubated in vitro. This 'protein-sparing' effect of exogenous BCAA, also seen with leucine alone[55], is probably operative to some extent in vivo.

However, in states of stress the major source of BCAA for peripheral oxidation may be muscle protein. In such conditions as major injury and sepsis there is a functional glucose intolerance, due in part to insulin resist-ance[56,57], which also prevents quantitative utilization of fatty acids[57]. In the trauma-septic state it is recognized that skeletal muscle plays a major role in fuel metabolism, due to hepatic failure and multiple systems organ failure[48,58-65].

Increased muscle protein catabolism[64] provides most of the BCAA, which is oxidized to BCKA. This concept is supported by many studies[58,59,62,64,66-68]. The organ distribution of enzyme activity in the rat (Table 1.2) suggests that a major proportion of BCKA are exported from muscle[54], enzyme distribution in man indicates a major, but less extensive, export of BCKA from muscle (Table 1.2). It is likely that stress increases the use of BCAA as a calorie source for muscle. However, it remains to be determined what effect stress has on the proportioning of BCKA (produced in muscle) for local use and for export to other organs. For example, Duff, *et al.*[66] measured arteriovenous differences across human calf muscle of septic or post-operative (non-septic) patients (Table 1.3). In the septic groups alanine production and phenylalanine release were greatly elevated, indicating increases in both BCAA oxidation and muscle proteolysis, respectively. However, BCAA output was depressed. It would be interesting to know the proportion of BCAA that could have been accounted for by measuring BCKA output by the muscle (see Refs 174, 175).

Table 1.3 Plasma amino acid flux across human calf muscle (nanomoles/g)

Amino acids	Patient group	
	Septic	*Non-septic post-op.*
BCAA*	− 34†	− 128
Phenylalanine	− 28	− 22
Alanine	− 208	− 202

*BCAA − branched-chain amino acids
†Negative value signifies net release of amino acid.
(From Ref. 66)

The observations outlined above have prompted attempts to prevent or even reverse the muscle catabolism of stress by the intravenous infusion of solutions enriched with BCAA. Some are of the opinion that this treatment acts by decreasing protein catabolism[59,60,65,69-73]; others disagree[10,39,55,74]. Certain caveats must attend some of the conclusions regarding decreased catabolism following administration of BCAA. For example, it has been pointed out that changes in plasma amino acid concentrations, or arteriovenous differences, are subject to certain errors. Non-muscle components (skin, feet, bone) of a rat hindlimb preparation can account for 40% of protein synthetic activity[75]. It is difficult to relate this quantitatively to arteriovenous differences, but there may be significant non-muscle contribution. Wahren's group observed that changes in human plasma amino acid concentrations following BCAA infusion suggested not decreased catabolism but a general redistribution of amino acids into tissues of the body[76-78]. Also, decreased efflux of BCAA from a perfused limb could easily be due to increased efflux of BCKA, i.e. not necessarily a decrease in BCAA oxidation[3,51,53,54,66]. Finally, the observations of Aoki, *et al.*[79] and McCormick and Webb[80] indicate that quantitative assessment of amino acid flux must take into account the erythrocyte- and peptide-amino acid fractions of whole blood.

Several studies have attempted to resolve some of the many complex factors integrating intracellular and whole-body BCAA metabolism. Figure 1.4 presents a hypothetical model of leucine metabolism and its compartments, based on several sources[10,55,81,82]. Discrete pools seem to exist for protein synthesis ('leucine$_1$') and for oxidation ('leucine$_2$'). The precursor pool for oxidation receives input preferentially from extracellular leucine, and probably communicates minimally with ('leucine$_1$'). Proteolysis may contribute a small proportion to the oxidative pool. The pool involved in protein synthesis does receive major input from extracellular leucine, but mechanisms exist for efficient recycling of amino acids derived from endogenous protein[10,55]. Suboptimal protein (amino acid) intake or supply increases the efficiency of reutilization of endogenous protein (leucine$_1$)[10,82] and transfers exogenous amino acids (leucine) rapidly to the oxidative pool[55]. An excess of protein increases further the channelling of exogenous protein to (leucine$_2$).

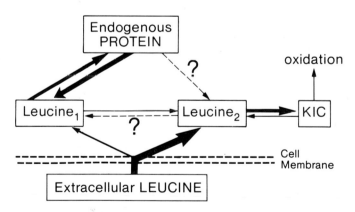

Figure 1.4   Intracellular compartmentation of L-leucine (hypothetical) (KIC = ketoisocaproate)
(Adapted from Refs. 10,55,81,82)

BRANCHED-CHAIN AMINO ACIDS AND ANABOLISM

The BCAA, especially leucine, have been shown to directly stimulate protein synthesis in muscle, an effect suggested to be distinct from BCAA oxidation for energy metabolism[7,10,84,85]. Buse, *et al.*[86] injected BCAA intraperitoneally into rats; increased polysome formation was found to occur in skeletal muscle. The suggested mode of action was an increase in acylated RNA, resulting in a stimulation of peptide chain initiation[85,86]. Many other studies from Busc's group have supported the contention that BCAA are anabolic for protein synthesis in muscle[13,87]. This may represent a 'fuel' effect (use of BCAA as fuel to drive anabolic reactions) and/or a 'signal'[41,85,88,89]. 'The crucial question . . . is whether an increase in the rate of oxidation of leucine controls, or is controlled by, protein synthesis'[39].

Moldawer, et al.[10] reviewed their own and other data on the effects of BCAA infusion, and concluded that BCAA do not merely decrease muscle protein catabolism or increase synthesis per se; rather, the BCAA seem to improve the efficiency of reutilization of intracellular amino acids derived from protein catabolism.[55].

Muscle proteolysis has been observed to be in part modulated by the intra-cellular redox state[40,90] ($NADH/NAD^+$). Further studies of the effects of BCAA on redox potential might yield additional clues as to their mode of action, or suggest a means of enhancing the anabolic functions of BCAA.

Regardless of the mechanisms involved, the evidence for an effect of BCAA on muscle protein metabolism is convincing, although much of it has been obtained in vitro. Also, some experiments have not been performed in models or preparations representing states of stress, e.g., trauma or sepsis, and it is in these states where most investigators feel there may be a clinical benefit associated with the use of BCAA. Some investigators[74,90a], do not believe that evidence warrants attributing to BCAA a role in the regulation of protein synthesis. Taken together, the above evidence can be seen to contain two elements. One is the use of BCAA as a peripheral-based calorie source. The second concerns their possible role in promoting protein anabolism and/or decreasing catabolism, again in muscle. The physiological significance of each element remains to be determined; one may prove to be predominant, but both probably interact to achieve the observed effects depending on existing metabolic conditions.

BRANCHED-CHAIN AMINO ACIDS AND NEUROTRANSMITTER METABOLISM

Consciousness is a function of the interrelated activities of the cerebral hemispheres and the brain stem and diencephalon[91]. Coma may be defined as brain failure, which can involve any or all of these areas of the brain. Functional or metabolic disturbances may be a cause, and affect the brain diffusely. Pathophysiologic processes associated with encephalopathy and coma, include cerebral oedema, hyperaemia, and ischaemia/anoxia[91-94]. Impaired oxygen consumption and reduced cerebral blood flow are symptomatic and exacerbating factors in brain failure. Since central neuro-transmitters modulate cerebral blood flow and metabolism[91,95-97], deranged neurotransmitter metabolism is thought to play a causative and complicating role in encephalopathy.

The specific organic causes of coma are not known for certain[93,94]. Impaired liver function can lead to hepatic encephalopathy (HE), including coma. Liver disease can result in altered ureagenesis[98-101], after resulting in hyperammonaemia. Many investigators feel that ammonia plays an important, possibly primary role in precipitating HE[93,98-100, 102, 103]. This is the basis for one aspect of treatment of HE, which is the reduction of blood ammonia by any of several means including neomycin, lactose, and/or lactulose[99-101,103,104]. Table 1.4 lists some of the proposed 'toxins' active in hepatic failure. Since liver failure can leave a large share of ammonia removal to skeletal muscle[4,5,99], malnutrition and muscle-wasting (which is often seen in chronic liver disease) can exacerbate hyperammonaemia[48]. This view gives added significance to the repletion or maintenance of muscle mass by nutritional support, which in this light acquires prophylactic as well as therapeutic value.

Table 1.4 Proposed 'toxins' in hepatic failure

Ammonia
Fatty Acids
Mercaptans
'False neurotransmitters' (gut flora)
Amino acids (ratios)

More recently Fischer, Freund and others have made significant conceptual advances in attempting to elucidate the mechanism of HE. It had been observed that in patients with hepatic failure and/or HE, plasma amino acid patterns are characterized by decreased levels of the BCAA, while one or more of the aromatic amino acids (phenylalanine, tyrosine, tryptophan) are increased[64,105-110]. Other amino acids are also often increased in plasma – usually methionine and glutamine[107,109,111]. The aminograms can vary with the degree of liver disease: in acute hepatic failure all plasma amino acids may be elevated[110,112,113]. Not all investigators report the specific pattern, decreased ratio ('R') of BCAA/aromatic amino acids, in liver disease[112-115]. Liver disease may be too heterogeneous an entity to enable characterization by plasma aminograms alone, since many other factors can affect this parameter[113,114,116]. Correlations with HE have been reported for other 'R' designations or plasma components (see Figure 1.5).

One of the more prominent hypotheses to explain HE may be termed the 'BCAA hypothesis'. In the latest version[117-119], it is proposed that the hyperammonaemia of cirrhosis, by way of insulin and glucagon interactions, results in the characteristic pattern of plasma amino acid levels – a decreased ratio (R) of BCAA/aromatic amino acids. Since the BCAA and aromatic amino acids compete at the blood–brain barrier for a common entry site[120], the decreased R-value permits unusually large amounts of the aromatic amino acids to enter the cerebrospinal fluid. There, the high levels of ammonia favour glutamine production. The transport system that mediates efflux of glutamine from the brain also mediates influx of the aromatic and other amino acids, thus perpetuating the cycle.

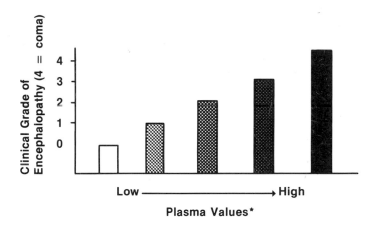

Figure 1.5 Representation of reported relationships between hepatic encephalopathy and selected plasma constituents. * Ratios: AAA/BCAA; TRP/BCAA; TRP/NAA; Concentrations: ammonia; octopamine (AAA = aromatic amino acids; BCAA = branched chain amino acids; TRP = tryptophan; NAA = neutral amino acids)

The elevated brain levels of aromatic amino acids are thought to result in the abnormal production of 'false neurotransmitters', such as octopamine and phenylethanolamine, which directly cause central nervous system dysfunction. False neurotransmitters are also postulated to originate in the gut, by bacterial action[121,122]. In addition, increased brain levels of tryptophan result in altered co-ordination of production of the neurotransmitters norepinephrine and serotonin, thereby contributing to HE. Many observations support the BCAA–HE hypothesis[112,115,123,124,124a]. Dietary administration of protein or amino acids rich in aromatic amino acids can alter the cerebral concentrations and metabolism of both amino acids and neurotransmitters[95,109,117,125–132]. This has led to reports of effects of dietary manipulations of aromatic amino acids on behaviour and/or cerebral function, including aggression[133], hyperkinesis in children[134], and effects on memory[135], pain [136], and appetite regulation[137].

With this background, reports have indicated that the value of the ratio

(R), BCAA/aromatic amino acids, correlates well with the presence or grade of HE[108,111,117,138,139]. Figure 1.5 summarizes this concept, and gives R as any of several published factors reported to show a correlation with the grade of HE. Others have not been able to substantiate this observation[110–112,115,124,141,142], finding often that R correlates with the histologic severity of liver disease, but not with HE.

Table 1.5 Plasma amino acid concentrations and transport constants for the blood–brain barrier

Amino acid	Normal plasma concentration (μm/l)	K_m (mM)	K_m (app) (mM)	Plasma concentration in hepatic failure (μm/l)
Leucine	172	0.23	0.76	154
Isoleucine	85	0.23	0.78	71
Valine	348	0.46	1.60	255
Phenylalanine	83	0.15	0.49	280
Tyrosine	88	0.16	0.72	305
Methionine	35	0.19	0.74	460

(Sources: Refs. 140,145,150)

It has been suggested that in liver disease the alterations in amino acid metabolism are the result of a functional derangement in the permeability of the blood–brain barrier (BBB)[118,143]. There are unquestionably alterations in amino acid transport across the BBB in this state. However, the sharing of a common carrier by the BCAA and aromatic amino acids[120] is itself sufficient to explain the observed transport anomalies. Many studies have documented the competitive nature of this amino acid transport system; a relative excess of one or more amino acid of the carrier system will affect transport rates of the other amino acids[117,118,144–149]. (For example, see Figure 1.6.) Since plasma contains all amino acids the competitive permutations will alter the individual affinity constants (K_m) to new values (K_m(app)) (Table 1.5),

Table 1.6 Biphasic transport constants for the blood–brain barrier

Amino acid	K_{m1} (mM)	K_{m2} (mM)
Leucine	0.32	1.86
Isoleucine	0.36	2.14
Valine	0.28	2.24
Phenylalanine	0.23	*
Tyrosine	N.D.	N.D.†
Methionine	N.D.	N.D.

*No K_{m2} for this amino acid; † N.D., not determined. (Ref. 145)

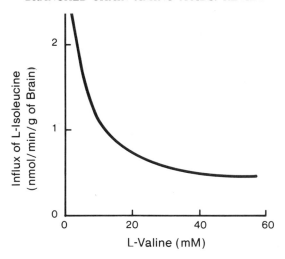

Figure 1.6 Influence of plasma valine concentration on influx of isoleucine into mouse brains *in vivo*
(From Ref. 144)

which obviously will alter characteristics of transport across the BBB. An added complication is the fact that several of the amino acids, including the BCAA, have biphasic affinity constants[145,151] (Table 1.6), which may suggest 'compartmental' amino acid transport across the BBB[151]. Thus, the innumerable potential interactions of amino acids with their BBB transport system obviate the need for postulating 'deranged permeability of the BBB. Further, a study of amino acid transport in dogs given a surgical portacaval shunt could detect no HE and no change in permeability even though plasma amino acid levels were altered in the expected fashion[148]. Similar conclusions were reached in a study using rats[148a].

Some drugs have been suggested to alter cerebral BBB permeability as a result of changes in cerebral blood flow[96]. Central adrenergic neurons (responding to catecholamine neurotransmitters – produced from aromatic amino acids) may regulate cerebral blood flow[96,97], and thereby influence 'permeability' indirectly. One might consider whether normalization of plasma (thereby brain) amino acid levels results in normalization of blood flow, and in this manner affects consciousness. This explanation carries the BCAA–HE hypothesis one step further. The hypothesis – that HE results at least in part from altered patterns of plasma amino acid levels, in particular the BCAA/aromatic amino acid ratio – has led to what may seem a logical suggestion for treatment. Correction, or 'normalization', of plasma amino acid patterns has been attempted by oral or intravenous administration of formulations relatively rich in BCAA and low in aromatic amino acids[76,106,111,131,138,139,152]. Others have not been successful in treating HE with BCAA-enriched solutions[110,153].

13

The BCAA hypothesis of HE implicates plasma (and thereby brain) levels of ammonia as the primary etiologic agent[117,118]. Rational and primary treatment of HE and hyperammonaemia would thus involve reduction of blood ammonia. Administration of BCAA, in the context of the hypothesis, would seem to be a distal approach to the problem. Further, several reports have indicated that infusion of BCAA resulted in reduction of blood ammonia concurrently with clinical improvement[106,115,131,152]. The solution infused (F080) was high in BCAA content (35% of total), low in aromatic amino acids[108,138]. However F080 also has less glycine and more arginine than the control solution; these two amino acids are involved in ammonia metabolism, but in different ways[100]. The question then follows: was clinical improvement due to beneficial changes in brain neurotransmitters following 'normalization' of plasma amino acid patterns, or was improvement in mental status (HE) due in part to the reduction of ammonia, which probably was related to the decreased glycine and increased arginine content?

The BCAA hypothesis of HE has been both attacked[112,153-157] (and see ref. no. 143: pp. 141–142) and defended[158]. The complexity of factors affecting HE makes it difficult to assign biological effects to a single variable. Zieve[93] has evaluated most of the theories of the etiology of HE. His approach is a global one; it is suggested that a synergism among potential toxins (Table 1.4) forms the etiologic base. His framework has not yet addressed the problem at

Table 1.7 Summary of clinical investigations of branched-chain amino acid infusions

Reference	Subject selection	Number infused*	Infusion duration
76	Healthy volunteer	6	150 min
78	Cirrhosis; healthy	?,6	150 min
106	Hepatic encephalopathy (HE)	15	6 days
108	Cirrhosis; hepatitis	8,3	Not given
131	HE	3	Not given
138	Septic coma	5	Not given
139	Cirrhosis, some HE	55	8–14 days
161	Surgical stress	5	5 days
162	HE	4	5–14 days
163	Cirrhosis, HE	16	Not given
164	HE	33	4–26 days
165	Post-operative	16	5 days
166†	Liver disease	6	to 7 days
167	HE	21	6 days
168	HE	32	5–14 days
169†	Liver disease	6	Not given
170†	HE	4	to 7 days
171	Cirrhosis	7	24 h
172	HE; hepatitis	4	3 h/treatment
173	HE	10	1–23 days

*Number of subjects infused with solutions containing branched-chain amino acids (BCAA) as 35% or more of total amino acids; includes 100%, i.e. only three BCAA. † Conventional solution supplemented with 5% valine

a molecular-cellular level, as do Fischer and colleagues[117]. If BCAA do confer clinical benefit in HE, elucidation of the mode of action will probably require consideration of all the theories.

ANIMAL AND CLINICAL STUDIES

Animals have been employed for studying the effects of enteral and parenteral BCAA supplementation. Rats, dogs, monkeys, and rabbits have been used in models involving sepsis, stress, 'post-injury catabolism', and portacaval shunt. End-points have included: nitrogen retention, bodyweight changes, plasma and muscle aminograms[70], protein turnover[160], and neurological function[115,152,153]. A discussion of this topic is presented by Dr Freund (Chapter 3).

Clinical studies of BCAA infusion have involved humans in the following categories; healthy volunteers, major surgical stress, septic coma and liver disease. Duration of infusion among the studies ranged from 150 min to 26 days. Efficacy and safety were established in 265 patients who received infusions of BCAA as 35% to 100% of total amino acids. The relevant references are presented in Table 1.7.

SUMMARY

The branched-chain amino acids (BCAA) have several unique qualities that render them theoretically useful in therapy of certain disease states. It has been proposed that in stressful clinical conditions, e.g. trauma or sepsis, administration of supplemental doses of BCAA may be of value. Benefit would be due to their use as a peripheral calorie source and to their tendency to promote protein synthesis.

A second area of interest is associated with hepatic coma. It has been proposed that the altered plasma levels of amino acids associated with this state initiate encephalopathy. The mechanism may involve increased levels of aromatic amino acids in the brain, resulting in an imbalance of proper neurotransmitters. Infusion of solutions higher in BCAA may correct the plasma and brain abnormalities, resulting in reversal of the comatose state. This hypothesis is the most contested aspect of BCAA. Regardless of the mechanisms, BCAA infusions have proven valuable in therapy of stress.

Continuing investigation of the use of BCAA supplementation in well-defined stress, sepsis and multiple organ failure is necessary to define metabolic mechanisms which underlie the observed results. It is also necessary to define the patient groups which will obtain maximum benefit from BCAA supplementation, as well as optimal levels of BCAA content. In all of the studies reviewed here, animal and human, infusion of high concentrations of

BCAA was apparently safe. There were no reports of adverse reactions associated with any infusions.

ACKNOWLEDGEMENTS

I gratefully acknowledge the advice and support of Dr Hugh Tucker, the co-ordinating skills of Ms Trina Palm and of Ms Frances Moore, and the patience and encouragement of my wife, Barbara.

References

1 Adibi, S. (1970). Metabolism of branched-chain amino acids in altered nutrition. *Metabolism*, **25**, 1287
2 Adibi, S. (1980). Roles of branched-chain amino acids in metabolic regulation. *J. Lab. Clin. Med.*, **95**, 475
3 Hutson, S. and Harper, A. (1981). Blood and tissue branched-chain amino and α-keto acid concentrations: effect of diet, starvation, and disease. *Am. J. Clin. Nutr.*, **34**, 173
4 Wahren, J., Felig, P. and Hagenfeldt, L. (1976). Effect of protein ingestion on splanchnic and leg metabolism in normal man and in patients with diabetes mellitus. *J. Clin. Invest.*, **57**, 987
5 Sherwin, R. (1978). Effect of starvation on the turnover and metabolic response to leucine. *J. Clin. Invest.*, **61**, 1471
6 Elwyn, D., Fürst, P., Askanazi, J. and Kinney, J. (1981). Effect of fasting on muscle concentrations of branched-chain amino acids. In M. Walser and J. Williamson (eds.). *Metabolism and Clinical Implications of Branched Chain Amino and Keto Acids.* pp. 547–552. (New York: Elsevier/North Holland)
7 Hutson, S., Zapalowski, C., Cree, T. and Harper, A. (1980). Acid metabolism in skeletal muscle. *J. Biol. Chem.*, **255**, 2418
8 Adibi, S. and Drash, A. (1970). Hormone and amino acid levels in altered nutritional states. *J. Lab. Clin. Med.*, **76**, 722
9 Ghisolfi, J., Charlet, P., Ser, N., Salvayre, R., Thouvenot, J. and Duole, C. (1978). Plasma free amino acids in normal children and in patients with proteincaloric malnutrition: fasting and infection. *Pediatr.* Res., **12**, 912
10 Moldawer, L., Sakamoto, A., Blackburn, G. and Bistrian, B. (1981). Alterations in protein kinetics produced by branched chain amino acid administration during infection and inflammation. In M. Walser and J. Williamson (eds.). *Metabolism and Clinical Implications of Branched Chain Amino and Keto Acids.* pp. 533–540. (New York: Elsevier/North Holland)
11 Tischler, M. and Goldberg, A. (1980a). Production of alanine and glutamine by atrial muscle from fed and fasted rats. *Am. J. Physiol.*, **238**, E487
12 Tischler, M. and Goldberg, A. (1980b). Leucine degradation and release of glutamine and alanine by adipose tissue. *J. Biol. Chem.*, **255**, 8074
13 Buse, M. (1981). *In vivo* effects of branched-chain amino acids on muscle protein synthesis in fasted rats. *Horm. Metabol. Res.*, **13**, 502
14 Hedden, M. and Buse, M. (1979). General stimulation of muscle protein synthesis by branched-chain amino acids. *Proc. Soc. Exp. Biol. Med.*, **160**, 410
15 Buse, M. and Reid, S. (1975). Leucine. A possible regulator of protein turnover in muscle. *J. Clin. Invest.*, **56**, 1250
16 Krebs, H. and Lund, P. (1974). Aspects of the regulation of the metabolism of the branched-chain amino acids. *Adv. Enzyme Regul.*, **15**, 375
17 Odessey, R. (1979). Amino acid and protein metabolism in the diaphragm. *Am. Rev. Respir. Dis.*, **119**, 107
18 Pek, S., Santiago, J. and Tai, T. (1978). L-Leucine-induced secretion of glucagon and insulin, and the 'Off-Response' to L-leucine *in vitro*. I. Characterization of the dynamics of secretion. *Endocrinology*, **103**, 1208

19 Malaisse, W. and Sener, A. (1981). Branched-chain amino and keto aoids: effects upon insulin secretion. In M. Walser and J. Williamson (eds.). *Metabolism and Clinical Implications of Branched Chain Amino and Keto Acids.* pp. 181–186. (New York: Elsevier/North Holland)

20 Ichihara, A., Noda, C. and Tanaka, K. (1981). Oxidation of branched-chain amino acids with special reference to their transaminase. In M. Walser and J. Williamson (eds.). *Metabolism and Clinical Implications of Branched Chain Amino and Keto Acids.* pp. 227–232. (New York: Elsevier/North Holland)

21 Khatra, B., Chawla, R., Sewell, C. and Rudman, D. (1977). Dehydrogenases in primate tissues. *Clin. Invest.*, **59**, 558

22 Paul, H. and Adibi, S. (1976). Assessment of effect of starvation, glucose, fatty acids and hormones on decarboxylation of leucine in skeletal muscle of rat. *J. Nutr.*, **106**, 1079

23 Sketcher, R., Fern, E. and James, W. (1974). The adaptation in muscle oxidation of leucine to dietary protein and energy intake. *Br. J. Nutr.*, **31**, 333

24 Veerkamp, J. and Wagenmakers, A. (1981). Branched-chain 2-oxo acid metabolism in human and rat muscle. In M. Walser and J. Williamson (eds.). *Metabolism and Clinical Implications of Branched Chain Amino and Keto Acids.* pp. 163–168. (New York: Elsevier/North Holland)

25 Mitch, W., Walser, M. and Sapir, D. (1981). Nitrogen sparing induced by leucine compared with that induced by its keto analogue, α-ketoisocaproate, in fasting obese man. *J. Clin. Invest.*, **67**, 553

26 Hauschildt, S. and Brand, K. (1980). Effects of branched-chain α-keto acids on enzymes involved in branched-chain α-keto acid metabolism in rat tissues. *J. Nutr.*, **110**, 1709

27 Mimura, T., Yamada, C. and Swendseid, M. (1968). Influence of dietary protein levels and hydrocortisone administration on the branched-chain amino acid transaminase activity in rat tissues. *J. Nutr.*, **95**, 493

28 Hauschildt, S., Luthje, J. and Brand, K. (1981). Influence of dietary nitrogen intake on mammalian branched-chain α-keto acid dehydrogenase activity. *J. Nutr.*, **111**, 2188

29 Shinnick, F. and Harper, A. (1977). Effects of branched-chain amino acid antagonism in the rat on tissue amino acid and keto acid concentrations. *J. Nutr.*, **107**, 887

30 Potter, D., Sullivan, S. and Cox, R. (1980). Rhythmic variations of valine and leucine decarboxylation in rat diaphragm. *Metabolism*, **29**, 435

31 Goodlad, G., Tee, M. and Clark, C. (1981). Leucine oxidation and protein degradation in the extensor digitorum longus and soleus of the tumor-bearing host. *Biochem. Med.*, **26**, 143

32 Li, J. and Goldberg, A. (1976). Effects of food deprivation on protein synthesis and degradation in rat skeletal muscle. *Am. J. Physiol.*, **231**, 441

33 Paul, H. and Adibi, S. (1978). Leucine oxidation in diabetes and starvation: effects of ketone bodies on branched-chain amino acid oxidation *in vitro. Metabolism*, **27**, 185

34 Goldberg, A., Tischler, M., DeMartino, G. and Griffin, G. (1980). Hormonal regulation of protein degradation in skeletal muscle. *Fed. Proc.*, **39**, 31

35 Frick, G., Tai, L., Blinder, L., and Goodman, H. (1981). L-Leucine activates branched-chain keto acid dehydrogenase in rat adipose tissue. *J. Biol. Chem.*, **256**, 2618

36 Jefferson, L., Li, J. and Rannels, S. (1977). Regulation by insulin of amino acid release and protein turnover in the perfused rat hemicorpus. *J. Biol. Chem.*, **252**, 1476

37 May, M., Mancusi, V., Aftring, R. and Buse, M. (1980). Effects of diabetes on oxidative decarboxylation of branched-chain keto acids. *Am. J. Physiol.*, **239**, E215

38 Waymack, P., DeBuysere, M. and Olson, M. (1980). Studies on the activation and inactivation of the branched-chain α-keto acid dehydrogenase in the perfused rat heart. *Biol. Chem.*, **255**, 9773

38a Rhead, W., Dubiel, B. and Tanaka, K. (1981). The tissue distribution of isovaleryl-CoA dehydrogenase in the rat. In M. Walser and J. Williamson (eds.). *Metabolism and Clinical Implications of Branched-Chain Amino and Keto Acids.* pp. 47–52. (New York: Elsevier/North Holland)

39 Lindsay, D. B. (1980). Amino acids as energy sources. *Proc. Nutr. Soc.*, **39**, 53

40 Buse, M., Herlong, H. and Weigand, D. (1976). The effect of diabetes, insulin and the redox potential on leucine metabolism by isolated rat hemidiaphragm. *Endo.*, **98**, 1166

41 Chua, B., Siehl, D. and Morgan, H. (1980). A role for leucine in regulation of protein turnover in working rat hearts. *Am. J. Physiol.*, **239**, 510

42 Freund, H., Yoshimura, N. and Fischer, J. (1979). Effect of exercise on postoperative nitrogen balance. *J. Appl. Physiol.*, **46**, 141

43 Morgan, H., Chua, B., Fuller, E. and Siehl, D. (1980). Regulation of protein synthesis and degradation during *in vitro* cardiac work, *Am. J. Physiol.*, **238**, 431

44 Rennie, M., Edwards, R., Krywawych, S., Davies, C., Halliday, D., Waterlow, J. and Millward, D. (1981). Effect of exercise on protein turnover in man. *Clin. Sci.*, **61**, 627

45 Odessey, R. and Goldberg, A. (1972). Oxidation of leucine by rat skeletal muscle. *Am. J. Physiol.*, **223**, 1376

46 Ben Galim, E., Hruska, K., Bier, D., Matthews, D. and Haymand, M. (1980). Branched-chain amino acid nitrogen transfer to alanine *in vivo* in dogs. *J. Clin. Invest.*, **66**, 1295

47 Garber, A., Karl, I. and Kipnis, D. (1976). Alanine and glutamine synthesis and release from skeletal muscle. *J. Biol. Chem.*, **251**, 836

48 Lockwood, A., McDonald, J., Reiman, R., Gelhard, A., Laughlin, J., Duffy, T. and Plum, F. (1979). The dynamics of ammonia metabolism in man. *J. Clin. Invest.*, **63**, 449

49 Yamamoto, H., Aikawa, T., Matsutaka, H., Okuda, T. and Ishikawa, E. (1974). Inter-organal relationships of amino acid metabolism in fed rats. *Am. J. Physiol.*, **226**, 1428

50 Harper, A. and Zapalowski, C. (1981). Interorgan relationships in the metabolism of the branched chain amino acids. In M. Walser and J. Williamson (eds.). *Metabolism and Clinical Implications of Branched Chain Amino and Keto Acids*. pp. 195–204. (New York: Elsevier/North Holland)

51 Odessey, R. and Goldberg, A. (1979). Leucine degradation in cell-free extracts of skeletal muscle. *Biochem. J.*, **178**, 475

52 Odessey, R., Khairallah, E. and Goldberg, A. (1974). Origin and possible significance of alanine production by skeletal muscle. *J. Biol. Chem.*, **249**, 7623

53 Hutson, S. and Harper, A. (1978). Blood and tissue branched-chain and α-keto acid concentrations: effect of diet, starvation, and disease. *Am. J. Clin. Nutr.*, **34**, 173

54 Livesey, G. and Lund, P. (1980). Enzymic determination of branched-chain amino acids and 2-oxoacids in rat tissues. *Biochem. J.*, **188**, 705

55 Schneible, P., Airhart, J. and Low, R. (1981). Differential compartmentation of leucine for oxidation and for protein synthesis in cultural skeletal muscle. *J. Biol. Chem.*, **256**, 4888

56 Ryan, N., George, B., Egdahl, D. and Egdahl, R. (1974a). Chronic tissue insulin resistance following hemorrhagic shock, *Ann. Surg.*, **180**, 402

57 Ryan, N., Blackburn, G. and Clowes, G. (1974b). Differential tissue sensitivity to elevated endogenous insulin levels during experimental peritonitis in rats. *Metabolism*, **23**, 1081

58 McMenamy, R., Birkhahn, R., Oswald, G., Reed, R., Rumph, C., Vaidyanath, N., Yu, L., Cerra, F., Sorkness, R. and Border, J. (1981). Multiple systems organ failure: I. The basal state. *J. Trauma.*, **21**, 99

59 McMenamy, R., Birkhahn, R., Oswald, G., Reed, R., Rumph, C., Vaidyanath, N., Yu, L., Sorkness, R., Cerra, F. and Border, J. (1981). Multiple systems organ failure: II. The effect of infusion of amino acids and glucose. *J. Trauma.*, **21**, 228

60 Moyer, E., McMenamy, R., Cerra, F., Reed, R., Yu, L., Chenier, R., Caruana, J. and Border, J. (1981). Multiple systems organ failure: III. Contrasts in plasma amino acid profiles in septic trauma patients who subsequently survive and do not survive – effects of intravenous amino acids. *J. Trauma.*, **21**, 263

61 Moyer, E., Border, J., Cerra, F., Caruana, J., Chenier, R. and McMenamy, R. (1981). Multiple systems organ failure: IV. Imbalances in plasma amino acids associated with exogenous albumin in the trauma-septic patient. *J. Trauma.*, **21**, 543

62 Moyer, E., Border, J., McMenamy, R., Caruana, J., Chenier, R., and Cerra, F. (1981). Multiple systems organ failure: V. Alterations in the plasma protein profile in septic trauma – effects of intravenous amino acids. *J. Trauma.*, **21**, 645

63 Siegel, J., Giovannini, I., Coleman, B., Cerra, F. and Nespoli, A. (1981). Death after portal decompressive surgery. *Arch. Surg.*, **116**, 1330

64 Cerra, F., Siegel, J., Coleman, B., Border, J. and McMenamy, R. (1980). Septic autocannibalism. *Ann. Surg.*, **192**, 570

65 Milewski, P., Threlfall, C., Heath, D., Holbrook, I., Wilford, K. and Irving, M. (1982). Intracellular free amino acids in undernourished patients with or without sepsis. *Clin. Sci.*, **62**, 83

66 Duff, J., Viidik, T., Marchuk, J., Holliday, R., Finley, R., Groves, A. and Woolt, L. (1979). Femoral arteriovenous amino acid differences in septic patients. *Surgery*, **85**, 344

67 Long, C., Birkhahn, R., Geiger, J. and Blakemore, W. (1981). Contribution of skeletal muscle protein in elevated rates of whole body protein catabolism in trauma patients. *Am. J. Clin. Nutr.*, **34**, 1087

68 Birkhahn, R., Long, C., Fitkin, D., Geiger, J. and Blakemore, W. (1980). Effects of major skeletal trauma on whole body protein turnover in man measured by L-(1, ^{14}C)-leucine. *Surgery* , **88**, 294

69 Chua, B., Siehl, D. and Morgan, H. (1979). Effect of leucine and metabolites of branched-chain amino acids on protein turnover in heart. *J. Biol. Chem.*, **254**, 8358

70 Freund, H., Yoshimura, N., Lunetta, L. and Fischer, J. (1978). The role of the branched-chain amino acids in decreasing muscled catabolism *in vivo*. *Surgery*, **83**, 611

71 Freund, H., Yoshimura, N. and Fischer, J. (1980). The role of alanine in the nitrogen conserving quality of the branched-chain amino acids in the postinjury state. *J. Surg. Res.*, **29**, 23

72 Lindberg, B. and Clowes, G. (1981). An experimental method for study of liver blood flow and metabolism in intact animals. *J. Surg. Res.*, **31**, 156

73 O'Donnell, T., Clowes, G., Blackburn, G., Ryan, T., Benotti, P. and Miller, J. (1976). Proteolysis associated with a deficit of peripheral energy duel substrates in septic man. *Surgery* , **80**, 192

74 Millward, D., Garlick, P., Nnanyelugo, D. and Waterlow, J. (1976). The relative importance of muscle protein synthesis and breakdown in the regulation of muscle mass. *Biochem. J.*, **156**, 185

75 Preedy, V. and Garlick, P. (1981). Rates of protein synthesis in skin and bone, and their importance in the assessment of protein degradation in the perfused rat hemicorpus. *Biochem. J.*, **194**, 373

76 Eriksson, S., Hagenfeldt, L. and Wahren, J. (1981). A comparison of the effects of intravenous infusion of individual branched-chain amino acids on blood amino acid levels in man. *Clin. Sci.*, **60**, 95

77 Hagenfeldt, L., Eriksson, S. and Wahren, J. (1980). Influence of leucine on arterial concentrations and regional exchange of amino acids in healthy subjects. *Clin. Sci.*, **59**, 173

78 Hagenfeldt, L. and Wahren, J. (1980). Experimental studies on the metabolic effects of branched-chain amino acids. *Acta Chir. Scand. Suppl.*, **498**, 88

79 Aoki, T., Brennan, M., Muller, W. and Cahill, G. (1974). Amino acid levels across normal forearm muscle: whole blood *vs.* plasma. *Adv. Enzyme Regul.*, **12**, 157

80 McCormick, M. and Webb, K. (1982). Plasma free, erythrocyte free and plasma peptide amino acid exchange of calves in steady state and fasting metabolism. *J. Nutr.*, **112**, 276

81 Matthews, D., Bier, D., Rennie, M., Edwards, R., Halliday, D., Millward, D. and Clugston, G. (1981). Regulation of leucine metabolism in man: a stable isotope study. *Science*, **214**, 1129

82 Nissen, S. and Haymond, M. (1981). Effects of fasting on flux and interconversion of leucine and α-ketoisocaproate *in vivo*. *Am. J. Physiol.*, **241**, E72

83 Motil, K., Matthews, D., Bier, D., Burke, J., Munro, H. and Young, V. (1981). Whole-body leucine and lysine metabolism: response to dietary protein intake in young men. *Am. J. Physiol.*, **240**, E712

84 Fulks, R., Li, J. and Goldberg, A. (1975). Effects of insulin, glucose and amino acids on protein turnover in rat diaphragm. *J. Biol. Chem.*, **250**, 290

85 Li, J. and Jefferson, L. (1978). Influence of amino acid availability on protein turnover in perfused skeletal muscle. *Biochim. Biophys. Acta*, **544**, 351

86 Buse, M., Atwell, R. and Mancusi, V. (1979). *In vitro* effect of branched-chain amino acids on the ribosomal cycle in muscles of fasted rats. *Horm. Metab. Res.*, **11**, 289

87 Buse, M. and Weigand, D. (1977). Studies concerning the specificity of the effect of leucine on the turnover of proteins in muscles of control and diabetic rats. *Biochim. Biophys. Acta*, **475**, 81

88 Sener, A. and Malaisse, W. (1981). The stimulus–secretion coupling of amino acid-induced insulin release: insulinotropic action of branched-chain amino acids at physiological concentrations of glucose and glutamine. *Eur. J. Clin. Invest.*, **11**, 455

89 Seglen, P., Gordon, P. and Poli, A. (1980). Amino acid inhibition of the autophagic/lyso-

somal pathway of protein degradation in isolated rat hepatocytes. *Biochim. Biophys. Acta,* **630**, 103

90 Tischler, M. (1980). Is regulation of proteolysis associated with redox-state changes in rat skeletal muscle? *Biochem. J.,* **192**, 963

90a Millward, D. and Waterlow, J. (1978). Effect of nutrition on protein turnover in skeletal muscle. *Fed. Proc.,* **37**, 2283

91 Spielman, G. (1981). Coma: a clinical review. *Heart Lung,* **10**, 700

92 Levy, D., Bates, D., Caronna, J., Cartlidge, N., Knill-Jones, R., Lapinski, R., Singer, B., Shaw, D. and Plum, F. (1981). Prognosis in nontraumatic coma, *Ann. Int. Med.,* **94**, 293

93 Zieve, L. (1981). The mechanism of hepatic coma. *Hepatology,* **1**, 360

94 Schenker, S., Breen, K. and Hoyumpa, A. (1974). Hepatic encephalopathy: current status. *Gastroenterology,* **66**, 121

95 Heffner, T., Hartman, J. and Seiden, L. (1980). Feeding increases dopamine metabolism in the rat brain. *Science,* **208**, 1168

96 Preskorn, S., Irwin, G., Simpson, S., Friesen, D., Rinne, J. and Jerkovitch, G. (1981). Medical therapies for mood disorders alter the blood-brain barrier. *Science,* **213**, 469

97 Peroutka, S., Moskowitz, M., Reinhard, J. and Snyder, S. (1980). Neurotransmitter receptor binding in bovine cerebral microvessels. *Science,* **208**, 610

98 Koen, H., Okuda, K., Musha, H., Tateno, Y., Fukuda, N., Mutasumoto, T., Shisido, F., Rikitake, T., Iinuma, T., Kurisu, A. and Arimizu, N. (1980). A dynamic study of rectally absorbed ammonia in liver cirrhosis using (^{13}N) ammonia and a positron camera. *Dig. Dis. Sci.,* **25**, 842

99 Weber, F., Fresard, K. and Lolly, B. (1982). Effects of lactulose and neomycin on urea metabolism in cirrhotic subjects. *Gastroenterology,* **82**, 213

100 Rudman, D., Galambos, J., Smith, R., Salam, A. and Warren, D. (1973). Comparison of the effect of various amino acids upon the blood ammonia concentration of patients with liver disease. *Am. J. Clin. Nutr.,* **26**, 916

101 Uribe, M., Berthier, J., Lewis, H., Mata, J., Sierra, J., Garcia-Ramos, G., Acosta, J. and Dehesa, M. (1981). Lactose enemas plus placebo tablets *vs.* neomycin tablets plus starch enemas in acute portal systemic encephalopathy. *Gastroenterology,* **81**, 101

102 Walshe, J., DeCarli, L. and Davidson, C. (1958). Some factors influencing cerebral oxidation in relation to hepatic coma. *Clin. Sci.,* **17**, 11

103 Weber, F. (1981). Therapy of portal-systemic encephalopathy: the practical and the promising. *Gastroenterology,* **81**, 174

104 Orlandi, F., Freddara, M., Candelaresi, M., Morettini, A., Corazza, G., DeSimone, A., Dobrilla, G., and Cavallini, G. (1981). Comparison between neomycin and lactulose in 173 patients with hepatic encephalopathy. *Dig. Dis. Sci.,* **26**, 498

105 Cascino, A., Cangiano, C., Calcaterra, V., Rossi-Fanelli, F. and Capocaccia, L. (1978). Plasma amino acids imbalance in patients with liver disease. *Am. J. Dig. Dis.,* **23**, 591

106 Egberts, E., Hamster, W., Jurgens, P., Schumacher, H., Fondalinski, G., Reinhard, U. and Schomerus, H. (1981). Effect of branched chain amino acids on latent portal-systemic encephalopathy. In M. Walser and J. Williamson (eds.). *Metabolism and Clinical Implications of Branched Chain Amino and Keto Acids.* pp. 453−463. (New York: Elsevier/North Holland)

107 Fischer, J., Yoshimura, N., Aguirre, A. James, J., Cummings, M., Abel, R. and Deindoerfer, F. (1974). Plasma amino acids in patients with hepatic encephalopathy. *Am. J. Surg.,* **127**, 40

108 Fischer, J., Rosen, H., Ebeid, A., James, J., Keane, J. and Soeters, P. (1976). The effect of normalization of plasma amino acids on hepatic encephalopathy in man. *Surgery,* **80**, 77

109 Knell, A., Davidson, A., Williams, R., Kantamaneni, B. and Curzon, G. (1974). Dopamine and serotonin metabolism in hepatic encephalopathy. *Br. Med. J.,* **1**, 549

110 McCullough, A., Czaja, A., Jones, J. and Go, V. (1981). The nature and prognostic significance of serial amino acid determinations in severe chronic active liver disease. *Gastroenterology,* **81**, 645

111 Sketcher, R., Fern, B. and James, W. (1974). The adaption in muscle oxidation of leucine to dietary protein and energy intake. *Br. J. Nutr.,* **31**, 333

112 Mans, A., Saunders, A., Kirsch, E. and Biebuyck, J. (1979). Correlation of plasma and

brain amino acid and putative neurotransmitter alterations during acute hepatic coma in the rat. *J. Neurochem.*, **32**, 285

113 Shaw, S. and Lieber, C. (1978). Plasma amino acid abnormalities in the alcoholic. *Gastroenterology*, **74**, 677

114 Siassi, F., Wang, M., Kopple, J. and Swendseid, M. (1977). Plasma tryptophan levels and brain serotonin metabolism in chronically uremic rats. *J. Nutr.*, **107**, 840

115 Smith, A., Rossi-Fanelli, F., Freund, H. and Fischer, J. (1979). Sulfur-containing amino acids in experimental hepatic coma in the dog and the monkey. *Surgery*, **85**, 677

116 Anderson, G. and Blendis, L. (1981). Plasma neutral amino acid ratios in normal man and in patients with hepatic encephalopathy: correlations with self-selected protein and energy consumption. *Am. J. Clin. Nutr.*, **34**, 377

117 James, J., Jeppson, B., Ziparo, V. and Fischer, J. (1979). Hyperammonaemia, plasma amino acid imbalance, and blood-brain amino acid transport: a unified theory of portal-systemic encephalopathy. *Lancet*, 772

118 James, J. and Fischer, J. (1981). Transport of neutral amino acids at the blood-brain barrier. *Pharmacology*, **22**, 1

119 Anonymous. (1980). Hepatic encephalopathy: a unifying hypothesis. *Nutr. Reviews*, **38**, 371

120 Oldendorf, W. and Szabo, J. (1976). Amino acid assignment to one of three blood-brain barrier amino acid carriers. *Am. J. Physiol.*, **230**, 94

121 Fischer, J. and James, J. (1972). Treatment of hepatic coma and hepatorenal syndrome. *Am. J. Surg.*, **123**, 222

122 Schafer, D. and Jones, E. (1982). Hepatic encephalopathy and the γ-aminobutyric-acid neurotransmitter system. *Lancet*, **1**, 18

123 Bloxam, D. and Curzon, G. (1978). A study of proposed determinants of brain tryptophan concentration in rats after portocaval anastomosis of sham operation. *J. Neurochem.*, **31**, 1255

124 Wustrow, Th., van Hoorn-Hickman, R., van Hoorn, W., Vinik, A., Fischer, M. and Terblanche, J. (1981). Acute hepatic ischaemia in the pig – the changes in plasma hormones, amino acids and brain biochemistry. *Hepato-gastroenterol.*, **28**, 143

124a Langer, M., Masala, A., Alagna, S., Rassu, S., Madeddu, G., Solinas, A. and Chiandussi, L. (1981). Growth hormone, (GH) secretion in hepatic encephalopathy. *Clin. Endocrinol.*, **14**, 189

125 Fernstrom, J. (1981). Effects of precursors on brain neurotransmitter synthesis and brain functions. *Diabetologia*, **20**, 281

126 Arnold, M. and Fernstrom, J. (1981). L-Tryptophan injection enhances pulsatile growth hormone secretion in the rat. *Endocrinology*, **108**, 331

127 Heroux, E. and Roberge, A. (1981). Different influences of two types of diets commonly used for rats on a series of parameters related to the metabolism of central serotonin and noradrenaline. *Can. J. Physiol. Pharmacol.*, **58**, 108

128 James, J., Hodgman, J., Funovics, J., Yoshimura, N. and Fischer, J. (1976). Brain tryptophan, plasma free tryptophan and distribution of plasma neutral amino acid. *Metabolism*, **25**, 471

129 Kamata, S., Okada, A., Watanabe, T., Kawashima, Y. and Wada, H. (1980). Effects of dietary amino acids on brain amino acids and transmitter amines in rats with a portacaval shunt. *J. Neurochem.*, **35**, 1190

130 Pardridge, W. (1981). Transport of nutrients and hormones through the blood-brain barrier. *Diabetologia*, **20**, 246

131 Weiser, M., Riederer, P. and Kleinberger, G. (1978). Human cerebral free amino acids in hepatic coma. *J. Neur. Transm.*, Suppl. **14**, 95

132 Ashely, D. and Curzon, G. (1981). Effects of long-term low dietary tryptophan intake on determinants of 5-hydroxytryptamine metabolism in the brains of young rats. *J. Neurochem.*, **37**, 1385

133 Thurmond, J., Kramarcy, N., Lasley, S. and Brown, J. (1980). Dietary amino acid precursors: effects on central monoamines, aggression, and locomotor activity in the mouse. *Pharmacol. Biochem. Behav.*, **12**, 525

134 Arnold, L. and Nemzer, E. (1982). New evidence on diet in hyperkinesis. *Pediatr.*, **69**, 250

135 Wurtman, R. (1981). The effects of nutritional factors on memory. *Acta Neurol. Scand.*, Suppl. **64**, 145

21

136 Seltzer, S., Marcus, R. and Stoch, R. (1981). Perspectives in the control of chronic pain by nutritional manipulation. *Pain*, **11**, 141

137 Wurtman, R., Wurtman, J. and Fernstrom, J. (1980). Composition and method for suppressing appetite for calories as carbohydrates. US Patent No. 4 210 637

138 Freund, H., Ryan, J. and Fischer, J. (1978). Amino acid derangements in patients with sepsis: treatment with branched chain amino acid rich infusions. *Ann. Surg.*, **188**, 423

139 Holm, E., Striebel, J., Meisinger, E., Haux, R., Langhans, W. and Becker, H. (1978). Amino acids mixtures for parenteral feeding in liver insufficiency. *Infusionsther.*, **5**, 274

140 Chase, R., Davies, M., Trewey, P., Silk, D. and Williams, R. (1978). Plasma amino acid profiles in patients with fulminant hepatic failure treated by repeated polyacrylonitrile membrane hemodialysis. *Gastroenterology*, **75**, 1033

141 Wardle, E. and Williams, R. (1980). Depressed uptake of serotonin by platelets in hepatic encephalopathy. *Biochem. Med.*, **24**, 223

142 Ono, J., Hutson, D., Dombro, R., Levi, J., Livingstone, A. and Zepra, R. (1978). Tryptophan and hepatic coma. *Gastroenterology*, **74**, 296

143 Jeppsson, B., Freund, H., Gimmon, Z., James, J., von Meyenfeldt, M. and Fischer, J. (1981). Blood-brain barrier derangement in sepsis: cause of septic encephalopathy? *Am. J. Surg.*, **141**, 136

144 Daniel, P., Pratt, O. and Wilson, P. (1977). The exclusion of L-isoleucine or of L-leucine from the brain of the rat, caused by raised levels of L-valine in the circulation, and the manner in which this exclusion can be partially overcome. *J. Neurol. Sci.*, **31**, 421

145 Benjamin, A., Verjee, Z. and Quastel, J. (1980). Kinetics of cerebral uptake processes *in vitro* of L-glutamine, branched-chain L-amino acids, and L-phenylalanine: effects of ouabain. *J. Neurochem.*, **35**, 67

146 Binek-Singer, P. and Johnson, T. (1981). The inhibition of brain protein synthesis following leucine or valine injection can be prevented. *Biochem. Biophys. Res. Commun.*, **103**, 1209

147 Fernstrom, J. and Faller, D. (1978). Neutral amino acids in the brain: changes in response to food ingestion. *J. Neurochem.*, **30**, 1531

148 Huet, P., Pomier-Layrargues, G., Duguay, L. and du Souich, P. (1981). Blood-brain transport of tryptophan and phenylalanine: effect of portacaval shunt in dogs. *Am. J. Physiol.*, **241**, G163

148a Sarna, G., Bradbury, M. and Cavanagh, J. (1977). Permeability of the blood-brain barrier after portocaval anastomosis in the rat. *Brain Res.*, **138**, 550

149 Pardridge, W. (1977). Kinetics of competitive inhibition of neutral amino acid transport across the blood-brain barrier. *J. Neurochem.*, **28**, 103

150 Cremer, J. (1981). Nutrients for the brain: problems in supply. *Early Hum. Dev.*, **5**, 117

151 Samuels, S. and Schwartz, S. (1981). Compartmentation in amino acid transport across the blood brain barrier. *Neurochem. Res.*, **6**, 755

152 Freund, H., Krause, R., Rossi-Fanelli, F., Smith, A. and Fischer, J. (1978). Amino acid-induced coma in normal animals; prevention by branched-chain amino acids. *Gastroenterology*, **74**

153 Zieve, L., Onstad, G., Doizaki, W., Timmerman, W. and Palm, S. (1980). High brain concentrations of phenylalanine, tryptophan and methionine do not cause coma in rats or dogs, *Gastroenterology*, **79**, (Abstr.)

154 Baker, A. (1979). Amino acids in liver disease: a cause of hepatic encephalopathy? *J. Am. Med. Assoc.*, **242**, 355

155 Zieve, L. (1979). Amino acids in liver failure. *Gastroenterology*, **76**, 219

156 Hawkins, R. (1981). Blood-brain barrier during portal-systemic encephalopathy. *Lancet*, **ii**, 302

157 Hawkins, R. (1982). The blood-brain barrier in encephalopathy. *Lancet*, **1**, 398

158 James, J., Freund, H. and Fischer, J. (1979). Amino acids in hepatic encephalopathy. *Gastroenterology*, **77**, 421

159 Preskorn, S., Irwin, G., Simpson, S., Friesen, D., Rinne, J. and Jerkovich, (1981). Medical therapies for mood disorders alter the blood-brain barrier. *Science*, **213**, 469

160 Blackburn, G., Moldawer, L., Usui, S., Bothe, A., O'Keefe, S. and Bistrian B. (1979). Branched chain amino acid administration and metabolism in starvation, injury and infection. *Surgery*, **86**, 307

161 Blackburn, G., Desai, S., Keenan, R., Bentley, B., Moldawer, L. and Bistrian, B. (1981). Clinical use of branched-chain amino acid enriched solutions in the stressed and injured patient. In M. Walser and J. Williamson (eds.). *Metabolism and Clinical Implications of Branched Chain Amino and Keto Acids*. pp. 521−526. (New York: Elsevier/North Holland)

162 Ferenci, P., Funovics, J. and Wewalka, F. (1978). Therapy of hepatic encephalopathy, modification of the plasma aminogram using AA infusions. *Chir. Form Exp. Klin.*, 183

163 Fiaccadori, F., Ghinelli, F., Pelosi, G., Sacchini, D., Vaona, G., Zeneroli, M., Rocchi, E., Santunione, V., Gibertini, P. and Ventura, E. (1980). Selective amino acid solutions in hepatic encephalopathy treatment. *Ric. Clin. Lab.*, **10**, 411

164 Fischer, J., Freund, H., Rosen, H., Yoshimura, N., Bradford, R. and Sofio, C. (1978). Effects of F080 in clinical hepatic encephalopathy: result of a phase I study. *Gastroenterology*, **75**, 963

165 Freund, H., Hoover, H., Atamian, S. and Fischer, J. (1979). Infusion of the branched chain amino acids in postoperative patients. *Ann. Surg.*, **190**, 18

166 Jellinger, K., Riederer, P., Rausch, W. and Kothbauer, P. (1978). Brain monoamines in hepatic encephalopathy and other types of metabolic coma. *J. Neur. Trans.*, Suppl 14, 103

167 Okada, A., Kamata, S., Kim, C. and Kawashima, Y. (1981). Treatment of hepatic encephalopathy with BCAA-rich amino acid mixture. In M. Walser and J. Williamson (eds.). *Metabolism and Clinical Implications of Branched Chain Amino and Keto Acids*. pp. 447−452. (New York: Elsevier/North Holland)

168 Rakette, S., Fischer, M., Reimann, H. and von Sommoggy, S. (1981). Effects of special amino acid solutions in patients with liver cirrhosis and hepatic encephalopathy. In M. Walser and J. Williamson (eds.). *Metabolism and Clinical Implications of Branched Chain Amino and Keto Acids*. pp. 419−425. (New York: Elsevier/North Holland)

169 Riederer, Von P., Jellinger, K., Rausch, W., Kleinberger, G. and Kothbauer, P. (1978). Zur biochemie der hepatischen enzaphalopathien. *Z. Gastroenterol.*, **16**, 768

170 Riederer, P. (1980). Oral and parenteral nutrition with L-valine: mode of action. *Nutr. Metabol.*, **24**, 209

171 Rossi-Fanelli, F., Angelico, M., Cangiano, C., Cascino, A., Capocaccia, R., DeConciliis, D., Riggio, O. and Capocaccia, L. (1981). Effect of glucose and/or branched-chain amino acid infusion on plasma amino acid imbalance in chronic liver failure. *J. Paren. Ent. Nutr.*, **5**, 414

172 Watanbe, A., Higashi, T. and Nagashima, H. (1978). An approach to nutritional therapy of hepatic encephalopathy by normalization of deranged amino acid patterns in serum. *Acta Med. Okayama*, **32**, 427

173 Leon, I., Martinez, J., Martin, P., Pombo, M., Leon, P. and Perez, A. (1978). Parenteral nutrition in patients with hepatic encephalopathy. *Rev. Clin. Esp.*, **151**, 129

174 Elia, M. and Livesey, G. (1981). Branched chain amino acid and oxo acid metabolism in human and rat muscle. In M. Walser and J. Williamson (eds.). *Metabolism and Clinical Implications of Branched Chain Amino and Keto Acids*. pp. 257−262. (New York: Elsevier/North Holland)

175 Abumrad, N., Patrick, L., Rannels, S. and Lacy, W. (1981). Branched chain amino acids, and α-keto isocaproate balance across human forearm muscle. In M. Walser and J. Williamson (eds.). *Metabolism and Clinical Implications of Branched Chain Amino and Keto Acids*. pp. 317−322. (New York: Elsevier/North Holland)

2
Intracellular and plasma branched-chain amino acid interrelationships

P. FÜRST, A. ALVESTRAND, J. BERGSTRÖM, J. ASKANAZI, D. ELWYN AND J. KINNEY

Many studies have documented abnormal plasma concentrations of the free amino acids in patients with different catabolic disorders. The interpretation and the pathogenesis of these abnormal patterns, however, remain speculative. Certain abnormalities appear to be caused by the catabolic disease *per se*, whereas others, resembling those observed in subjects with low intake of protein and energy, may be related to malnutrition.

It is now well documented that during catabolism the distribution of free amino acids between the extracellular and intracellular compartments is altered[1-5]. Therefore the plasma concentrations do not necessarily reflect the intracellular concentrations. Skeletal muscle contains the largest pool of free amino acids. Determination of the free amino acid concentration in muscle is therefore of particular interest in the study of amino acid and protein metabolism in catabolic patients.

METHODS

In the past 20 years a relatively simple and safe technique for taking muscle biopsies has been developed in Stockholm[6] and subsequently introduced and applied at Columbia University, New York. In this procedure, a 0.5 cm incision in the thigh is made under local anaesthesia, permitting a needle biopsy of the quadriceps femoris muscle amounting to 30–50 mg of tissue[6,7]. Multiple biopsies can be taken in almost any situation. With the use of modern microanalytical techniques[6,7] this biopsy can be analysed for water[6], electrolytes[8], lipids[6,9] and free amino acids[10,11]. Similar measurements are made on a plasma sample and the amount of intracellular and extracellular water may then be calculated[6,11], based on the chloride method and assuming a normal (-87.2 mV) transmembrane potential. Intracellular concentrations of free amino acids or electrolytes are then estimated based on intra- and extracellular water distribution (see Table 2.1). In this calculation it is also assumed that none of the amino acids (except tryptophan) are bound to protein. These calculations are described extensively elsewhere[10,11]. Although the lack of measured values of membrane potential may lead to quantitative errors, it does not invalidate the qualitative conclusions. The problems using this calculation in severe catabolic patients have been discussed in detail[1,4,11,12].

Table 2.1 Measurements made on muscle biopsy

Direct measurements	*Derived parameters*
Water	Intracellular and extracellular
Electrolytes	water
Na^+, K^+, Mg^{++}, Cl^-	Intracellular concentrations of
Lipid	electrolytes and amino acids
Free amino acids	

THE UNIQUE INTRACELLULAR PATTERN OF CATABOLISM

Muscle free amino acid patterns have been measured to date in several catabolic conditions; in uraemia[1,5,13-15], diabetes mellitus[16], after major abdominal operations[17,18], or after total hip replacement[3], after severe injury[4,19], during sepsis[4], during experimental starvation[20] or semi-starvation[2,3] and during immobilization[2]. One important finding is that the concentrations of muscle free amino acids change in response to changes in diet and physiological or pathological states. This is shown for selected amino acids in Figure 2.1; branched-chain amino acids (BCAA) are exemplified by valine.

Each catabolic condition appears to have its own unique and reproducible intracellular pattern. Valine, like the other BCAA, increases uniformly in trauma and is unchanged in diabetes. Glutamine by contrast decreases considerably in trauma, starvation and diabetes but is unchanged in uraemia. Changes in plasma free amino acid concentrations may parallel those in muscle, but frequently differ either quantitatively or qualitatively. These results indicate that the intracellular free amino acid pattern of muscle is characterized by a 'unique pattern of catabolism' distinctive for the disease or condition studied.

FREE AMINO ACID POOLS AND THEIR ROLE IN PROTEIN SYNTHESIS AND BREAKDOWN

The interpretation of changes in muscle free amino acids must rest on their role in protein synthesis and breakdown. The total body pool of free amino

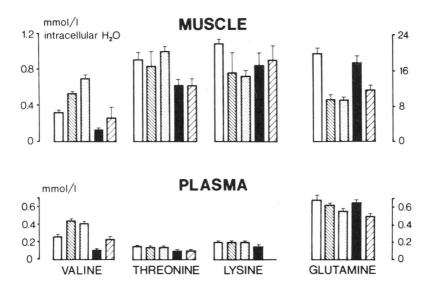

Figure 2.1 Free amino acids in different catabolic states

acids is about 120 g, only 1% of the total protein pool, and it amounts to about 30% of the daily protein turnover which is assumed to be about 400 g per day[21]. These figures suggest that there are sufficient amounts of free amino acids in the body pool for at least 7 h at the normal rate of protein synthesis. However, when the breakdown rates of the individual amino acids are considered the above calculation is misleading. Table 2.2 exemplifies the distribution of selected amino acids between extra- and intracellular water and the sum of essential and non-essential amino acids in the respective compartments. As shown in the table the BCAA reveal similar concentrations in cell and plasma, whereas the non-essential amino acids are highly concentrated in cells as compared to plasma. The differences between extra- and intracellular contents of valine, isoleucine and leucine are mainly due to differences in size of the two compartments and only slightly influenced by different concentrations. Considering the distribution of amino acids in mixed tissue protein (Table 2.2, column 4)[22], it is evident that it is very different from that in the free amino acid pool (Table 2.2, column 3). Consequently, this means that there is a large difference in availability of individual amino acids for protein synthesis. One may calculate this difference as the length of time protein synthesis could continue before it uses up all the free amino acids in the total body pool, assuming that the rate of protein synthesis is 400 g per day. As calculated in Table 2.2 (column 5), the availability of the BCAA is considerably less than those of non-essential amino acids or lysine. This means that alterations in rates of protein synthesis or breakdown in any organ or tissue cannot continue for more than a few minutes unless there are compensatory changes in other organs, involving changing rates of net protein synthesis, or amino acid catabolism or of dietary amino acid supply.

Table 2.2   The estimated distribution of selected amino acids (AA) as well as essential and non-essential AA between extra- and intracellular water. The calculation of AA turnover time is based on the availability of individual free AA in the total body water assuming a daily protein turnover of 400 g/day and considering the distribution of individual AA in mixed tissue protein

Amino acid	Free AA concentration (g/70 kg man)			Mixed tissue protein concentration (g AA/400 g)	AA turnover (h)
	Intra-cellular (1)	Extra-cellular (2)	Total body (3)	(4)	(5)
Valine	0.9	0.7	1.6	20	1.9
Leucine	0.7	0.4	1.1	33	0.8
Isoleucine	0.4	0.2	0.6	21	0.7
Lysine	3.2	0.7	3.9	37	2.5
Total essential	9.5	3.4	12.9	180	2.0
Glutamine	69	2.1	71	31	55
Serine	2.2	0.2	2.4	15	3.8
Total non-essential	105	4.5	110	220	12
Total AA	115	7.8	123	400	7

BCAA IN INJURY

In Figure 2.2 changes in intracellular muscle free and plasma free concentrations of BCAA are illustrated in different catabolic conditions. Bedrest (immobilization)[2], starvation[20], semi-starvation[2,3], post-operative trauma[3,18], severe injury[4,19], and onset of sepsis[14] are conditions associated with negative nitrogen balance which is presumed to originate in large part from net protein degradation in skeletal muscle[23]. In the above conditions the observed negative nitrogen balance varies in extent from -1 g N per day on bedrest to 20–40 g N or more per day in severe injury and sepsis. This means that in each of these conditions there is a net transport of amino acids out of muscle to other tissues.

The patterns of changes in amino acid concentrations in muscle and plasma show many similarities during catabolism. In all cases there is an increase in BCAA aromatic amino acids and methionine and a decrease in glutamine and in basic amino acids. There is, however, a gradual response, with minimal changes seen in muscle with bedrest and maximal changes in

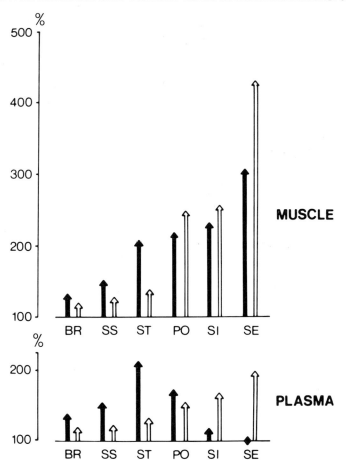

Figure 2.2 Branched-chain amino acids (↑) and aromatic amino acids and methionine (⇑) as affected by catabolism in muscle and plasma. % changes from normal. BR = Bedrest; SS = semi-starvation; ST = starvation; PO = post-operative trauma; SI = severe injury; *SE* = sepsis

sepsis. Obviously, the alterations observed in these conditions tend to follow a pattern according to the specifications of the different amino acid transport system. The BCAA together with phenylalanine, tyrosine and methionine are transported by the leucine-preferring system (L-system)[24]. As shown in Figure 2.2 in the milder conditions increases of muscle BCAA are proportionally greater than those of the aromatic amino acids and methionine, while in severe injury and sepsis, the increases of muscle aromatic amino acids and methionine are greater than for BCAA. It is also notable that in the milder conditions, plasma concentrations of BCAA rise proportionally to intracellular concentrations, but in severe catabolic states increased intracellular BCAA concentrations are not accompanied with similar rises in plasma.

BCAA IN URAEMIC CATABOLISM

In the untreated uraemic patients the findings of high plasma concentrations of citrulline, low plasma concentrations of serine, leucine and isoleucine and low plasma and muscle concentrations of threonine, valine, lysine and histidine confirm the existence of a typical free amino acid pattern in plasma and muscle of patients with advanced renal failure[5,15]. In patients treated with low protein diet supplemented with an amino acid mixture of essential amino acids in the proportions recommended by Rose[13,14] some of the abnormalities were found to be corrected. However, the abnormal distribution of leucine and isoleucine persisted with low plasma and normal muscle intracellular concentrations, while muscle valine concentrations were still reduced by about 30% even though the regimen provided 3–4 times the minimum requirement of valine (Figure 2.3). These abnormalities in the BCAA pattern resemble those observed in rats fed a low protein diet containing an excess of leucine[25] thus suggesting the existence of a BCAA antagonism in chronic uraemia[1,15]. Antagonism of BCAA has earlier been described[26] showing that excess or deficiency of one of the BCAA might change the distribution of the others. This phenomenon was associated with deterioration of growth and diminished amino acid utilization. The BCAA antagonism observed with depletion of the extra- and intracellular valine pools together with abnormal distribution of leucine and isoleucine indicate that uraemic patients may require BCAA in different proportions to non-uraemic individuals.

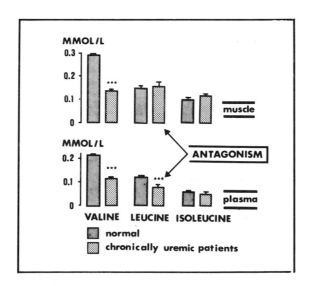

Figure 2.3 Branched-chain amino acid antagonism in uraemia[5]. Results are mean ± SEM. Significant differences between healthy controls and patients are indicated; *** p 0.001

HYPOTHESES AND INTERPRETATIONS

Changes of intracellular amino acid pattern in injury

Increased tissue catabolism alone cannot explain the observed intracellular amino acid pattern, since certain amino acids show no change or reveal a decreased concentration. However, the selected changes seen for BCAA, the aromatic amino acids and methionine may be accounted for, if it is postulated that the concentration of the L-system in muscle is rate limiting and that higher efflux requires higher intracellular concentrations. This helps to explain the finding that intracellular concentrations as well as intra–extracellular transmembrane gradients increase with increasingly negative nitrogen balance. In most severe catabolic states, net proteolysis in muscle is much greater and there are fourfold increases in muscle concentrations of the aromatic amino acids and methionine[4,19]. The liver's ability to handle this outflow will be taxed, leading to increases in plasma concentrations. If the L-transport system is limiting in severe catabolic states, the high muscle concentrations of aromatic amino acids and methionine will competitively reduce the outflow of BCAA, thus plasma concentrations of BCAA will decrease with respect to those in muscle.

Changes of intracellular amino acid pattern in uraemia

The pathogenesis of the abnormal uraemic intracellular pattern (BCAA antagonism) remains speculative. The decreased IC/EC gradients of leucine and isoleucine and the unaltered valine gradient suggest that the membrane transport or permeability of leucine and isoleucine is changed, either as a direct effect of uraemia or secondary to disturbed amino acid metabolism or endocrine abnormalities. Selective depletion of free valine in uraemia would require that this amino acid is preferentially catabolized. It is generally believed that the same transaminase is involved in the degradation of all BCAA[27] and that a single dehydrogenase complex controls the rate-limiting decarboxylation of the branched-chain keto acids (BCKA)[28]. However, the demonstration of metabolic defects involving only one or two of the BCAA in the human[29] suggests that decarboxylation of BCKA may occur by more than one enzymatic mechanism[30]. There is also the possibility that the affinity of the individual BCKA to one or more enzymes of dehydrogenase complex is due to change in pH[31] or presence of high concentrations of one or several metabolites which accumulate in the uraemic state[32]. An interesting observation is that diabetic rats fail to show depression in valine and isoleucine when fed a diet high in leucine and it has been suggested that insulin plays a role in the BCAA antagonism[33] in uraemia. Insulin secretion is usually enhanced in an attempt to compensate for the well-known uraemic insulin resistance[34,35]. Leucine is a potent agonist for insulin secretion, thus the high leucine supply (cf. Rose formula) may further stimulate insulin secretion. In contrast to the impairment of insulin action with respect to glucose metabolism, sensitivity to the effect of insulin on BCAA metabolism in tissues appears to be normal[36]. It is, thus, possible that hyperinsulinaemia may contribute to the abnormal BCAA pattern. This is supported by a recent finding which shows that the effect of insulin on BCKA is different in normal individuals and in the uraemic patients[37].

The obvious question is whether low intracellular valine and BCAA antagonism *per se* could limit protein synthesis. Although tRNA charging should be adequate even at low concentrations (Harper, personal communication), compartmentation in the cell resulting in lower amino acid concentration at the site of protein synthesis than in the tissues as a whole would reduce charging and inhibit protein synthesis[38].

THE UNIQUE ROLE OF BCAA

The BCAA, particularly leucine, stimulate protein synthesis and inhibit protein breakdown in muscle *in vitro*[28,39,40]. The BCAA are not catabolized to any extent by the liver, but are primarily oxidized by the skeletal muscle[28,39]. As a consequence, arterial concentrations of BCAA rise more after a protein meal than do the other amino acids which are largely absorbed by the liver[41]. There is an increase in the rate of muscle protein synthesis after a meal[42] which is presumably the result of this increased supply of BCAA. In support of this is the finding by several investigators that infusion of BCAA in human subjects improves nitrogen balance[43,44]. Thus, it seems probable that the effect of dietary protein in increasing muscle protein synthesis is mediated by an increased concentration of BCAA. One might expect the reverse to hold true, i.e. that a decrease in muscle protein synthesis or an increase in catabolism would be mediated by a decreased concentration of BCAA in intra- and extracellular compartments. However, the opposite occurs in trauma, in which increased concentrations are seen in both muscle and plasma. This suggests that changes in BCAA do not mediate but rather result from changes in protein synthesis.

It is suggested that BCAA inhibit the flux of amino acids from muscle, effectively decreasing muscle breakdown[28,39,40,45] which may result in improved nitrogen balance[43,44]. This could be explained if the rate of BCAA oxidation determines the rate of overall protein catabolism[28,45]. There are also speculations that in sepsis muscle oxidizes BCAA to compensate for an energy deficit caused by the reduced intracellular oxidation of glucose[46] and that in sepsis there is an energy deficit due to the inadequate supply of metabolic fuel which may force the utilization of amino acid for energy production[47].

These theories assume reduced intracellular BCAA concentration while in contrast, direct measurements revealed considerably increased BCAA levels in the cells. Furthermore, in recent investigations we were also able to demonstrate that severe trauma and sepsis were associated with tissue depletion of energy-rich compounds and the cellular energy level was decreased[48,49]. The combination of a tissue energy deficit and the observed accumulation of intracellular BCAA thus suggests an impairment of substrate utilization, rather than lack of fuel availability to the cell.

CONCLUSIONS AND FUTURE ASPECTS

None of these results support that the administration of BCAA only may specifically effect protein synthesis. BCAA together with other amino acids are undoubtedly of nutritional importance and it is possible that the beneficial results observed in the short term may be related to a specific action of leucine. Nevertheless, it is slowly becoming clear that nutrition at the height of the flow phase after injury requires further detailed studies of the influence of nutrients on the altered metabolism at this time. The imbalance of cellular amino acids in post-operative catabolism may be one of the fundamental inadequacies in cell nutrition adversely affecting normal protein synthesis. This indicates that proportions and minimum requirements based on data from healthy subjects cannot be applied directly in post-operative catabolism and also suggests that such patients should be supplied with amino acids in other proportions. A better understanding of the factor affecting protein synthesis and breakdown is needed not only from the theoretical standpoint, but has important practical implications if it can lead to a reduction of catabolic losses and an increase of protein gains during repletion therapy.

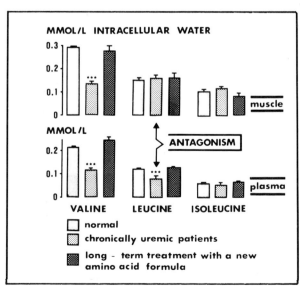

Figure 2.4 The effect of nutrition on branched-chain amino acid antagonism in uraemia[5,15] other details as Figure 2.4

Recent experiences in conservative treatment of chronic uraemia may serve as an example. Based on information on free amino acids in plasma and muscle a new amino acid formula (NAF) was studied in which the proportions of essential amino acids were altered and tyrosine was added[5,15]. Results of metabolic studies show that the essential amino acid abnormalities in plasma and muscle typical for uraemia can be corrected by nutritional means[5,50] (Figure 2.4). Longterm treatment with the new formula resulted in improved nitrogen balance[51], possibly due to correction of BCAA antagonism and the repletion of the free amino acid pools which were low and possibly limiting for protein synthesis. The results, thus, suggest that normalization of amino acid pools may improve nitrogen utilization. Although the observed difference in nitrogen balance may be small when expressed in grams of nitrogen per day, the difference in the cumulative balance may become great and of obvious clinical importance in longterm treatment.

ACKNOWLEDGEMENTS

This work was supported by grants from the Swedish Medical Research Council (project no. B82-17X-04219-098) and by Public Health Service grants GM 14546 and HL 239-75 from the National Institutes of Health by Contract DA-49-193-MD-2552, with the US Army Research and Development Command.

References

1 Bergström, J., Fürst, P., Norée, L-O. and Vinnars, E. (1978). Intracellular amino acids in muscle tissue of patients with chronic uremia. The effect of peritoneal dialysis and infusion of essential amino acids. *Clin. Sci. Mol. Med.*, **54**, 51

2 Askanazi, J., Elwyn, D. H., Kinney, J. M., Gump, F. E., Michelsen, C. B., Stinchfield, F. E., Fürst, P., Vinnars, E. and Bergström, J. (1978). Muscle and plasma amino acids after injury: the role of inactivity. *Ann. Surg.*, **188**, 797

3 Askanazi, J., Fürst, P., Michelsen, C. B., Elwyn, D. H., Vinnars, E., Gump, F. E., Stinchfield, F. E. and Kinney, J. M. (1980). Muscle and plasma amino acids after injury: hypocaloric glucose *vs* amino acids infusion. *Ann. Surg.*, **191**, 465

4 Askanazi, J., Carpantier, Y. A., Michelsen, C. B., Elwyn, D. H., Fürst, P., Kantrowitz, L. R., Gump, F. E. and Kinney, J. M. (1980). Muscle and plasma amino acids following injury. Influence of intercurrent infection. *Ann. Surg.*, **192**, 78

5 Alvestrand, A., Bergström, J. and Fürst, P. (1982). Plasma and muscle free amino acids in uremia: influence of nutrition with amino acids. *Clin. Nephrol.* (In press)

6 Bergström, J. (1962). Muscle electrolytes in man. *Scand. J. Clin. Lab. Invest.*, **68**, 7

7 Bergström, J. (1975). Percutaneous needle biopsy of skeletal muscle in physiological and clinical research. *Scand. J. Clin. Lab. Invest.*, **35**, 609

8 Bergström, J. and Fridén, A-M. (1975). The effect of hydrochlorothiazide and amiloride administered together on muscle electrolyte in normal subjects. *Acta Med. Scand.*, **197**, 415

9 Fröberg, S. O. (1972). Methodological, experimental and clinical studies. *Thesis*, Uppsala

10 Bergström, J., Fürst, P., Norée, L-O. and Vinnars, E. (1974). Intracellular free amino acid concentration in human muscle tissue. *J. Appl. Physiol.*, **36**, 693

11 Bergström, J., Alvestrand, A., Fürst, P., Hultman, E., Sahlin, K., Vinnars, E. and Widström, A. (1976). Influence of severe potassium depletion and subsequent repletion with

potassium on muscle electrolytes, metabolites and amino acids in man. *Clin. Sci. Mol. Med.*, **51**, 589

12 Bergström, J., Fürst, P., Holmström, B., Vinnars, E., Askanazi, J., Elwyn, D., Michelsen, C. B. and Kinney, J. (1981). Influence of injury and nutrition on muscle water and electrolytes. *Ann. Surg.*, **193**, 810

13 Alvestrand, A., Bergström, J., Fürst, P., Germanis, G. and Widstam, U. (1978). Effect of essential amino acid supplementation on muscle and plasma free amino acids in chronic uremia. *Kidney Intern.*, **14**, 323

14 Fürst, P., Ahlberg, M., Alvestrand, A. and Bergström, J. (1978). Principles of essential amino acid therapy in uremia. *Am. J. Clin. Nutr.*, **31**, 1949

15 Fürst, P., Alvestrand, A. and Bergström, J. (1980). Effects of nutrition and catabolic stress on intracellular amino acid pools in uremia. *Am. J. Clin. Nutr.*, **33**, 1387

16 Roch-Nordlund, A. E., Alinder, A., Ahlberg, M., Fürst, P. and Werner, G. (1974). Nitrogen metabolism in diabetic patients. *Acta Endocrinol.*, **77**, 190

17 Vinnars, E., Bergström, J. and Fürst, P. (1975). Influence of postoperative state in the intracellular free amino acids in human muscle tissue. *Ann. Surg.*, **163**, 665

18 Fürst, P., Bergström, J., Kinney, J. M. and Vinnars, E. (1977). Nutrition in postoperative catabolism. In J. E. Richards and J. M. Kinney (eds.). *Nutritional Aspects of Care of the Critically Ill*, p. 387. (Edinburgh, London: Churchill Livingstone)

19 Fürst, P., Bergström, J., Chao, L., Larsson, J., Liljedahl, S-O., Neuhäuser, M., Schildt, B. and Vinnars, E. (1979). Influence of amino acid metabolism in severe trauma. *Acta Chir. Scand.*, **494**, 136

20 Elwyn, D., Fürst, P., Askanazi, J. and Kinney, J. M. (1981). Effect of fasting on muscle concentrations of branched chain amino acids. In M. Walser and J. R. Williamson (eds.). *Metabolism and Clinical Implication of Branched Chain Amino and Keto-Acids*, p. 547. (New York: North Holland, Elsevier)

21 Munro, H. N. and Thompson, R. S. T. (1953). Influence of glucose on amino acid metabolism. *Metab. Clin. Exp.*, **2**, 354

22 Fürst, P., Jonsson, A., Josephson, B. and Vinnars, E. (1970). Distribution in muscle and liver vein protein of ^{15}N administered as ammonium acetate to man. *J. Appl. Physiol.*, **29**, 307

23 Kinney, J. M. (1976). Surgical diagnosis, patterns of energy, weight and tissue changes. In A. W. Wilkinson and E. P. Cuthbertson (eds.). *Metabolism and the Response to Injury.* p. 121. (Kent: Portman Medical)

24 Christensen, H. N. (1969). Some special kinetic problems of transport. *Adv. Enzymol.*, **32**, 1

25 Shinnic, F. L. and Harper, E. A. (1977). Effects of branched-chain amino acid antagonism in the rat on tissue amino acid and ketoacid concentrations. *J. Nutr.*, **107**, 887

26 Harper, A. E. (1964). Amino acid toxicities and imbalances. In H. N. Munro and J. B. Allison (eds.). *Mammalian Protein Metabolism.* Vol. II, pp. 87–134. (New York: Academic Press)

27 Ichihara, A. and Koyama, E. (1966). Transaminase of branched-chain amino acids. I. Branched chain amino acids alpha-ketoglutarate transaminase. *J. Biochem.*, **59**, 160

28 Odessey, R. and Goldberg, A. L. (1972). Oxidation of leucine by rat skeletal muscle. *Am. J. Physiol.*, **22**, 1376

29 Budd, M. A., Tanaka, K., Holmes, L. B., Efron, M. L., Crawford, J. D. and Isselbacher, K. J. (1967). Isovaleric acidemia. Clinical features of a new genetic defect of leucine metabolism. *N. Engl. J. Med.*, **277**, 321

30 Connelly, J. L., Danner, D. J. and Bowden, J. A. (1968). Branched-chain alpha-keto-acid metabolism. I. Isolation, purification and partial characterization of bovine liver alpha-ketoisocaproic: alpha-keto-beta-methylvaleric acid dehydrogenase. *J. Biol. Chem.*, **243**, 1198

31 Randle, P. J. (1981). Discussion In M. Walser and J. R. Williamson (eds.). *Metabolism and Clinical Implications of Branched-Chain Amino and Ketoacids.* p. 619. (New York: Elsevier/North Holland)

32 Bergström, J. and Fürst, P. (1979). Uraemic toxins. In W. Drukker, F. M. Parsons and J. F. Maher (eds.). *Replacement of Renal Function by Dialysis.* pp. 334–368. (The Hague: Martinus Nimhoff)

33 Clark, A., Yamada, C. and Swendseid, M. (1968). Effect of L-leucine on amino acid levels in

ADVANCES IN CLINICAL NUTRITION

plasma and tissue of normal and diabetic rats. *Am. J. Physiol.*, **215**, 1324

34 DeFronzo, R. A., Nadres, R., Edgar, P. and Walker, W. G. (1973). Carbohydrate metabolism in uremia. A review. *Medicine*, **52**, 469
35 DeFronzo, R. A., Alvestrand, A., Smith, D., Hendler, R., Hendler, E. and Wahren, J. (1981). Insulin resistance in uremia. *J. Clin. Invest.*, **67**, 563
36 DeFronzo, R. A. and Felig, P. (1980). Amino acid metabolism in uremia: insights gained from normal and diabetic man. *Am. J. Clin. Nutr.*, **33**, 1378
37 Schander, P., Mattaei, H. V., Henning, H. V., Scheler, F. and Langenbeck, U. (1981). Blood levels of branched-chain α-ketoacids in uremia: effect of an oral glucose tolerance test. *Klin. Wochenschr.*, **59**, 845
38 Waterlow, J. C., Garlick, P. J. and Millward, D. J. (1978). *Protein-Turnover in Mammalian Tissues and in the Whole Body.* p. 656. (Amsterdam: Elsevier/North-Holland)
39 Buse, M. B. and Reid, M. J. (1975). Leucine, possible regulator of protein turnover in muscle. *J. Clin. Invest.*, **58**, 1250
40 Fulks, R. M. and Goldberg, A. L. (1975). Effects of insulin, glucose and amino acid on protein turnover in rat diaphragm. *J. Biol. Chem.*, **250**, 290
41 Elwyn, D. H., Hamendra, C. P. and Shoemaker, W. V. (1968). Amino acid movements between gut, liver and periphery in unanesthetized dogs. *Am. J. Physiol.*, **215**, 1260
42 Waterlow, J. C. (1969). The assessment of protein nutrition and metabolism in the whole animal with special reference to man. In H. N. Munro (ed.). *Mammalian Protein Metabolism.* p. 325. (New York: Academic Press)
43 Freund, H. R., Ryan, J. A. and Fischer, J. E. (1978). Amino acid derangements in patients with sepsis: treatment with branched-chain amino acid rich infusion. *Ann. Surg.*, **188**, 423
44 Freund, H. R., Yoshimura, N., Lunetta, L. and Fischer, J. E. (1978). The role of the branched-chain amino acids in decreasing muscle catabolism *in vivo. Surgery*, **83**, 611
45 Odessey, R., Khairallah, E. Z. and Goldberg, A. (1974). Origin and possible significance of alanine production by skeletal muscle. *J. Biol. Chem.*, **249**, 7623
46 O'Donnell, T. F. Jr., Clowes, G. H. A. Jr. and Blackburn, G. L. (1976). Proteolysis associated with a deficit of peripheral energy fuel substrates in septic man. *Surgery*, **80**, 192
47 Ryan, N. T., Blackburn, G. L. and Clowes, G. H. A. Jr. (1974). Differential tissue sensitivity to elevated endogenous insulin levels during experimental peritonitis in rats. *Metabolism*, **23**, 1081
48 Liaw, K. Y., Askanazi, J., Michelsen, C. B., Kantrowitz, L. R., Fürst, P. and Kinney, J. M. (1980). Effect of injury and sepsis on high-energy phosphates in muscle and red cells. *J. Trauma*, **20**, 755
49 Liaw, K. Y., Askanazi, J., Michelsen, C. B., Fürst, P., Elwyn, D. H. and Kinney, J. M. (1982). Effect of postoperative nutrition on muscle high energy phosphates. *Ann. Surg.* **195**, 12
50 Alvestrand, A., Ahlberg, M., Bergström, J. and Fürst, P. (1981). The effect of nutritional regimens on branched-chain amino acid (BCAA) antagonism in uremia. In M. Walser and J. R. Williamson (eds.). *Metabolism and the Clinical Implications of Branched-Chain Amino and Ketoacids.* p. 547. (New York: North-Holland/Elsevier)
51 Alvestrand, A., Ahlberg, M., Bergström, J. and Fürst, P. (1982). Clinical results of long-term treatment with low protein diet and a new amino acid preparation in chronic uremic patients. *Clin. Nephrol.* (In press)

36

3
Branched-chain amino acid supplementation in the injured rat model

H. R. FREUND, Z. GIMMON and J. E. FISCHER

INTRODUCTION

Post-injury or post-operative catabolism is associated with negative nitrogen balance, muscle protein breakdown and weight loss. Depending on the severity of the injury, body protein is catabolized at a rate of 75–150 g/day, a loss of up to 300–600 g of lean body mass per day. Recent studies have attributed special anti-catabolic effects to the branched-chain amino acids (BCAA): valine, leucine and isoleucine. *In vitro* experiments have demonstrated that the BCAA serve as energy substrate for the muscle, participate in gluconeogenesis through alanine and glutamine, and play a role in regulating muscle protein degradation and synthesis[1–4]. Fulks *et al.*[2] and Buse and Reid[1] demonstrated that the BCAA play a crucial role in regulating nitrogen balance in muscle by decreasing protein degradation and increasing protein synthesis in a dose-dependent fashion. Odessey *et al.*[4] suggested that BCAA inhibit the flux of amino acids from the muscle, thus effectively decreasing muscle breakdown. Recently, Sapir, Walser *et al.*[5–7] demonstrated prolonged improvement in nitrogen conservation by the administration of alpha keto-analogues of the BCAA during fasting and hepatic encephalopathy. Sherwin[8] reported a less negative nitrogen balance in fasted obese subjects infused with leucine.

The following series of studies were undertaken to elucidate the *in vivo* role of the BCAA in preventing or minimizing post-injury lean body mass breakdown.

EXPERIMENTAL STUDIES

A standard experimental model of moderate operative injury was utilized. Sprague-Dawley rats weighing 300–370 g underwent laparotomy and jugular vein cannulation, while under pentobarbital anaesthesia. Following surgery, animals were placed in metabolic cages and infused for a total of 96 h. Water was allowed *ad libitum*. Intake and output, bodyweight change, and nitrogen balance were recorded daily. After 96 h of infusion the animals were killed by decapitation, and blood, muscle, and liver were harvested immediately and frozen for amino acid determination performed by a Beckman 121-MB amino acid analyser, on the supernate of protein or plasma deproteinized by the addition of 5% sulfosalicylic acid. In some experiments plasma glucose, insulin, albumin, and liver function tests were determined.

Investigation of nitrogen-conserving effect of BCAAs

Initially the nitrogen-conserving quality of four amino acid formulations differing mainly in their branched-chain amino acid concentrations were compared. All groups (10–12 rats per group) received isocaloric amounts of 5% (50 g l^{-1}) dextrose at 3 cal/100 g of bodyweight/24 h. Hypocaloric glucose was used in an attempt to avoid masking the effects of amino acids. All groups received adequate amounts of electrolytes, minerals, vitamins and trace elements. Group I did not receive any amino acids. All the solutions used in groups II to V contained 3% (30 g l^{-1}) amino acids, and were infused at 0.5 g/100 g of bodyweight/24 h. Group II received Freamine*, containing 22% BCAAs (i.e. 22% of infused amino acids), group III received a specially designed amino acid formulation containing 35% branched-chain amino acids, group IV received an amino acid solution containing a total of 52% BCAAs, and group V received only the three BCAAs in equimolar quantities (100% BCAA).

The control post-operative rat undergoing anaesthesia, laparotomy and jugular vein cannulation and infused with 5% dextrose has a mean negative nitrogen balance of 200–300 mg N/day. The mean cumulative 4 days and day-to-day nitrogen balances are summarized in Figure 3.1. All groups demonstrated negative nitrogen balance following injury and the use of hypocaloric solutions. All four groups of animals which received amino acids were less negative in respect to nitrogen balance than the control group with 5% dextrose only ($p<0.001$). Among the four groups receiving amino acids the 35% BCAA and 100% BCAA groups were significantly less negative than the 22% and the 52% BCAA groups ($p<0.001$). This was true both for the mean cumulative 4-day nitrogen balance and for days 1, 2, and 4 (Figure 3.1). All five groups of animals lost weight during the 4-day period of the experiment.

Comparing plasma and muscle amino acid patterns of the various groups of animals, the groups receiving 5% dextrose, 22% BCAA, 35% BCAA, and 52% BCAA had very similar patterns, with most of the amino-acid levels elevated. This pattern reverted almost completely to normal with the infusion

*McGaw, Glendale, California

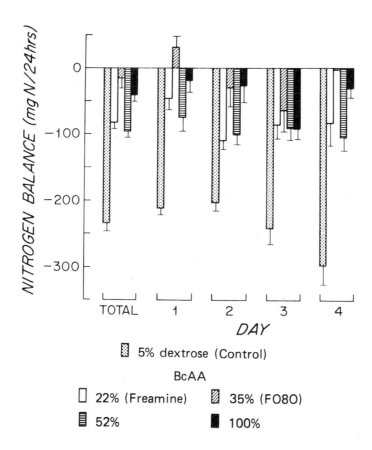

Figure 3.1 Nitrogen balance (mean ± SEM) in injured rats infused with amino acid formulations containing 22%, 35%, 52% or 100% (as % of total amino acids infused) BCAA in 5% dextrose (Courtesy of Freund *et al.* (1978). The role of the branched chain amino acids in decreasing muscle catabolism *in vivo. Surgery*, **83**, 611)

of 100% BCAA, where only the three BCAAs and alanine were elevated (Figures 3.2 and 3.3).

This set of experiments was performed with only 5% dextrose as the caloric source in an attempt to single out the effects of the BCAA without masking it with excess calories. Subsequent experiments were directed at elucidating the effectiveness of BCAAs and hypertonic glucose, mimicking the clinical situation of hypertonic glucose-amino acid infusions in post-operative patients.

Effect of maintenance levels of carbohydrate on protein-sparing action of BCAAs

Four groups of rats undergoing anaesthesia, laparotomy and jugular vein

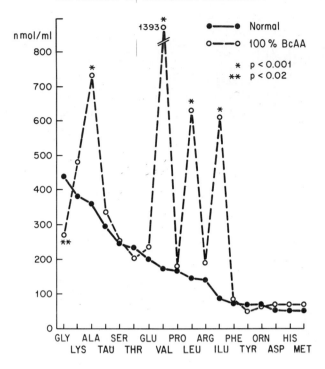

Figure 3.2 Plasma amino acid patterns in injured rats receiving a 100% (as % of total amino acids infused) BCAA formulation in 5% dextrose. Note the almost normal amino acid pattern except for elevated levels of leucine, valine, isoleucine and alanine (Courtesy of Freund *et al.* (see Fig. 3.1)

cannulation were infused with isocaloric amounts of 24% dextrose at a dose of 17 cal/100 g bodyweight/24 h.

Group I did not receive any amino acids. All the solutions used in groups II–IV contained 3.4% (34 g l^{-1}) amino acids and differing only as to their amino acid formulation, principally in the content of BCAA, and were infused at 0.62 g/100 g of bodyweight/24 h.

Group II received Freamine, containing 22% BCAA, group III received a specially designed amino acid formulation, containing 35% BCAA, group IV received a solution containing only the three BCAAs: valine 1.4%, leucine 1.0% and isoleucine 1.0% (BCAA comprising 100% of total amino acids infused).

All three solutions produced significantly better nitrogen balance ($p<0.001$) and less weight loss ($p<0.01$–0.001) than the group receiving 24% dextrose only (Figure 3.4). The animals receiving amino acid solutions in 24% dextrose (II–IV) were all in nitrogen equilibrium or mild positive nitrogen balance, with no significant difference between the three groups (Figure 3.4). Even the infusion of only the three BCAAs in 24% dextrose resulted in

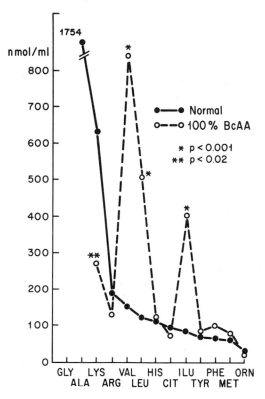

Figure 3.3 Muscle amino acid patterns in injured rats receiving a 100% (as % of total amino acids infused) BCAA formulation in 5% dextrose. Note the almost normal amino acid pattern except for elevated levels of leucine, valine, isoleucine and alanine (Courtesy of Freund *et al.* (see Fig. 3.1))

nitrogen equilibrium similar to the two other more balanced amino acid solutions.

Comparing the plasma and muscle amino acid patterns in the groups receiving amino acids in 24% dextrose with aminograms of normal animals, the groups receiving 22% and 35% BCAA show a very similar pattern, with most of the amino acids present at increased levels probably being the result of muscle protein catabolism to satisfy energy requirements. Unlike the other two groups receiving amino acid infusions, the group receiving 100% BCAA showed an almost normal plasma and muscle amino acid pattern except for marked increases in the concentrations of valine, leucine, isoleucine and alanine. This near normal pattern in animals infused with a completely unbalanced amino acid formulation containing only the three BCAAs may be the result of the complete inhibition of amino acid efflux from the muscle by pharmacological doses of the BCAA so that the muscle depends almost solely on the exogenous supply of BCAAs to satisfy its metabolic require-

41

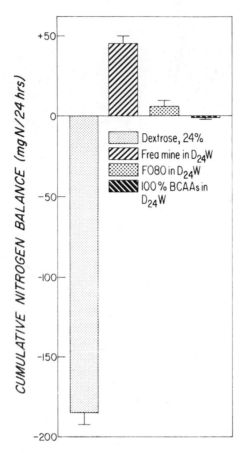

Figure 3.4 Nitrogen balances (mean ± SEM) in injured rats infused with amino acid formulations containing 22%, 35% or 100% (as % of total amino acid infused) BCAA in 24% dextrose. Note nitrogen equilibrium achieved by the infusion of BCAA only (Courtesy of Freund *et al.* (1980). The effect of branched chain amino acids and hypertonic glucose infusion on post-injury catabolism in the rat. *Surgery*, **87**, 401)

ments, similar to the *in vitro* studies reported by Odessey *et al.*[4].

Comparing this with earlier studies, the addition of maintenance amounts of carbohydrate (17 kcal/100 g rat/24 h) does not appear to result in significantly greater protein sparing than infusions of the branched-chain amino acids with 5% dextrose (3.0 kcal/100 g rat/24 h), suggesting that in the presence of pharmacologic doses of branched-chain amino acids, little additional protein-sparing may be expected with carbohydrate.

The role of alanine and gluconeogenesis in nitrogen-sparing action of BCAAs
Trying to determine and separate the role of alanine and gluconeogenesis in this nitrogen-conserving quality of the BCAA, we infused the injured rat model with alanine alone, comparing it to the infusion of BCAA alone and

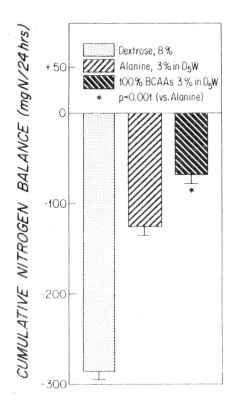

Figure 3.5 Nitrogen balances (mean±SEM) in injured rats infused with either 8% dextrose, 3% alanine in 5% dextrose, or 3% BCAA in 5% dextrose (Courtesy of Freund *et al.* (1980) The role of alanine in the nitrogen conserving quality of the branched chain amino acids in the post-injury state. *J. Surg. Res.*, **29**, 23)

isocaloric amounts of glucose only. Group I received 8% dextrose with no amino acids, isocaloric with group II receiving 3% alanine in 5% dextrose, and group III receiving 3% BCAA (valine, leucine, and isoleucine in equal amounts) in 5% dextrose (100% BCAA).

Anaesthesia, laparotomy and jugular vein cannulation in the control group receiving 8% dextrose resulted in mean daily negative nitrogen balance of 286 ± 10 mg N (range 262–330 mg N/24 h) and weight loss of 13.8 ± 1% of their bodyweight. The group receiving 100% BCAAs (group III) showed a mean daily negative nitrogen balance of 68 ± 10 mg N (range 46–110 mg N/24 h) and weight loss of 12.7 ± 3% bodyweight, while the group receiving alanine (group II) showed a mean daily negative nitrogen balance of 126 ± 11 mg N (range 71–161 mg N/24 h) and weight loss of 11.9 ± 1% body-weight. Thus both alanine and 100% BCAA had a significantly better nitrogen-conserving quality as compared with the control group receiving isocaloric 8% dextrose ($p<0.001$) with the 100% BCAA solution being superior to alanine ($p<0.001$) (Figure 3.5). That the 100% BCAA group exhibits a decisively better nitrogen balance than the alanine group, prompts the conclusion that only part of the improved nitrogen-sparing quality of the 100% BCAAs solution documented in this study is due to the production of alanine and its utilization for gluconeogenesis.

In further investigating the role of the BCAAs in post-injury metabolism, the extent of the contribution of each of the three BCAAs (leucine, valine, and isoleucine) to this improved nitrogen-conserving quality of the BCAAs was examined.

Mechanisms of nitrogen sparing of BCAAs

Using the same rat model, four groups of animals were infused with isocaloric amounts of 5% dextrose at a dose of 3 kcal/100 g bodyweight/24 h. Group I did not receive any amino acids but additional 1.5% glucose to a total of 6.5%; group II received a 1.5% solution of valine, group III, a 1.5% solution of isoleucine, group IV, a 1.5% solution of leucine, and group V, 1.5% alanine. All four amino acid containing solutions demonstrated a similarly improved nitrogen-conserving quality when compared to the group receiving 6.5% dextrose only ($p<0.001$) (Figure 3.6). The last group of experiments were repeated and 5 μCi[^{14}C]-tyrosine was added to the infusate for the last 24 hours to permit calculation of liver and muscle protein synthesis rate, and rate of tyrosine flux[9]. Our results demonstrated that the fractional synthesis rate of muscle was increased significantly only by the infusion of valine and unaltered by the infusion of alanine, isoleucine, and leucine (Table 3.1). The fractional synthesis rate of liver protein was significantly increased in all groups receiving BCAA compared to the glucose or alanine group (Table 3.1). Total body protein breakdown rate assessed by tyrosine flux was decreased to a similar extent in all groups receiving amino acids as compared to animals receiving glucose only (Table 3.1).

Table 3.1 Fractional synthesis rates in liver and muscle protein and total body protein breakdown rate in injured rats receiving valine, leucine, isoleucine or alanine

	Group I (6.5% dextrose)	Group II (1.5% alanine in 5% dextrose)	Group III (1.5% isoleucine in 5% dextrose)	Group IV (1.5% leucine in 5% dextrose)	Group V (1.5% valine in 5% dextrose)
Fractional synthesis rate of liver (%/day)	8.2±1.5	8.7±2.5	12±2.0*	12±3.0*	12±7*
Fractional synthesis rate of muscle (%/day)	1.2±0.5	1.0±0.2	1.5±0.4	1.1±0.5	2.3±1.4*
Breakdown rate (μmol tyr/hour)	63±18	33±11*	33±5*	38±9*	32±9*

*$p<0.05$

44

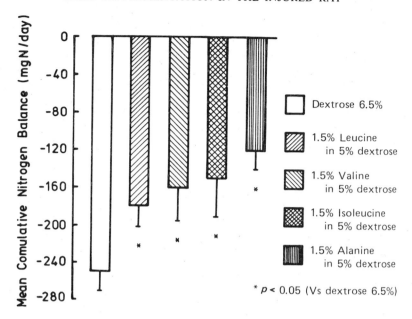

Figure 3.6 Nitrogen balances (mean±SEM) in injured rats infused with dextrose only (6.5%) or 5% dextrose with either 1.5% valine, 1.5% leucine, 1.5% isoleucine or 1.5% alanine (Courtesy of Freund *et al.* (1981). Nitrogen sparing mechanisms of singly administered branched chain amino acids in the injured rat. *Surgery*, 90, 237)

These results point to the unique effect of valine on muscle protein synthesis rate, and to similar beneficial effects of all three BCAAs on fractional synthesis rate of liver protein. All BCAAs, as well as alanine, caused a reduction in total body protein breakdown. The nitrogen-sparing effect of the BCAA thus seems to be in part the result of increased protein synthesis in liver and muscle, and reduction in protein degradation.

These preliminary results suggested that the nitrogen-conserving quality of amino acid solutions in the post-operative rat might be improved by increasing the amount of branched-chain amino acids. However, it became important to determine the minimum amount of BCAA needed to exert their beneficial effect so as to create a balanced amino acid formulation containing enough BCAA combined with adequate amounts of all other essential and non-essential amino acids.

The BCAA's enriched and balanced amino acid formulation

Aminosyn*, a balanced amino acid formulation, was modified to contain 40%, 45% and 50% BCAA. The 4.25% amino acid formulation was mixed with either 5% or 25% dextrose and adequate amounts of electrolytes, minerals, vitamins and trace elements.

*Abbott Labs., North Chicago, Illinois

Table 3.2 Mean daily nitrogen balances (mg N/day) of six groups of injured rats receiving balanced amino acid formulations containing 40%, 45% or 50% BCAA in 5 or 25% dextrose (w/v)

	40%* BCAA	45%* BCAA	50%* BCAA
5% (w/v) Dextrose	− 44 ± 7	− 10 ± 13	− 16 ± 12
25% (w/v) Dextrose	+ 48 ± 11	+ 86 ± 7	+ 89 ± 16

*% of total amino acids infused

The results of nitrogen balances as summarized in Table 3.2 point towards the 45–50% BCAA-containing solutions as the most effective. Studying the plasma amino acid patterns, the 40% and 45% BCAA-containing formulations maintained an almost normal pattern of essential and non-essential amino acids, including alanine and glutamine. Exceptions were the increases in plasma levels of the BCAA (up to 160–260%) which were, however, increased to a lesser degree in the 40% and 45% groups compared with the 50% group.

Plasma albumin levels in the 45% BCAA groups (5% and 25% glucose) were maintained at 2.8–2.9 g/100 ml compared to significantly lower levels of 2.6–2.7 g/100 ml in the 40% and 50% BCAA groups.

COMMENTS

The metabolic response to injury is characterized by an extensive and complex neuroendocrine activation resulting in rapid breakdown of muscle protein as substrate for energy production and gluconeogenesis, resulting in the well-known post-injury negative nitrogen balance and weight loss. Skeletal muscle is the major site of protein breakdown after injury. Muscle protein breakdown in addition to being wasteful leads also to relative deficiency of essential amino acids like leucine, isoleucine and valine which are selectively and extensively utilized by the muscle as an energy source and substrate for gluconeogenesis. This relative deficiency might have a bad effect on the availability of essential amino acids for protein synthesis. The muscle breakdown itself affects all skeletal muscle so that at a certain stage if catabolism continues and no nutritional support is afforded, vital functions like respiration, immunological activity and wound healing could be seriously affected. In recent years, there has been increased interest in reducing post-injury muscle catabolism by providing exogenous amino acids and calories. However, only very limited attention has been given to the qualitative amino acid requirements in the post-injury state.

The branched-chain amino acids (BCAA) valine, leucine and isoleucine are the only essential amino acids that are principally oxidized by skeletal muscle[1,3,4,10–13]. The rate of oxidation of BCAAs in muscle is stimulated by fasting, hormonal influence, stress, diabetes, and other conditions associated with muscle protein wasting and negative nitrogen balance. Oxidation of the BCAAs supplies energy to the muscle, and also nitrogen and possibly the carbon skeleton for the glucose alanine cycle and muscle glutamine synthesis[4,14-20]. Recent *in vitro* experiments have also ascribed regulatory functions to the BCAAs. Fulks *et al.*[2] suggested that the BCAAs promoted protein synthesis and reduced protein degradation in a dose-dependent fashion. Odessey *et al.*[4] showed the BCAAs to inhibit net protein breakdown in muscle and to stimulate *de novo* synthesis of alanine and glutamine, thus offering energy substrates for liver, kidney and gut. Buse and Reid[1] suggested that leucine might act as a regulator of protein turnover in skeletal muscle by inhibiting protein degradation and promoting protein synthesis. In human studies, Sapir *et al.*[5,6] and Walser *et al.*[7] achieved prolonged and improved nitrogen conservation during fasting and during hepatic encephalopathy by the administration of alpha-keto analogues of the BCAAs. Sherwin[8] infused leucine to fasting obese subjects with a resultant improvement in nitrogen balance but unchanged 3-methylhistidine excretion suggesting stimulation of muscle protein synthesis. It thus seemed reasonable to assume that these three special properties of the BCAAs, namely being an energy substrate for skeletal muscle, substrate for gluconeogenesis through alanine and glutamine and having a regulatory function in muscle protein degradation and synthesis, could be utilized for the reduction of post-injury protein catabolism.

Rats subjected to injury consisting of laparotomy and jugular vein cannulation demonstrated that the nitrogen-conserving quality of amino acid solutions administered in the post-operative period might be improved by increasing the amount of BCAAs. Furthermore, the infusion of a solution consisting only of the three BCAAs resulted in nitrogen equilibrium and near-normal plasma and muscle amino acid patterns, suggesting inhibition of amino acid efflux from the muscle. The exogenous supply of BCAAs must have satisfied the metabolic requirements of the muscle. This protein-sparing quality was present both in animals receiving hypocaloric amounts of 5% dextrose and animals receiving 25% dextrose infusions. Moreover, the increased caloric intake as glucose did not result in significantly greater protein-sparing, suggesting that in the presence of pharmacological doses of BCAAs, little additional nitrogen-sparing can be achieved by adding carbohydrate. Additional experiments including infusions of alanine alone, suggest that only part of this improved nitrogen-sparing quality of the BCAAs is due to production of alanine and its utilization for gluconeogenesis. This nitrogen-sparing effect is probably a combination of the role of BCAA as an energy source for muscle, serving as a substrate for alanine production and gluconeogenesis, and blocking amino acid efflux from muscle as suggested by Odessey *et al.*[4] in *in vitro* studies, and confirmed by the almost normal plasma and muscle amino acid patterns of animals infused with 100% BCAA, both with 5% and 24% dextrose solutions. Among the three BCAAs themselves, valine alone increased fractional synthesis rate of protein in both muscle and liver, while leucine and isoleucine positively affected fractional synthesis rate in liver only. All three BCAAs and alanine decreased protein breakdown rate.

CONCLUSIONS

Summarizing these animal studies the following conclusions concerning the post-injury state can be drawn.

Early nutritional support in the post-injury state will result in nitrogen equilibrium or even mildly positive nitrogen balance, suggesting that the 'obligatory catabolic phase' as viewed by Moore[21], can be minimized or even abolished by the provision of adequate amino acids and calories.

The infusion of only the three branched-chain amino acids in the immediate post-operative period is as effective in achieving nitrogen equilibrium as are other more balanced amino acid solutions.

The nitrogen-conserving quality of amino acid solutions in the post-injury period might be improved by increasing the amount of BCAAs.

Based on nitrogen balance data, plasma albumin levels and plasma amino acid patterns, we advocate the use of a balanced amino acid solution containing adequate amounts of essential and non-essential amino acids together with 45% BCAAs as the most appropriate amino acid formulation in the post-injury state.

References

1 Buse, M. G. and Reid, M. (1975). Leucine, a possible regulator of protein turnover in muscle. *J. Clin. Invest.*, **58**, 1251
2 Fulks, R. M., Li, J. B. and Goldberg, A. L. (1975). Effects of insulin, glucose and amino acids on protein turnover in rat diaphragm. *J. Biol. Chem.*, **250**, 280
3 Odessey, R. and Goldberg, A. L. (1972). Oxidation of leucine by rat skeletal muscle. *Am. J. Physiol.*, **223**, 1376
4 Odessey, R., Khairallah, E. A. and Goldberg, A. L. (1974). Origin and possible significance of alanine production by skeletal muscle. *J. Biol. Chem.*, **249**, 7623
5 Sapir, D. G., Owen, O. E., Pozefsky, T. *et al.* (1974). Nitrogen sparing induced by a mixture of essential amino acids given chiefly as their keto-analogues during prolonged starvation in obese subjects. *J. Clin. Invest.*, **54**, 974
6 Sapir, D. G. and Walser, M. (1977). Nitrogen sparing induced early in starvation by infusion of branched chain keto-acids. *Metabolism*, **26**, 301
7 Walser, M., Maddry, W. C. and Herlong, H. F. (1982). Ornithine salts of branched chain keto acids in the treatment of hepatic encephalopathy. In E. Holm (ed.). *Amino Acid and Ammonia Metabolism: Basic Data and Therapeutic Measures*, pp. 141–148. (Baden-Baden/Koln/New York: Witzstrock)
8 Sherwin, R. S. (1978). Effect of starvation on the turnover and metabolic response to leucine. *J. Clin. Invest.*, **61**, 1471
9 Garlick, P. J., Millward, D. J. and James, W. P. T. (1973). The diurnal response of muscle and liver protein synthesis *in vivo* in meal fed rats. *Biochem. J.*, **136**, 935
10 Buse, M. G. and Buse, J. (1967). Effect of free fatty acids and insulin on protein synthesis and amino acid metabolism of isolated rat diaphragms. *Diabetes*, **16**, 753
11 Manchester, K. L. (1965). Oxidation of amino acid by isolated rat diaphragm and the influence of insulin. *Biochim. Biophys. Acta*, **100**, 295

12 Miller, L. L. (1962). The role of the liver and non-hepatic tissues in the regulation of free amino acid levels in blood. In J. J. Holden (ed.). *Amino Acid Pools.* pp. 708–728. (Amsterdam: Elsevier)
13 Young, V. (1969). The role of skeletal and cardiac muscle in the regulation of protein metabolism. In H. N. Munro (ed.) *Mammalian Protein Metabolsim.* Vol. 4, pp. 585–674 (New York: Academic Press)
14 Felig, P. (1973). The glucose alanine cycle. *Metabolism*, **22**, 179
15 Felig, P., Pozefsky, T., Marliss, E. *et al.* (1970). Alanine: key role in gluconeogenesis. *Science*, **167**, 1003
16 Garber, A. J., Karl, I. E. and Kipnis, D. M. (1977). Metabolic interrelationship and factors controlling skeletal muscle protein degradation and the selective synthesis and release of alanine and glutamine. Clinical nutrition update – amino acids. *Am. Med. Assoc.*, pp. 10–20
17 Goldberg, A. L. and Odessey, R. (1972). Oxidation of amino acids by diaphragms from fed and fasted rats. *Am. J. Physiol.*, **223**, 1384
18 Marliss, E. B., Aoki, T. T., Pozefsky, T. *et al.* (1971). Muscle and splanchnic glutamine and glutamate metabolism in post-absorptive and starved man. *J. Clin. Invest.*, **50**, 814
19 Ruderman, N. B. and Berger, M. (1974). The formation of glutamine and alanine in skeletal muscle. *J. Biol. Chem.*, **249**, 5500
20 Wahren, J., Felig, P. and Hagenfeldt, L. (1976). Effect of protein ingestion on splanchnic and leg metabolism in normal men and in patients with diabetes mellitus. *J. Clin. Invest.*, **57**, 987
21 Moore, F. D. (1959). *Metabolic Care of the Surgical Patient.* (Philadelphia: W. B. Saunders)

4
Review of branched-chain amino acid supplementation in trauma

F. R. CERRA

INTRODUCTION

The metabolic response to surgical stress appears to induce alterations in carbohydrate, fat, and protein metabolism that differentiate it from the biochemistry and physiology of non-stressed, fasting man. As the severity and duration of the stress increases, so do the metabolic alterations. Some of these stress changes include increased fat mobilization and turnover[4], poorly suppressed gluconeogenesis, elevations in lactate and pyruvate with little change in their ratio[7], hyperglycaemia with varying degrees of insulin resistance[7,13], characteristic reductions or increases in plasma amino acid levels[6,7,9,12], increased negative nitrogen balance[14] and increased protein catabolism[3,20].

Current methods of nutritional support, when applied to these stressed conditions, have not produced the results that were observed in patients who were not stressed or simply starved. This acquired malnutrition of stress as well as the effects of stress on already malnourished patients, are thought to have a continuing impact on surgical morbidity and mortality[4,5,8,10,16,20,25].

With a better understanding of the stress response has come a realization that the principles of nutritional support are much different than those of non-stressed, fasting man. Total caloric needs increase, but the fuel mix to meet that need changes. There seems to occur an increasing reliance on fat and amino acids as energy sources with a reduced fractional need for glucose. It becomes apparent that standard nutritional support can achieve varying degrees of protein-sparing during stress[10,12,15,24], but that a reduction in lean body mass with its consequences is still the inevitable result. Once the stress has abated, however, standard, fasting principles again seem to apply.

The branched-chain amino acids (BCAA) are essential for protein synthesis to occur. They are catabolized primarily in peripheral tissue, especially muscle[17,22,23]. The enzyme activation for such catabolic functions is 3–10 times greater in muscle than liver, with activity in muscle increasing before that in liver during stress[22]. With stress states, the catabolism of BCAA increases, a phenomenon thought to occur mostly in the periphery with their utilization as an energy source[5,6,8,10]. The precise origin of this increased demand for BCAA is not known precisely, but is thought to reflect neuro-humeral modulation, by insulin[11,17,18,26]. A portion of the BCAAs produced from proteolysis in protein catabolism is released unchanged from muscle[19], accounting for the rising plasma levels seen in late septic stress. The increased utilization and turnover during stress explains the reduced plasma levels seen in minor stress. When given to non-stressed, fasting man, the plasma BCAA levels rapidly rise[26].

A possible metabolic regulator function has also been identified for the BCAAs, particularly leucine. Leucine can directly stimulate muscle protein synthesis in several animal models[17,18]. In other settings, the BCAA can reduce cardiac muscle and skeletal muscle proteolysis as well as stimulate visceral protein synthesis[11,17,18,22,23].

Thus, there are three potentially beneficial effects of increasing the BCAA load during the balanced amino acid support of stress: to serve as a caloric substrate; to stimulate protein synthesis; and to reduce protein catabolism.

Experimental data have shown a beneficial effect of the BCAA on nitrogen balance under stressed conditions through effects on both protein synthesis and degradation[5,11,15-18,22,23]. Because of the potential benefits of these effects to patients, prospective, blind studies are undertaken in surgical patients who were randomized to either standard total parenteral nutrition (TPN) or BCAA-rich TPN in the immediate post-operative period.

THE STUDY

Elective general surgery or multiple trauma patients aged 19–71 years were studied consecutively. All patients were previously well-nourished and were entered into the study within 16 h of the elective, abdominal surgery or multiple trauma. None of the patients had cirrhosis, hepatitis, pancreatitis, insulin-dependent diabetes mellitus, or a glomerular filtration rate of less than 30 ml/h as measured by a 24 h creatinine clearance assay. No patient received steroids. All patients had oxygen transport appropriately monitored and treated so that there was reasonable assurance that oxygen delivery was meeting demand.

Each patient was randomized to one of two nutritional treatment groups. Group I received a commercial 3.5% balanced amino acid solution* in 25% glucose at a rate to provide 1.0 g protein per kg/day and 35 glucose kcal/kg/day. The final solution was 15.5% BCAA (as % of total amino

*Travesol, Travenol Laboratories, Inc., Chicago, Illinois

Table 4.1 Nutritional support data*

	Branched-chain amino acids			Control		
	Day 0	Day 3	Day 6	Day 0	Day 3	Day 6
No. of patients	7	7	7	8	8	8
Glucose (g)	237.0 ±132.0	671.0 ±153.0	761.0 ±103.0	243.0 ±59.0	591.0 ±79.0	685.0 ±80.0
Amino acids (g)	18.40± 3.40	75.20± 13.00	88.00± 5.80	18.50± 4.80	62.70± 8.50	75.00± 7.00
(g/kg)	0.26± 0.07	1.03± 0.20	1.21± 0.15	0.24± 0.07	0.79± 0.13	0.96± 0.18
Amino acid nitrogen (g)	2.60± 0.50	10.80± 1.90	12.50± 0.80	3.10= 0.80	10.60± 1.40	12.60± 1.20

*Mean ±S.D.

acids). Group II received the same solution, but branched-chain-rich so that the final solution was 3.5% amino acids with 50% BCAA (as % of total amino acids). It was given in 25% glucose to provide 35 glucose kcal/kg/day and 1.0 g protein/kg/day.

No intravenous fat was given to either group I or II. No oral nutrition was given for the 7 days of the study. All patients received the same full complement of vitamins, minerals, and trace elements daily (Table 4.1 and 4.2).

Table 4.2 Amino acid input*

	Branched-chain amino acids			Control		
	Day 0	Day 3	Day 6	Day 0	Day 3	Day 6
Ala	1.60±0.20	5.10±0.90	6.00±0.40	3.90±1.00	12.50±1.70	15.00±1.40
Leu	3.10±0.60	12.50±2.20	14.70±0.97	1.20±0.30	3.90±0.50	4.60±0.40
Ileu	3.10±0.60	12.50±2.20	14.70±0.97	0.90±0.20	2.90±0.40	3.50±0.30
Val	3.10±0.60	12.50±2.30	14.70±0.97	0.90±0.20	2.90±0.40	3.50±0.30
Glut						
Pro	0.26±0.05	1.10±0.18	1.20±0.08	0.80±0.20	3.00±0.40	3.20±0.30
Phe	0.37±0.06	1.50±0.26	1.80±0.12	1.20±0.30	3.90±0.50	4.70±0.40
Tyr	0.20±0.01	0.10±0.01	0.11±0.10	0.17±0.02	0.25±0.03	0.30±0.03
Trp	0.11±0.02	0.44±0.08	0.52±0.03	0.33±0.09	1.13±0.11	1.40±0.13
Meth	0.35±0.06	1.40±0.20	1.70±0.11	1.11±0.30	3.60±0.50	4.40±0.40

*Mean±S.D.

Each patient was monitored on day 0, 3, and 7 with routine lab tests that included a complete blood count, electrolytes, urea, creatinine, SGOT, bilirubin, LDH and alkaline phosphatase. Fluid balance and daily weights were measured, as were serum albumin, prothrombin time and platelet count. Plasma aminograms were performed daily on an arterial blood sample drawn 30 min after stopping the infusion, so that the effects of the solution alone could be eliminated[7,9]. Daily urine collections were analysed for total nitrogen, urea, creatinine, free amino acids and 3-methylhistidine excretion. The 3-methylhistidine excretion was monitored as an index of the degree of existing protein catabolism, particularly in muscle[21].

Within 48 h of entering the study, each patient was receiving the desired input of glucose calories and amino acids, which was maintained throughout the study. The result was an isocaloric-isonitrogenous situation in which the principal difference between the groups was in the amount of BCAA received: 0.5 g/kg/day in group II compared to 0.155 g/kg/day in group I.

Because the patients were not fed orally and had prolonged ileus or pre-operative bowel preparation, bowel movement during the study was quite unusual. The various urinary outputs are summarized in Table 4.3. Creatinine and 3-methylhistidine excretion was the same on each day regardless of the study group. Urinary amino acid output did not differ between the groups, in spite of the 3-fold increase in input of BCAA. Although there was no glutamine input, there was a consistently large urinary output of glutamine in both groups (Figure 4.1). Very small amounts of the other amino acids were excreted relative to input. Balance studies of some amino acids are presented in Figures 4.2–5.

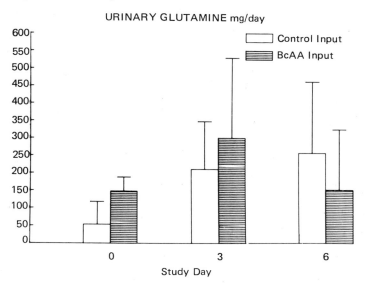

Figure 4.1 Histogram depicting the glutamine balance in the control and BCAA groups. During the stress phase (day 3 peak), there was increased glutamine excretion, even in the presence of no exogenous input

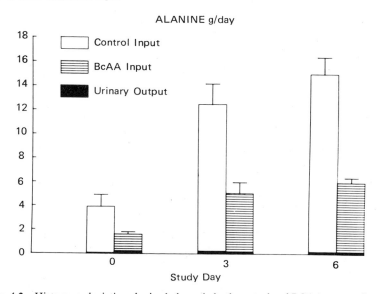

Figure 4.2 Histogram depicting alanine balance in both control and BCAA groups. In spite of a large exogenous input, there is little urinary excretion

Table 4.3 Output data summary*

	Branched chain amino acids			Control		
	Day 0	Day 3	Day 6	Day 0	Day 3	Day 6
No. of patients	7	7	7	8	8	8
Urinary nitrogen (g)	11.10 ± 8.60	7.60 ± 3.80	7.90 ± 3.90	5.90 ± 3.80	13.20 ± 4.00	10.70 ± 5.40
Creatinine (g)	1.40 ± 0.80	1.50 ± 0.60	1.50 = 0.60	1.20 ± 0.80	1.01 ± 0.50	1.30 ± 0.40
3-Methylhistidine (µg)	255.0 ± 197.0	181.0 ± 82.0	151.0 ± 53.0	195.0 ± 155.0	176.0 ± 80.0	165.0 ± 97.0

*Mean ± S.D.

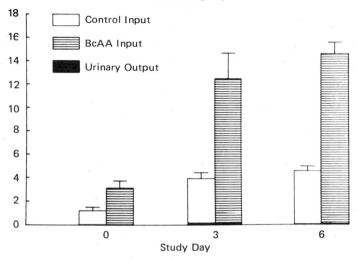

Figure 4.3 Histogram depicting leucine balance in both control and BCAA groups. In spite of a large exogenous input, there is little urinary excretion

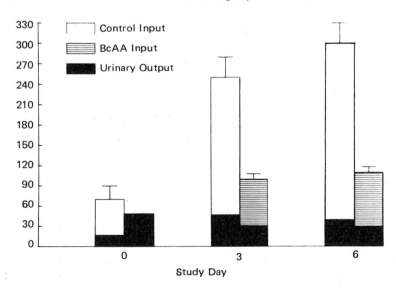

Figure 4.4 Histogram depicting tyrosine balance in both control and BCAA groups. In spite of a large exogenous input, there is little urinary excretion.

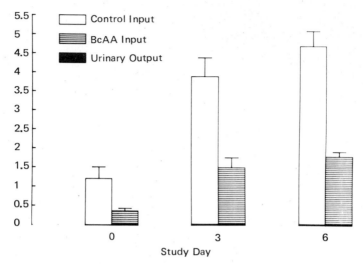

Figure 4.5 Histogram depicting phenylalanine balance in both control and BCAA groups. In spite of a large exogenous input, there is little urinary excretion

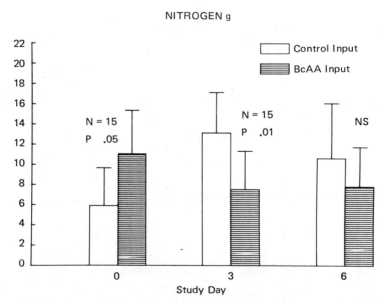

Figure 4.6 Histogram depicting urinary nitrogen excretion on Days 0, 3 and 6 for control (group I) and BCAA (Group II) support patients. With increasing isocaloric–isonitrogenous input, the urinary nitrogen excretion of group II is much lower on day 3

The differences in total urinary nitrogen output are depicted in Figure 4.6. The total urinary nitrogen continued to increase in group I as the nutritional input increased. In group II, however, the urinary nitrogen excretion fell by day 3 and remained low. The net nitrogen balance is summarized in Figure 4.7. The group II patients began with a much more negative balance ($p<0.05$). By day 3, however, they were in significant positive balance relative to the control group I ($p<0.01$). The group I patients achieved an equivalent positive nitrogen balance by day 6.

Table 4.4 Urinary free amino acid output (mg/day; mean±S.D.)

	Branched-chain amino acids			Control		
	Day 0	Day 3	Day 6	Day 0	Day 3	Day 6
Ala	42.0±36.0	54.0± 52.0	50.0± 38.0	21.0±17.0	68.0± 48.0	64.0± 46.0
Leu	7.0± 4.0	22.0± 14.0	11.0± 7.0	7.0± 5.0	18.0± 13.0	6.0± 5.0
Ileu	4.0± 4.0	8.0± 5.0	16.0± 29.0	0.4± 0.1	3.0± 3.0	2.0± 0.06
Val	7.0± 6.0	21.0± 15.0	25.0± 25.0	6.0± 4.0	15.0± 8.0	16.0± 10.0
Glut	147.0±40.0	300.0±230.0	153.0±173.0	53.0±64.0	210.0±137.0	258.0±205.0
Phe	40.0±60.0	30.0± 19.0	23.0± 11.0	10.0± 8.0	36.0± 11.0	33.0± 21.0
Tyr	31.0±22.0	40.0± 33.0	30.0± 24.0	17.0±16.0	49.0± 17.0	47.0± 32.0
Trp	32.0±37.0	42.0± 16.0	189.0±128.0	26.0±20.0	42.0± 12.0	40.0± 10.0
Meth	3.0± 1.5	11.0± 8.0	4.0± 2.0	2.0± 1.0	9.0± 7.0	8.0± 6.0

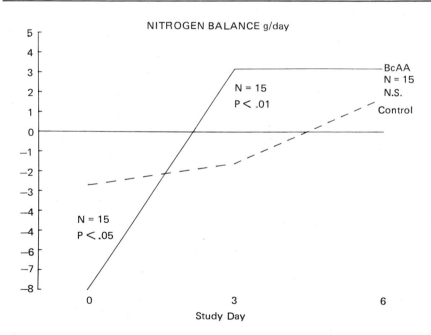

Figure 4.7 The net nitrogen balance is depicted on days 0, 3 and 6. By day 3, the 50% BCAA patients (group II) are in positive nitrogen balance. The control (group I) patients illustrate a metabolic stress response with conventional nutritional support in which nitrogen retention occurs after significant abatement of the stress response

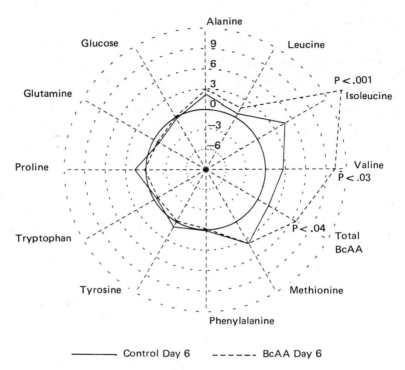

Figure 4.8 The plasma amino acid levels are compared. The inner dark circle is the mean value for each variable for the control group I on day 0. Each dotted circle represents three standard deviations from this control mean. The BCAA levels did not rise until day 6, when there was significant abatement of the stress response

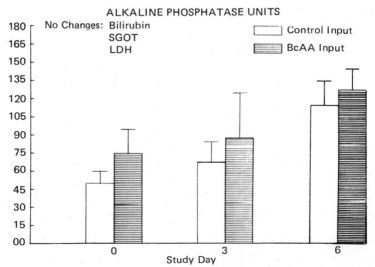

Figure 4.9 Histogram of the liver enzyme changes during stress TPN. The change was the same in both groups.

The plasma amino acid profile data are summarized in Table 4.5 and Figure 4.8. Except for the BCAA, there were no differences between the groups. The BCAA levels were also not different on day 0 or day 3. By day 6, however, the levels of isoleucine, valine and total BCAA were significantly higher in the group II patients ($p<0.001$, $p<0.03$, $p<0.04$ respectively).

The serum bilirubin, SGOT and LDH changed little in all patients. The alkaline phosphatase however, rose threefold in all patients, irrespective of whether or not they were group I or II. No changes of significance in the other measurements were noted (Figure 4.9).

CONCLUSION

These studies compare the effects of 50% BCAA-TPN with 15.5% BCAA-TPN in non-septic, non-cirrhotic, elective abdominal surgery or multiple trauma patients in the 7 post-operative days. With isocaloric-isonitrogenous intravenous input, the BCAA-rich group achieved positive nitrogen balance by the third day. Although the urinary nitrogen output was decreased in the BCAA-rich group by the third day the plasma BCAA levels did not increase until the sixth day. There was no change in urinary 3-methylhistidine excretion between the groups or within each group. The serum alkaline phosphatase level rose equally in both groups.

The control patients illustrated a typical metabolic stress response supported with standard TPN. Effective protein-sparing was present on day 3; with positive nitrogen balance occurring on day 6 after the stress response had abated. Plasma amino acid levels at that time were more typical of those of overnight fasting man. The addition of higher doses of BCAA (0.5 g/kg/day), however, induced a state of nitrogen retention during the stress response. The marked rise in plasma BCAA levels on day 6 in the 50% BCAA group is quite similar to that seen when BCAA are given to non-stressed man. These data, together with a reduction in urinary nitrogen output on day 3 at a time of increased input and no observed change in the 3-methylhistidine excretion is consistent with a shifting of the balance of protein synthesis and degradation in favour of synthesis in the presence of high BCAA infusion during stress. The site of increased synthesis and the use of BCAAs as an energy source or their effects on protein catabolism require further study.

Thus, the ability to favourably modulate the metabolic response to stress with alternate fuels has become a clinical reality. The development of new substrates, refining those currently available and a safe and effective application of nutritional support would be improved if a categorization of the stress level could be agreed. Some potential criteria are presented in Table 4.6.

Table 4.5 Plasma amino acids*

	Branched-chain amino acids			Control		
	Day 0	Day 3	Day 6	Day 0	Day 3	Day 6
Ala µM/l	282.0±163.0	256.0±141.0	387.0±158.0	240.0± 51.0	265.0± 68.0	350.0± 30.0
Leu µM/l	75.0± 16.0	99.0± 30.0	117.0± 53.0	77.0± 25.0	76.0± 31.0	92.0± 20.0
Ileu µM/l	26.0± 9.0	68.0± 27.0	96.0± 48.0	24.0± 5.0	45.0± 25.0	48.0± 9.0‡
Val µM/l	227.0±118.0	392.0±195.0	422.0±214.0†	184.0± 23.0	287.0±188.0	245.0±114.0
Glut µM/l	346.0±267.0	291.0±236.0	278.0±230.0	359.0±127.0	286.0±121.0	245.0±176.0
Pro µM/l	38.0± 11.0		83.0± 65.0	77.0± 37.0	118.0± 31.0	137.0± 24.0
Phe µM/l	71.0± 25.0	75.0± 38.0	71.0± 50.0	78.0± 44.0	81.0± 25.0	84.0± 36.0
Tyr µM/l	58.0± 19.0	33.0± 14.0	48.0± 18.0	50.0± 13.0	55.0± 12.0	61.0± 9.0
Trp µM/l	141.0± 87.0	107.0± 56.0	89.0± 39.0	127.0± 60.0	113.0± 50.0	102.0± 59.0
Meth µM/l	18.0± 7.0	21.0± 14.0	30.0± 14.0	15.0± 4.0	32.0± 5.0	30.0± 12.0
Glucose mM/l	196.0± 73.0	223.0±130.0	165.0± 83.0	162.0± 34.0	213.0± 85.0	155.0± 41.0

*Mean ±S.D.
†P<0.03 relative to day 0
‡P<0.011

Table 4.6 Metabolic response to stress

Level	Clinical prototype	Urinary		Glucagon* (pg/ml)	Plasma					
		Nitrogen (g/day)	3-Methyl-histidine (mmol/day)		Glucose† (mmol/l)	Lactate‡ (mmol/l)	Leucine (mmol/l)	Proline (mmol/l)	Phenyl-alanine* (mmol/l)	Methionine* (mmol/l)
0	Starvation	<5	<100	<20	5±2	10±5	120±10	200±20	60±15	10±5
1	Elective general surgery	5–10	130±20	50±9	10±1	1200±200	74±12	213±40	74±8	15±5
2	Multiple trauma	10–15	200±20	120±40	10±1	1200±200	74±12	213±40	74±8	15±5
3	Sepsis	>15	450±50	500±50	16±2	3000±500	180=30	350±50	124±17	105±25

*in absence of cirrhosis
†in absence of steroids, diabetes, pancreatitis
‡with a normal lactate/pyruvate ratio

References

1 Adibi, S. A. (1980). Roles of branched chain amino acids in metabolic regulation. *J. Lab. Clin. Med.*, **95**, 475

2 Alberti, K. G. M. M., Batstone, G. F., Foster, K. J. and Johnston, D. G. (1980). Relative role of various hormones in mediating the metabolic response to injury. *J.P.E.N.*, **4**, 141

3 Birkham, R. H., Long, C. L. and Litkin, B. S. (1980). Effects of major skeletal trauma on whole body protein turnover in man measured by ^{14}C-leucine. *Surgery*, **88**, 294

4 Blackburn, G. L. (1977). Lipid metabolism in infection. *Am. J. Clin. Nutr.*, **30**, 1321

5 Blackburn, G. L., Moldawer, L. L., Usui, S., Bethe, A., O'Keefe, S. J. D. and Bistrian, B. R. (1979). Branched chain amino acid administration and metabolism during starvation, injury, and infection. *Surgery*, **86**, 307

6 Border, J. T., Chenier, R. and McMenamy, R. H. (1976). Multiple systems organ failure: muscle fuel deficit with visceral protein malnutrition. *Surg. Clin. N. Am.*, **56**, 1147

7 Cerra, F. B., Siegel, J. H. and Border, J. R. (1979). Correlation between metabolic and cardiopulmonary measurement in patients after trauma, general surgery and sepsis. *J. Trauma*, **19**, 621

8 Cerra, F. B., Caprioli, J. and Siegel, J. H. (1979). Proline metabolism in sepsis, cirrhosis and general surgery: the peripheral energy deficit. *Ann. Surg.*, **190**, 577

9 Cerra, F. B., Siegel, J. H. and Border, J. R. (1979). The hepatic failure of sepsis: cellular *vs* substrate. *Surgery*, **86**, 409

10 Cerra, F. B., Siegel, J. H. and Coleman, B. (1980). Septic autocannibalism: a failure of exogenous nutritional support. *Ann. Surg.*, **192**, 570

11 Chua, B., Siehl, D. L. and Morgan, H. E. (1979). Effect of leucine and metabolites of branched chain amino acids on protein turnover in heart. *J. Biol. Chem.*, **254**, 8358

12 Clowes, G. H. A., Heidman, M. and Lindberg, B. (1980). Effects of parenteral alimentation on amino acid metabolism in septic patients. *Surgery*, **88**, 531

13 Clowes, G. H. A., O'Donnell, T. F., Blackbumb, V. and Maki, T. N. L. (1976). Energy metabolism and proteolysis in traumatized and septic man. *Surg. Clin. N. Am.*, **56**, 1169

14 Cuthbertson, D. and Tilstone, W. (1969). Metabolism during the post injury period. *Adv. Clin. Chem.*, **12**, 1

15 Freund, H. and Fischer, J. E. (1978). Nitrogen-conserving quality of the branched-chain amino acids: possible regulator effect of valine in postinjury muscle catabolism. *Surg. Forum*, **29**, 69

16 Freund, H., Yoshimure, N. and Fischer, J. E. (1980). The effect of branched chain amino acids and hypertonic glucose infusions on postinjury catabolism in the rat. *Surgery*, **87**, 401

17 Hutson, S. M., Cree, R. C. and Harper, A. E. (1980). Regulation of leucine and ketoisocaproic acid metabolism in skeletal muscle. *J. Biol. Chem.*, **255**, 2418

18 Li, J. B. and Jefferson, L. S. (1978). Influence of amino acid availability on protein turnover in perfused skeletal muscle. *Biochim. Biophys. Acta*, **544**, 351

19 Lindsay, D. B. (1980). Amino acids as energy sources. *Proc. Nutr. Soc.*, **39**, 53

20 Long, C. L., Jeevanandan, M. and Kinen, J. M. (1977). Whole body protein synthesis and catabolism in septic man. *Am. J. Clin. Nutr.*, **30**, 1340

21 Long, C. L., Haversberg, L. N., Young, V. R., Kinney, J. M. and Munro, H. N. (1975). Metabolism of 3-methylhistidine in man. *Metabolism*, **24**, 929

22 Odessey, R. and Goldberg, A. L. (1972). Oxidation of leucine by rat skeletal muscle. *Am. J. Phys.*, **223**, 1376

23 Sketcher, R. D., Fern, E. B. and James, W. P. T. (1974). The adaptation in muscle oxidation of leucine to dietary protein and energy intake. *Br. J. Nutr.*, **31**, 333

24 Tweedle, D. E. (1980). Metabolism of amino acids after trauma. *J.P.E.N.*, **4**, 165

25 Wedge, J. H., De Campos, R., Kerr, A., Smith, R., Farrell, R., Ilic, V. and Williamson, D. H. (1976). Branched-chain amino acids, nitrogen excretion and injury in man. *Clin. Sci. Mol. Med.*, **50**, 393

26 Wolfe, B. M. (1980). Substrate-endocrine interactions and protein metabolism. *J.P.E.N.*, **4**, 188

5
The importance of study design to the demonstration of efficacy with branched-chain amino acid enriched solutions

L. L. MOLDAWER, M. M. ECHENIQUE, B. R. BISTRIAN, J. L.
DUNCAN, R. F. MARTIN, E. M. St. LEZIN AND G. L. BLACKBURN

INTRODUCTION

The host response to injury is characterized by disturbances in protein metabolism which result in an enhanced appearance of nitrogen, sulphur, and phosphorus in the urine. Increased catabolism of muscle protein due either to inhibition of protein synthesis or enhanced protein degradation increases the availability of endogenous amino acids to sustain protein synthesis and support increased energy expenditure[1].

After most forms of severe injury or infection, there is also increased protein synthesis, presumably due to increases in the synthetic capacity of such visceral tissues as hepatic non-secretory, acute phase, and leukocytic proteins[2]. In the absence of dietary protein intake, the source of amino acids for this increased anabolism by visceral tissues is skeletal muscle and connective tissue[3]. Simultaneously, enhanced amino acid oxidation and urinary nitrogen excretion following injury or infection is due to a constant proportion of amino acids being degraded to meet the host's increased energy requirements. The enhanced amino acid oxidation observed after injury can be attributed to an increased branched-chain amino acid (BCAA) oxidation by skeletal muscle. After severe injury or infection, the obligatory energy requirements are not met entirely by the oxidation of carbohydrate and fat fuels. Utilization of carbohydrate in the citric acid cycle is regulated, in part, by the pyruvate dehydrogenase complex and during experimental infection[4], its activity is reduced. Although fat remains the primary fuel for skeletal muscle, its utilization is also inhibited during severe injury or infection. Increased insulin levels antagonize the lipolytic effects of catecholamines[5] and also foster the re-esterification of free fatty acids within the adipocyte thereby decreasing free fatty acid release and oxidation. Furthermore, the reduction in carnitine acyltransferase activity during sepsis[6] and the impairment in hepatic ketogenesis[7] contribute to the change in fuel availability which is met by increased branched-chain amino acid oxidation.

We have reported previously in a study of 18 patients following surgery that patients who failed to develop ketonaemia on a carbohydrate-free intake excreted more nitrogen in the urine[8,9]. In a similar series of trauma patients, Williamson showed that failure to develop hyperketonaemia after fasting was also associated with the greatest rate of skeletal protein breakdown as measured by urinary 3-methylhistidine excretion[10]. In both of these studies, there was a positive correlation between plasma levels of ketone bodies and BCAAs.

It is now clear that the oxidation of branched chain amino acids occurs almost exclusively extrahepatically. Skeletal muscle, due to its relative mass and the presence of enzymes for BCAA degradation, is the major site of oxidation[11]. If the metabolic response to injury is characterized by increased utilization of branched-chain amino acids for fuel by skeletal muscle, then a portion of the protein catabolic response to injury may be due to inadequate levels of BCAAs to support protein synthesis. The increased degradation of skeletal muscle would then represent an endogenous mechanism by which skeletal muscle 'autocannibalizes' to meet energy requirements and as an additional consequence also provides glucogenic precursors for other tissues. Under these conditions, dietary administration of BCAAs at rates equivalent to their increased oxidation would offset these losses and restore concentrations necessary to support optimal protein synthesis.

BCAA supplementation, or the use of BCAA-enriched amino acid formulaes, would then be most beneficial in injured or stressed patients in whom BCAA oxidation, and therefore requirements for replacement, are increased to a point where usual protein intakes of standard crystalline amino acid solutions will not provide sufficient BCAAs. For example, a severely stressed patient losing 15–20 g/day of nitrogen in the urine on a protein-free regimen is catabolizing 95–125 g/day protein. Assuming that BCAAs compose 19% of protein by weight, the appearance of 15–20 g/day urinary nitrogen represents the oxidation of 18–24 g/day of BCAAs. Isotopic studies performed by our group[12] and by Long and his associates[13] with L-[l-^{14}C]-leucine have confirmed that leucine oxidation rates in severely stressed patients are between 1.5 and 2.5 mmol/h or 5–9 g/day. To offset these increased losses of BCAAs to oxidation, 95–125 g of a standard crystalline amino acid solution containing 20% BCAAs (1.35–1.75 g/kg day for a 70 kg man) would have to be provided. If a BCAA-enriched solution in which the contribution of BCAAs to the total amount is increased to 50%, the BCAAs required to offset oxidation could be met by administration of only 40–60 g of amino acid (0.60–0.85 g/kg day). Nitrogen balance in an isonitrogenous comparison of less than 95 g of amino acids would be expected to favour BCAA-enriched formulae over a standard amino acid solution.

These calculations are based on the assumption that BCAA oxidation and requirements play a regulatory role in amino acid kinetics and that the requirement for amino acids in the critically ill patient is really a requirement for BCAAs. Although these assumptions appear contrary to standard ideas about dietary amino acid requirements, there is ample *in vitro* evidence demonstrating a regulatory role for the BCAAs, especially leucine, in skeletal protein dynamics. More importantly, the metabolic milieu associated with severe injury or infection is one of increased amino acid availability and flux with dietary amino acid intake making a relatively small contribution to the total amino acid appearance.

Unlike the malnourished patient, the stressed individual has an ample supply of amino acids available for protein synthesis as a result of increased protein breakdown. In stressed patients, the protein deficit appears to result from greater proportions of free amino acids being oxidized rather than being reutilized for protein synthesis[13,14]. The putative goal of amino acid therapy in the critically ill patient should be not only to provide the substrate for net protein synthesis but also to offset the increased proportion of free amino acids that are oxidized.

Maximal benefits with BCAA-enriched solutions would be most noticeable in stressed patients whose total protein intake is restricted because of fluid overload or hepatic or renal failure. The beneficial effect of BCAA administration would also be anticipated in patients whose nitrogen losses exceed 15 g/day, thus necessitating very high intakes of standard amino acid solution.

EXPERIMENTAL EVIDENCE

Studies of the efficacy of BCAA-enriched amino acid solutions need to consider both the degree of stress and the total amino acid intake of the patients. The unstressed individual receiving 1–1.5 g protein/kg day should do as well with standard crystalline amino acid solutions as with BCAA-enriched formulae. In moderately stressed individuals excreting 10–15 g N/day, the improved utilization of BCAA-enriched solutions may only be identified if moderate protein intakes are used. Only in the severely stressed patient are isonitrogenous comparisons of branched-chain-enriched and standard amino acid solutions likely to show a difference with usual protein intakes (1.0–1.5 g/kg day).

In a recent report, Cerra and his associates[15] presented 15 elective general surgery or multiple trauma patients receiving a total parenteral nutrition regimen delivering 35 glucose kcal/kg day and 145 mg nitrogen/kg day of complete crystalline amino acids. The amino acids were given as either a 15.6% BCAA solution or as a 50% BCAA solution. Nitrogen balance and

Table 5.1 Summary of input and output data with branched-chain amino acid enriched feedings

	Branched-chain amino acid enriched feeding			Standard crystalline amino acid feeding		
	Day 0	3	7	Day 0	3	7
Nitrogen intake g/day	2.6	10.8	12.5	3.1	10.6	12.6
Glucose intake kcal/day	806	2281	2587	826	2009	2329
Urinary nitrogen excretion g/day	11*	8*	8	6*	13*	11
Estimated nitrogen balance g/day	−8.4*	+2.8*	+4.5	−2.9*	−2.4*	+1.6
3-Methylhistidine excretion µg/day	255	181	151	195	176	165

Reprinted with permission, Dr Frank Cerra
*$p<0.05$, versus other feeding at same time

3-methylhistidine excretion were evaluated over 7 days following surgery or trauma. As shown in Table 5.1, although the patients receiving the BCAA-enriched amino acid formula were significantly more catabolic prior to initiation of the total parenteral nutrition formula, urinary nitrogen excretion was significantly less after 3 days on the BCAA-enriched formula when compared to patients receiving the standard amino acid diet. Institution of both regimens produced a small but insignificant reduction in urinary 3-methylhistidine excretion, suggesting that differences in muscle protein breakdown between patients receiving the standard crystalline amino acid and BCAA-enriched formulae were insignificant.

The authors concluded that the early nitrogen retention associated with administration of BCAA-enriched formulae probably reflected a stimulation in protein synthesis. However, the results could be attributed either to an increase in protein synthesis, a reduction in non-muscle protein breakdown, or some combination of the two. Close evaluation of the patient population reveals a moderate degree of stress.

We have proposed the use of a catabolic index to evaluate the degree of metabolic stress experienced by a patient[16]. The primary variables affecting the magnitude of the catabolic response are the nutritional status of the patient, the level of dietary intake, and the degree of stress. The particular value of the catabolic index is the integration of these parameters into one number:

$$\text{Catabolic index (CI)} = \text{UUN} - (0.5\,\text{N}_{in} + 3)$$

A score of −5 to 0 indicates mild stress; 0 to +5, moderate stress; and +5 to +10, severe stress.

A major benefit of the catabolic index is that it relies on urine urea measurements, which are both easy to obtain and inexpensive. The catabolic index also moderates the impact of high nitrogen intake on urine urea nitrogen production by fractional reduction. The impact of low nitrogen intakes is minimized as well by addition of a constant $+3$ g of nitrogen.

An evaluation of Cerra's data with the catabolic index on day 0, when the study was initiated, demonstrates a mean value of $+3.0$ reflecting a moderate degree of stress. Coupled with the fact that the patients were receiving only 60–80 g/day of amino acid (1.0 g/kg day), these patients would be expected to benefit from a BCAA-enriched formula in that they were receiving moderate amounts of dietary amino nitrogen and had an appreciable degree of stress.

Table 5.2 Patient characteristics

Patient	Diagnosis	Anthropometric measurements (% of standard)			Serum albumin (g/dl)	Catabolic index
		Triceps skinfold	Arm muscle circumference	Mean arm circumference		
1	Metastatic ovarian carcinoma	57%	72%	68%	3.1	−2.8
2	Ovarian cancer, pelvic abscess	50%	70%	65%	2.3	−0.5
3	Sepsis, renal failure, GI bleed	46%	92%	85%	2.9	+0.5
4	Crohn's disease, short bowel syndrome, retro-peritoneal bleeding	117%	87%	90%	2.6	+2.3
5	Acute pancreatitis	58%	94%	88%	3.2	+2.1
6	Diverticulitis, intra-abdominal abscess, small bowel fistula	96%	89%	85%	2.6	−0.3

In addition to Cerra's work, we have completed two studies investigating BCAA-enriched formulae in adults in hospital. In the initial study, Echenique investigated six protein-malnourished patients who required total parenteral nutrition. The clinical diagnoses and nutritional indices are included in Table 5.2. The patients were given a total parenteral regimen containing both carbohydrate and lipid. Two amino acid sources were evaluated, a standard crystalline amino acid formula containing either 15.6% or 24% BCAAs and one enriched with BCAA to a final concentration of 50%. The diets were administered over 4 to 7 days and the patients were then crossed over to receive the other diets. The order of infusions was randomly selected and attempts were made to keep the diets isonitrogenous.

Results are presented in Table 5.3. In five of six patients, amino acid intake exceeded 1.3 g/kg day but nitrogen balance did not differ significantly when

Table 5.3 Amino acid intake and nitrogen balance in mildly stressed patients receiving adequate dietary intakes

Patient	Branched-chain amino acid enriched diet		Standard crystalline amino acid formula	
	Amino acid intake (g/kg day)	Nitrogen balance (g/day)	Amino acid intake (g/kg day)	Nitrogen balance (g/day)
1	1.9	+0.4	1.6	+0.8
2	1.6	−1.9	2.2	+1.7
3	0.9	−3.5	0.7	−3.5
4	1.4	−3.2	1.3	−3.5
5	1.9	−0.7	2.0	+1.0
6	2.0	+2.7	1.6	+1.7
Mean ±S.E.	1.6±0.2	−1.0±0.9	1.6±0.2	−0.3±1.0

the patients were receiving the standard crystalline amino acid diet or the BCAA-enriched formula.

In these six patients, use of a BCAA-enriched formula as part of a total parenteral regimen failed to improve nitrogen retention over that observed with a standard crystalline amino acid formula. In retrospect, the failure to observe marked improvements in nitrogen balance can be attributed to the fact that the patients were not very catabolic prior to starting the feeding regimen and their dietary amino acid intake was not restricted. Based upon a catabolic index of approximately 0, the six individuals could only be classified as a group as minimally stressed. Furthermore, as evidenced in Table 5.3, dietary protein intake averaged 1.6 g/kg day. Under these conditions, BCAA oxidation was offset by use of the standard crystalline amino acid formula and BCAA supplementation provided quantities far in excess of that required to replace losses. The degree of metabolic stress and dietary protein intake in these patients differed substantially from Cerra's study where the patients were less catabolic at the beginning of the study and received considerably more protein. The findings confirm the need to identify those patients who will benefit most from BCAA supplementation.

Many patients with dietary protein restriction and considerable catabolic stress can be found in the surgical intensive care unit. Nutritional support in this critically ill population is difficult due to both general fluid restriction and associated organ failure. These patients often undergo considerable stress due to infection or sepsis coupled with respiratory, renal, and hepatic dysfunction.

However, studies which estimate net protein catabolism from urinary excretion are difficult to undertake in a critical care setting. 3–4 days on a constant dietary intake are required to attain steady state urine nitrogen excretion. Because of the heterogeneity of this patient population, crossover studies must be used to minimize patient variability. Since crossover studies are needed and since attainment of new steady states in urine nitrogen excretion requires several days, nitrogen balance studies evaluating two different diets must last at least a week. In the critical care setting, it is recognized that the clinical course of a patient is rarely constant over such a period and as a result, large numbers of patients are necessary to reduce the variability inherent in such a study. Nevertheless, it is this patient population in which BCAA-enriched solutions may be most beneficial.

We have investigated an alternative approach to studying changes in protein metabolism in the critically ill patient. Instead of estimating net protein catabolism based on urinary nitrogen excretion, we have utilized a continuous infusion of L-[l-^{14}C]-leucine and have measured rates of plasma amino acid flux and oxidation.

The assumptions and limitations inherent to a continuous infusion of radiolabelled amino acid have been discussed extensively elsewhere[17]. However, central to the measurements is the assumption that the kinetics of the individual tracer amino acid reflect accurately the behaviour of amino acids in general. Work from this laboratory[12] and from Long et al.[13] reported good correlation between leucine oxidation rates and total urine nitrogen excretion in severely injured patients when both measurements could be accurately obtained. Such a correlation confirms that most amino acids are oxidized at rates similar to the isotopic leucine and that the kinetics of leucine in particular can be used as an estimate for whole body protein turnover.

Unlike nitrogen balance measurements, infusions of isotopically labelled amino acids offer two additional benefits. The first and perhaps most important in the critical care setting is the rapidity with which isotopic steady states can be achieved. We reported previously that in a group of 13 patients immediately after operation, an isotopic steady state in the plasma compartment was achieved within 10 h of starting an isotopic infusion and within 4 h of changing a dietary regimen[12]. Similar results have been reported in healthy young adults during fasting and refeeding periods[18]. The rapidity with which measurements can be made is in direct contrast to nitrogen balance studies which require days.

Secondly, the use of isotopic labels also permits derivation of whole body protein synthesis and breakdown rates. As Waterlow reported fifteen years ago[19], improvements in nitrogen balance can be achieved through an increase in protein synthesis, a reduction in protein breakdown, or a combination of the two. Using an isotopically labelled amino acid permits an estimation of whether the changes in protein balance are obtained through alterations in protein synthesis or breakdown.

Using this technique, Echenique et al.[20] in this laboratory assessed the value of BCAA-enriched solutions in five critically ill individuals in the intensive care unit requiring total parenteral nutrition. Two complete feeding solutions were compared: one containing 15.6% of the amino acids as BCAAs and the other containing 50% BCAAs. The diets were isonitrogenous but total protein and caloric intakes were dictated by the nutritional status of the patient and the presence of organ failure. The average amino acid and caloric intake in the five patients was 0.9 g/kg day and 24 kcal/kg day, respectively.

Urine urea nitrogen excretion was collected so that each patient's catabolic index could be calculated. The average value in the five patients was + 1.7 indicating a moderate degree of stress. Based upon the limited protein intake that these patients could tolerate and the moderate degree of stress they experienced, we predicted that BCAA fortification would offer some additional benefit in reducing net protein catabolism.

The patients were assigned randomly to one feeding regimen for 24 h. During the last 10 h, 50 μCi of L[l-^{14}C]-leucine was added to the nutrient infusions and rates of leucine kinetics evaluated as previously described[12].

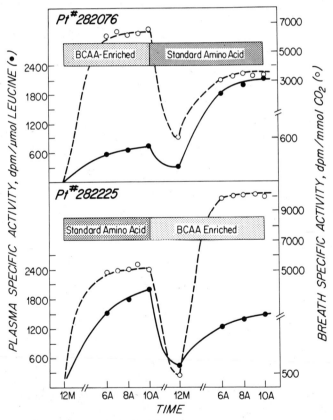

Figure 5.1 Plateau values in plasma and breath in 48 h crossover studies. Patients were randomized to receive total parenteral nutrition regimen containing either 15.6% or 50% BCAAs for 24 h. During the subsequent 24 h the patients were crossed over to the other regimen. 50 μCi L-[l-^{14}C]-leucine was added to the infusion during the last 10 h. Data from two patients show that a steady state was achieved in both plasma and breath regardless of the order of infusions

The patients were then crossed over to the other feeding regimen for the subsequent 24 h. An additional 50 μCi L-[l-^{14}C]-leucine were added to the second nutrient infusion during the last 10 h and leucine kinetics were estimated. In this manner, each patient served as his own control. Two days was a sufficiently brief period to assume that the clinical course of the patient was relatively unchanged. Three of the patients received the standard amino acid formula first and the remaining two patients received the BCAA-enriched formula initially.

As demonstrated in Figure 5.1, isotopic steady states in blood and expired breath were achieved in these critically ill patients after 24 h on the diet and 10 h of isotopic infusions. It appeared that the order of nutrient administration had little effect on the attainment of isotopic steady states after background enrichments were subtracted and turnover rates did not depend upon the order of infusions. However, it was also clear that the BCAA-enriched feedings altered the dilution of [^{14}C]-leucine in the plasma compartment and increased the appearance of [^{14}C]-carbon dioxide in the expired breath.

Whole body leucine kinetics are presented in Figure 5.2. The BCAA-enriched feeding significantly increased plasma leucine flux and oxidation. However, leucine balance, the difference between intake and oxidation, became significantly more positive when the BCAA-enriched feeding was

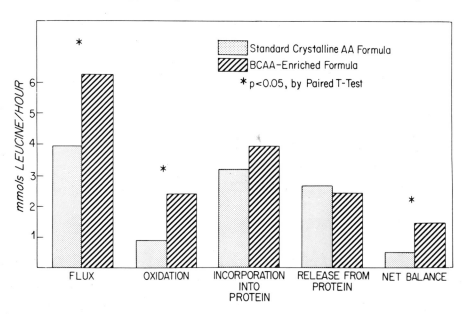

Figure 5.2 Leucine kinetics and isotope balance with the two feeding formulae. Rates of leucine flux, oxidation, incorporation into and release from protein, and balance were determined in five patients receiving either standard amino acid formulae or one enriched to 50% BCAAs. Leucine flux and oxidation were significantly increased with the BCAA-enriched formula.

employed. The retention of dietary leucine in the body can only be interpreted as an increase in amino acid retention as protein equivalent similar to an improvement in nitrogen balance. Although the findings are consistent with increased protein retention with the BCAA-enriched feeding, the mechanism of nitrogen sparing is unclear. Increases in leucine incorporation into whole body protein and reductions in its release from protein breakdown were observed but neither of the changes was significant.

In summary, preliminary evidence suggests a role for BCAA-enriched formulae in the stressed patient. Close attention to study design will be necessary if the efficacy of one solution over another is to be demonstrated. The optimal patient in which to evaluate such enriched solutions may be the moderately to severely stressed individual who as a result of fluid intolerance or organ failure has restricted dietary intake. Furthermore, in such patient populations, nitrogen balance measurements may be of limited value due to the time required to perform such studies. Under these circumstances, the use of isotopic amino acid infusions provides the opportunity to evaluate the efficacy of a feeding solution rapidly.

ACKNOWLEDGEMENTS

Supported in part by grants GM-24206, GM-22691 and GM-24401, awarded by the National Institute of General Medical Sciences, DHHS. The authors gratefully acknowledge the assistance of Travenol Laboratories, Deerfield, Illinois.

References

1 Blackburn, G. L. and O'Keefe, S. J. D. (1977). Protein sparing therapy during stress and injury. *Proc. Western Hemisph. Nutrit. Congr.* Vol. IV, p. 220. (Chicago: American Medical Association)
2 Powanda, M. C. (1977). Changes in body balances of nitrogen and other key nutrients: description and underlying mechanisms. *Am. J. Clin. Nutr.*, **30**, 1254
3 Wannemacher, R. W., Jr., Dinterman, R. E., Pekarek, R. S. *et al.* (1975). Urinary amino acid excretion during experimentally induced sandfly fever in man. *Am. J. Clin. Nutr.*, **28**, 110
4 Ryan, N. T., Blackburn, G. L., Clowes, G. H. A. *et al.* (1974). Differential tissue sensitivity to elevated endogenous insulin levels during experimental peritonitis in rats. *Metabolism*, **23**, 1081
5 Blackburn, G. L. (1977). Lipid metabolism in infection. *Am. J. Clin. Nutr.*, **30**, 1321
6 Border, J. R., Burns, G. P., Rumph, C. *et al.* (1970). Carnitine levels in severe infection and starvation: a possible key to the prolonged catabolic state. *Surgery*, **68**, 175
7 Beisel, W. R. and Wannemacher, R. W., Jr. (1980). Gluconeogenesis, ureagenesis and ketogenesis during sepsis. *J. Parent. Enter. Nutr.*, **4**, 277
8 Miller, J. D. B., Bistrian, B. R., Blackburn, G. L. *et al.* (1977). Failure of postoperative infection to increase nitrogen excretion in patients maintained on peripheral amino acids. *Am. J. Clin. Nutr.*, **30**, 1523
9 Miller, J. D. B., Blackburn, G. L., Bistrian, B. R. *et al.* (1977). Effect of deep surgical sepsis on protein-sparing therapies and nitrogen balance. *Am. J. Clin. Nutr.*, **30**, 1528
10 Williamson, D. H., Farrell, R., Kerr, A. *et al.* (1977). Muscle protein catabolism after injury in man as measured by urinary excretion of 3-methylhistidine. *Clin. Sci.*, **52**, 527

11 Shinnick, F. L. and Harper, A. E. (1976). Branched chain amino acid oxidation by isolated rat tissue preparations. *Biochim. Biophys. Acta*, **437**, 477

12 O'Keefe, S. J. D., Moldawer, L. L., Young, V. R. *et al.* (1981). The influence of intravenous nutrition on protein dynamics following surgery. *Metabolism*, **30**, 1150

13 Birkhahn, R. H., Long, C. L., Fitkin, K. *et al.* (1980). Effects of major skeletal trauma on whole body protein turnover in man measured by L-(l-^{14}C)-leucine. *Surgery*, **88**, 294

14 Sakamoto, A., Moldawer, L. L., Palombo, J. D. *et al.* (1982). Alterations in tyrosine and protein kinetics produced by injury and branched chain amino acid administration. *Clin. Sci.*, (in press)

15 Cerra, F. B., Upson, D., Angelico, R. *et al.* (1982). Branched chains support postoperative protein synthesis. *Surgery*, **92**, 192

16 Bistrian, B. R. (1979). A simple technique to estimate severity of stress. *Surg. Gyn. Obstet.*, **148**, 675

17 Waterlow, J. C., Garlick, P. J. and Millward, D. J. (1978). *Protein Turnover in Mammalian Tissues and in the Whole Body*. (New York: Elsevier/North-Holland)

18 Garlick, P. J., Clugston, G. A., Swick, R. W. *et al.* (1980). Diurnal pattern of protein and energy metabolism in man. *Am. J. Clin. Nutr.*, **33**, 1983

19 Waterlow, J. C. (1968). Observations on the mechanism of adaptations to low protein intakes. *Lancet*, **2**, 1063

20 Echenique, M. M., Bistrian, B. R., Moldawer, L. L. *et al.* (1982). Improvement in amino acid utilization in the critically ill with parenteral formulas enriched with branched chain amino acids. (in review)

6
Branched-chain amino acids in surgically stressed patients

J. TAKALA, J. KLOSSNER, J. IRJALA AND S. HANNULA

INTRODUCTION

The idea of reducing post-injury catabolism by branched chain amino acids (BCAA) is based largely on experimental studies[1-7]. The potential benefits from BCAA have not been confirmed clinically.

This controlled clinical study in progress was designed to test whether the response to total parenteral nutrition (TPN) during severe stress is modified by increasing the proportion of BCAA in the nitrogen supply.

PATIENTS

All the patients were treated and studied in the surgical critical care unit. The results obtained with 10 catabolic post-operative or severely traumatized patients are reported (Table 6.1). All the patients selected for the study excreted a minimum of 0.16 g of urea nitrogen per kg of body weight per day in the urine at the start of the study.

The study lasted 6 consecutive days. On day 0 the patients received infusions with a maximum glucose concentration of 5%. The baseline urinary excretion of urea nitrogen was measured. The sum of urea excretion and the change in the body pool of urea nitrogen during day 0 (the urea appearance), referred to as UrP_0, was used as a rough estimate of the severity of the catabolism (Table 6.2). The body pool of urea nitrogen was assumed to be equal to $0.6 \times$ (initial body weight + weight change) $\times$ serum urea nitrogen. The patients were randomized to receive the TPN with either 50% or 15% of the infused amino acids as BCAA. Both TPN-programmes were isonitrogenous with a constant daily supply of 0.24 g of nitrogen per kg of bodyweight. The TPN was started on day 1 and continued for 5 days. The

Table 6.1   Patients included in the study

Age	Sex	Diagnosis
Branched-chain amino acid group		
70	Male	Retroperitoneal abscess
68	Male	Intestinal strangulation
31	Male	Multiple trauma of the thorax and abdomen
33	Male	Multiple trauma of the extremities
29	Male	Gunshot injury of the thorax and abdomen
28	Male	Multiple trauma of the thorax and abdomen
Control group		
36	Male	Acute pancreatitis
43	Male	Acute pancreatitis
72	Male	Tension pneumothorax
35	Male	Acute pancreatitis

energy supply was determined individually from the Harris-Benedict formula and the estimated degree of catabolism. A fraction of 20–50% of the energy input was infused as fat. The energy input was increased gradually, but maintained constant during days 4 and 5 (Table 6.2). All the nutrients except the fat emulsion were premixed and infused at a constant rate 23 h a day. The 1 h stop each morning was used for analysis of the post-prandial amino acid concentrations. Another plasma sample for analysis of the plasma amino acid concentrations was taken daily 3–4 hours from the commencement of the infusion, and was used for analysis of the plasma amino acid concentrations. The urinary excretion of amino acids was measured from a diurnal specimen.

Table 6.2   Urea appearance and energy input

	UrP_0 (g/kg)	Energy input days 4 and 5 (kcal/kg)
BCAA	0.18 ± 0.06	37.6 ± 9.1
Control	0.26 ± 0.14	43.5 ± 7.8

The efficacy of the TPN was evaluated by measurements of the nitrogen balance and serum concentrations of prealbumin. The nitrogen balance was measured on days 1, 4 and 5 and the prealbumin at the start and at the end of the TPN.

The measurements of the nitrogen balance were corrected for the accumulation of urea. Thus the nitrogen balance was calculated to be equal to the difference between the nitrogen input and the sum of nitrogen excreted in the urine and the change in the body pool of urea nitrogen. The statistical analysis of the difference between and within the groups was done by the analysis of variance and covariances with a split-plot design including repeated measures. The amino acid data were analysed by Student's *t*-test.

RESULTS

Nitrogen balance

The mean nitrogen balances on days 1, 4 and 5 are given in Table 6.3. The data were analysed by using either the UrP_0 or the actual input of energy or both as covariates and without covariates. No significant differences between the treatment alternatives were demonstrated.

Table 6.3 Nitrogen balance (g/kg day) mean ± S.D.

Treatment	Day 1	Day 4	Day 5
Corrected for the accumulation of urea			
BCAA	− 0.09 ± 0.09	− 0.10 ± 0.09	− 0.10 ± 0.10
Control	− 0.09 ± 0.10	− 0.14 ± 0.12	− 0.06 ± 0.11
Without correction for the accumulation of urea			
BCAA	− 0.05 ± 0.12	− 0.12 ± 0.09	− 0.10 ± 0.09
Control	− 0.07 ± 0.10	− 0.14 ± 0.12	− 0.06 ± 0.14

Serum concentration of prealbumin

The mean serum concentrations of prealbumin are shown in Table 6.4. Analysed identically with the nitrogen balance data, no differences between the treatment alternatives were noted.

Table 6.4 Serum concentration of prealbumin

Treatment	Serum prealbumin prior to TPN (g/l)	Serum prealbumin after TPN (g/l)
BCAA	0.140 ± 0.090	0.115 ± 0.069
Control	0.135 ± 0.086	0.113 ± 0.038

Plasma concentrations of BCAA

The concentrations of plasma BCAA were low at the start of the study in both groups. Compared to the fasting values at day 0, postprandial levels of BCAA increased in the BCAA group. On the contrary the concentrations remained low or decreased in the control group (Table 6.5). The difference observed between the groups was significant for valine and isoleucine on day 5.

Table 6.5 Plasma concentrations of BCAA

Treatment	Day 0 (μmol/l)	Significance of difference between treatments	Day 5 post prandial (μmol/l)	Significance of difference between treatments
BCAA				
leucine	91 ± 50	NS	134 ± 32	NS
isoleucine	40 ± 24	NS	78 ± 10	$p < 0.05$
valine	149 ± 69	NS	392 ± 54	$p < 0.05$
Control				
leucine	134 ± 63		110 ± 25	
isoleucine	45 ± 35		54 ± 1	
valine	220 ± 79		209 ± 42	

Urinary excretion of BCAA and 3-methylhistidine

The urinary excretion of BCAA was low in both treatment alternatives. No significant difference between the groups was observed (Table 6.6). The urinary excretion of 3-methylhistidine was lower in the BCAA-group on day 1. No other significant differences were noted (Table 6.6).

COMMENTS

An attempt was made to take into account the metabolic state of the patients by estimating the urea appearance and the energy expenditure. A criticism of the validity of these estimates is clearly justified, hence the differences in the primary catabolism and in the actual intake of energy were taken into account in the statistical analysis.

This study did not demonstrate any benefit from an increased supply of BCAA. The results are however, preliminary. The analysis of the nitrogen balance demonstrated that estimating the changes in the body pool of urea was essential since the retention of urea in the body may be considerable in the early post-traumatic phase.

The results suggest that the validity of comparative balance studies, requiring several days, in severely stressed patients may be questioned, as considerable metabolic alterations inevitably occur during the observation period in such a heterogeneous group of patients.

Table 6.6 Urinary excretion of BCAA and 3-methylhistidine

Treatment	Day 0 (μmol/day)	Significance of difference between treatments	Day 1 (μmol/day)	Significance of difference between treatments	Day 5 (μmol/day)	Significance of difference between treatments
BCAA						
leucine	59 ± 80	NS	80 ± 106	NS	132 ± 225	NS
isoleucine	0	NS	39 ± 49	NS	60 ± 104	NS
valine	74 ± 110	NS	133 ± 155	NS	187 ± 329	NS
3-methylhistidine	167 ± 125	NS	177 ± 34	$p < 0.05$	93 ± 107	NS
Control						
leucine	23 ± 40		47 ± 63		51 ± 14	
isoleucine	5 ± 8		12 ± 21		51 ± 14	
valine	33 ± 37		54 ± 62		74 ± 16	
3-methylhistidine	156 ± 132		311 ± 64		124 ± 61	

References

1 Blackburn, G. L. *et al.* (1979). Branched chain amino acid administration and metabolism during starvation, injury, and infection. *Surgery*, **86**, 307

2 Buse, M. G. *et al.* (1979). *In vitro* effect of branched chain amino acids on the ribosomal cycle in muscles of fasted rats. *Horm. Metab. Res.*, **11**, 289

3 Chua, B. *et al.* (1979). Effect of leucine and metabolites of branched chain amino acids on protein turnover in heart. *J. Biol. Chem.*, **254**, 8358

4 Hutson, S. M. *et al.* (1980). Regulation of leucine and α-ketoisocaproic acid metabolism in skeletal muscle. *J. Biol. Chem.*, **255**, 2418

5 Jefferson, L. S. *et al.* (1974). Insulin in the regulation of protein turnover in heart and skeletal muscle. *Fed. Proc.*, **33**, 1098

6 Jefferson, L. S. *et al.* (1979). Regulation by insulin of amino acid release and protein turnover in the perfused rat hemicorpus. *J. Biol. Chem.*, **252**, 1476

7 Li, J. B. and Jefferson, L. S. (1978). Influence of amino acid availability of protein turnover in perfused skeletal muscle. *Biochim. Biophys. Acta*, **544**, 351

7
Selective nutritional support for hepatic regeneration: experimental and clinical experiences with branched-chain amino acids

H. JOYEUX, B. SAINT-AUBERT, C. ASTRE, P. C. ANDRIGUETTO,
P. VIC, C. HUMEAU AND Cl. SOLASSOL

Hepatic regeneration (HR) is a transient process of the liver which occurs when liver mass is reduced by 10% or more and ceases when the initial liver mass has been restored[4]. The reserve capacity of this phenomenon is extraordinary since only 5% of the total liver mass (e.g. the caudate lobe) can serve to reconstitute the entire initial liver mass[21]. The rapidity with which this process occurs, dictates that in the second post-operative week regeneration has assumed such 'biological priority' as to eventually assure complete restoration of the resected mass[3].

However, liver regeneration is a fragile process. This temporary activity is governed by two variables: residual liver mass after partial hepatectomy and the metabolic status of the host. If the remaining liver mass is less than 30% of total initial mass, induced DNA synthesis is retarded, or even halted when the mass falls below 10%[29,32]. Marked metabolic perturbations occur after partial hepatectomy. The intensity of these abnormalities varies with the extent of liver resection and includes: changes in glycogen reserves, mobilization of fat stores and protein breakdown. The overall effect on remaining hepatocytes may inhibit cell division thereby impeding the regenerative process.

Since 1974, our laboratory[9] has studied liver regeneration in an effort to identify the stimuli which will produce the best survival following 90–95% partial hepatectomy[10]. The only artificial means of support which extended survival of animals (4 days) after total or subtotal hepatectomy was total parenteral nutrition[11–14].

We have tried to stimulate the metabolic processes supporting regeneration by modifying nutritional factors. These factors play a vital role in 'hepatic assistance' following total or sub-total hepatectomy[11]. Our goal was to influence cell synthesis through selective nutritional support in animals with metabolic disturbances due to total or subtotal hepatectomy[11,21]. Our results have since been applied to patients with severe surgical hepatic insufficiency following extensive liver resection.

SELECTIVE NUTRITIONAL SUPPORT IN SURGICAL HEPATIC INSUFFICIENCY

Our goal was to establish a formula with a specific association of amino acids for the surgical patient with hepatic insufficiency.

Materials and methods

Two experimental dog models[11] were used to determine amino acid disturbances related to surgical hepatic insufficiency. Amino acids were analysed in both serum and urine. The first model consisted of total hepatectomy, the second subtotal. Both groups received total parenteral nutrition which included glucose, electrolytes and clotting proteins in the form of fresh complete plasma collected from donor animals. The nutritional solution (formula I) contained the following (per kg per 24 h): H_2O; 65 ml; glucose; 7.5 g; electrolytes (mEq/kg bodyweight 24 h): Na^+; 3.26; K^+; 2.13; Cl^-; 3.12; Mg^{2+}; 0.74; Ca^{2+}; 0.42; P, 1.06; acetate, 1.12; lactate, 1.06; SO_4, 0.13. Ten ml blood samples were taken on heparin for amino acid determinations immediately before surgery and every 24 hours thereafter. Five percent sulfosalicylic acid supernatants were analysed by a Technicon TSM amino acid analyser[25].

Serum and urine amino acid perturbations

Total plasma amino acid concentration increased from day 0 to day 3 after hepatectomy, reaching a 2-fold overall increase on day 2. Two groups of amino acids were found to be of particular interest: aromatic amino acids (AAA) were elevated in all dogs to about four times the normal control value. Of the branched chain amino acids (BCAAs), leucine and isoleucine (leu, ileu) decreased and valine (val) remained unchanged. In spite of a distinct hyperaminoacidaemia, isoleucine decreased to 20% of the control value. The normal BCAA/AAA ratio (3–5) decreased to as low as 0.4–0.5. No differences were noted for total versus 95% hepatectomized dogs (Figure 7.1).

Urinary amino acids were excreted in increasing quantities from day 0 to day 3. Gram quantities of amino acids were present in the urine.

Creating a solution specifically formulated for hepatic surgical insufficiency

We reduced the concentration of aromatic amino acids and methionine since

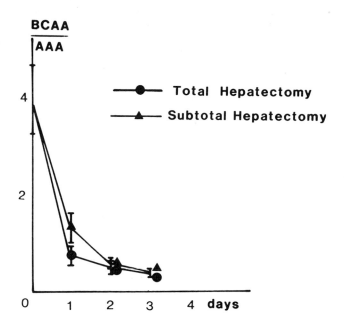

Figure 7.1 Variation of BCAA/AAA ratio after total and subtotal hepatectomy in the dog. (BCAA: branched chain amino acids; AAA: aromatic amino acids)

they may constitute a source of brain complications due to acute hepatic insufficiency. We therefore limited their concentration during the liver regeneration process. Whereas a high plasma concentration of amino acids can be drawn upon during protein synthesis, a decreased plasma concentration of branched amino acids or an imbalanced plasma amino acid pool can inhibit liver protein synthesis. In fact, it has been shown in rats[6] that 1.0–1.5 hours after hepatectomy there is a stimulation of the A amino acid transport system during early and mid-S phase of DNA synthesis. At this time the L system (leucine-preferring) is set in action (which also transports isoleucine and valine). For these reasons we gave BCAAs beyond what was required to restore normal plasma equilibrium so as to ensure an available supply when needed (first week after hepatic resection in dog and patient).

Because of a hyperaminoacidaemia we did not see the need to administer non-essential amino acids during the early regeneration period. The essential non-branched-chain amino acids were given in the usual recommended proportions with phenylalanine, methionine and tryptophan in small amounts. The concentration of threonine was increased slightly. The majority of the amino acids given were BCAA (Table 7.1).

Table 7.1 Composition of selective amino acid solutions

	Definitive composition of the selective amino acid solution*	Proposed selective amino acid solution for surgical hepatic insufficiency
Isoleucine (g/l)	14	22.40
Leucine (g/l)	6.42	12.80
Lysine (acetate) (g/l)	3.20	3.20
Methionine (g/l)	0.80	1.00
Phenylalanine (g/l)	0.70	1.00
Threonine (g/l)	2.00	2.00
Tryptophan (g/l)	0.40	0.40
Valine (g/l)	12.80	12.80
Arginine (g/l)	0.60	0.60
Cysteine (g/l)	0.22	–
Total nitrogen	4.79 g/l	6.27 g/l

*Roger Bellon Laboratories

The final formula

We were not able to create as stable a solution as desired due to the low solubility of the BCAAs. Leucine in the presence of large amounts of valine and isoleucine presented the greatest problem and its concentration had to be reduced. The basic amino acid arginine was added to adjust the pH (7.0) of the solution and for its role in generating urea (Table 7.1).

THE IMPORTANCE OF SELECTIVE NUTRITIONAL SUPPORT DURING LIVER REGENERATION FOLLOWING 65% HEPATECTOMY IN THE DOG

In order to test the effect of our formula we chose an easily reproducible experimental model with only minor post-operative complications but which would serve to compare the effects of different parenteral diets.

The experimental model
A 65% hepatectomy was performed[1].

Anaesthesia
A Teflon catheter was introduced into a leg vein of the animal. Anaesthesia was induced with sodium thiopental at 20 mg/kg. Anaesthesia was maintained by dextromoramide and chlorprothixène as needed. During the operation the animal was infused with 5% Ringer–lactate–glucose and received oxygen–nitrogen protoxyde (50/50) through a tracheal tube. At the end of surgery the animal was placed on an adjustable oxygen–air respirator. At this time, if the animal did not regain consciousness it received 20 mg doxapram chlorhydrate i.v.

Central venous catheter

The internal jugular vein was catheterized through a 1 cm incision on the lateral surface of the neck and the catheter was then passed through the sternocleidomastoid muscle. The catheter (35 cm long and with an internal diameter of 1.5 mm) was threaded down the internal jugular vein to the superior vena cava. The external portion of the catheter was passed subcutaneously to exit at a point 5 cm beyond the original site of penetration, to avoid infection and allow intravenous feeding while the animal was able to move about freely in its metabolic cage.

Surgical technique

Hepatic resection was performed through a mid-line incision and section of the triangular and coronary ligaments of the liver. The II, III, IV and V lobes (Couinaud's classification) were excised, amounting to 65% of total liver mass. The pedicles associated with these lobes were ligated and divided. Each pedicle consisted of a branch of the portal vein, a branch of the hepatic artery and the corresponding bile duct. After ligating and sectioning the hepatic veins, the resected portion was removed as a single block. Blood loss with this technique does not exceed 50 ml. The various layers of the abdominal wall were then closed without a drain. The average duration of surgery was 1 h 15 min.

Animals

Seventy-five middle-aged healthy mongrel dogs of both sexes were divided randomly into three groups, each of which received a different type of nutrition during the post-operative phase following 65% hepatectomy. Vitamins were given i.m. and i.v. in the same amount to all three groups.

Group I: fed orally (25 animals)

This group was fed *ad libitum*. These animals received a solution of electrolytes and glucose in water intravenously during the first 48–60 post-operative hours to compensate any losses due to surgery and to maintain a metabolic balance during this phase. The volume delivered was identical to that given to animals undergoing 95 or 100% hepatectomy (formula I) (see Table 7.2).

Group II: parenteral nutrition with all amino acids (25 animals)

This group was fed exclusively by the parenteral route from the first post-operative day and received the same electrolyte, glucose and water ration as group I. In addition, Group II also received a balanced mixture of all amino acids (E/T ratio: 3.22; formula II – Table 7.2 and Table 7.3).

Group III or the selective nutrition group (25 animals)

The only difference in this group was the kind of amino acids administered.

This formula contained only essential amino acids of which 82.5% were branched-chain (Table 7.2 and Table 7.3).

Table 7.2 Nutrition after 65% hepatectomy in the dog

Nutrients	Group I Formula I	Group II Formula II	Group III Formula III
	(units/kg bodyweight 24 h)		
H_2O (ml)	65	65	65
Dextrose (g)	7.5	7.5	7.5
Nitrogen (g)	0	0.24*	0.24†
Na^+ (mEq)	3.26	3.26	3.26
K^+ (mEq)	2.13	2.13	2.13
Cl^- (mEq)	3.12	3.12	3.12
Mg^{2+} (mEq)	0.74	0.74	0.74
Ca^{2+} (mEq)	0.42	0.42	0.42
P (mEq)	1.06	1.06	1.06
Acetate (mEq)	1.12	1.12	1.12
Lactate (mEq)	1.06	1.06	1.06
(mEq)	0.13	0.13	0.13

*All amino acids
† Specific amino acids with 82.5% BCAA

Table 7.3 Amino acid solutions given after 65% hepatectomy in the dog (group II: all amino acids, group III: specific amino acids)

	Solution containing all amino acids	Solution containing specific amino acids
Essential amino acids		
Isoleucine	5.1	14
Leucine	13.9	6.42
Lysine	19.5	(acetate) 3.20
Methionine	9.4	0.80
Phenylalanine	12.5	0.70
Threonine	5	2.00
Tryptophan	2.6	0.40
Valine	12.5	12.80
Non-essential amino acids		
Alanine	9.5	0
Arginine	25	0.60
Aspartic acid	4	0
Cystine-cysteine	1.5	0.22
Glutamic acid	5	0
Glycine	9	0
Histidine	5	0
Ornithine	3.4	0
Proline	8.0	0
Serine	1.3	0
Tyrosine	0.25	0
Citrulline	3	0
Total nitrogen (g/l)	25	4.79
Percentage BCAA		82.5 %

Surveillance of liver regeneration

In addition to routine clinical and laboratory follow-up, clotting factors were monitored since they reflect hepatic synthetic activity and mitotic index was also followed[1].

The purpose of the study was to follow liver regeneration step by step from day 1 to day 12 post-hepatectomy. Our first procedure consisted of a series of liver biopsies for all three groups. For this, animals were killed on post-operative days 1–11. The results show strong regenerative activity on days 2–5. Five animals were killed on each of these days and liver specimens were examined by both light and electron microscopy[30].

Results

Animals were randomly assigned to each group without preference as to breed, age or sex. 65% hepatectomy was well tolerated by all animals. All responded to hepatectomy in the same manner except that dogs in groups II and III presented mild indications of transient hepatic insufficiency.

Evaluation of hydration, electrolytes, BUN, blood glucose and serum protein
Blood tests did not show very significant changes. The BUN for group II decreased slightly and rose above 0.10 for two animals in group III on the 5th and 7th post-operative days. Glucose remained within normal limits for all three groups, was more or less constant and showed little deviation compared to normal values.

We did not observe post-hepatectomy hypoglycaemia and therefore considered the exogenous glucose supply to be adequate.

Low blood protein was recorded over the first 4 post-operative days for group I. This could be due to inadequate exogenous protein. Protein values tended to fall into the normal range once the dogs began to eat again. Group II protein values remained constant whereas group III values were slightly subnormal on post-operative days 6 and 7. Serum bicarbonate remained normal throughout the entire post-operative period for all three groups reflecting a balanced water and electrolyte parenteral solution.

Sodium, potassium and chloride remained unchanged for all three groups, any losses being compensated by the infused source.

Nitrogen balance
We only compared groups II and III since these were the only ones to receive nitrogen parenterally. Both groups (II and III) showed a nitrogen deficit during the first 10 post-operative days. This deficit was slightly more pronounced for group II during the first 3 and last 4 days (Figure 7.2).

Clotting factors
Clotting factors were markedly low, particularly the vitamin K dependent factors. Factor II was invariably in the low normal range for group III but

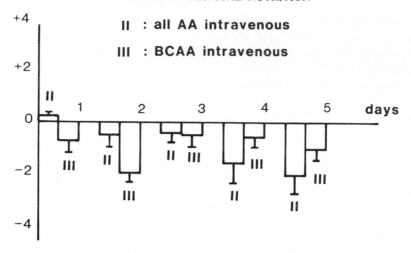

Figure 7.2 Nitrogen balance after 65% hepatectomy in the dog (AA: amino acids, BCAA: branched chain amino acids)

tended to rise after the 10th post-operative day for groups I and II. A sharp fall in factor V (proaccelerin) on day 2, was followed by a progressive rise, climbing more quickly in group I than groups II and III.

Factors VII and X were lower in group III although group II values were also very low. In the control group, these values rose progressively to eventually reach normal.

Factor VII was not significantly depressed for group II. The relatively low values observed during the first post-operative week for groups I and III returned to normal thereafter.

Both fibrinogen and Quick's test remained unchanged for all three groups.

Plasma amino acid profile and BCAA/AAA ratio
Plasma BCAAs, particularly valine and isoleucine, were significantly higher for group III compared to groups I and II.

The BCAA/AAA ratio was also significantly higher in group III compared to groups I and II (Figure 7.3).

Histomorphology
The morphological differences observed for the three groups are subjective since they could not be quantified. All three groups demonstrated massive steatosis on the 1st post-operative day but this began to regress on day 3. The steatosis was slightly less for groups II and III and disappeared almost completely by day 4 which was not the case for group I.

There was a slight increase in cell organelles for all three groups with, again, a difference for group I where slightly more organelles were observed.

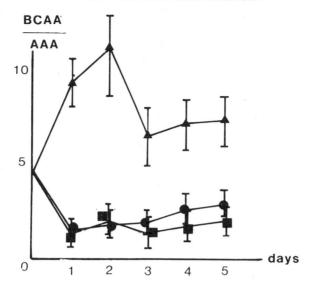

Figure 7.3 Variation of BCAA/AAA ratio after 65% hepatectomy in the dog. ■ – group I, oral alimentation; ● – group II, all amino acids, intravenous; ▲ – group III, BCAA intravenous

Thus, cellular kinetics were essentially the same for all groups except for minor differences which may only be subjective. This finding will eventually have to be confirmed quantitatively.

Variation of the mitotic index

Mitotic index was calculated according to Weibel's method[31] using the technique described by Humeau[8]. At microscopic examination, mitoses appeared normal and were in nests.

After 65% hepatectomy in the dog the mitotic index peak occurs at day 3, and 4. Mean values for each group from day 2 to day 5 show a statistically significant difference between groups III and I. There is no significant difference between groups I and II (Table 7.4).

Table 7.4 Mean mitotic index from day 2 to 5

Group	Mitotic index	Significantly different from group I*
Group I	0.673 ± 0.40	–
Group II	0.829 ± 0.80	0
Group III	1.098 ± 0.09	+

*Wilcoxon test: $0 = NS$; $+ = p < 0.01$

Table 7.5 Mean mitotic index per day (M ± SEM)

Group	Day 2 (n=5)	Day 3 (n=5)	Day 4 (n=5)	Day 5 (n=5)
I	0.475 ± 0.100*	0.728 ± 0.008 †	0.714 ± 0.060 ‡	0.774 ± 0.060
II	0.602 ± 0.120	0.918 ± 0.220	1.077 ± 0.170	0.719 ± 0.110
III	0.891 ± 0.160*	0.974 ± 0.07 †	1.562 ± 0.230 ‡	0.965 ± 0.100

Wilcoxon test: *, $p = 0.05$; †, $p = 0.02$; ‡, $p = 0.01$

Mean values in each group on each day (Table 7.5) reveal highly significant differences between the three groups and confirm an overall increase of mitotic index in group III animals (Figure 7.4).

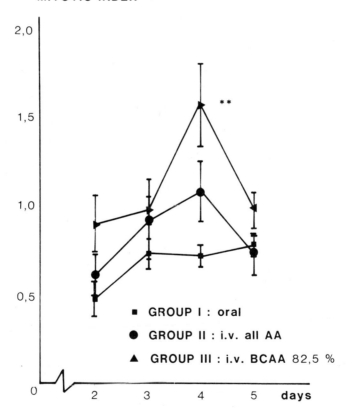

Figure 7.4 Variation of mitotic index after 65% hepatectomy in the dog (number of mitoses per 100 counted hepatocytes)

CLINICAL APPLICATION WITH TEN CASES OF EXTENSIVE HEPATIC RESECTION (70–90%)

Having demonstrated the benefit of selective nutritional support for surgical hepatic insufficiency in animals following 65% hepatectomy, we decided to apply our findings to the human situation.

Table 7.6   Selective nutritional support (units/kg 24 h)

H_2O	28.5 ml
Dextrose	3.6 g
Nitrogen	0.07 g
Na	0.71 mmol
K	0.68 mmol
Ca	0.05 mmol
Mg	0.03 mmol
Cl	1.12 mmol
Acetate	0.36 mmol
Sulphate	0.03 mmol
Lactate	0.06 mmol
Phosphate	0.05 mmol
Trace elements	+

Selective parenteral nutrition in man

Selective parenteral nutrition was restricted to the post-operative period only (Table 7.6). All patients received the same pre-operative nutrition (i.v.) during the 3 days leading up to surgery. The protein charge consisted of a balanced solution of all amino acids (25 g nitrogen/l).

Patients

Ten liver tumour patients (2 benign and 8 malignant) received post-operative nutritional support following excision of 70–90% of their liver. The percent of total liver mass which was resected was determined by three-dimensional tomoscintiscans[15]. Excision was performed as described above (ligation of afferent and efferent blood supply prior to resection). In two cases, metastatic nodules were also removed from the remaining liver mass.

Follow-up

Clotting factors

As shown in Figure 7.5, there was a significant difference in post-hepatectomy proaccelerin values. Patients receiving selective nutrition (group I) experienced a lesser decrease in proaccelerin activity which remained at 40% compared with 18% activity for the group receiving normal (all amino acids) nutrition (group II). Moreover, the rebound followed the same pattern with an 85% recovery on day 5 for selective nutrition compared to the 48% control value. On day 10, patients who had received a selective parenteral nutrition had normal proaccelerin values while control subjects were still at 50%. The same pattern was observed for prothrombin values.

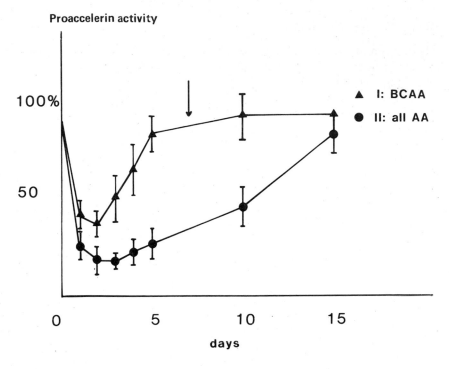

Figure 7.5 Proaccelerin activity after trisegmentectomy in patients. Selective nutrition was given from day 1 to day 7 (see vertical arrow). From day 8 to day 15 nutrition was similar in group II and group I

Plasma amino acid levels
Patients who received non-selective nutrition after hepatectomy (group II) presented the expected changes: increased tyrosine (Figure 7.6) and phenylalanine (Figure 7.7), slightly elevated leucine (Figure 7.8) and lesser concentrations of amino acids valine and isoleucine (Figures 7.9 and 7.10). Overall this led to a marked fall in the BCAA/AAA ratio (Figure 7.11). Some of the values were very low e.g. 1 on day 3 and 1.6 by day 15. The amino acid which varied the most was methionine (Figure 7.12). On day 3 it rose to 10-fold its base value and did not return to normal until day 12.

In the group receiving the solution enriched with branched amino acids (group I), infusion led to normal levels of isoleucine, phenylalanine, tyrosine and methionine. Leucine varied only slightly. Valine was abnormally high as was the BCAA/AAA ratio. This treatment was discontinued on day 7 and by day 10 all amino acids and the BCAA/AAA were normal except for perhaps isoleucine which remained low (Figures 7.6–7.12).

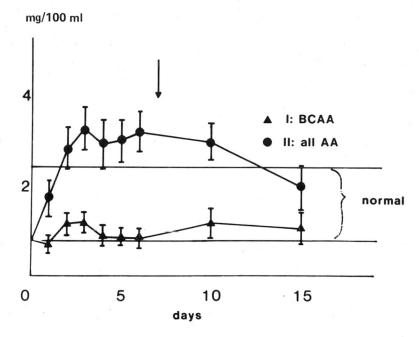

Figure 7.6 Plasma tyrosine level after trisegmentectomy in patients. Selective nutrition was given from day 1 to day 7 (see vertical arrow). From day 8 to day 15 nutrition was similar in group II and group I

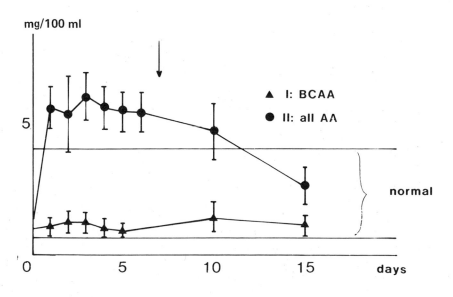

Figure 7.7 Plasma phenylalanine level after trisegmentectomy in patients

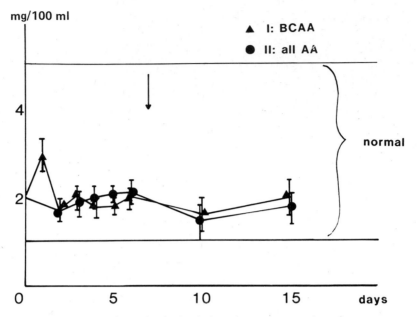

Figure 7.8   Plasma leucine level after trisegmentectomy in patients

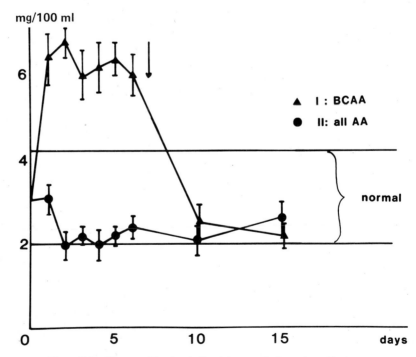

Figure 7.9   Plasma valine level after trisegmentectomy in patients

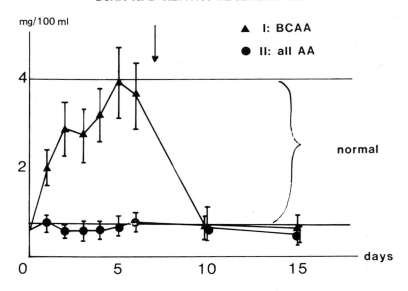

Figure 7.10 Plasma isoleucine level after trisegmentectomy in patients

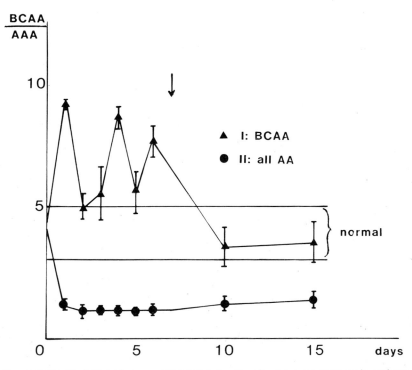

Figure 7.11 Variation of plasma BCAA/AAA ratio after trisegmentectomy in patients

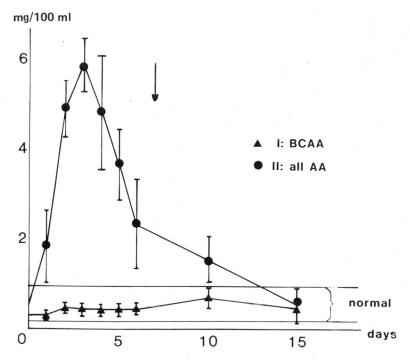

Figure 7.12   Plasma methionine level after trisegmentectomy in patients

DISCUSSION

Nutritional considerations during liver regeneration

Most of the studies dealing with nutrition during liver regeneration have employed rats which underwent 68% hepatic resection followed by an oral diet[7]. Stirling has shown that low calorie, low nitrogen pre-operative diets (complete fasting for 48 h or half-fasting for 14 days prior to surgery) retards peak mitotic activity and lowers the maximum mitotic index from 26 to 16 per 1000 hepatocytes[26]. It is nevertheless remarkable that clearly atrophied hepatocytes with ultrastructural changes such as decreased rough endo-plasmic reticulum, retain the capability to hypertrophy and proliferate in response to parenchymal resection[23-28]. Although the role played by the administration of proteins was not confirmed by Montecuccoli[18], its import-ance was upheld by the studies of MacGowan and Fausto in 1975[16]. Thus, fasting in the absence of protein exerts a depressive, but not suppressive effect on the regenerative process. Short has shown the major role of six amino acids: isoleucine, threonine, valine, lysine, methionine and tryptophan[22]. Weinbren and Dowling have been the only proponents of the major role played by nutrition (glucose) following extensive resection leading to severe hepatic insufficiency[33].

Therefore, nutritional support is essential to liver regeneration even if detoxification of excess aromatic amino acids, leading to the production of false neurotransmitters[5], is envisioned[2]. In their article on hepatotrophic substances we feel that Starzl and Terblanche should have placed greater emphasis on the role of nutritional factors[24]. The present authors began promoting this belief in 1974[9] which, in 1977, led us to formulate a totally new concept for nutrition as supportive liver therapy[10].

The concept and role of selective amino acid solutions

The development of such a solution necessitated that we establish the metabolic consequences related to severe surgical hepatic insufficiency. This led to the formulation of our original amino acid solution which consisted of a very high concentration of BCAA.

A study of 75 dogs divided into control groups and one test group confirmed the benefits of the above solution on the mitotic index which is presently the most representative measure of liver regeneration. Confirmation in human trials is still forthcoming but the recent development of a quantitative evaluation of liver regeneration in man[15] may help evaluate this type of therapy.

Other nutritional supports for the liver

Several enteral (Hepatic Aid, Travasorb Hepatic) and parenteral nutritional formulae are available and have been proposed for the treatment of medical hepatic insufficiency[17-19]. These solutions restore normal amino acid levels in plasma[20-27], create a positive nitrogen balance and ameliorate electroencephalographic signs. However, there is no publication to date which attests to their benefit during regeneration of the 'insufficient' liver. Certain findings[20] suggest an anticatabolic effect for BCAAs on muscle protein turnover, but no proof has been provided as yet.

Perspectives

What remains to be done is to compare the different nutritional therapies to ascertain whether enteral or portal vein infusion would be more effective than a systemic approach or whether lipids should also be given. Metabolic studies in animals and patients undergoing partial hepatectomy should offer valuable information towards reaching this goal.

ACKNOWLEDGEMENTS

The authors are grateful to H. Viguier, M. Brissac, M. Cros and C. Legrand for skilful technical assistance and to R. Lodise for translation of the manuscript. This work was supported by grant INSERM no. 807008.

References

1 Andriguetto, P. C., Vic, P., Saint-Aubert, B., Quijano, F., M'Jahed, A., Piccolboni, D., Astre, C. and Joyeux, H. (1982). Branched-chain amino acids induce increased mitotic index in the liver of 65% hepatectomized dogs. (In preparation)

2 Asanuma, Y., Malchesky, P., Smith, J., Zawicki, I., Werynski, A., Carey, W., Ferguson, D., Hermann, R., Kayashima, K. and Nose, Y. (1981). Chronic ambulatory liver support by membrane plasmapheresis with on-line detoxification. *Trans. Am. Soc. Artif. Int. Org.*, **27**, 416

3 Blumgart, L. H., Leach, K. G. and Karran, S. J. (1971). Observations on liver regeneration after right hepatic lobectomy. *Gut*, 12,292

4 Bucher, N. L. R. (1967). Experimental aspects of hepatic regeneration. *N. Engl. J. Med.*, **277**, 686

5 Fischer, J. E., Funovics, J. M., Aguirre, A., James, J. H., Keane, J. M., Westorp, R. I. C., Yoshimura, N. and Westman, T. (1975). The role of plasma amino acids in hepatic encephalopathy. *Surgery*, **78**, 276

6 Guidotti, G. G., Borghetti, A. F. and Gazzola, G. C. (1978). The regulation of amino acid transport in animal cells. *BBA*, **515**, 329

7 Higgins, G. M. and Anderson, R. M. (1931). Experimental pathology of the liver. Restoration of the liver of the white rat following partial surgical removal. *Archiv. Pathol.*, **12**, 186

8 Humeau, C., Vic, P., Arnal, F., Mathieu Daude, P., Vlahovitch, B. and Sentein, P. (1980). L'activité mitotique dans les gliomes. Comparaison entre gliomes bénins et malins. *Biol. Cell*, **38**, 3, 179

9 Joyeux, H., Yakoun, M., Ould-Said, H., Brissac, C. and Solassol, Cl. (1974). Regeneration hépatique après hépatectomie partielle sous nutrition parenterale. In C. Romieu, C. Solassol, H. Joyeux and B. Astruc (eds.). *Proceedings of the International Congress on Parenteral Nutrition*. pp. 460–479. (Montpellier: France)

10 Joyeux, H., Joyeux, A., Raoux, Ph., Brissac, C., Yakoun, M., Blanc, F., Fourcade, J. and Solassol, Cl. (1977). Hepatic assistance after subtotal or total hepatectomy in the dog: a new concept. *Trans. Am. Soc. Artif. Int. Organs*, **23**, 683

11 Joyeux, H., Joyeux, A., Raoux, Ph., Brissac, C., Blanc, F. and Solassol, Cl. (1977). Troubles métaboliques de l'anhépatie expérimentale et nutrition parentérale. *Ann. Anesth. Franç.* **18**, 939

12 Joyeux, H., Joyeux, A., Ould-Said, H., Carretier, M., Yakoun, M., Brissac, C., Blanc, F. and Solassol, Cl. (1979). Artificial liver support with total parenteral nutrition after total hepatectomy in the dog. *Acta Chir. Scand.*, **494** (Suppl.), 164

13 Joyeux, H., Carretier, M., Ould Said, H., Saint-Aubert, B., Brissac, C., Walkin, P. and Solassol, Cl. (1980). Artificial liver in surgical hepatic insufficiency. *Am. Soc. Artif. Int. Org.*, Abstracts 9, 73

14 Joyeux, H., Joyeux, A., Carretier, M., Raoux, Ph., Ould Said, H., Saint-Aubert, B., Blanc, F., Rouanet, Cl., Pujol, H. and Solassol, Cl. (1980). Suppléance hémostatique après hépatectomie totale ou subtotale chez le chien. *J. Chir. Paris*, **117**, 199

15 Joyeux, H., Saint-Aubert, B., Andriguetto, P. C., M'Jahed, A., Yakoun, M., Solassol, Cl., Collet, H., Faurous, P. and Suquet, P. (1982). Evaluation de l'index pondéral hépatique (IPH), par tomoscintigraphie tridimensionnelle, au cours de la régénération hépatique après hépatectomie de 70 à 90%. Etude expérimentale et clinique. *Gastroenterol. Clin. Biol.* (submitted)

16 MacGowan, J. and Fausto, N. (1975). Ornithine decarboxylase activity and delayed DNA synthesis in the regenerating liver of protein deprived rats. *Fed. Proc. 59th Annual Meeting*, Abstract 3616, p. 858

17 Martin, D. J. (1981). Formulation of enteral feedings in hepatic insufficiency. *Nutr. Support Services*, **1**, 34

18 Montecuccoli, G., Novello, F. and Stirpe, F. (1972). Effect of protein deprivation on DNA synthesis in resting and regenerating rat liver. *J. Nutr.*, 102, 507

19 Quercia, R. A. (1981). Malnutrition and nutritional support in hepatic failure. *Nutr. Support Services*, **1**, 22

20 Rossi-Fanelli, F., Angelico, M., Cangiano, C., Cascino, A., Capocaccia, R., Deconciliis, D., Riggio, O. and Capocaccia, L. (1981). Effect of glucose and/or branched chain amino

acid infusion on plasma amino acid imbalance in chronic liver failure. *J. Parent. Ent. Nutr.*, **5**, 414
21 Saint-Aubert, B. (1982). *Limites de la Régénération Hépatique.* Thèse No. 77. (University of Montpellier I (School of Medicine))
22 Short, J., Armstrong, N. B., Gaza, D. J. and Lieberman, L. (1975). Hormones and amino acids and the control of nuclear DNA replication in liver. In R. Lesch and W. Reutter (eds.). *Liver Regeneration after Experimental Injury.* pp. 296–308. (New York: Stratton International Medical)
23 Silmes, A. M. and Dalman, P. (1974). Nucleic acid and polyamine synthesis in the rat during short term protein deficiency: response of the liver to partial hepatectomy. *J. Nutr.*, **104**, 47
24 Starzl, T. E. and Terblanche, J. (1979). Hepatotrophic substances. In H. Popper and F. Schaffner (eds.). *Progress in liver diseases.* pp. 135–151. (New York: Grune and Stratton)
25 Stein, W. H. and Moore, S. (1954). The free amino acids in human plasma. *J. Biol. Chem.*, **211**, 915
26 Stirling, G. A., Bourne, L. D. and Marsh, T. (1975). Effect of protein deprivation and a reduced diet on the regenerating rat liver. *Br. J. Exp. Pathol.*, **56**, 502
27 Striebel, J. P., Holm, E., Kutz, H. and Storz, L. W. (1979). Parenteral nutrition and coma therapy with amino acids in hepatic failure. *J. Parent. Ent. Nutr.*, **3**, 240
28 Talaricio, K. S., Feller, D. D. and Neville, E. D. (1971). Mitotic response to various dietary conditions in the normal and regenerating rat liver. *Proc. Soc. Exp. Biol. Med.*, **136**, 381
29 Tuczeck, H. V. and Rabes, M. H. (1971). Verlust der proliferations föhigkeit der hepatozyten nach subtotaler hepatektomie. *Experientia*, **27**, 526
30 Vic, P. (1982). Critères morphologiques de la régénération après hépatectomie. *Thèse de Biologie humaine.* Faculté de Médecine de Montpellier
31 Weibel, E. R. (1969). Stereological principes for morphometry in electron microscopic cytology. *Int. Rev. Cytol.*, **235**, 26
32 Weinbren, K. and Woodward, E. (1964). Delayed incorporation of [^{32}P]-orthophosphate into deoxyribonucleic acid of rat liver after subtotal hepatectomy. *Br. J. Exp. Pathol.*, **5**, 442
33 Weinbren, K. and Dowling, F. (1972). Hypoglycaemia and the delayed proliferative response after subtotal hepatectomy. *Br. J. Exp. Pathol.*, **53**, 78

8
The effect of branched-chain amino acids on body protein breakdown and synthesis in patients with chronic liver disease

J. D. HOLDSWORTH, M. B. CLAGUE, P. D. WRIGHT AND
I. D. A. JOHNSTON

INTRODUCTION

Abnormal patterns of plasma amino acids are common in patients with liver disease[1-5] and can be corrected by the administration of both branched-chain enriched amino-acid solutions[6-9] or the three branched-chain amino acids (BCAA) alone[10,11]. These solutions have been reported as valuable in hepatic encephalopathy[6,7,9] in restoring the profile of plasma amino acids towards normal[6,8-11] and improving nitrogen balance[6,7,11]. There is also evidence that nitrogen losses are reduced following injury in patients or animals with normal liver function given branched-chain amino acids. The effect recorded appears to be out of proportion to the nitrogen content of the infusions[13]. O'Keefe et al.[15] investigated the effect of oral BCAA on body protein metabolism in liver disease and reported that turnover, breakdown and synthesis were diminished.

A study was planned to examine the effect of intravenous BCAA on body protein metabolism in patients with chronic liver disease.

METHODS

Body protein metabolism can be measured by a variety of techniques in which the kinetics of labelled amino acids are analysed[16]. Leucine was selected for this study and constant rate infusions of sodium [14C]bicarbonate and L[1-14C]leucine[17], were given over 4 h with all measurements carried out on plasma samples[18]. The 14C-labelled substances were separated on the day of collection and counted in a Packard Tri-carb liquid scintillation counter. Plasma leucine was determined separately on an amino-acid autoanalyser.

12 patients with a histologically proven stable chronic liver disease were allocated at random to one of two treatment groups. All patients underwent an initial control study during which nutrition was provided by an oral feed, followed by a second study when nutrition was given intravenously as either branched-chain amino acids with dextrose (Group A, $n=6$) or dextrose alone (Group B, $n=6$). Details of the patients and their nutritional intake are given in Tables 8.1 and 8.2. No subject had obvious encephalopathy and patients with infection, recent variceal bleeding, surgery or blood transfusion were excluded. Permission was granted by the Isotope Advisory Panel and Local Ethical Committee for two studies (25 μCi per study) on patients over the age of 45 years. Informed consent was obtained from each patient.

Table 8.1 The clinical details of the patients

Group	A ($n = 6$)	B ($n = 6$)
Nutrition	BCAA + dextrose	Dextrose
Age (mean ± SD)	57 ± 7	55 ± 13
Sex		
male	4	2
female	2	4
Childs grade*		
A	1	2
B	2	2
C	3	2

*Pugh modification of Childs' Classification[19]

Table 8.2 The composition and volume of the oral and intravenous nutrition ($kg^{-1} h^{-1}$)

Nutrition	Oral feed*	4% BCAA† + 10% dextrose	10% dextrose
Calories	0.64	0.06‡ + 0.58	0.64
Nitrogen (mg)	3.84	2.11 + 0.0	0.0
Volume (ml)	0.80	0.48 + 1.46	1.60

Intravenous fluids were prepared in 3 l bags and administered peripherally, the rate controlled by an IMED 922H volumetric infusion pump.
*Clinifeed 400 chocolate (Roussel Laboratories Ltd., North End Road, Wembley Park, Middlesex).
† A 4% solution of L-BCAA, composition: leucine 13.82 g/l, isoleucine 13.82 g/l and valine 12.35 g/l (Travenol Laboratories Ltd., Caxton Way, Thetford, Norfolk).
‡ Allowance made for calories derived from amino acids.

Following an overnight fast the oral feed was commenced at 06.30, given as small aliquots each half-hour. At 08.00, having taken blood for background radioactivity, the [14C]bicarbonate infusion was started, continuing for 1½ h with blood sampling during the last ½ h to establish plasma [14C]bicarbonate activity. The isotope infusion was then changed to L[1-14C]leucine, continued for a further 2½ h, blood being drawn during the final hour to determine [14C]bicarbonate and [14C]leucine activity. At 12.00 the oral feed and isotope infusion were discontinued and the intravenous nutrition commenced. The i.v. feeding continued for 24 h before the second metabolic study was performed in exactly the same way as the first, between 08.00 and 12.00.

The body is considered as having a single, but compartmentalized pool of free leucine. As leucine is not synthesized in the body, the amino acid can only enter this pool by oral or intravenous intake (I) or by the breakdown of protein (B). Leucine leaves this pool by either becoming incorporated into new protein (S) or by being oxidized. In a steady state, when the size of the pool remains constant, the following equation must apply to leucine turnover (Q),

$$Q = B + I = S + C$$

The same applies to the turnover (q) of the radiolabelled leucine, but if the amount incorporated into protein is considered to be so diluted that a negligible quantity returns to the free pool during the course of a short infusion (i.e. $b = 0$) then,

$$q = i$$

Assuming that the unlabelled leucine behaves in a similar manner to the labelled then,

$$Q \text{ (mmol/h)} = \frac{i \quad \text{(dpm/h)}}{SA_{plateau} \text{ (dpm/mmol)}}$$

where $SA_{plateau}$ is the specific radioactivity or concentration of labelled to unlabelled leucine at equilibrium. The infusion rate of L[1-14C] is known and the specific radioactivity in plasma at equilibrium can be measured. Calculating turnover (Q) thus, the release of leucine from protein (B) can be determined knowing the intake of leucine (I).

The [14C]bicarbonate activity in the plasma is measured during the bicarbonate infusion (i_{bicarb}) and combining this with the level of [14C]bicarbonate during the [14C]leucine infusion the entry rate of $^{14}CO_2$ into the bicarbonate pool from labelled leucine can be determined. Leucine catabolism (C) is then calculated and so the rate of leucine incorporated into protein synthesis (S) is derived.

From these values for leucine the rate of body protein breakdown and synthesis (g/kg day) can be calculated assuming the average leucine content of body protein to be 8%[20].

Table 8.3 Body leucine metabolism (g/kg day) and plasma leucine (μmol/l)

Study	1 (Control) Oral	2 Intravenous
Group A		
Turnover (Q)	0.24 ± 0.06	0.47 ± 0.06‡
Intake (I)	0.06	0.16
Breakdown (B)	0.18 ± 0.06	0.31 ± 0.06‡
Catabolism (C)	0.04 ± 0.02	0.14 ± 0.04 †
Synthesis (S)	0.20 ± 0.05	0.33 ± 0.05 ‡
Plasma leucine	79 ± 31	208 ± 47 ‡
Group B		
Turnover (Q)	0.29 ± 0.07	0.24 ± 0.07*
Intake (I)	0.06	0.0
Breakdown (B)	0.23 ± 0.07	0.24 ± 0.07
Catabolism (C)	0.06 ± 0.03	0.04 ± 0.02*
Synthesis (S)	0.23 ± 0.05	0.20 ± 0.05
Plasma leucine	77 ± 27	57 ± 20†

* $p \leqslant 0.05$, † $p \leqslant 0.01$, ‡ $p \leqslant 0.001$.

RESULTS

Leucine metabolism and plasma leucine (Table 8.3) were similar for both groups during the control studies. In patients given branched-chain amino acids (Group A), metabolism altered significantly during the second study; all measurements increased, whereas in the dextrose only group (B) turnover fell in association with small changes in synthesis and breakdown. The difference in turnover between the groups is highly significant ($p < 0.001$).

Table 8.4 Body protein metabolism (g/kg day)

Study	1 (Control)	2	Change (2 − 1)
Group A			
Breakdown (B)	2.28 ± 0.78	3.85 ± 0.74 †	+ 1.57 ± 0.47
Synthesis (S)	2.47 ± 0.64	4.06 ± 0.61 †	+ 1.59 ± 0.24
Balance (S − B)	+ 0.19 ± 0.20	+ 0.21 ± 0.51	+ 0.02 ± 0.47
Group B			
Breakdown (B)	2.81 ± 0.90	2.95 ± 0.80	+ 0.14 ± 0.35
Synthesis (S)	2.76 ± 0.59	2.49 ± 0.63	− 0.27 ± 0.36
Balance (S − B)	− 0.05 ± 0.41	− 0.46 ± 0.24*	− 0.41 ± 0.25

* $p \leqslant 0.01$, † $p \leqslant 0.001$, between studies 1 and 2.

Protein synthesis and breakdown (Table 8.4) were both increased after branched-chain administration (Group A) and as both increased by a similar amount, protein balance was maintained. In group B the small rise in break-down and fall in synthesis resulted in a negative balance. The groups have been compared by calculating the change in protein metabolism from the first to the second study for each patient (Table 8.4), differences were highly significant ($p < 0.001$) for both protein breakdown and synthesis but not balance.

DISCUSSION

It has been suggested that the administration of large amounts of leucine may invalidate measurements of protein metabolism using L[1-^{14}C]leucine[21], but since the quantity of leucine involved in protein breakdown and synthesis must remain fixed in relation to the total amino acids used for these processes, the method is valid under these circumstances. The same cannot be said of leucine turnover and catabolism; they must not be converted into values for body protein, and for these reasons leucine metabolism has been presented in full so that the relative contribution of each measurement to turnover can be assessed. The large changes in Group A must be due to the presence of the branched-chain amino acids and not dextrose because in Group B, where only dextrose was given, small changes occurred.

Dextrose alone induced a pattern of changes that can be interpreted as protein metabolism adapting to an absence of dietary amino acids. The changes in Group A confirm the previously reported nitrogen-sparing effect of branched-chain amino acids[7,12-14,22], but are at variance with those of O'Keefe et al.[15], who used a different amino acid mixture given orally to measure protein metabolism and expressed their results in g/day. The increase in synthesis was not entirely unexpected since animal work has shown that leucine increased synthesis while decreasing breakdown in isolated skeletal muscle[23,24], and BCAA can also increase total liver nitrogen[13], while branched-chain keto acids stimulate albumin synthesis[25]. The main interest of this study was the associated rise in protein breakdown, which has not been reported previously.

There are several explanations of these findings. If the *in vitro* effects of leucine on skeletal muscle[23,24] occur *in vivo*, then infusing large quantities of BCAA may have increased muscle protein synthesis primarily. The demands of a high level of synthesis may induce increased breakdown in other tissues or organs of the body. More work is required to examine the mechanism behind these changes. The changes in synthesis and breakdown are, however, of a similar order suggesting a cause and effect relationship. The increased plasma leucine in the branched-chain group suggests that a similar change may have taken place in skeletal muscle cells where branched-chain amino acids are mainly metabolized.

The exact role of branched-chain amino acids in the management of liver disease is uncertain. This study suggests that any benefits in terms of synthesis and balance may be counteracted by an increase in breakdown.

The calorie intakes in this study were extremely low for technical reasons and further work is required to establish the effect of differing calorie intakes.

Further studies are planned to examine the effect of conventional amino-acid solutions enriched with branched-chain amino acids on body protein turnover, in patients with chronic liver disease.

SUMMARY

Protein turnover has been measured in two groups of six patients with stable chronic liver disease. One group was given an infusion of branched-chain amino acids with dextrose and the other dextrose alone. Branched-chains with dextrose increased both protein breakdown and synthesis while maintaining protein balance. Dextrose alone, in small amounts, caused only small changes in breakdown and synthesis, the net effect being a negative balance.

Acknowledgements

J. D. Holdsworth was in receipt of a Research Fellowship awarded by Travenol Laboratories Limited.

We wish to thank Dr Gordon Dale, Mr Jim Bonham, Mr Ian Smeaton and Dr Mike Keir for technical assistance and Dr Oliver James for clinical co-operation.

References

1 Iber, F. L., Rosen, H., Levenson, S. M. and Chalmers, T. C. (1957). The plasma amino acids in patients with liver failure. *J. Lab. Clin. Med.*, **50**, 417

2 Richmond, J. and Girdwood, R. H. (1962). Observations on amino acid absorption. *Clin. Sci.*, **22**, 301

3 Iob, V., Coon, W. W. and Sloan, M. (1967). Free amino acids in liver, plasma and muscle of patients with cirrhosis of the liver. *J. Surg. Res.*, **7**, 41

4 Fischer, J. E., Yoshimura, N., Aguirre, A., James, J. H., Cummings, M. G., Abel, R. M. and Diendorfer, F. (1974). Plasma amino acids in patients with hepatic encephalopathy. *Am. J. Surg.*, **127**, 40

5 Rosen, H. M., Yoshimura, N., Hodgman, J. H. and Fischer, J. E. (1977). Plasma amino acid patterns in hepatic encephalopathy of differing etiology. *Gastroenterology*, **72**, 483

6 Fischer, J. E., Rosen, H. M., Ebeid, A. M., James, J. H., Keane, J. M. and Soeters, P. B. (1976). The effect of normalisation of plasma amino acids on hepatic encephalopathy in man. *Surgery*, **80**, 77

7 Fischer, J. E. (1981). The etiology of hepatic encephalopathy. *Acta Chir. Scand. Suppl.*, **507**, 50

8 Egberts, E. H., Hamster, W., Jurgens, P., Schumacher, H., Fondalinski, G., Reinhard, U. and Schomerus, H. (1981). Effect of branched chain amino acids on latent portal-systemic

encephalopathy. In Walser, M. and Williamson, J. R. (eds.) *Metabolism and Clinical Implications of Branched Chain Amino and Keto Acids.* pp. 453−463. (Amsterdam: Elsevier−North Holland)

9 Okada, A., Kamata, S., Kim, C. W. and Kawashima, Y. (1981). Treatment of hepatic encephalopathy with BCAA-rich amino acid mixture. In Walser, M. and Williamson, J. R. (eds.) *Metabolism and Clinical Implications of Branched Chain Amino and Keto Acids.* pp. 447−452. (Amsterdam: Elsevier−North Holland)

10 Wahren, J., Eriksson, S. and Hagenfeldt, L. (1981). The influence of branched chain amino acids and keto acids on arterial concentration and brain exchange of amino acids in man. In Walser, M. and Williams, J. R. (eds.) *Metabolism and Clinical Implications of Branched Chain Amino and Keto Acids.* pp. 471−488. (Amsterdam: Elsevier−North Holland)

11 Okita, M., Watanabe, A. and Nagashima, H. (1981). Treatment of liver cirrhosis with branched chain amino acid supplemented diet. *Gastroenterol. Jap.*, **16**, 389

12 Freund, H., Hoover, H. C., Atamian, S. and Fischer, J. E. (1979). Infusion of the branched chain amino acids in postoperative patients. *Ann. Surg.*, **190**, 18

13 Blackburn, G. L., Moldawer, L. L., Usui, S., Bothe, A., O'Keefe, S. J. D. and Bistrian, B. R. (1979). Branched chain amino acid administration and metabolism during starvation, injury and infection. *Surgery*, **86**, 307

14 Freund, H., Yoshimura, N. and Fischer, J. E. (1980). The effect of branched chain amino acids and hypertonic glucose infusions on post injury catabolism in the rat. *Surgery*, **87**, 401

15 O'Keefe, S. J. D., Abraham, R. R., Davis, M. and Williams, R. (1981). Protein turnover in acute and chronic liver disease. *Acta Chir. Scand.* Suppl., **507**, 99

16 Waterlow, J. C., Garlick, P. J. and Millward, D. J. (1978). Summary of methods of measuring total protein turnover. In *Protein Turnover in Mammalian Tissues and in the Whole Body.* pp. 327−338. (Amsterdam: Elsevier−North Holland)

17 Clague, M. B. and Keir, M. J. (1982). A modified technique for measuring whole body protein metabolism in surgical patients using $L(1~^{14}C)$leucine. *Clin. Sci. Mol. Med.*, (in press)

18 Clague, M. B., Keir, M. J. and Wright, P. D. (1981). Determination of the oxidation rate of ^{14}C-labelled substances in surgical patients by assessment of $^{14}CO_2$ production without the need to collect expired air. *Clin. Sci. Mol. Med.*, **60**, 233

19 Pugh, R. N. H., Murray-Lyon, I. M., Dawson, J. L., Pietroni, M. C. and Williams, R. (1973). Transection of the oesophagus for bleeding oesophageal varices. *Br. J. Surg.*, **60**, 646

20 Block, R. J. and Weiss, K. W. (1956). *Amino Acid Handbook.* (Illinois: Charles C. Thomas)

21 Desai, S. P., Moldawer, L. L., Bistrian, B. R., Blackburn, G. L., Bothe, A. and Schulte, R. D. (1981). Amino acid and protein in humans using $L(U~^{14}C)$tyrosine and $L(1~^{14}C)$leucine. In Walser, M. and Williamson, J. R. (eds.) *Metabolism and Clinical Implications of Branched Chain Amino and Keto Acids.* pp. 307−312. (Amsterdam: Elsevier−North Holland)

22 Sherwin, R. S. (1978). Effect of starvation on the turnover and metabolic response to leucine. *J. Clin. Invest.*, **61**, 1471

23 Buse, M. G. and Reid, S. S. (1975). Leucine; a possible regulator of protein turnover in muscle. *J. Clin. Invest.*, **56**, 1250

24 Goldberg, A. L. and Chang, T. W. (1978). Regulation and significance of amino acid metabolism in skeletal muscle. *Fed. Proc.*, **37**, 2301

25 Kirsch, R. E., Frith, L. O. and Saunders, S. J. (1976). Stimulation of albumin synthesis by keto analogues of amino acids. *Biochim. Biophys. Acta*, **442**, 437

Section 2
Nutritional Support in Renal Failure

9
Nutritional therapy for patients with acute renal failure

J. D. KOPPLE AND E. I. FEINSTEIN

Patients with acute renal failure have widely varying metabolic and nutritional states. Some patients have no evidence of negative protein balance, and they have normal plasma electrolyte concentrations and fluid and acid-base status. In general, these patients have no severely catabolic underlying illnesses. These patients are usually not oliguric, and the cause of their renal failure is typically an isolated non-catabolic event, such as a nephrotoxic reaction to a drug. On the other hand, most patients with documented acute renal failure have some degree of net protein catabolism and have altered fluid, electrolyte, or acid-base status[1,2]. These patients often have positive water balance, hyperkalaemia, hyperphosphataemia, hypocalcaemia, azotaemia, hyperuricaemia, and metabolic acidosis with a large anion gap.

In some patients with acute renal failure, net protein degradation is massive[1,2] and they may have net protein losses of 150–250 g/day. These patients are more likely to have acute renal failure due to shock, or sepsis. The marked net breakdown of protein can promote a more rapid increase in plasma potassium, phosphorus, hydrogen ion, and products of nitrogen metabolism. Since in non-uraemic humans, wasting and malnutrition can impair wound healing and immune function[3,4] and increase morbidity and mortality, it is likely that the profound catabolic response of many patients with acute renal failure will have the same adverse consequences. Indeed, morbidity and mortality of patients with acute renal failure continues to be high despite the recent advances in medical care[1,2,5,6].

Prior to the mid 1960s many workers advocated marked or total restriction of protein and administration of small amounts of carbohydrates and fats to reduce the catabolic response of these patients[7,8]. However, with the development of modern methods for parenteral nutrition, it was natural to examine whether administration of amino acids and other nutrients, could improve nutritional status and ameliorate the severity of uraemic toxicity, or even modify the course of acute renal failure, particularly in patients who are unable to eat. It was recognized that the uraemic patient has intolerance to many nutrients and that special amino-acid formulations might be required.

Abel and associates used amino-acid formulations specifically designed for patients with acute renal failure[9-11]. They published a series of reports concerning intravenous administration of hypertonic solutions which contained glucose and the eight essential amino acids without histidine. They reported that serum potassium, phosphorus and magnesium often fell, and that serum urea nitrogen (SUN) stabilized or decreased. These investigators carried out a prospective double-blind study in which 53 patients with acute renal failure were assigned randomly to receive intravenous nutrition with solutions providing either amino acids with hypertonic glucose or isocaloric quantities of hypertonic glucose alone[10]. The total energy intake averaged 1426 and 1641 kcal/day with the two solutions. The amino-acid intake with the essential amino-acid solution was approximately 16 g/day. The recovery of renal function was significantly more frequent in the patients receiving the essential amino acids and glucose. Overall survival was slightly and not significantly greater in these patients. However, there was significantly increased survival with the essential amino-acid solution in those patients with more severe renal failure, as indicated by the need for dialysis therapy, and in those with severe complications, such as pneumonia and generalized sepsis. There was a tendency, which was also not significant, for recovery of renal function to occur more rapidly in the group receiving essential amino acids and glucose.

Although several other studies suggested that in patients with acute renal failure, total parenteral nutrition with essential and non-essential amino acids might improve survival[12-14], the lack of adequate controls or a prospective randomized design made the results inconclusive. In contrast, Leonard, Luke and Siegel carried out a randomized prospective study in patients with acute renal failure who were given infusions of either 1.75% L-essential amino acids and 47% dextrose or 47% dextrose alone[1]. Patients who were able to eat or tolerate tube feedings were excluded from the study, and many of the patients had severe associated illnesses. The authors reported that the rate of rise in SUN was significantly less in the group receiving essential amino acids. However, mean nitrogen balance was about 10 g/day negative with both treatment groups, and there was no difference in the rate of recovery of renal function or survival between the two treatments.

The observation that parenteral nutrition with glucose and amino acids (and minerals and vitamins) might improve recovery of renal function or survival in patients with acute renal failure, in whom negative nitrogen balance is common, and morbidity and mortality are very high, stimulated us to carry out a prospective controlled study comparing intravenous treatments with glucose alone or with essential or essential and non-essential amino

Table 9.1 Nutritional intake in patients with acute renal failure*

		Treatment group	
	Glucose	Glucose + EAA †	Glucose + ENAA‡
Prescribed intake			
Essential amino acids (g/day)	0	21.0	21.2
Non-essential amino acids (g/day)	0	0	20.9
Actual intake			
Total energy infused (kcal/day)	2678 ± 744**	2265 ± 598	2445 ± 720
Total nitrogen intake (g/day)	0	2.32 ± 0.45	5.34 ± 0.66

*Does not include supplemental glucose or amino acids given during dialysis (see text).

†Provided through the courtesy of Abbott Laboratories, Inc., and contains (g/day): histidine 1.95; isoleucine 2.10; leucine 3.30; lysine 2.43; methionine 3.30; phenylalanine 3.30; threonine 1.50; tryptophan 0.75; and valine 2.40.

‡Aminosyn^R, courtesy of Abbott Laboratories, Inc., and contains (g/day): histidine 1.25; isoleucine 3.06; leucine 3.99; lysine 3.06; methionine 1.70; phenylalanine 1.87; threonine 2.21; tryptophan 0.68; valine 3.40; alanine 5.44; arginine 4.16; glycine 5.44; proline 3.65; serine 1.78; and tyrosine 0.37.

**Mean ± standard deviation.

acids[2]. 29 men and one woman with acute renal failure who were unable to receive adequate oral nutrition were studied. They received parenteral nutrition with glucose (seven patients), glucose and 21 g/day of essential amino acids (EAA, 11 patients), or glucose 21.2 g/day of essential and 20.9 g/day of non-essential amino acids (ENAA, 12 patients) (Table 9.1). When patients underwent haemodialysis, they received a supplemental infusion into the venous side of the shunt which provided 280–330 g of glucose alone or in combination with either 21 g of EAA or 42.1 g of ENAA, depending upon the parenteral nutrition regimen to which the patient was assigned. The grand mean (±SD) of the energy intake was 2678 ± 744 kcal/day with the glucose intake, 2265 ± 598 kcal/day with the EAA intake, and 2445 ± 720 kcal/day with the ENAA intake and did not differ among the three regimens. Studies were conducted with all three intravenous regimens in the patients treated at Los Angeles County/University of Southern California Medical Center and with the EAA and ENAA infusions in the patients at Veterans Administration Wadsworth Medical Center.

The results indicated that there was no significant difference in recovery of renal function or survival between those patients receiving glucose alone, glucose and EAA, or glucose and ENAA (Table 9.2). In those patients in whom shock or sepsis was the cause of renal failure, only 17% recovered renal function or survived. In contrast, in patients in whom renal failure was due to other causes, 83% recovered from their renal failure and 67% survived to leave the hospital. Although there was a tendency for greater recovery of renal function and survival in the patients receiving glucose and EAA, there was a slightly lower incidence of hypotension or sepsis as the cause of renal failure in this group (Table 9.2).

Table 9.2 Characteristics of patients with acute renal failure

		Treatment group	
	Glucose	Glucose + EAA	Glucose + ENAA
No. of subjects	7	11	12
Age (years)	40 ± 19*	55 ± 23	38 ± 21
Duration of TPN (days)	17 ± 13	6.9 ± 2.7	6.3 ± 2.3
Cause of renal failure (recovered renal function, survived)			
Hypotension	5 (0, 0)†	4 (1, 1)	7 (2, 2)
Sepsis	—	1 (0, 0)	1 (0, 0)
Rhabdomyolysis	1 (1, 1)	2 (2, 2)	3 (1, 1)
Antibiotics	1 (1, 1)	2 (2, 1)	1 (1, 0)
Radiocontrast material	—	1 (1, 1)	—
Unknown	—	1 (1, 1)	—
Total	7 (2, 2)	11 (7, 6)	12 (4, 3)

*Mean ± standard deviation.
†In the parentheses, the first number indicates the number of patients who recovered from renal failure, the second number indicates those who survived.

With each treatment regimen, the patients were often very catabolic and wasted. The grand mean of the urea nitrogen appearance was 10.4 ± 5.9 g/day in the glucose group, 6.7 ± 7.2 g/day in the EAA group, and 14.0 ± 8.0 g/day in the ENAA group (EAA vs ENAA, $p < 0.05$). Estimated total nitrogen output in the three groups is shown in Table 9.3. One may assess the degree of negative nitrogen balance in these patients by subtracting estimated mean total nitrogen output from the average nitrogen intake (Table 9.3). These values do not take into consideration the nitrogen intake

Table 9.3 Urea nitrogen appearance and estimated nitrogen balance in patients with acute renal failure

		Treatment group	
	Glucose	Glucose + EAA	Glucose + ENAA
No. of subjects	7	11	11
Urea nitrogen appearance* (g/day)			
Maximum	14.9 ± 5.6†	10.1 ± 10.0††	20.5 ± 12.3
Mean‡	10.4 ± 5.9	6.7 ± 7.2 ††	14.0 ± 8.0
Estimated mean total nitrogen output ‡,** (g/day)	12.0 ± 5.7	8.4 ± 7.0††	15.5 ± 7.7
Mean nitrogen intake minus estimated mean nitrogen output ‡ (g/day)	12.0 ± 5.7***	6.2 ± 7.0‡‡	10.2 ± 7.7***

*Calculated during the interdialytic interval as previously described[15].
†Mean ± standard deviation.
‡Grand mean of the mean values for the entire period of study in each patient.
**Calculated from the equation, total nitrogen output (g/day) = 0.97 (urea nitrogen appearance) + 1.93[15].
††Probability that values do not differ from the glucose + ENAA group; $p < 0.05$.
‡‡Probability that estimated nitrogen balance does not differ from zero; $p < 0.02$.
***Probability that estimated nitrogen balance does not differ from zero; $p < 0.005$.

Table 9.4 Final serum protein concentrations after parenteral nutrition in patients with acute renal failure

Serum proteins	Normal	Glucose	Glucose + EAA	Glucose + ENAA
No. of subjects	49	5	7	9
Total protein (g/dl)	7.43 ± 0.63*	6.1 ± 1.0†	5.7 ± 0.9‡	5.0 ± 0.6‡
Albumin (g/dl)	5.06 ± 0.35	2.8 ± 0.7‡	2.8 ± 0.4‡	2.8 ± 0.7‡
Transferrin (mg/dl)	309 ± 39	127 ± 53‡	131 ± 62‡	161 ± 44‡

*Mean ± standard deviation.
Probability that serum concentrations are not different from normal values,
  †$p < 0.005$, ‡$p < 0.001$.

from infusion of blood and blood products or the nitrogen output from wound drainage, respiration, flatus, and the integument. The data indicate that nitrogen balance was often severely negative. However, several patients were in only slightly negative balance or even neutral or possibly slightly positive balance. These patients were found only in the EAA and ENAA groups.

The final values for serum total protein, albumin and transferrin were each significantly below normal in each treatment group (Table 9.4). There was no consistent pattern of abnormality for serum IgG, IgA, IgM, C3 or C4. Serum protein concentrations were often low at the onset of treatment with the three intravenous regimens and did not fall significantly during the course of treatment in any group. However, the low concentrations of these proteins persisted even though many patients received infusions of blood and blood products during the period of study.

Final plasma total essential amino acids, total non-essential amino acids, and total amino acids (Figure 9.1) were significantly reduced in each treatment group as compared to normal men ingesting normal diets. This finding was observed in the patients given the EAA or ENAA infusion even though their blood was drawn for amino-acid analyses while they were receiving infusions of the amino acids, while blood was sampled from the normal subjects after an overnight fast. The alterations in plasma concentrations of individual amino acids at the termination of total parenteral nutrition were varied. For the essential amino acids, there were, in general, decreased concentrations of histidine, isoleucine, leucine, lysine, threonine, and valine; methionine and phenylalanine were normal or increased. For the plasma non-essential amino acids, arginine, citrulline, and tyrosine were most clearly decreased; there were low normal to normal levels of alanine, asparagine, glutamine, serine, and ornithine; and normal to increased concentrations were observed for aspartate, glutamate, glycine, cystine, and N^{τ}-methyl-histidine (3-methylhistidine).

The results of this study indicate that patients with acute renal failure who

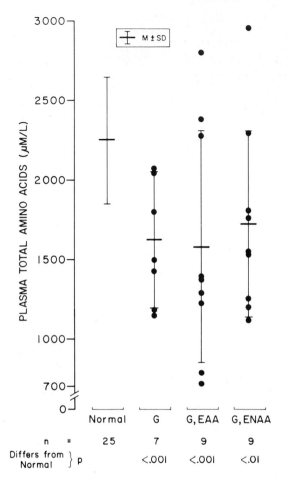

Figure 9.1 Plasma total amino-acid concentrations in normal subjects and patients with acute renal failure who received total parenteral nutrition which provided glucose (G), glucose and the nine essential amino acids (G, EAA), or glucose, the nine essential amino acids and six non-essential amino acids (G, ENAA). The circles indicate values in individual patients at the end of parenteral nutrition in each patient. The thicker horizontal line indicates the mean value for each group, and the brackets represent one standard deviation

are unable to receive adequate nourishment through the gastrointestinal tract have a high incidence of wasting and malnutrition, particularly if shock or sepsis is the cause of renal failure. There was no clear-cut improvement in outcome with any nutritional treatment. Although the patients who received glucose and essential amino acids displayed a tendency toward greater recovery of renal function and survival, there was a lower incidence of hypotension or sepsis as the cause of renal failure in this group (Table 9.2). Moreover, the high urea nitrogen appearances, negative nitrogen balances, and low concentrations of serum proteins and plasma amino acids suggest that with each treatment regimen the patients were often hypercatabolic and wasted.

Since malnutrition is associated with impaired immune function and wound healing[3,4], it is possible that the catabolic state of many of the uraemic patients may have contributed to their high morbidity and mortality. Hence, we decided to assess whether a higher nitrogen intake might reduce morbidity or mortality, presumably by improving nutritional status.

A study was therefore carried out in which patients were assigned in random order to receive one of two parenteral nutrition regimens[16]. With one regimen, the goal was to provide a constant amount of the nine essential amino acids (EAA, 21 g/day, 2.3 g amino acid nitrogen/day). With the other regimen, we attempted to administer a quantity of essential and non-essential amino acids (ENAA) which was varied according to the urea nitrogen appearance so that the daily nitrogen intake exceeded the urea nitrogen appearance by 2.0 g N/day. Nitrogen intake, however, was not allowed to exceed a maximum of 15 g/day. The daily intake of ENAA was adjusted according to the urea nitrogen appearance every 2 or 3 days. During the study, patients treated with either regimen did not receive nutrition through the gastrointestinal tract.

The EAA solution (Nephramine, McGaw Laboratories) and the ENAA solution (Aminosyn 10%) were similar or identical to those used in the first study[2]. The ratio of essential to non-essential amino acids in the ENAA solutions was 1:1. All patients received hypertonic glucose, minerals and vitamins with both treatment regimens. Attempts were made to provide about 35 kcal/kg/day for all patients.

At present, 11 patients have been studied; five were treated with the EAA regimen and six received the ENAA infusions. In ten patients, the acute renal failure was caused by shock; eight of these patients were trauma victims. Three EAA patients recovered renal function and two survived. Two ENAA patients recovered renal function and none survived. Mean daily energy intake in this ongoing study was not different with the two treatments. The ENAA group received more nitrogen than the patients receiving EAA (11.3 ± 1.9 vs 2.3 ± 0.3 g/day, $p < 0.01$), and the urea nitrogen appearance was significantly greater with the ENAA regimen. Moreover, the protein nitrogen balance, calculated as nitrogen intake minus the urea nitrogen appearance, was slightly but not significantly more positive with the ENAA intake (ENAA vs EAA, -2.9 ± 4.0 g N/day vs -5.2 ± 2.9 g N/day, p:NS).

These preliminary data suggest that a regimen of parenteral nutrition that provides larger quantities of ENAA as compared to about 21 g/day of EAA is associated with increased urea nitrogen appearance, no marked improvement in nitrogen balance, and no greater survival. Given the current survival data, it is unlikely that if the sample size were four to five times larger there would be significantly greater recovery of renal function or survival with the

ENAA treatment. The sample size was too small to assess whether recovery of renal function or survival is improved with the EAA regimen. Moreover, the increased fluid load and greater urea nitrogen appearance with the ENAA infusion may increase dialysis requirements and hence the cost of medical therapy with this regimen.

The experience in acutely uraemic rats may provide some insight into the question of optimal nutritional therapy for patients with acute renal failure. Flugel-Link and co-workers in our laboratory examined urea nitrogen appearance, protein synthesis and degradation, and amino acid metabolism in the hemicorpus of rats with acute uraemia[17]. Studies were carried out in ten rats with acute renal failure caused by bilateral nephrectomy and ten sham-operated controls. Rats were fasted for 30 h after surgery except for two gavages with sodium bicarbonate, and the posterior hemicorpus was then perfused for 2 h with a modified Krebs–Hensleit buffer containing 3% albumin and 30% bovine red cells. Protein synthesis was assessed by uptake of [^{14}C]phenylalanine into muscle protein; protein degradation, by dilution of [^{14}C]phenylalanine with unlabelled phenylalanine from degraded protein.

The uraemic rats were more catabolic as evidenced by greater urea nitrogen appearances (198 ± 50 mg nitrogen/30 h vs 151 ± 35 in sham-operated controls, $p < 0.05$). In the perfused hemicorpus of these animals, there was a slight but not significant reduction in protein synthesis. However, protein degradation was significantly greater in the uraemic rats as compared to the control animals, 233 ± 42 vs 170 ± 36 nmol phenylalanine released/h/g hemicorpus, $p < 0.001$. There was also greater net release of phenylalanine, tyrosine, alanine, total non-essential amino acids, total amino acids, potassium and phosphorus from the perfused hemicorpus of the uraemic rats and greater release of citrulline from the hemicorpus of the control animals.

These observations suggest that in the hemicorpus of the acutely uraemic rat there is enhanced protein degradation and release of amino acids. While some of the released amino acids are catabolized *in situ*, others appear to flow in increased quantities into the general circulation. The finding of low plasma and intracellular muscle amino acids in the acutely uraemic rats suggests that there is even a greater propensity to degrade the amino acids than to release them from the hemicorpus. Indeed, the increased urea nitrogen appearance of the uraemic rats suggests that these animals have enhanced gluconeogenesis of amino acids in the liver. This possibility is supported by studies from other investigators which demonstrate increased uptake and degradation of amino acids and enhanced production of glucose and urea in the liver of acutely uraemic rats[18,19].

The foregoing experiments in rats may explain why the patients with acute renal failure did not seem to benefit from greater quantities of intravenous amino acids. The animal studies suggest that the enhanced hepatic gluconeogenesis was a primary cause of the patients' hypercatabolism. Their low plasma amino-acid concentrations might actually be protective by reducing the amino-acid substrates available for accelerated gluconeogenesis and ureagenesis in the liver. The infusion of greater quantities of amino acids may, by increasing the concentrations of amino acids flowing into the liver, enhance the rate of synthesis of urea and possibly other potentially toxic metabolites.

The dilemma, then, is how to adequately nourish the patient with acute renal failure whose tissues seem to be primed to catabolize infused amino acids. It is possible that patients may become more anabolic with infusion of solutions which contain a different ratio of essential to non-essential amino acids or altered proportions of individual amino acids. There may be benefits to adding more branched-chain amino acids or ketoacid or hydroxyacid analogues to the intravenous solutions. The administration of anabolic hormones may be advantageous. Since catabolic processes in liver, muscle and possibly other organs seem to be enhanced in acute uraemia, research is indicated to develop methods to reduce the circulating levels of catabolic hormones or decrease the activity of enzymes which degrade amino acids or proteins or form urea or other metabolic products. If we are able to reduce these catabolic processes, it must still be demonstrated that this will lead to less wasting, malnutrition and morbidity or mortality in the patient with acute renal failure.

SUMMARY

A large proportion of patients with acute renal failure are hypercatabolic and show evidence for wasting and malnutrition. Patients with acute renal failure who have impaired function of the gastrointestinal tract are at particular risk for these complications. Mortality is high, particularly if shock or sepsis is the cause of acute renal failure. Treatment with different quantities and types of intravenous amino acid solutions does not seem to cause dramatic improvement in the severity of the catabolic response or the rate of recovery of renal function or survival. The data suggest that more research is necessary to examine whether patients may benefit from other combinations of amino acids or their analogues, and from the use of anabolic hormones. If clinical procedures can be developed for reducing the increased catabolic processes in tissues, it would be important to examine whether such techniques are beneficial to the patient.

ACKNOWLEDGEMENTS

This work was supported in part by a grant from Abbott Laboratories, Inc., and a contract with the USPHS-NIH AM 3-2210 and NIH Am 2-7931.

References

1 Leonard, C. D., Luke, R. G. and Siegel, R. R. (1975). Parenteral essential amino acids in acute renal failure. *Urology*, **6**, 154
2 Feinstein, E. I., Blumenkrantz, M. J., Healy, H., Koffler, A., Silberman, H., Massry, S. G. and Kopple, J. D. (1981). Clinical and metabolic responses to parenteral nutrition in acute renal failure. A controlled double-blind study. *Medicine*, **60**, 124
3 Bozzetti, F., Terno, G. and Longoni, C. (1975). Parenteral hyperalimentation and wound healing. *Surg. Gynecol. Obstet.*, **141**, 712
4 Law, D. K., Dudrick, S. J. and Abdou, N. I. (1973). Immunocompetence of patients with protein-calorie malnutrition. The effects of nutritional repletion. *Ann. Intern. Med.*, **79**, 545
5 Berne, T. V. and Barbour, B. H. (1971). Acute renal failure in general surgical patients. *Arch. Surg.*, **102**, 594
6 Stott, R. B., Cameron, J. S., Ogg, C. S. and Bewick, M. (1972). Why the persistently high mortality in acute renal failure? *Lancet*, **2**, 75
7 Bull, G. M., Joekes, A. M. and Lowe, K. G. (1949). Conservative treatment of anuric uremia. *Lancet*, **2**, 229
8 Blagg, C. R., Parsons, F. M. and Young, B. A. (1962). Effects of dietary glucose and protein in acute renal failure. *Lancet*, **1**, 608
9 Abel, R. M., Abbott, W. M. and Fischer, J. E. (1972). Intravenous essential L-amino acids and hypertonic dextrose in patients with acute renal failure. *Am. J. Surg.*, **123**, 631
10 Abel, R. M., Beck, C. H., Jr., Abbott, W. M., Ryan, J. A., Jr., Barnett, G. O. and Fischer, J. E. (1973). Improved survival from acute renal failure after treatment with intravenous essential L-amino acids and glucose. *N. Engl. J. Med.*, **288**, 695
11 Abel, R. M., Abbott, W. M., Beck, C. H., Jr., Ryan, J. A., Jr. and Fischer, J. E. (1974). Essential L-amino acids for hyperalimentation in patients with disordered nitrogen metabolism. *Am. J. Surg.*, **128**, 317
12 Baek, S. M., Makabali, G. G., Bryan-Brown, C. W., Kusek, J. and Shoemaker, W. (1975). The influence of parenteral nutrition on the course of acute renal failure. *Surg. Gynecol. Obstet.*, **141**, 405
13 McMurray, S. D., Luft, F. C., Maxwell, D. R., Hamburger, R. J., Futty, D., Szwed, J., Lavelle, K. J. and Kleit, S. A. (1978). Prevailing patterns and predictor variables in patients with acute tubular necrosis. *Arch. Intern. Med.*, **138**, 950
14 Milligan, S. L., Luft, F. C., McMurray, S. D. and Kleit, S. A. (1978). Intra-abdominal infection and acute renal failure. *Arch. Surg.*, **113**, 467
15 Kopple, J. D. (1981). Nutritional therapy in kidney failure. *Nutr. Rev.*, **39**, 193
16 Feinstein, E. I., Kopple, J. D., Silberman, H. and Massry, S. G. (1983). Parenteral nutrition with increased nitrogen intake in the treatment of hypercatabolic acute renal failure. *Kidney Int.* (In press)
17 Flugel-Link, R. M., Salusky, I., Jones, M. and Kopple, J. (1983). Altered muscle protein and amino acid metabolism in rats with acute renal failure. *Kidney Int.* (In press)
18 Lacy, W. W. (1969). Effect of acute uremia on amino acid uptake and urea production by perfused rat liver. *Am. J. Physiol.*, **216**, 1300
19 Fröhlich, J., Schölmerich, J., Hoppe-Seyler, G., Maier, K. P., Talke, H., Schollmeyer, P. and Gerok, W. (1974). The effect of acute uremia on gluconeogenesis in isolated perfused rat livers. *Eur. J. Clin. Invest.*, **4**, 453

10
Can low protein diet retard the progression of chronic renal failure?

J. BERGSTRÖM, M. AHLBERG and A. ALVESTRAND

INTRODUCTION

Patients with advanced renal failure who exhibit symptoms of uraemia may experience symptomatic relief and a decrease in blood urea concentration when the dietary protein intake is reduced. Treatment with low protein diet (LPD) in chronic uraemia may significantly prolong life and postpone the start of chronic dialysis. Marked protein restriction (15–20 g protein/day) may be required to control uraemia when glomerular filtration rate is reduced to 2–10% of normal. At such low protein intakes essential amino acids (EAA) or their N-free keto analogues (KAA) have to be given with the diet to fulfil nutritional requirements and prevent depletion of body protein[1,2].

It has been suggested earlier that protein reduction not only affords symptomatic relief in patients with chronic renal failure but may also improve renal function[3] or slow the rate of progression[4]. Mitch et al.[5] reported that, in most cases of chronic renal insufficiency, the reciprocal of serum creatinine declines linearly with time as renal failure progresses. This observation makes it possible to estimate the effect of therapy. Using this method Walser et al.[6] showed that the progression of the renal insufficiency was halted or retarded in several patients when treatment with low protein diet supplemented with KAA was instituted. Similar results were reported by Barsotti et al.[7] who measured creatinine clearance repeatedly in their patients.

Prompted by these observations we decided to make a retrospective analysis of patients with chronic renal failure treated with protein-restricted diet (15–20 g/day) supplemented with EAA or KAA to find out whether the progression of renal failure had been influenced by this treatment.

123

MATERIAL AND METHODS

To evaluate the progression of renal failure the inverse of serum creatinine was plotted against time. If the progression of renal disease is altered the slope of the line (regression coefficient) will change. In order to include a patient in the study we decided that the reciprocal of creatinine should decline in an approximate linear fashion with time before the diet was changed and that we should have at least eight creatinine determinations over a period of 200 days before as well as during LPD. Of all patients treated with LPD + EAA or KAA over the period 1975–1980, 17 fulfilled these requirements. The diagnoses were as follows: chronic glomerulonephritis in two patients, chronic pyelonephritis in three patients, analgesic nephropathy in two patients, congenital renal dysplasia in one patient and polycystic kidney disease in nine patients.

Before the patients were switched to LPD they had either no protein restriction or were ordered a diet containing 60 or 40 g protein per day. The prescribed LPD diet contained 15–20 g protein per day[8] and was supplemented orally in 14 cases with EAA (Aminess[R], AB Vitrum or a modified EAA preparation[9], and in three cases with a KAA preparation containing the keto analogues of valine, isoleucine, leucine and phenylalanine, the hydroxy analogue of methionine and the amino acids threonine, lysine, histidine and tryptophan (Ketoperlen[R], Pfrimmer AG).

All patients received aluminium hydroxide gel or tablets before as well as during treatment with LPD and the dose was adjusted to keep the plasma phosphate concentration within the normal range (below 1.8 mmol/l). Oral calcium carbonate 1–3 g per day was given to most of the patients to prevent a negative calcium balance. Apart from a minor dose of vitamin D_3 (250 IU) included in a multivitamin preparation no patients received vitamin D or its analogues.

The LPD which contained about 500 mg Na^+ per day was generally supplemented with NaCl powder (2–4 g per day) to improve palatability. $NaHCO_3$ was given to prevent acidosis. We aimed to keep the plasma bicarbonate concentration at or above 22 mmol/l. Furosemide was given as required to prevent sodium overload and control hypertension. Additional antihypertensive therapy was prescribed when indicated.

Each patient was checked at our out-patient clinic during the initial observation period as well as during treatment with LPD, usually by the same physician. The principles for control of fluid and electrolyte balance and blood pressure were the same in both periods.

RESULTS

Figure 10.1 shows the regression lines of the reciprocal of serum creatinine versus time in the 17 patients before and after introduction of the low protein

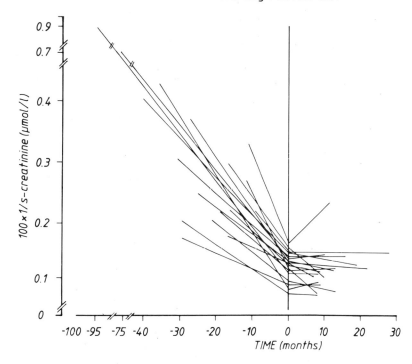

PROGRESSION OF RENAL INSUFFICIENCY BEFORE AND
DURING TREATMENT WITH 20g PROTEIN DIET

Figure 10.1 Regression lines of the reciprocal of the serum creatinine vs. time in 17 patients before and during treatment with LPD supplemented with EAA or KAA. The patients were switched to LPD at time 0

diet. In most of the cases the slope of the regression line (regression coefficient) decreased markedly after switching to the low protein diet. The mean regression coefficient ±SEM before LPD was −0.230 ± 0.026 and during LPD −0.022 ± 0.020 ($p<0.001$). A typical curve in an individual case is shown (Figure 10.2). In eight cases the improvement in renal function appeared to be associated with a decrease in serum phosphate, whereas in three cases progression of renal disease was retarded in spite of a significant increase in serum phosphate. The plasma phosphate concentration was slightly lower during LPD (1.47 ± 0.29 mmol/l) than before (1.63 ± 0.24 mmol/l; $p<0.052$). Plasma calcium was significantly higher during LPD (2.34 ± 0.11 mmol/l) than before (2.22 ± 0.15 mmol/l; $p<0.001$) but the Ca × P product was not significantly different. Improved blood pressure control may also influence the progression of renal disease. However, in the present material blood pressure control was not improved by low protein diet.

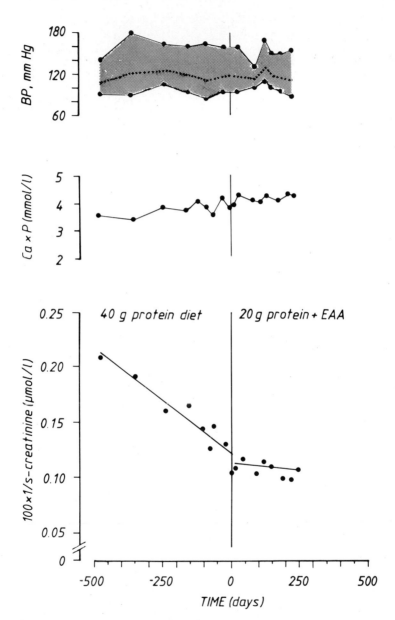

F, 48 yrs; Polycystic kidney disease

Figure 10.2 The reciprocal of the serum creatinine vs. time in a patient treated with LPD + EAA. Blood pressure was not different and Ca × P product was higher during LPD than before

126

DISCUSSION

The results of this study are in keeping with previous observations that LPD may slow the progression of chronic renal failure. However, our results have to be evaluated with caution due to the fact that the study was retrospective and not controlled. In most cases clinical and laboratory check-ups were more frequent during treatment with LPD than before. There is, thus, a possibility that improved patient care during LPD with regard to control of hypertension, fluid and electrolyte balance, infections, etc., might have affected the results, although we found no evidence that this was the case. Furthermore, evaluation of renal function by following the reciprocal of serum creatinine is questionable since a decrease in muscle mass due to protein depletion and inactivity may reduce creatinine production and extra renal elimination of creatinine may be enhanced with the severity of renal failure[10]. However, the creatinine clearance data which was available in some of our patients (Figure 10.3) supported the validity of the conclusions drawn from the curves of reciprocal of serum creatinine with time.

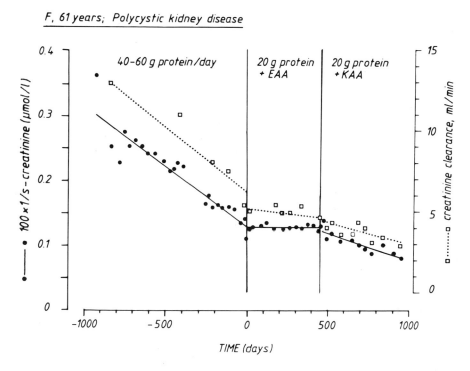

Figure 10.3 The reciprocal of the serum creatinine and creatinine clearance, respectively, vs. time. In this patient progression of chronic renal failure was halted during LPD + EAA but renal function deteriorated further during another period on LPD + KAA

The mechanisms by which LPD may slow the progression of renal failure in man have not been elucidated. Since the phosphate content in low protein diets is generally low, it was suggested that improved control of serum phosphate might be involved[6]. In support of this explanation are reports in subtotally nephrectomized rats[11] and in rats with nephrotoxic serum nephritis[12], that a low phosphate diet may prevent deterioration of renal function. However, Barrientos et al.[13] could not find that improved control of serum phosphate by low phosphate diet and 12 g aluminium hydroxide per day had any effect on progression of renal failure in patients with chronic renal disease; the protein contents of their low phosphate and control diets were not given. Our own data do not support the suggestion that an improved control of serum phosphate by LPD is the only mechanism by which LPD exerts its effect on the progression of renal failure, since in some of our patients LPD had an effect in spite of enhanced phosphate retention.

Hostetter et al.[14] have speculated that the final common outcome of chronic renal disease, i.e. predictable progression from mild renal insufficiency to end-stage renal failure independently of the aetiology of the initial injury, depends on some critical loss of renal mass. Compensatory raised glomerular pressures and flows influence the ultimate sclerotic destruction of those nephrons which survive the initial renal insult. In partially nephrectomized rats severe restriction of dietary protein intake could partly prevent glomerular hyperfiltration and reduce the structural abnormalities in remaining glomeruli[15]. This effect was apparently not dependent on improved phosphate control since the phosphate content of the low protein diet and control diet were comparable. These data suggest that protein restriction per se or some factor associated with protein restriction other than phosphate is responsible. It should be pointed out that the effect of protein restriction in preventing glomerular hyperfiltration in remaining nephrons was found in rats with experimental renal failure but that there is no evidence that the same mechanism is operating in human renal disease. If this was the case one might expect that switching to more marked protein restriction would result in an immediate decrease in glomerular filtration rate and, thus, an increase in serum creatinine, a phenomenon not observed by us or reported by others.

In conclusion, available clinical data from different groups, although anecdotal or not well-controlled, strongly suggest that progression of end-stage renal failure in man may be retarded or halted by LPD supplemented with EAA or KAA. This may have important clinical implications in the future, especially if it can be shown that this principle of treatment is operative in early renal failure as well. Prospective, randomized studies are now required to confirm the present findings and to evaluate by which mechanisms protein restriction affects the progression of renal failure.

References

1 Bergström, J., Fürst, P. and Noree, L.O. (1975). Treatment of chronic uremic patients with protein-poor diet and oral supply of essential amino acids. I. Nitrogen balance studies. *Clin. Nephrol.*, **3**, 187
2 Walser, M. (1975). Ketoacids in the treatment of uremia. *Clin. Nephrol.*, **3**, 180
3 Levin, D. M. and Cade, R. (1965). Metabolic effects of dietary protein in chronic renal failure. *Ann. Intern. Med.*, **4**, 642
4 Kluthe, R., Oeschlen, D., Quirin, H. and Jedinsky, H. J. (1971). Six years experience with a special low-protein diet. In Kluthe, R., Berlyne, G. and Burton, B. (eds.) *Uremia, International Conference on Pathogenesis, Diagnosis and Therapy.* (Stuttgart: Georg Thieme Verlag)
5 Mitch, W. E., Walser, M., Buffington, G. A. and Lemann, J., Jr. (1976). A simple method of estimating progression of chronic renal failure. *Lancet*, **2**, 1326
6 Walser, M., Mitch, W. E. and Collier, V. U. (1979). The effect of nutritional therapy on the course of chronic renal failure. *Clin. Nephrol.*, **11**, 66
7 Barsotti, G., Guiducci, A., Ciardella, F. and Giovannetti, S. (1981). Effects on renal function of a low-nitrogen diet supplemented with essential amino acids and ketoanalogues and of hemodialysis and free protein supply in patients with chronic renal failure. *Nephron*, **27**, 113
8 Noree, L. O. and Bergström, J. (1975). Treatment of chronic uremic patients with protein-poor diet and oral supply of essential amino acids. II. Clinical results of long-term treatment. *Clin. Nephrol.*, **3**, 195
9 Fürst, P., Alvestrand, A. and Bergström, J. (1980). Effects of nutrition and catabolic stress on intracellular amino acid pools in uremia. *Am. J. Clin. Nutr.*, **33**, 1387
10 Mitch, W. E., Collier, V. U. and Walser, M. (1980). Creatinine metabolism in chronic renal failure. *Clin. Sci.*, **58**, 327
11 Ibels, L. S., Alfrey, A. C., Haut, L. and Huffer, W. E. (1978). Preservation of function in experimental renal disease by dietary restriction of phosphate. *N. Engl. J. Med.*, **298**, 122
12 Karlinsky, M. L., Haut, L., Buddington, B. *et al.* (1980). Preservation of renal function in experimental glomerulonephritis. *Kidney Int.*, **17**, 293
13 Barrientos, A., Arteaga, J., Rodicio, J. L., Alvarez Ude, F., Alcazar, J. M. and Ruilope, L. M. (1982). Role of control of phosphate in the progression of chronic renal failure. *Min. Elec. Metab.*, **7**, 127
14 Hostetter, T. H., Rennke, H. G. and Brenner, B. M. (1982). Compensatory renal hemodynamic injury: a final common pathway of residual nephron destruction. *Am. J. Kidn. Dis.*, **1**, 310
15 Hostetter, T. H., Olson, J. L., Rennke, H. G. *et al.* (1981). Hyperfiltration in remnant nephrons: a potentially adverse response to renal ablation. *Am. J. Physiol.*, **241**, 85

11
Experience with prolonged intradialysis hyperalimentation in catabolic chronic dialysis patients

D. V. POWERS AND A. J. PIRAINO

Protein and caloric deficits can develop easily during chronic dialysis-therapy (CDT)[1-7]. Amino acid loss from plasma to dialysate and accidental blood loss during haemodialysis occur[4,8,9]. Altered taste sense[10-11] and suppressed appetite result in inadequate food intake which is not improved easily in our experience, by dietary advice. Intercurrent stressful events, such as infection, gastrointestinal bleeding, cardiac arrythmia, etc., cause increased protein catabolic rates[12-13]. Such periods of weight loss frequently result in a failure to return to the original weight.

At this stage of significant bodyweight reduction, symptoms of anorexia and fatigue, muscular wasting and a slow steady weight loss characterize the clinical picture of the chronic dialysis patient who is 'failing to thrive'. Increasing dialysis time in the belief that these are indications of uraemia can lead to further deterioration. Haemodialysis is a catabolic stimulus and can cause a further loss of already depleted body protein. The clinical course described is commonly seen in patients who have been on chronic haemodialysis for a number of years. In an attempt to prevent the progression of such a course we have over the last 5 years prescribed intradialysis hyperalimentation with the hope of reversing weight loss[14].

METHODS OF ADMINISTRATION AND RATIONALE OF THERAPY

Intradialysis hyperalimentation consists of infusing during haemodialysis into the venous line a solution of glucose and amino acids. The solution contains glucose at a concentration of 250 g/l and varying concentrations (5.1–10%) of amino acids. The majority of patients received a general amino acid

Table 11.1 Weight profile for weight gaining patients. Weight record of 150 treatments in 19 intradialysis hyperalimentation series of 14 chronic dialysis patients on dialysis, 12 months or more prior to hyperalimentation, who sustained weight loss pre-treatment and gained weight during treatment. Weight recorded only when patient was oedema-free and at estimated dry weight

Patient	Start of dialysis Weight (kg)	Date	Measured at specified number of months before hyperalimentation						Weight of patients on dialysis (kg) Measured after various numbers of treatments											Ideal weight* (kg)
			18	12	6	3	2	1	0	15	30	45	60	75	90	105	120	135	150	
B.B.	53.5	9–71		71.7	67.6	65.8				55.3	58.0		60.3	60.8		59.9			63.5	68.0
M.C.	46.3	5–75	45.8	44.9	44.9	43.5	43.1		42.6	46.3	46.3		46.3	45.8	46.7					49.0
J.D.1	63.9	7–76	65.3	61.7	59.9		60.3		58.0	58.5	58.5	61.2	61.2	61.2	61.2		59.4	59.4	59.4	60.3
J.D.2	63.9	7–76		56.2		51.7		49.4	49.9	51.2	52.2	53.5	54.0							60.3
J.D.3	63.9	7–76				53.5		53.1	51.7	51.7	52.2	52.6	54.4	54.8	55.3	55.3				60.3
E.D.	99.8	4–74	63.5				51.7	52.2	52.2	52.6	54.8	54.4	54.4							Amputee
R.G.	59.9	3–78			72.1	68.5	68.0	64.4	64.9	65.3	67.6	72.1	74.8							71.2
M.G.1	61.2	1–75			54.8		54.0		44.0		50.8	49.9								59.9
M.G.2	61.2	1–75	54.0	54.0	56.7				50.3	50.8	52.6	56.2	58.5	61.2	59.4		60.8			59.9
R.H.1	54.4	11–72		56.7		55.3	54.4		50.8	53.5	54.8	55.3	55.3	61.2						54.4
R.H.2	54.4	11–72						53.1	50.8	52.6	54.4	54.8	55.3							54.4
V.H.†	68.5	2–80				68.0	65.8		57.6	57.2	58.0	58.0	59.4	59.9	59.9					59.0
A.M.1	61.7	6–72		54.0		50.8		49.0	42.6	44.0	49.4	50.8	51.2	49.9	50.8					54.4
A.M.2	61.7	6–72	50.8		51.2	49.0		47.2	42.2	43.1	44.4	45.8	48.1	49.9	50.8					54.4
W.N.	63.0	1–75		60.3	57.6				53.5	52.6	51.2	52.2	52.2	56.7	58.5		59.0			64.4
M.M.	68.0	6–74	68.0	62.6					53.5		47.2				50.8			53.5		63.5
C.P.	74.4	8–77	72.6	69.4		62.6		55.3	60.8	61.7		63.9								
K.W.	51.2	7–75	48.5	47.2	44.0				39.9	44.4	44.0									51.7
M.G.	63.5	5–70	59.0		59.0		55.3					52.2	52.2	53.5	54.8	54.8	54.8	54.8		

*Ideal weight established from 1959 Metropolitan Height and Weight Standards
Patient V.H. had only 5 months of haemodialysis prior to hyperalimentation

solution, namely 8.5% Freamine II (McGaw Laboratories, Irvine, Ca.) or Travasol 8.5% or 10% (Travenol Laboratories, Deerfield, Ill.)*. A few patients received an essential amino acid solution, Nephramine 5.1% (McGaw Laboratories, Irvine, Ca.), together with oral L-histidine. 800 ml of solution is usually infused during each haemodialysis at a rate of 200 ml/h. Patients are haemodialysed in the usual way (3–4 times per week for 3–4.5 h) using Travenol 'RSP' machines and various positive and negative pressure dialysers.

The adverse symptoms noted most frequently have been nausea and vomiting during dialysis and afterwards rebound hypoglycaemia, which can be avoided by a snack before the end of the infusion period. Nausea and vomiting, on the other hand, seem to arise from two mechanisms: one is related to the development on dialysis of acute hypotension which responds to extra saline; the second is nausea related to a particular amino acid solution, and this can be relieved by switching to a different solution.

The rationale of our therapy has been to supply calories and amino acids simultaneously at a time when chronic dialysis patients are best able to utilize these nutrients, i.e., during a dialysis treatment[15–17]. At the same time the administered fluid can be removed by ultrafiltration (and possibly also by solute drag). Moreover, if amino acid antagonisms generated by uraemia and poor nutrition are present and suppressing appetite[18–19], the dialysis procedure together with a new supply of amino acids may enable the body to correct amino acid deficiencies and imbalances.

RESULTS OF INTRADIALYSIS HYPERALIMENTATION: WEIGHT CHANGE RESPONSES

We have not treated 25 weight-losing CDT patients on dialysis 12 months or more prior to hyperalimentation dialysis. From 30 to 225 hyperalimentation treatments have been administered per patient, at a rate of 10–15 infusions per month. Of these 25 patients, 14 have had a weight gain; two, no further weight loss and ten continued to lose weight[†]. Table 11.1 shows the rate of weight changes during 19 series of treatments in the 14 patients who gained weight.

The most striking clinical observation noted during the treatment of these patients has been the development of a voracious appetite, in marked contrast to a previously depressed appetite. During periods of rapid weight gain non-protein energy consumption in a few patients reached 60 cal/kg bodyweight 24 h. Such improvement was not sustained however, usually lasting between 10 and 30 treatments (1–3 months). The onset of improved appetite was also quite variable, sometimes not occurring until the forty-fifth to sixtieth treatment, and in some patients never occurring.

The relationship between weight gain and the degree of appetite improve-

*Synthamin in U.K.
†One patient had both a weight-losing and subsequent weight-gaining series of treatments

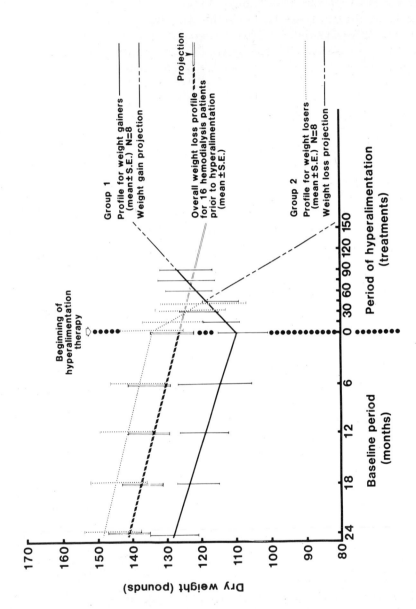

Figure 11.1 Comparison of response to hyperalimentation: change in dry weight compared for two groups of chronic dialysis patients

ment is an important clinical observation. The amount of amino acids supplied with intradialysis infusion was not large, usually less than 70 g/week (400 ml of 8.5% amino acid solution, with one-third loss into the dialysate, during three infusions per week). On the assumption that there is no glucose loss into a glucose-containing dialysate, which is unlikely with high blood sugar levels during a 25% glucose infusion, then only about 2000 extra calories per week are supplied. The correction of amino acid imbalances operating at the level of appetite control may have occurred as a result of this therapy. The formulation of an amino acid mixture better suited for the correction of the amino acid abnormalities of catabolic renal failure may produce even better results.

Cessation of weight loss appeared to be more directly related to the nutritive quality of the infusion. Appetite improvement was not noticeable and estimation of dietary intake (fraught with errors in an out-patient setting) did not show any change as compared to pre-intradialysis hyperalimentation. From 3 to 8 g of unbound amino acids, and an equal amount of bound amino acids, may be lost per dialysis treatment into the bath solution[4,5,8,9]. Kopple has shown recently, however, that an intradialysis infusion may result in the retention by the body of most of the amino acids infused. During an infusion of 39.5 g amino acids and 200 g glucose, unbound amino acid dialysate losses increased only from 8 g without the infusion to 12 g with the infusion. The net effect is an increase of 105 g of protein per week[20] which is a significant contribution to a restricted diet.

HYPERPARATHYROIDISM AND ADVANCED METABOLIC BONE DISEASE IN CHRONIC DIALYSIS PATIENTS (Figure 11.1)

Our original report[14] described 16 catabolic patients, on dialysis 24 months or more, who lost weight, prior to the onset of hyperalimentation dialysis, of 14.5 ± 2.37 lb (6.58 ± 1.07 kg). A general amino acid solution plus glucose was administered for 60 treatments over 20 weeks. Eight patients (group 1) responded with a new gain in weight of $10.5\pm2.1\%$. Average weight gain was 0.18 lb (0.08 kg) per treatment. Eight patients (group 2) continued to lose weight at an average of 0.50 lb (0.23 kg) per treatment and all but one of these patients have subsequently died (Fig. 11.1).

Table 11.2 displays the pre-dialysis plasma chemistry values of these two groups and two other groups. Group 3 comprised 25 weight-stable patients who were not given hyperalimentation. Group 14 consisted of five patients with much less dialysis experience of 1–3 months and then given intradialysis hyperalimentation with essential amino acids and glucose and orally supplemented with histidine. The chemistry values of the control group (group 3) are compared to the other groups. Group 2 patients displayed both significantly higher plasma alkaline phosphatase values, indicative of advanced metabolic bone disease, and higher plasma parathyroid hormone

Table 11.2 Baseline and treatment data for groups 1, 2 and 14 as compared to group 3 (control patients)

	Group 3	Baseline comparisons			Treatment comparisons		
		Group 1	Group 2	Group 14	Group 1	Group 2	Group 14
Albumin (g/100 ml)	3.86±0.457	NS	NS	3.52±0.436 p<0.01	NS	NS	3.64±0.23 p<0.05
Globulin (g/100 ml)	3.186±0.848	NS	NS	NS	NS	NS	3.67±0.373
Blood urea nitrogen (mg/100 ml)	87.43±26.55	NS	NS	77.52±21.43 p<0.05	NS	NS	
Creatinine (mg/100 ml)	13.07±4.35	NS	NS	NS	NS	NS	7.41±2.13 p<0.001
Total iron binding capacity (μg/100 ml)	264.28±63.34	NS	NS	239.13±47.63 p<0.05	NS	NS	218.25±50.18 p<0.05
Alkaline phosphatase (IU/l)	169.19±187.31	116.192±187.31 p<0.02	383.4±207.99 p<0.05	NS	116.11±74.7 p<0.02	353.9±237.2 p<0.01	NS
Parathyroid hormone (pg/ml)	4059.44±3197.31	2731.8±2445.67 p<0.02	5802.4±3026 p<0.02	1306.67±649.44 p<0.001	2307.86±1827.0 p<0.05	NS	2086.67±1348.92 p<0.02
Haemoglobin (g/100 ml)	8.19±4.54	NS	NS	7.24±0.916 p<0.01	NS	NS	7.02±0.832 p<0.01
Haematocrit (%)	25.69±5.38	22.47±5.37 p<0.02	NS	23.18±2.72 p<0.001	22.47±5.56 p<0.02	NS	22.76±2.46 p<0.01
Sodium (mEq/l)	138.5±5.49	NS	NS	140.09±3.31 p<0.01	NS	NS	NS

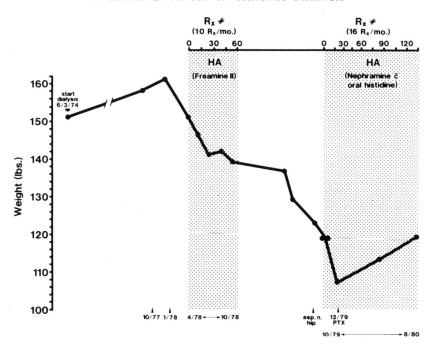

Figure 11.2 Weight profile: M.M.

levels at the start of hyperalimentation. This group had severe proximal muscle myopathy and neuropathy, related to advanced secondary hyperparathyroidism. The plasma aminograms of this group were more abnormal than those who gained weight in group 1. Post-intradialysis hyperalimentation values of phenylalanine, methionine, leucine and isoleucine were significantly higher than group 1. These amino acids have been implicated in the generation of amino acid imbalances.

In uraemia parathyroid hormone may be a neurotoxin[21], and cause severe myopathy[22] and promote skeletal muscle protein breakdown in uraemic rats[23]. The one patient of this advanced hyperparathyroid group alive today was a treatment failure during intradialysis hyperalimentation before parathyroidectomy; a similar course of treatment given after parathyroidectomy resulted in weight gain (Figure 11.2). The effect of parathyroidectomy may be important for weight gain, as we have observed similar weight gain, unaccompanied by hyperalimentation, following parathyroidectomy (Figure 11.3).

REPRODUCIBILITY OF WEIGHT GAIN

Prolonged intradialysis hyperalimentation can be effective in promoting weight gain in patients with wasting but without parathyroid disease (group 1

137

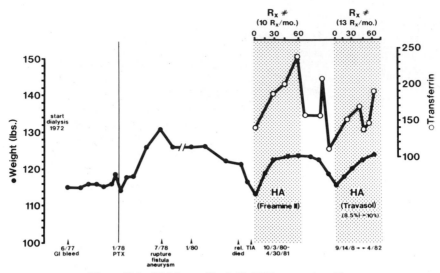

Figure 11.3 Weight profile: R.H. PTX = parathyroidectomy

and group 14 patients, Figure 11.1) (patients in Table 11.1). Moreover, in some patients weight gain is reproducible and can be achieved when different commercially available amino acid solutions are used. This is depicted in the weight profiles during two periods of treatment in patient R.H. (Figure 11.3) and in patient A.M. (Figure 11.4). The reproducibility of weight gain seems to rule out any placebo effect. The fall-off in weight which can occur when dialysis hyperalimentation stops is shown in Figure 11.5. In seven treatment series in six patient, there was weight loss prior to hyperalimentation followed by weight gain during infusion treatment with subsequent weight loss after treatment.

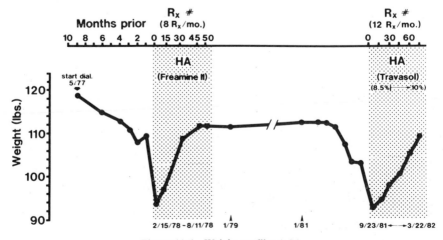

Figure 11.4 Weight profile: A.M.

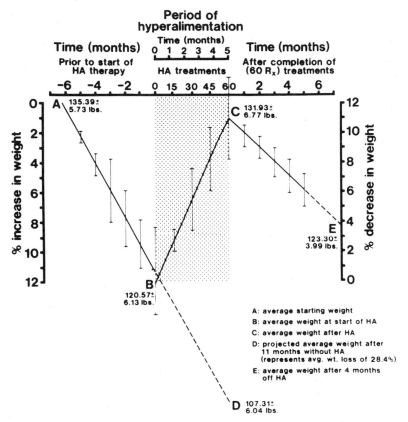

Figure 11.5 Percentage weight change (mean ± S.E.) in response to initiation and termination of hyperalimentation therapy in chronic dialysis patients ($N = 7$). Average rate of weight change: AB = −3.27% of bodyweight changes per month ($r = 0.94$) BC = +2.13% of bodyweight changes per month ($r = 0.98$) CE = −0.90% of bodyweight changes per month ($r = 0.95$) (r = correlation coefficient of line). A *vs.* D ($p < 0.01$); A *vs.* B ($p > 0.05$) ns; C *vs.* E ($p > 0.05$) ns

This sequence of weight changes confirms the effect of hyperalimentation dialysis.

Our clinical observations can be summarized as follows.

(1) Intradialysis hyperalimentation can reverse continuing weight loss and the symptoms of anorexia, lassitude and muscular weakness which accompany the wasting syndrome during chronic dialysis.

(2) Such treatment is ineffective in patients with advanced metabolic bone disease due to hyperparathyroidism.

(3) Even low-grade intercurrent stressful events (infection, etc.) can delay the onset of steady weight gain during a prolonged treatment period. In this circumstance cessation of weight loss is significant until the stressful event is resolved.

139

(4) Cessation of treatment following a period of weight gain may be followed by further weight loss.

(5) The best weight gain occurs when, during a treatment period, appetite is improved.

RATIONALE FOR FORMULA CHANGE: AMINO ACID IMBALANCE AND APPETITE SUPPRESSION

The association of amino acid imbalance with depressed food intake was first suggested by Rose in 1931[24]. An amino acid imbalance, as defined by Harper *et al.*[19] is the excess presence in the body of an essential amino acid along with the reduced concentration of a different amino acid which is rate-limiting for growth. Experimental animals fed a diet containing excess leucine display a reduced food intake proportional to the amount of leucine in the diet. When valine and isoleucine are added to this diet, the imbalance is removed and normal food intake is resumed[25]. Low plasma levels of some amino acids correlate well with reduced food intake, while the concentration of these amino acids is even lower in muscle and brain tissue[26].

The growth inhibition produced by an imbalanced amino acid diet in experimental animals can be overcome by forced feeding of the imbalanced diet[19]. Theoretically at least, the catabolic renal patient could similarly avoid protein inadequacy by a regime of total parenteral nutrition and dialysis. It would seem more practical however, if appetite stimulation and weight gain could be achieved by removal of amino acid imbalances and this should be the objective in prescribing amino acid formulations.

There are many amino acid abnormalities found in untreated uraemia. These include high plasma concentrations of methionine and phenylalanine accompanied by low levels of tyrosine (due to reduced phenylalanine hydroxylase activity)[27,28], low plasma histidine concentrations due to impaired synthesis[29,30] and reduced levels of valine, leucine, lysine and threonine[27,31]. Plasma serine levels may be low due to disturbance of the renal pathway by which it is synthesized from glycine[32,33]. Urea cycle amino acids citrulline, ornithine and arginine are elevated as well as taurine, proline, glycine, alanine and glutamic acid[5,8,34–37]. These plasma amino acid abnormalities in untreated uraemia are generally reflected in muscle levels. However, low plasma leucine and serine and high plasma methionine and glutamic acid concentrations may be associated with normal muscle values[27].

A low protein diet plus essential amino acids (Rose formula) and histidine did not change the branched-chain amino acid (BCAA) abnormalities in plasma and muscle[38]. However, a new formula with decreased concentrations of isoleucine and leucine and increased valine given along with a low protein diet returned plasma and muscle BCAA concentrations to normal. These investigators felt a BCAA antagonism existed with valine as the limiting amino acid. Well-nourished haemodialysis patients have been reported to have normal plasma and muscle concentrations of valine, tyrosine and threonine and low plasma leucine. Elevated plasma and muscle concentrations of phenylalanine and the urea cycle amino acids, citrulline, arginine and ornithine remained abnormal in well-nourished dialysis patients. In addition, in these patients, concentrations of plasma and muscle histidine, muscle lysine and plasma tyrosine and methionine were elevated, along with plasma elevations of some non-essential amino acids[40].

We have examined pre- and post-dialysis plasma aminograms of 25 dialysis patients whose weight was stable. Post-dialysis levels of phenylalanine, methionine, leucine and isoleucine were higher compared to the pre-dialysis levels. Post-dialysis levels of histidine, lysine, serine, proline, glycine, alanine, arginine, 3-methylhistidine, cystine, citrulline, tyrosine, taurine, aspartic acid and ornithine were lower than pre-dialysis. The elevations of post-dialysis phenylalanine, methionine, leucine and isoleucine suggest concentrations in excess of requirements, despite losses on dialysis[14].

Chami et al.[41] infused separately eight essential amino acids between dialysis treatments and reported a slower rate of removal for phenylalanine, valine, leucine, isoleucine and threonine from dialysis patients compared to healthy controls. We have infused during dialysis (group 14) a solution of eight essential amino acids (Nephramine)* plus glucose, to poorly nourished patients and there was a significant increase in concentrations of these five amino acids after dialysis[14].

Phenylalanine excess inhibits protein synthesis[42]. Methionine, according to Harper, is the most toxic of all amino acids[19]. In rats administered high loads of methionine, serine and glycine concentrations are decreased[43]. In the metabolism of methionine to cysteine and cystine, serine is necessary at the cystathionine synthetase step[19].

Plasma serine concentrations were abnormally low in our two most protein-depleted patient groups (group 2 and group 14) and did not improve with hyperalimentation. Cystine values were normal before dialysis but dropped to low or subnormal values in all our patient groups[14].

In contrast to the elevated non-essential amino acid concentrations in our well-nourished weight-stable patients (group 3) and in those patients reported by Alvestrand et al.[40], severe non-essential amino acid deficits were present in our group 14 patients, i.e., those patients not far removed in time from a period of severe, acute decrease in renal function (Figure 11.6). These latter patients had not been given a general amino acid but an essential amino acid solution in case non-essential amino acids might exacerbate their uraemic symptoms. Unfortunately, we were unaware at that time of their large non-essential amino acid deficits.

*McGaw Laboratories, Irvine, California. Essential amino acid solution was supplemented with 1.2 g per treatment of oral L-histidine.

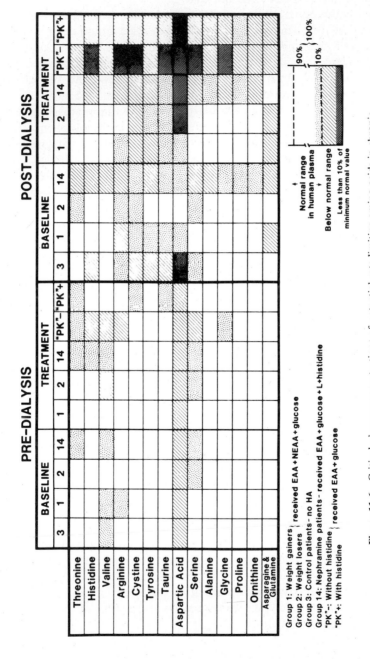

Figure 11.6 Critical plasma concentrations of potential rate-limiting amino acids in chronic dialysis patients

Group 1: Weight gainers ⎱ received EAA+NEAA+glucose
Group 2: Weight losers ⎰
Group 3: Control patients - no HA
Group 14: Nephramine patients - received EAA+glucose+L+histidine
"PK"-: Without histidine ⎱ received EAA+glucose
"PK"+: With histidine ⎰

Plasma arginine levels in these group 14 patients were abnormally low before and after infusion and in groups 1 and 2 (general amino acid solution patients) were in the normal range. All these values were significantly lower than in our weight-stable control patients (group 3). In contrast to reports of elevated ICM and plasma arginine concentrations in non-dialysed and dialysed CRF patients[35,40] we relate this difference to the weight loss of these patients (Groups 1, 2 and 14). Very low plasma arginine levels have been reported in sepsis[44]. Moreover, urea has been reported to be a selective inhibitor of arginine succinate lyase; in uraemic rats the activity of two other cycle enzymes, transcarbamylase in liver and arginine synthetase in kidney, is likewise reduced[28,45]. Removal of uraemic inhibition by dialysis may, as Farrell and Hone suggest, result in increased urea cycle activity and consequently further increase arginine requirements[3]. We conclude a general amino acid solution is superior to a formula containing only essential amino acids when given to catabolic chronic dialysis patients. We suggest that the best results in such patients may be produced by a general amino acid solution containing all essential amino acids but lower concentrations of phenyl-

Table 11.3 Recommendations for formula changes in the hyperalimentation of catabolic chronic dialysis therapy patients

	Amount contained in Freamine II (mg/100 ml)	Recommended change
Essential amino acids		
L-Threonine	340	→
L-Phenylalanine	480	▼
L-Methionine	450	▼
L-Histidine	240	△
L-Lysine	620	→
L-Valine	560	△
L-Leucine	770	▼
L-Isoleucine	590	▼
L-Tryptophan	130	→
Non-essential amino acids		
L-Serine	500	△
L-Proline	950	→
Glycine	1700	→
L-Alanine	600	→
L-Arginine	310	△
L-Tyrosine	–	△
L-Cysteine	20	→
L-Aspartic acid	–	?
L-Cystine	–	?

Should be increased, △; should be decreased, ▼; no change recommended →

alanine, methionine, leucine and isoleucine and a greater concentration of valine. Histidine, arginine and serine concentrations should be increased and tyrosine should be included. Among the non-essential amino acids, alanine, glycine and proline should not be changed. Cystine supplementation may or may not be necessary depending on its synthesis from methionine; the need for aspartic acid is likewise uncertain (Table 11.3).

References

1 Schaeffer, G., Heinze, V., Jontofosohn, R., Katz, N., Rippich, T. H., Schafer, B., Sudoff, A., Zimmerman, W. and Kluthe, R. (1975). Amino acid and protein intake in RDT patients: a nutritional and biochemical analysis. *Clin. Nephrol.*, **3**, 228
2 Bianchi, R., Mariani, G., Giuseppina, T. M. and Carmassi, F. (1978). The metabolism of human serum albumin in renal failure on conservative and dialysis therapy. *Am. J. Clin. Nutr.*, **31**, 1615
3 Farrell, P. C. and Hone, P. W. (1980). Dialysis-induced catabolism. *Am. J. Clin. Nutr.*, **33**, 1417
4 Tepper, J., Vanderhem, B. K., Tuma, G. J., Arisz, L. and Donker, A. J. M. (1978). Loss of amino acids during hemodialysis: quantitative and qualitative investigations. *Clin. Nephrol.*, **10**, 16
5 Young, G. A. and Parsons, F. M. (1970). Plasma amino acid imbalance in patients with chronic renal failure on intermittent dialysis. *Clin. Chim. Acta*, **27**, 491
6 Maillet, C. and Garber, A. J. (1980). Skeletal muscle amino acid metabolism in chronic uremia. *Am. J. Clin. Nutr.*, **33**, 1343
7 Kopple, J. D. and Swendseid, M. E. (1975). Protein and amino acid metabolism in uremic patients undergoing maintenance dialysis. *Kidney Int.*, **7** (Suppl.), 64
8 Rubini, M. E. and Gordon, S. (1968). Individual plasma-free amino acids in uremics: effect of hemodialysis. *Nephron*, **5**, 339
9 Kopple, J. D., Swendseid, M. E., Shinaberger, J. H. and Umezana, C. Y. (1973). The free and bound amino acids removed by hemodialysis. *Trans. Am. Soc. Artif. Organs*, **19**, 309
10 Burge, J. C., Park, H. S., Witlock, C. P. and Schemmel, R. A. (1979). Taste acuity in patients undergoing long-term hemodialysis. *Kidney Int.*, **15**, 49
11 Mahajan, S. K., Prasad, A. S., Lambujon, J., Abassi, A. A., Briggs, W. A. and McDonald, F. D. (1980). Improvement of uremia hypogeusia by zinc: a double blind study. *Am. J. Clin. Nutr.*, **33**, 1517
12 Grodstein, G. P., Blumenkrantz, M. J. and Kopple, J. D. (1980). Nutritional and metabolic response to catabolic stress in uremia. *Am. J. Clin. Nutr.*, **33**, 1411
13 Giordano, C., DeSanto, N. G. and Senatore, R. (1978). Effects of catabolic stress in acute and chronic renal failure. *Am. J. Clin. Nutr.*, **31**, 1561
14 Piraino, A. J., Firpo, J. J. and Powers, D. V. (1981). Prolonged hyperalimentation in catabolic chronic dialysis therapy patients. *J. Parenteral Enteral Nutr.*, **5**, 463
15 Fürst, P., Bergström, J., Josephson, B. and Noree, L-O. (1970). The effect of dialysis and administration of essential amino acids on plasma and muscle synthesis. Studies with ^{15}N in uraemic patients. *Proc. Eur. Dial. Transplant. Assoc.*, **7**, 175
16 DeFronzo, R. A. and Alvestrand, A. (1980). Glucose intolerance in uremia: site and mechanism. *Am. J. Clin. Nutr.*, **33**, 1438
17 Metcoff, J., Lindeman, R., Baxter, D. and Pederson, J. (1978). Cell metabolism in uremia. *Am. J. Clin. Nutr.*, **30**, 1627
18 Giordano, C., Depascale, C., Pluvio, M., DeSanto, N. G., Fella, A., Esposito, R., Capasso, G. and Pota, A. (1975). Adverse effects among amino acids in uremia. *Kidney Int.*, **7** (Suppl.), 306
19 Harper, A. E., Benevange, N. J. and Wohlhueter, R. M. (1970). Effects of ingestion of disproportionate amounts of amino acids. *Physiol. Rev.*, **50**, 428

20 Wolfson, M., Jones, M. R. and Kopple, J. D. (1982). Amino acid losses during dialysis with infusion of amino acids and glucose. *Kidney Int.*, **21**, 500
21 Avram, M. M., Iancu, M., Morrow, P., Feinfeld, D. and Huatuco, A. (1979). Uremic syndrome in man: new evidence for parathormone as a multisystem neurotoxin. *Clin. Nephrol.*, **11**, 59
22 Mallete, L. E., Patten, B. M. and Engel, W. K. (1975). Neuromuscular disease in secondary hyperparathyroidism. *Ann. Intern. Med.*, **82**, 474
23 Garber, A. J. (1976). Abnormalities of carbohydrate metabolism in chronic uremia. In B. B. Mackey (ed.). *Proceedings of the 9th Annual Contractors' Conference of the Artificial/ Chronic Uremia Program, National Institute of Arthritis, Metabolism and Digestive Diseases.* pp. 5−8. (Bethesda, Maryland: US Department of Health, Education and Welfare, National Institutes of Health)
24 Rose, W. C. (1931). Feeding experiments with mixtures of highly purified amino acids. I. The inadequacy of diets containing 19 amino acids. *J. Biol. Chem.*, **94**, 155
25 Rogers, Q. R. and Leung, M. G. (1977). The control of food intake: when and how are amino acids involved. In M. Kare and O. Maller (eds.). *The Chemical Senses and Nutrition.* pp. 213−249. (New York: Academic Press)
26 Peng, Y., Tews, J. K. and Harper, A. E. (1972). Amino acid imbalance, protein intake and changes in rat brain and plasma amino acids. *Am. J. Physiol.*, **222**, 314
27 Bergström, J., Fürst, P., Noree, L-O. and Vinnars, E. (1978). Amino acids in muscle tissue of patients with chronic uraemia: effect of peritoneal dialysis and infusion of essential amino acids. *Clin. Sci. Mol. Med.*, **54**, 51
28 Swendseid, M. E., Wang, M., Vyhmeister, I., Chan, W., Siassi, F., Tam, C. F. and Kopple, J. D. (1976). Amino acid metabolism in the chronically uremic rat. *Clin. Nephrol.*, **3**, 240
29 Fürst, P. (1972). ¹⁵N studies in severe renal failure. II. Evidence for the essentiality of histidine. *Scand. J. Clin. Lab. Invest.*, **30**, 307
30 Swendseid, M. E. and Kopple, J. D. (1975). Evidence that histidine is an essential amino acid in normal and chronically uremic man. *J. Clin. Invest.*, **55**, 881
31 Bergström, J., Fürst, P., Noree, L-O. and Vinnars, E. (1975). Intracellular free amino acids in uremic patients as influenced by amino acid supply. *Kidney Int.*, **7**, 345
32 Pitts, R. F., Damian, A. C. and MacLeod, M. G. (1970). Synthesis of serine by rat kidney *in vivo* and *in vitro*. *Am. J. Physiol.*, **219**, 584
33 Fukuda, S. and Kopple, J. D. (1980). Uptake and release of amino acids by the kidney of dogs made chronically uremic with uranyl nitrate. *Min. Electrolyte Metabol.*, **3**, 248
34 Giordano, C. (1972). Diet in uremia. *Proceedings of the European Dialysis and Transplant Association*, **9**, 419
35 Fürst, P., Alvestrand, A. and Bergström, J. (1980). Effects of nutrition and catabolic stress on intracellular amino acid pools in uremia. *Am. J. Clin. Nutr.*, **33**, 1387
36 McGale, E. H. F., Pickford, J. C. and Aber, G. M. (1972). Quantitative changes in plasma amino acids in patients with renal disease. *Clin. Chim. Acta*, **38**, 395
37 Counahan, R., El-Bishti, M., Cox, B. D., Ogg, C. S. and Chantler (1976). Plasma amino acids in children and adolescents on hemodialysis. *Kidney Int.*, **10**, 471
38 Alvestrand, A., Bergström, J., Fürst, P., Germanis, G. and Widstam, U. (1978). The effect of essential amino acid supplementation on muscle and plasma free amino acids in chronic uremia. *Kidney Int.*, **14**, 323
39 Alvestrand, A., Ahlberg, M., Bergström, J. and Fürst, P. (1981). Effects of nutritional regimens on branched chain amino acid (BCAA) antagonism in uremia. In M. Walsen and J. Williamson (eds.). *Metabolism and Clinical Implications of Branched-Chain Amino and Keto Acids.* pp. 605−613. (New York: Elsevier/North Holland)
40 Alvestrand, A., Bergström, J. and Fürst, P. (1979). Intracellular free amino acids in patients treated with regular haemodialysis. *Proc. EDTA*, **16**, 129
41 Chami, J., Reidenberg, M. M., Wellner, D., David, D. S., Rubin, A. L. and Stenzel, K. H. (1978). Pharmacokinetics of essential amino acids in chronic dialysis patients. *Am. J. Clin. Nutr.*, **31**, 1652
42 Roscoe, J. P., Eaton, M. D. and Choy, G. C. (1968). Inhibition of protein synthesis in Krebs 2 ascites cells and cell-free systems by phenylalanine and its effect on leucine and lysine in the amino acid pool. *Biochem. J.*, **109**, 507

43 Klavins, J. V. (1965). Pathology of amino acid excess. V. effects of methionine on free amino acids in serum. *Biochim. Biophys. Acta*, **104**, 554
44 Freund, H. R., Ryan, J. A. and Fischer, J. E. (1978). Amino acid dearrangements in patients with sepsis: treatment with branched chain amino acids rich infusion. *Am. Surg.*, **188**, 423
45 Menyhart, J. and Grof, J. (1977). Urea as a selective inhibitor of arginosuccinate lyase. *Eur. J. Biochem.*, **75**, 405

Section 3
Special Care Enteral Feedings

12
Enteral feeding in liver failure

J. E. WADE, M. ECHENIQUE AND G. L. BLACKBURN

PROTEIN METABOLISM

The normal liver is responsible for several functions indispensable for normal amino acid metabolism and homeostasis. These functions include the degradation of several amino acids for gluconeogenesis, conversion of amino acids to urea and glutamate, formation of several non-essential amino acids or enzymes, and the removal of amino acids from plasma[1]. Regulation of plasma amino acid concentrations appears important to body homeostasis, normal organ function, protein metabolism and to prevent metabolic acidosis. The gluconeogenetic and catabolic enzyme pathways are essential to this effort.

Most amino acids other than the branched-chains are cleared from circulation and are metabolized by the liver. Amino transferases, responsible for the metabolism of branched-chain amino acids (BCAAs), are primarily concentrated in skeletal muscle, kidney, brain, adipose and other peripheral tissues but not in the liver[2]. During stress, injury, and/or infection, BCAA will be oxidized by skeletal muscle producing ATP, CO_2, alanine, glutamine and water. The BCAAs contribute their nitrogen moiety and pyruvate supplies the carbon chain for the formation of alanine and glutamine, which are transported to the liver and kidneys for glucose production via gluconeogenesis and urea production[3]. BCAAs are important not only as substrates for energy production but as a source of dispensable amino acids and as regulators of protein metabolism[3]. The latter is important for redistribution of amino acids from skeletal muscle to visceral tissues for protein synthesis[4].

Trauma, stress, and infection are characterized by an increase in energy expenditure, increased excretion of urine urea nitrogen and by increased skeletal muscle breakdown[5]. Oxidation of BCAA is also increased in skeletal muscle, with other amino acids released into plasma[4,6]. Plasma amino acid levels in this state show an elevation of aromatic and sulphur-containing amino acids with normal or decreased levels of BCAAs[7,8] (Table 12.1).

Table 12.1 Plasma amino acid levels in normal subjects and in patients with hepatic disease*

Amino acid	Normal subjects (μg/ml)	Chronic liver disease ‡ (μg/ml) †
Valine	30.1 ± 1.9	14.8 ± 1.5
Leucine	18.5 ± 1.2	8.26 ± 0.9
Isoleucine	10.0 ± 0.7	3.80 ± 0.69
Threonine	14.5 ± 1.8	9.2 ± 0.69
Methionine	3.7 ± 0.2	10.86 ± 3.5
Phenylalanine	8.2 ± 1.1	22.84 ± 1.4
Tryptophan	9.0 ± 1.0	12.65 ± 2.2
Glutamate	7.7 ± 0.8	29.35 ± 7.8
Alanine	33.6 ± 2.2	24.28 ± 2.22
Tyrosine	11.3 ± 1.3	25.48 ± 3.1

*Adapted from Rosen, H.M. et al. (1977)[8]
†Encephalopathic patients, meeting two of the four following criteria: (1) serum bilirubin 2.5 mg/100 ml, (2) SGOT 80 i.u. per 100 ml, (3) prothrombin time, 2.5 s prolonged, (4) serum albumin 3.0 g/100 ml
‡receiving 20 g protein intake

LIVER DYSFUNCTION

Liver dysfunction may be produced by cirrhosis, sepsis, shock, hepatitis, surgery, or trauma. In late stage of liver failure or during high protein intake, there is decreased conversion of ammonia to urea causing the high plasma ammonia levels characteristic of cirrhosis and hepatic encephalopathy[9,10]. In addition, the liver is unable to remove many amino acids from plasma, producing an imbalance comprising an elevation of aromatic and sulphur-containing amino acids. It has been postulated that when these amino acids, rather than the BCAAs, are transported across the blood–brain barrier, an increase in the production of false neurotransmitters occurs in the brain. This may be one of the mechanisms of hepatic encephalopathy[7].

The administration of BCAAs has been shown to decrease skeletal muscle breakdown, thereby limiting the amount of ammonia available to be converted to urea[4,6]. Decreasing the production of ammonia may be a factor in improving hepatic encephalopathy. Formulas containing 35–50% of the amino acids as branched-chain have been shown to produce a recovery of encephalopathic coma state by several hours, improve nitrogen balance and perhaps change mortality rate[7].

Herlong et al. performed a double-blind crossover comparison of BCAA (68 μmol/l) vs. ornithine salts of the respective keto acids (34 μmol/l) over a 7–10 day period in eight patients with chronic portal-systemic encephalopathy. Patients that received the ornithine salts of the alpha keto acids had a significant improvement in the electroencephalographic abnormalities and clinical grade of encephalopathy compared to patients receiving BCAA solutions. More work will have to be done to explore this therapeutic possibility in this group of patients[11].

Some individuals with severe liver dysfunction may be unable to convert phenylalanine to tyrosine and/or methionine to cystine. Since tyrosine and cysteine are present in low concentrations in most feeding formulas there may be an increased requirement for these two amino acids in patients with severe liver dysfunction[12].

VITAMIN DEFICIENCIES IN LIVER DISEASE

Poor nutritional status is characteristic of patients with liver dysfunction. The causes include decreased intake, abnormalities in metabolism, diets high in carbohydrate and low in protein, vitamins, minerals and trace elements. Decreased intake and imbalanced diet are probably the principal causes of nutritional deficiencies in liver disease[13].

Several vitamins are present in abnormally low amounts in patients with liver disease. Leevy *et al.*[14] studied the levels of several water-soluble vitamins associated with different types of liver dysfunction and found reduced serum levels in 32% of alcoholic patients with normal livers, in 44% of patients with fatty livers, and in 49% with cirrhosis. Low plasma levels of thiamine, riboflavin, pyridoxine, and nicotinic acid were found in patients with cirrhosis. Low levels of other enzymes like vitamin B_{12}, pantothenic acid or biotin were also noted.

Halsted *et al.*[15] studied the mechanisms of altered folate levels in alcoholic patients and reported that a combination of dietary folate deficiency plus prolonged alcohol intake could result in intestinal malabsorption of water-soluble substances and account for the poor nutrition of these individuals. Folic acid is the vitamin most frequently deficient, serum levels being low in 30% of patients with normal liver, 40% in individuals with fatty liver, and 47% with cirrhosis.

Clinical changes which can be caused by vitamin deficiencies include: peripheral neuropathy secondary to thiamine deficiency, glossitis and chelosis due to low riboflavin or nicotinic acid levels, pellagra and Wernicke's encephalopathy from thiamine and nicotinic acid deficiencies and megaloblastic anaemia due to vitamin B_{12} and folate deficiencies[13].

Deficiencies of vitamins A, D, E, and K occur frequently in patients with hepatic disease. Deficiency of vitamin K may produce bleeding abnormalities that can be corrected by supplementation of this vitamin. Vitamin A deficiency may produce night blindness and vitamin D deficiency may cause bone disorders[16,13]. These conditions too, can be treated with supplementation of the respective vitamins.

INTESTINAL FUNCTION IN LIVER DISEASE

Malabsorption of fats, D-xylose, thiamine, and folic acid have been described in alcoholic patients with and without liver disease and in patients with cirrhosis, hepatitis, and other causes of liver dysfunction. Steatorrhea is a common finding[13].

Ethanol ingestion has been shown to produce a decreased activity of jejunal lactase and thymidine kinase plus a reduced oxygen consumption in jejunal slices in rats[17]. These changes may play a role in the abnormal absorption found after acute or chronic ethanol consumption. Marin et al.[18] reviewed autopsy protocols and histological sections of 154 patients with Laennec's cirrhosis and reported a 28% incidence of pancreatic abnormalities. In addition, they performed a prospective study of small bowel absorptive capacity in 20 patients (50% of these patients were found to have alcoholic steatorrhea). Several studies of small bowel absorption and function were performed. They yielded normal radiologic findings, normal Schilling's tests, normal small bowel biopsies, and normal D-xylose tests. Their results indicated that steatorrhea in cirrhotic patients is due not to small bowel dysfunction but to neomycin administration or pancreatic insufficiency.

Fast et al.[19] studied fat absorption in patients with alcoholic cirrhosis using radioactive iodinated glycerol licoleate in a test meal containing 25 μCi I[131] and 0.5 g fat/kg bodyweight. Blood samples were examined at 2, 3, 4, and 6 h and assayed for radioactive content. Twelve of 19 patients showed abnormal fat absorption curves. D-xylose tests were performed to assess small bowel function and were found normal. They conclude that the absorption defect in patients with cirrhosis is not due to the small bowel dysfunction but could be a manifestation of pancreatic insufficiency.

Deficiencies of bile salts in the intestinal lumen have been implicated as a cause of fat malabsorption in cirrhotic patients. Summerskill et al.[20] studied seven patients who had a malabsorption syndrome associated with hepatic disease but had neither clinical jaundice nor bile in the urine. They conclude that a 'quantitative deficiency of bile' is not a likely cause of fat malabsorption since jaundice was not present in these patients and normal amounts of bile were measured in duodenal aspirates. Additionally, excretion of urobilinogen was found to be normal in jaundiced patients with steatorrhea. Abnormal bowel flora have been suggested as a cause of malabsorption in cirrhosis. Other possibilities sometimes mentioned are increases in hydrostatic pressure within the portal venous and lymphatic systems and pancreatic insufficiency. In the Summerskill study, steatorrhea was not corrected with antibiotics, making the abnormal bowel flora hypothesis less likely as a cause of fat malabsorption. Only two of the patients in this study had ascites, the others had no evidence of portal hypertension making increases in hydrostatic pressure an unsatisfactory explanation for malabsorption. Pancreatic insufficiency was not identified in these individuals by clinical radiographic or chemical finding. Autopsy performed in two cases revealed no pancreatic abnormalities.

Therapeutic trials using broad-spectrum antibiotics, human bile therapy and pancreatic exocrine replacement have not yielded any significant improvement in fat absorption and they have not elucidated any specific cause of malnutrition.

In conclusion, specific causes of steatorrhea and malabsorption have not been determined. Different mechanisms are probably at work in different patients. Correction of malnutrition by means of diets high in protein and low in fats (especially long-chain fatty acids), replacement of vitamins and electrolytes, and abstinence from ethanol continue to be very important aspects of the management of these patients.

FLUID AND ELECTROLYTE ABNORMALITIES IN PATIENTS WITH LIVER DISEASE AND ASCITES

Ascites arises from the local imbalance of hydrostatic and osmotic forces across the splanchnic and hepatic beds. The capacity of the lymphatic system to absorb ascitic fluid determines the presence or absence of ascites[21]. Increased reabsorption of sodium in cirrhotic patients is characterized by (1) increased proximal tubular reabsorption of sodium due to diminished glomerular filtration rates and diminished activity of naturetic hormones and (2) increased sodium reabsorption in the distal tubules due to high aldosterone levels. The abnormal liver may be unable to inactivate aldosterone, increasing its effect on sodium reabsorption and potassium excretion[22].

Medical management of ascites involves restriction of sodium and free water and the use of diuretics to increase their elimination. A restriction which limits intake to 20 mEq of sodium per day or, in some cases, 10 mEq per day may be necessary in the initial management. After ascites has been corrected, sodium intake can be increased slowly to approximately 40 mEq per day. The necessity for free water restriction is controversial; some authors recommend restriction only if excessive water retention or hyponatraemia is present[22]. The use of diuretics can cause several complications including hyperuricaemia, hypo- or hyperkalaemia, metabolic acidosis or alkalosis, hypomagnesaemia, hypocalcaemia, diarrhoea, and glucose imbalance[22].

Soler et al.[23] studied whole body potassium measurements of 55 cirrhotic patients with different stages of liver disease. They found that cirrhotic patients are not always depleted of potassium stores and that the aetiology of cirrhosis influences the level of deficiency. Alcoholic cirrhosis produces low whole body potassium levels and many patients with chronic active hepatitis show normal body potassium stores. In addition, they demonstrated that serum potassium determinations may be misleading, normal values being

ADVANCES IN CLINICAL NUTRITION

present in patients with and without potassium depletion. Severe liver dysfunction and ascites had a very high incidence of decreased whole body potassium stores.

ENTERAL FEEDING

Evaluation of a patient's nutritional status and requirements, in the face of organ failure and its subsequent limitations on substrate delivery is crucial in designing the nutritional requirements. The decision to feed enterally in hepatic failure requires a careful assessment of: gastrointestinal (GI) function, the degree of organ failure, electrolyte and fluid requirements, overall nutritional status and nutrient needs, the ability of the liver to tolerate increasing levels of protein and the method of delivery.

ORAL FEEDING

Anorexia[24] and encephalopathy are primary factors in the reduced oral intake. The degree of encephalopathy can affect the patient's ability to choose appropriate foods, and diminish his/her ability to consume nutrients[25]. Clinical staff play a significant role in encouraging oral intake. Calculations of protein and energy consumed, are required to demonstrate the adequacy or inadequacy of oral intake.

The nutrient requirements of patients with liver disease have been described[13]. In general, alterations in the protein source and amount and in the sodium and fluid allocation, increased energy needs, and changes in the consistency of food tend to yield unpalatable diets. The oral delivery of a specific nutrient intake requires the ingenuity of clinical dietitians, the development of realistic meal plans, support from the food service in preparation and presentation of the prescribed diet and monitoring by the clinical staff of the patient's intake.

Investigators have tested a variety of protein sources. Fischer et al.[7] have recommended the use of protein sources high in BCAA. The prescribed protein sources include casein or milk protein to be followed by increasing amounts of poultry and fish. Red meats are limited or excluded because of their content of aromatic amino acids. Rudman et al.[26,27] have suggested limiting the ingestion of ammoniogenic protein sources. Greenberger et al.[28] have proposed the use of vegetable protein diets to provide approximately 40 g of protein from low ammoniogenic amino acid sources.

The practical problem associated with each of these dietary regimens is the limitation on patient long-term compliance due to the limited choices of food. Another problem is the volume of food that must be consumed in order to provide an adequate energy intake. Because the majority of people who have liver disease experience some degree of anorexia, it is often difficult to gain the patient's co-operation in the consumption of sufficient amounts of nutrients. Small meals taken frequently throughout the day are the best tolerated. In our experience it is generally necessary to supplement the patient's caloric intake with a variety of oral feeding modules (i.e. Polycose, Sumacal, Microlipid and safflower oil). In some instances a branched-chain-enriched oral formula, taken as a liquid or pudding, has been used to increase calorie and protein intake. This type of modular feeding, although beneficial during the convalescent stages of recovery, appears inadequate during the acute phase of illness; as the patient's ability to co-operate is often too limited and therapeutic interventions or restrictions may inhibit adequate consumption. Depending on the patient's actual intake of nutrients (as assessed by calculation of protein and energy) as well as the degree of anorexia and/or encephalopathy present, an enteral method of 'forced feeding' may be necessary.

FORMULA DIETS

A variety of commercial formulas afford the opportunity to feed a wide range of patients enterally[29]. Two formulas containing a branched-chain-enriched amino acid profile are currently available specifically for patients with hepatic disease (Tables 12.2 and 12.3). Their clinical benefit requires further evaluation but limited clinical evidence[30,31] is available. Modular feeding in hepatic failure is also possible[32].

Recently Travasorb Hepatic has been used on a limited basis in our institution and further investigation as to its benefit in the treatment of hepatic failure will be conducted. Travenol Hepatic is a defined formula diet which contains approximately 10.6% of total calories as L-amino acids, 77% as carbohydrate and 12% as fat. This formula has an altered amino acid composition consisting of approximately 50% BCAA with a low aromatic and ammoniogenic amino acid content. This formula provides approximately 1.1 cal per ml and 100% of the RDA for vitamins and minerals when given in an amount approximately equal to 2300 calories. At recommended dilution the approximate osmolality is 550 mOsmol/l; osmolarity is 690 mOsmol/kg of water.

In certain clinical conditions our preference is to use formula diets that contain no electrolytes. This is especially true in renal failure or hepato-renal failure. Electrolyte imbalances (hypernatraemia, hypercalcaemia and/or hyperkalaemia) generally necessitate the use of low electrolyte or electrolyte-free solutions. In organ failure, the addition of electrolytes as individual modular components may be beneficial.

In the adaptive and convalescent stages of feeding it may be important to provide adequate mineral and micronutrient supplementation in order to optimize the utilization of nitrogen. The provision of vitamin and minerals in the Travenol product may potentiate improved utilization of its nitrogen source.

Table 12.2 Nutrient comparison of enteral formulas designed for liver disease

	Travasorb Hepatic	Hepatic Aid
Energy distribution		
Protein (% total calories)	10.6 (L-amino acids)	10.3 (L-amino acids)
Carbohydrate (% total calories)	77.4 (glucose oligosaccharides, sucrose)	69.9 (maltodextrins, sucrose)
Fat (% total calories)	12.0 (MCT, sunflower oil)	19.7 (Soy oil, mono- and diglycerides)
Caloric density (kcal/ml)	1.1	1.6
% branched-chain amino acids	50	36
Osmolarity (mOsmol/l)	550	950
Vitamins	RDA (per 2268 kcal)	None
Minerals	(per 2268 kcal)	(per 2240 kcal)
Sodium (mEq)	40	20
Potassium (mEq)	60	Negligible
Calcium (mEq)	40	Negligible
Magnesium (mEq)	32.5	Negligible
Phosphorus (mEq)	9.9	Negligible
Chloride (mEq)	40.5	Negligible
Zinc (mEq)	15	Negligible
Iodine (mEq)	150	Negligible
Manganese (mEq)	2.5	Negligible
Iron (mEq)	18	Negligible
Copper (mEq)	2.0	Negligible

The caloric density of Travasorb Hepatic may need to be increased. Many patients require fluid restriction and, therefore, caloric concentrations greater than 1.1 cal/ml. One method of increasing the caloric content of this formula would be to decrease the amount of free water used to prepare the solution. Although the osmolarity of the solution would then be increased, the rate at which the formula is provided could be decreased.

A second formula designed specifically for liver disease is Hepatic Aid by McGaw Laboratories. This branched-chain-enriched formula provides approximately 10.3% of its total calories as protein, 70% as carbohydrate, and 20% as fat. This product provides approximately 36% of its nitrogen source as BCAAs. The solution has no vitamins, is essentially electrolyte-free and has an osmolarity of 900 mOsmol/l. When used on a longterm basis a vitamin mineral supplement must be added.

The addition of vitamin and mineral supplements is no easy task. Current formulations with adequate nutrient composition are in tablet form and therefore difficult to crush. When used in conjunction with small bore feeding tubes, the potential for tube obstruction is increased. Currently there

Table 12.3 Amino acid composition of selected formulas (g/100 g total amino acids)

	Organ failure		1.0 cal/ml meal replacements		2.0 cal/ml meal replacements		1.0 cal/ml defined formula diets		Protein modules	
	Travasorb Hepatic	Hepatic Aid	Ensure	Isocal	Isocal HCN	Magnacal	Criticare HN	Vivonex HN	Casec	Pro-Mix
Branched-chain amino acids										
Isoleucine	16.7	11.3	4.5	5.0	5.1	5.3	5.3	4.2	5.1	5.0
Leucine	20.0	13.9	8.7	9.2	8.9	8.9	9.1	6.6	8.9	11.1
Valine	13.3	10.5	5.0	6.7	6.4	6.4	6.7	4.6	6.4	5.3
Essential amino acids										
Lysine	11.7	7.7	6.9	6.9	7.6	7.0	7.6	4.9	7.6	9.2
Methionine	1.3	1.2	2.6	2.3	2.5	2.3	4.0	4.6	2.5	3.2
Phenylalanine	1.0	1.2	4.8	4.8	4.9	5.0	4.3	7.1	4.9	3.7
Tryptophan	1.0	0.8	1.2	1.7	1.1	1.7	1.4	1.3	1.1	2.2
Threonine	3.7	5.7	4.1	3.8	3.7	3.9	4.4	4.2	3.7	5.0
Non-essential amino acids										
Alanine	5.7	9.6	3.1	3.1	3.3	2.9	3.2	5.2	3.3	4.8
Arginine	11.7	7.6	3.7	3.6	3.6	3.1	3.6	4.1	3.6	3.0
Asparagine	–	–	–	–	–	NA	–	–	–	–
Asparate Acid	–	–	7.6	7.9	6.7	6.8	7.1	11.0	6.7	10.3
Cystine	–	–	0.5	0.5	0.3	0.3	0.4	–	0.3	3.6
Glutamine	–	–	–	–	–	–	–	18.2	–	–
Glutamic Acid	–	–	22.9	20.8	21.3	20.7	20.7	–	21.3	15.8
Glycine	3.3	11.3	2.0	2.3	1.8	1.9	2.1	9.8	1.7	2.0
Histidine	3.3	3.0	2.4	2.6	2.8	3.5	2.7	2.3	2.8	2.2
Hydroxyproline	–	–	–	–	–	0.2	–	–	–	–
Proline	7.3	10.1	9.8	9.5	10.4	10.4	9.6	6.9	10.4	5.4
Serine	–	6.3	5.5	4.8	4.5	4.9	5.5	4.2	4.5	4.4
Tyrosine	–	–	4.9	4.5	5.1	4.8	2.3	0.8	5.1	3.6

157

are no available liquid preparations providing the RDA of vitamins in combination with the desired minerals and trace elements.

Travasorb Hepatic may be taken orally or given by tube. (Hepatic Aid is also available as an instant pudding.) Both of these organ failure solutions require reconstitution with water. The actual mixing of each product is similar, however, the Travasorb Hepatic appears to remain in solution longer. This may be of importance when providing the formula via small bore feeding tubes.

The ability of 40 g of Travasorb Hepatic to normalize amino acids in liver patients is shown in Table 12.4. Use of a standard crystalline amino acid solution providing 70 g protein (Aminosyn) greatly alters this important amino acid profile.

Once a patient is able to tolerate 40 g of protein and his or her fluid and electrolyte status is stable, consideration is given to changing the enteral prescription to a casein-based replacement formula (i.e. Magnacal, Isocal, Ensure). These formulations are less costly and as the patient's condition improves, are as well tolerated as the branched-chain-enriched products.

CYSTEINE AND TYROSINE DEFICIENCY

Recently, Rudman et al.[12] have reported failure of four parenterally fed cirrhotic patients to achieve nitrogen balance when given solutions containing limited concentrations of cystine and tyrosine. Plasma levels of these two amino acids were less than 30% of normal fasting levels. In two patients supplemented with oral cystine (0.3 g TID) and tyrosine (0.3 g TID), plasma levels returned to normal and nitrogen balance became positive. It was concluded therefore that hepatic dysfunction reduced the liver's ability to synthesize cystine from methionine and tyrosine from phenylalanine. Therefore, nutritional repletion (based upon such indices as nitrogen balance and albumin synthesis) was inhibited by a deficiency of these two amino acids. It is of interest to note that Travasorb Hepatic and Hepatic Aid are devoid of cystine and tyrosine (Table 12.4). However rather than providing tyrosine to all hepatic failure patients most of whom have elevated tyrosine levels, it would be more reasonable to identify the minority of hepatic failure patients who require these amino acids. The measurement of urea nitrogen production and of nitrogen balance will identify poor utilization of intake in this group.

TUBE FEEDING

In many instances, the use of small silastic feeding tubes is necessary. The benefits of tube feeding and the procedures for insuring its safety are well documented[33,34]. During initial stabilization of the acutely ill patient with coagulopathies and potential nasoenteric bleeding sites the various

Table 12.4 Plasma amino acid profile following administration of Travasorb Hepatic

Amino acid	Baseline concentration* (µmol/ml)	Concentration day 7† (µmol/ml)	Concentration day 13‡ (µmol/ml)
Leucine	0.092	0.087	0.090
Isoleucine	0.005	0.050	0.063
Valine	0.235	0.186	0.210
Taurine	<0.005	0.020	0.061
Aspartate	<0.005	<0.005	0.012
Threonine	0.123	0.136	0.203
Serine	0.088	0.063	0.054
Asparagine	0.027	0.027	0.025
Glutamate	0.012	0.021	0.007
Glutamine	0.310	0.367	0.450
Glycine	0.304	0.211	0.238
Alanine	0.183	0.166	0.160
Tyrosine	0.063	0.059	0.040
Phenylalanine	0.152	0.075	0.054
Methionine	0.044	0.018	0.015

*Aminosyn 3.5% (Abbott) infusing at a rate of 80 ml/h/24° (70 g protein)

†Half-strength Travasorb Hepatic (Travenol) infusing at a rate of 65 ml/h/24° via nasogastric feeding tube (23.4 g protein)

‡Full-strength Travasorb Hepatic (Travenol) infusing at a rate of 65 ml/h/24° via nasogastric feeding tube (47 g protein)

techniques for tube insertion must be evaluated carefully. If a stylet is used to aid in a passage of the tube, it is important to insure that the tip of the stylet is locked into the distal tip of the feeding tube prior to its passage. Otherwise, during insertion, the stylet tip may cause damage.

Fluid restriction, gastrointestinal function, protein intolerance, alterations in liver function all dictate the rate, concentration and adjustments in the tube feeding prescription. Once a patient can take oral fluids, tube feeding is stopped.

SUMMARY

The design and selection of enteral formulas and diet prescriptions for patients with hepatic disease should be based on the degree of liver dysfunction present, the presence and degree of encephalopathy, the presence and degree of ascites and the type of electrolyte and acid base imbalances associated with the different therapeutic modalities used in treating these patients.

Individuals with significant liver dysfunction and encephalopathy may benefit from solutions containing a high concentration of BCAAs (Table 12.2). These will return plasma amino acid levels to normal (Table 12.4), with a dietary intake of 40 g of amino acids, a decrease in protein breakdown and therefore a decrease in ammonia production.

In the presence of hyperaldosteronism, oedema, and ascites, a formula low in sodium and with a moderate amount of potassium may be needed. In addition, a high density formula greater than 1.5 cal/ml may be of benefit to patients for whom fluid restriction is required due to severe hyponatraemia and significant water retention[35].

The enteral route is the preferred method of feeding when the gastro-intestinal tract is functional. Clinical, therapeutic and practical consider-ations must be evaluated when designing and implementing enteral nutrition prescriptions. Although branched-chain-enriched enteral formulas and/or diets high in BCAA and/or low in ammonia content show promise in the treatment of hepatic dysfunction, further prospective, randomized studies are needed.

ACKNOWLEDGEMENTS

Supported in part by grant GM-24206 awarded by the National Institute of General Medical Sciences, DHHS

References

1 McMurray, W. C. (1977). *Essentials of Human Metabolism: Nitrogen Metabolism*, pp. 210–276. (New York: Harper & Row)
2 Khatra, B. S., Chawla, R. K. and Sewell, L. W. (1977). Distribution of branched-chain and keto acid dehydrogenase in primate tissues. *J. Clin. Invest.*, **59**, 558
3 Adibi, S. A. (1976). Metabolism of branched chain amino acids in altered nutrition. *Metabolism*, **25**, 1287
4 Blackburn, G. L., Moldawer, L. L., Usui, S. *et al.* (1979). Branched chain amino acid administration and metabolism during starvation, injury, and infection. *Surgery*, **86**, 307
5 Cuthbertson, D. P. (1979). The metabolic response to injury and its nutritional implications: retrospect and prospect. *J. Parent. Ent. Nutr.*, **3**, 108
6 Freund, H., Yoshimura, N. and Fisher, Y. (1980). The role of alanine in the nitrogen conserving quality of the branched-chain amino acids in the post-injury state. *J. Surg. Res.*, **29**, 23
7 Fischer, J. E., Rosen, H. M. and Ebeid, A. M. (1976). The effect of normalization of plasma amino acids on hepatic encephalopathy in man. *Surgery*, **80**, 77
8 Rosen, H. M., Yoshimura, N., Hodgeman, J. M. and Fischer, J. E. (1977). Plasma amino acid patterns in encephalopathy of differing etiology. *Gastroenterology*, **72**, 483
9 Rudman, D., DiFuleo, T. J. and Galambos, J. T. (1973). Maximal rates of excretion of urea in normal and cirrhotic subjects. *J. Clin. Invest.*, **52**, 2441
10 Lockwood, A. H., McDonald, J. M. and Reiman, R. E. (1979). The dynamics of ammonia metabolism in man. Effects of liver disease and hyperammonemia. *Am. Soc. Clin. Invest.*, **63**, 449
11 Herlong, H. F., Maddrey, W. C. and Walser, M. (1980). The use of ornithine salts of branched chain keto acids in portal systemic encephalopathy. *Ann. Intern. Med.*, **93**, 545
12 Rudman, D., Kutner, M., Ansley, J. *et al.* (1981). Hypotyrosinemia, hypocystinemia, and failure to retain nitrogen during total parenteral nutrition of cirrhotic patients. *Gastro-enterology*, **81**, 1025
13 Mezey, E. (1978). Liver disease and nutrition. *Gastroenterology*, **74**, 770
14 Leevy, C. M., Baker, H. and Tenhorne, W. (1965). B-complex vitamins in liver disease of the alcoholic. *Am. J. Clin. Nutr.*, **16**, 339
15 Halsted, C. H., Robles, E. A. *et al.* (1973). Intestinal malabsorption in folate-deficient alcoholics. *Gastroenterology*, **64**, 526

16 Morgan, A. G., Kellcher, J., Walker, B. E. *et al.* (1976). Nutrition in cryptogenic cirrhosis and chronic aggressive hepatitis. *Gut*, **17**, 113
17 Baraona, E., Pirola, R. and Lieber, C. (1976). Small intestinal damage and changes in cell population produced by ethanol ingestion in the rat. *Gastroenterology*, **66**, 226
18 Marin, G. A., Clark, M. L. and Senior, J. R. (1969). Studies of malabsorption occurring in patients with Laennec's cirrhosis. *Gastroenterology*, **56**, 727
19 Fast, B. B., Wolfe, S. J., Stormont, J. M. *et al.* (1959). Fat absorption in alcoholics with cirrhosis. *Gastroenterology*, **37**, 321
20 Summerskill, W. H. and Moertel, C. G. (1962). Malabsorption syndrome associated with anicteric liver disease. *Gastroenterology*, **42**, 380
21 Losowsky, M. S. and Scott, B. B. (1973). Ascites and oedema in liver disease. *Br. Med. J.*, **3**, 336
22 Frankes, J. T. (1980). Physiologic considerations in the medical management of ascites. *Arch. Int. Med.*, **140**, 620
23 Soler, N. G., Jain, S., James, H. *et al.* (1976). Potassium status of patients with cirrhosis. *Gut*, **17**, 152
24 Morgan, A. G., Kelleher, J., Walker, B. E. *et al.* (1976). Nutrition in cryptogenic cirrhosis and chronic aggressive hepatitis. *Gut*, **17**, 133
25 Hurlow, A. (1977). Diet in the treatment of liver disease. *J. Human Nutr.*, **31**, 105
26 Rudman, D., Smith, R. A., Salan, A. A. *et al.* (1973). Ammonia content of food. *Am. J. Clin. Nutr.*, **27**, 487
27 Rudman, D., Galambos, J. T., Smith, R. B. *et al.* (1973). Comparison of the effect of various amino acids upon the blood ammonia concentration of patients with liver disease. *Am. J. Clin. Nutr.*, **26**, 916
28 Greenberger, M. J., Carley, J., Schenker, S. *et al.* (1977). Effect of vegetable and animal protein diets in chronic hepatic encephalopathy. *Dig. Dis.*, **22**, 845
29 Bothe, A., Wade, J. E. and Blackburn, G. L. (1981). Enteral nutrition, an overview. In G. Hill (ed.). *Surgical Nutrition.* pp. 76–103. (London: Churchill Livingstone)
30 Freund, H., Yoshimura, N. and Fischer, J. E. (1979). Chronic hepatic encephalopathy: long term therapy with a branched chain amino acid enriched elemental diet. *J. Am. Med. Assoc.*, **242**, 347
31 Keohane, D. D., Attrill, H., Grimble, G. *et al.* (1982). Nutritional support of malnourished encephalopathic cirrhotic patients using a specially formulated enteral diet. *Gastroenterology*, **82**, 1098 (abstract)
32 Smith, J., Horowitz, J., Henderson, J. M. and Heymsfield, J. (1982). Enteral hyperalimentation in undernourished patients with cirrhosis and ascites. *Am. J. Clin. Nutr.*, **35**, 56
33 Kaminski, M. V. (1976). Enteral hyperalimentation. *Surg. Gynec. Obstet.*, **143**, 12
34 Griggs, B. A. and Hopper, M. C. (1979). Nasogastric tube feeding. *Am. J. Nursing,* **79**, 481
35 Conn, H. O. (1972). The national management of ascites. In H. Popper and F. Schanffner (eds.). *Progress in Liver Disease.* pp. 269–288. (New York: Grune & Stratton)

13
Nutritional support in the thermally injured patient

D. E. BEESINGER, K. GALLAGHER AND S. MANNING

THE HYPERMETABOLIC STATE

The burn patient has become the classic experimental model for the study of hypermetabolism and the catabolic state. Multiple investigators have documented increases of 50–100% over normal basal metabolic rates in patients with major cutaneous injuries, a situation that indicates special concern for the provision of adequate nutritional support.

Resting metabolic rate increases in a curvilinear fashion with increasing burn size. Rates are near normal for injuries involving less than 10% of total body surface area (TBSA) but reach a maximum of up to twice normal when the extent of the burn exceeds 40% TBSA[1]. This hypermetabolic response is marked by an increase in oxygen consumption, increased body temperature, protein wasting, and weight loss. Respiratory rate, carbon dioxide production, cardiac output, and wound blood flow increase at the same time. Catabolism of protein, fat and glycogen are increased similarly to supply an elevated glucose flow to the wound.

The onset and duration of this hypermetabolic state in burn patients is of considerable interest. Characteristically, oxygen consumption and energy expenditure increase quickly during the first 24 h after the burn and reach a maximum level after several days. The patient then maintains an elevated level until the wounds are almost healed[2].

Energy needs can be reduced early in the hospital course by surgical removal of eschar and skin coverage of the underlying soft tissue defect[3]. Autograft application is the preferred cover after early tangential or full thickness excision of the burn wound, but porcine xenograft or cadaver homograft also effectively reduces fluid loss and heat dissipation, thereby decreasing energy utilization.

163

The ambient temperature to which the patient is exposed also affects metabolic expenditure. A decrease of body metabolic rate (BMR) of approximately 10% is seen when patients are kept in a warm environment. When burn patients are allowed to control the ambient temperature for their individual comfort, they select an environmental temperature of between 28 and 35 °C[4].

Many theories are proposed for the aetiology of the hypermetabolic state in thermally injured patients. Body heat is lost at an increased rate due to radiation across the injured skin, and energy is required for the evaporation of water across an open wound[2]. Evaporative water loss can exceed $100 \, ml/m^2$ BSA per hour, and the heat of evaporation for water is $0.576 \, kcal/ml$. Energy for this process must be supplied by the patient when skin temperature is less than environmental temperature so that a large expenditure is required[5].

When patients are exposed to high environmental temperatures, however, BMR remains elevated, so evaporative cooling does not adequately explain this phenomenon. Multiple factors, many poorly understood, must be present to produce the hypermetabolic state of burn patients. It is postulated that changes in the body's hormonal milieu are responsible for elevated oxygen consumption since levels of catecholamines, corticosteroids, and glucagon are increased[3]. Debate in this area, however, centres on whether these hormonal changes are primary or secondary[6-8].

Despite debate, it is clear that the burn patient requires special nutritional supplementation to prevent weight loss and promote wound healing. The total number of calories required daily can be calculated using Curreri's formula:

$$kcal = (25 \times body \, wt.) + (40 \times \% \, TBSA \, burn)$$

An alternative formula has been advocated by Pruitt:

$$kcal = 2000 \, kcal + 2200 \, kcal/m^2 \, TBSA$$

Total daily protein requirement in grams can be obtained by multiplying the daily calorie need by 20 or 25%, depending upon the physician's preference, and dividing by 4, the approximate number of calories per gram of protein. An alternative method is the determination of protein catabolic rate by urea mass balance.

NUTRITIONAL ASSESSMENT

Periodic nutritional assessment is essential for the burn patient. The burn team should include a dietitian who understands the needs of these patients and has the resources available to assure that adequate calories in the correct form are given. Daily weights are required though often obtained with difficulty due to discomfort (Table 13.1).

Table 13.1 Curreri and Pruitt have derived formulae for determination of total calorie needs for the burn patient based on body weight and/or percentage of total surface area burned. Protein requirement is 20−25% of calculated calorie needs.

Total calories = (25 × body wt.) + (40 × % TBSA burn)

or

(g) = 2000 kcal + 2200 kcal/m² TBSA

$$\text{Protein (9)} = \frac{(0.25) \times (\text{total calories})}{4}$$

On the fifth day after injury and at 10 day intervals thereafter, laboratory determinations are made of blood urea nitrogen (BUN), serum albumin, transferrin and total lymphocyte count. Simultaneously, urine is collected for creatinine and urea nitrogen measurements. Skin testing is also performed routinely, with most major burn patients being anergic until most of the wounds are covered despite adequate nutritional support.

Serum albumin and transferrin correlate well with nutritional status in burn patients but total lymphocyte count is of less value. This determination usually correlates well with visceral protein in stable patients, but is affected by sepsis, blood administration, and intravascular volume shifts, so that it becomes the poorest indicator of nutritional status.

Nitrogen balance has been used in the past to reflect nutritional status. An inherent problem exists, however, in burn units where an aggressive surgical approach to the wound is used. Every 3−4 days, the patient is taken to the operating room for excision and wound coverage. Large volumes of blood and blood products are transfused, and the patient is without nutritional intake for about 12 h. Under these circumstances, daily calorie counts may frequently be inaccurate, further negating the accuracy of nitrogen balance.

A second method of nutritional assessment has been used recently in renal failure patients and may have application for the burn victim[9]. Nitrogen mass balance or urea kinetics is a means of determining net protein catabolism that is not dependent upon changes in intravascular volume, nitrogen intake, or the accuracy of calorie counts.

There is a relationship between the net rate of protein catabolism (PCR) and the net rate at which urea nitrogen will be generated (G). For each gram of protein catabolized and not matched by anabolic processes (net catabolism), 0.15 g of urea nitrogen will be produced which will remain in body water or be excreted (net urea generation rate). This level of net urea nitrogen production, however, occurs only when net protein catabolism exceeds 11 g/day, the obligatory amount that must be supplied to make up for nitrogen losses in stool and production of other nitrogen substances. This linear relationship above 11 g/day allows accurate estimation of the net rates of protein catabolism from the net rate of urea generation, which is computed from BUN values.

Mass balance principles are utilized in this determination based on the fact that whatever urea is produced from protein catabolism will be either excreted in the urine or distributed in body water and reflected by BUN. Sargent has developed a formula which allows computation of protein catabolic rate when urine and blood urea concentrations are known and which is not dependent upon knowledge of protein intake:

$$KrU = \frac{Uu}{BUN} \times \frac{Uv}{t}$$

where KrU is renal urea clearance (mg/ml), Uv is urinary volume (ml), and t is time interval of collection (min).

Urea generation rate (GU) is then determined by:

$$GU = KrU \times BUN$$

The constant relationship of urea genesis to protein catabolism then allows computation of protein catabolic rate (PCR):

$$PCR = (GU + 1.2) \times (9.35)$$

This method of determination of level of PCR has been used extensively in renal failure patients. It is valuable in diagnosing the severity of catabolic states and depicting the daily protein needs of the patient since intake need only exceed protein catabolism plus protein lost across the wound in order to produce positive nitrogen balance. In addition, knowledge of PCR allows discernment of the point at which protein catabolism starts to decrease so that attention to nutritional supplementation can be relaxed.

A study is currently under way to test the method in catabolic burn patients. Though results are preliminary, it appears that urea mass balance may be an easier and more logical method than nitrogen balance studies to follow the nutritional needs of these patients.

METHODS OF DELIVERY

After the nutritional needs of the burn patient have been determined, the most appropriate means of delivery of nutritional support must be selected. Since protein/calorie demands must be met, the physician must resort to whatever means are necessary to supply adequate nutritional supplementation.

Burn injuries of less than 30% TBSA rarely need nutritional supplements since their metabolic demands are less significant and the patient is usually able to consume adequate food when placed on a high calorie, high protein diet. Oral vitamins should be supplied, however, and zinc sulphate and ascorbic acid should be given since these are necessary for wound healing. Daily weights and calorie counts should be taken and when deficits are identified, further nutritional support must be added.

Exceptions to this generalization include patients who are of advanced age, who have catabolic disease processes prior to the accident, and who are malnourished before admission. Special consideration must also be given to those who sustain burns to the hands making self-feeding difficult and those with burns about the mouth who experience excessive pain with eating. Patients with inhalation injuries, even in the absence of skin involvement, also generally need supplementation.

When it is obvious that adequate protein and calories cannot be taken orally, tube feeding is required. Most patients with burns greater than 40% TBSA fall into this category. Tube feedings should be instituted as soon as cardiovascular haemodynamics are stable, usually 48–72 h after injury.

As soon as the nasogastric tube is removed, an 8 French feeding tube is placed through the nose into the stomach. Optimally, the tip of the tube should rest in the duodenum, but this may be difficult even with the newer, weighted tubes. Feeding should commence despite the inability to pass the tube beyond the stomach.

Continuous tube feeding has been found preferable to intermittent, bolus feeding[10]. Abdominal discomfort is less and patients are able to receive the calculated daily requirements more rapidly. In addition, the incidence of aspiration is less and patients are more likely to eat at meal times. A volumetric pump is also essential for continuous feedings since a gravity drip system is too imprecise and obstruction of the feeding tube can occur.

The selected feeding formula should be started at half-strength on the first day of feeding and given at a rate of 40 ml/h for 24 h. The tube is aspirated to check gastric residue every 4 h and bowel sounds are monitored simultaneously.

After satisfactory acceptance of the first day's feeding, the feeding formula is infused at full strength without altering the infusion rate. Again, gastric residue and bowel sounds are monitored. On day three, the infusion rate is increased to 660 ml/h, and thereafter the rate is increased daily by 20 ml/h, until the total intake meets the calculated requirements.

Tube feedings are continued throughout the hospital course up to the time that wounds are covered. At this time, metabolic demands are nearly normal and adequate oral intake is assured. If the patient has a non-abdominal surgical procedure, tube feedings should be restarted within 12 h postoperatively. Generally, the full-strength formulae will be well tolerated but the rate of infusion should be decreased to 40 ml/h initially.

The obvious problem with such a plan of nutritional supplementation is the changes that must be made on days of operation. In a burn unit where early excision and grafting of wounds is practised, there will, of necessity, be days when the needs are not fully met. This problem can be minimized, however, by the strict adherence to a rigid protocol for tube feeding.

Many feeding preparations are now available for the burn surgeon each of which is supposed to be the perfected formula. Ideally, the formula should supply a high calorie to volume ratio and should have low osmolality to avoid diarrhoea. Consideration must also be given to the relative cost of the preparation, and it should be easily prepared for use.

Preparations are available which provide 2 calories per ml, a factor of importance since the burn patient often has an expanded intravascular volume and limited myocardial reserve secondary to pre-existing heart disease or myocardial depression. It then becomes imperative to be able to deliver a large number of calories in a small volume of fluid.

The well-known problem with solutions with high calorie/volume ratios is the high osmolality of the preparation. One must choose a high caloric density product which has an osmolality of less than 600 mosm/kg of water to avoid disabling diarrhoea.

Special consideration must also be given to the constituents of the formula selected. Disagreement exists about the amount of protein that the burn patient should receive but the percentage of intake coming from protein is as important as the total number of calories supplied[11]. When inadequate protein supplementation is supplied, nitrogen balance cannot be achieved with perhaps decreased immunologic function, lowered serum protein values, greater numbers of septic episodes, and even greater mortality.

The percentage of total calories which must be supplied as protein is uncertain. It is apparent, however, that adequate dietary supplementation should include at least 20% of calories from protein, and it is possible that an even greater percentage is required. Contrary to previous reports, a total calorie to nitrogen ratio of at least 125:1 is necessary.

The stress of thermal injury combined with inadequate dietary fat administration can lead to essential fatty acid deficiency[12]. Abnormal levels of red blood cell membrane fatty acids, specifically linoleic, arachidonic, and descosahexanoic, polyunsaturated fatty acids which cannot be synthesized in the body have been documented, confirming the need for supplementation in the burn patient.

Administration of fat emulsion in these patients has been shown to prevent fatty acid deficiency and to correct it in patients who have been maintained on carbohydrate-rich, fat-free, parenteral nutrition. As with protein, however, it is not known how much fat emulsion should be administered. This information has been difficult to obtain since thermal injury stimulates a neuroendocrine environment which favours fatty acid utilization, thereby allowing rapid clearance of infusions.

Fat emulsion infusion rates of 3–5 g fat/kg body weight every 24 h appear to be safe and easily cleared in burn patients. A shift in respiratory quotient toward that of fat oxidation suggests that this amount of fat is utilized effectively. Intravenous fat emulsion is a satisfactory isotonic, high caloric, non-carbohydrate energy source for burn patients. Since the effect of excessively large volumes of fat emulsion on cellular membranes has not been adequately examined, it is advisable that no more than 35% of the total daily calories be supplied as fat, since fat may not be as effective in bringing about nitrogen sparing as glucose.

Equally important as the determination of how many and what kind of calories to administer is the way in which they should be delivered. Any protocol must identify which patients will receive tube feedings, the rates of administration, and the methods of monitoring progress. Such a protocol is now in use at Hermann Hospital, Houston, Texas, a 12 bed acute care burn centre with a yearly admission rate of about 200 adult victims of thermal, electrical and inhalation injuries.

Before January, 1981, there was no formal protocol for enteral nutritional support, and multiple feeding preparations were used, particularly those with a low calorie content (Ensure, Isocal, Precision, and Sustacal). A protocol was later developed which outlined the nutritional support for all patients with burns greater than 30% TBSA and described the plan of dilution and the daily volume of tube feedings. In addition, one preparation was selected which provided 2 cal/ml at an acceptable osmolality and a calorie to nitrogen ratio of about 125:1 (Travasorb MCT). The efficacy of nutritional management before and after institution of the protocol was evaluated by a retrospective review of 55 patients who required enteral tube feedings for nutritional support.

All adult admissions were divided into two groups. Group I were those treated before institution of the protocol and Group II after the introduction of Travasorb. They were then subdivided into three groups depending on size of burn, 10–40%, 41–70%, and greater than 70%. Patients were well-matched for age and burn size in pre- and post-protocol groups. Patients who required intravenous hyperalimentation for extended periods of time were excluded.

The daily caloric need was determined for each patient by Curreri's formula, kcal = $(25 \times$ kg body wt.$) + (40 \times$ % TBSA$)$. The protein need was determined as 20% of total calorie requirement. Comparison of Group I and Group II within each subgroup was then made to determine the effect of the protocol on the percentage of days when at least 90% of the predicted caloric and protein needs were supplied by oral and tube feedings.

Table 13.2 Achievement of calorie and protein requirements before and after protocol

	10–40%	Burn size 41–70%	≥71%
Group I (n = 22)			
Calorie ≥90%	63.6%	25.8%	27.8%
Protein ≥90%	43.9%	21.0%	21.4%
Group II (n = 39)			
Calorie ≥90%	68.7%	59.0%	71.7%
Protein ≥90%	49.2%	44.0%	36.8%

As expected for burns of 10–40%, total calorie needs were provided satisfactorily with either regimen. 63.6% in Group I and 68.7% in Group II. With injuries however of 41–70%, the picture was different. In Group I, adequate requirements were met only 25% of the time studied and this was increased to 59% in Group II. In burns greater than 70%, the goal was met on only 27.8% of the time studied before the protocol but in 71.7% of the time after the introduction of the strict protocol.

90% of daily protein requirement was met in 43.9% and 49.2%, respectively, of Group I and II patients who had 10–40% burns. In burns of 41–70%, 21% received >90% of protein need before protocol and 44% did so after protocol. In larger burns, >70%, an increase from 21.4% to 36.8% was seen (Table 13.2) in the study.

References

1 Boswick, J. A. Jr., Thompson, J. D. and Kershner, C. J. (1977). Critical care of the burned patient. *Anesthesiology*, **47**, 164
2 Liljedahl, S.-O. (1980). Treatment of the hypercatabolic state in burns. *Ann. Chir. Gynaecol.*, **69**, 191
3 Kauste, A. (1980). Parenteral and enteral nutrition of the thermally injured patient. *Ann. Chir. Gynaecol.*, **69**, 197
4 Pruitt, B. A. (1979). Metabolic changes and nutrition in burn patients. *Ann. Chir. Plast.*, **1**, 21
5 Barr, P. O., Brike, G., Liljedahl, S.-O. and Plantin, L. O. (1968). Oxygen consumption and water loss during treatment of burns with warm dry air. *Lancet*, **1**, 164
6 Wilmore, D. W. (1981). Glucose metabolism following severe injury. *J. Trauma*, **21**, 705
7 Blackburn, G. (1981). Metabolism and nutritional support. *J. Trauma*, **21**, 707
8 Curreri, P. W. (1981). Future areas of research in the metabolism of burned patients. *J. Trauma*, **21**, 711
9 Sargent, J. A. (1982). Urea mass balance: nutrition and treatment of the acutely ill patient. *Nutr. Support Services*, **2**, 33
10 Hiebert, J. M. *et al.* (1981). Comparison of continuous vs. intermittent tube feedings in adult burn patients. *J. Parent. Enteral Nutr.*, **5**, 73
11 Alexander, J. W. *et al.* (1980). Beneficial effects of aggressive protein feeding in severely burned children. *Ann. Surg.*, **192**, 505
12 Wilmore, D. W., Moylan, J. A., Helmkamp, G. M. and Pruitt, B. A. Jr. (1973). Clinical evaluation of a 10% intravenous fat emulsion for parenteral nutrition in thermally injured patients. *Ann. Surg.*, **178**, 503

14
Breast milk in neonatal care

H. C. BØRRESEN

INTRODUCTION

Surveys in Germany towards the end of the 19th century showed that survival in the first year of life was largely dependent on successful breast feeding[17,18]. It became widely recognized that breast milk protects infants against gastro-intestinal and respiratory infections[17]. Only the last two decades have produced evidence which can replace speculations as to why this is so[2,15,16,31]. Mammals suckling their young evidently provide more than ideal nourishment. Indeed primates may need breast milk and close physical contact with the mother for several years[12]. The human infant being no exception is probably entitled to enjoy breast feeding and 'single child privileges' for 4 years or more lest profound psychological needs be frustrated and physiological requirements be inadequately met[20]. Furthermore, in prehistory natural breast feeding may have mediated ecological equilibrium between food resources and population growth.

HUMAN MILK AS NOURISHMENT

Breast milk is singularly suited to sustain balanced accumulation of the constituents of the lean body mass. In primates the rapid growth and maturation of the central nervous system entail rigorous nutritional requirements[11,38]. The protein of human milk supplies tyrosine while the phenylalanine content is so low that even infants with hyperphenylalaninaemia tolerate breast feeding to cover up to 80% of their energy need[21]. The cysteine content is high enough to ensure adequate intake of sulphur amino acids without excessive methionine load. Breast milk also contains free amino acids, notably glutamic acid and the amino acid derivatives taurine[13] and carnitine[8]. The infant's ability to synthesize the latter two compounds is low[9,13]. Exogenous taurine may be needed by the growing brain as well as skeletal and heart muscle[13]. The human neonate conjugates bile acids preferentially with taurine but resorts to glycine if the diet is taurine-deficient[13]. It

171

remains to be clarified whether this change in bile acid physiology impairs fat absorption or aggravates diarrhoea in the sick neonate[13,32,33].

Exogenous L-carnitine is instrumental for successful transition from intra-uterine glucose oxidation to postnatal utilization of fat to cover 40–50% of the energy need[9,28]. Carnitine deficiency limits the ability to increase fatty acid oxidation and ketone body production necessary to cope with glucose shortage[28].

Breast milk supplies fatty acids essential for the assembly of myelin and other cell membranes and also for prostaglandin synthesis. Human milk is a rich source of arachidonic acid (C22:4ω6) which is more effective than its precursor linoleic acid (C18:2ω6) in preventing deficiency of essential fatty acids[11]. The fat of breast milk is easily digested. It is absorbed efficiently even at the low concentration of cholic acid conjugates (about 1 mmol/l) in the neonatal gut[14]. The bile acid activated lipase characteristic of primate milk appears to be responsible[14,37]. This fits nicely with the reduced fat absorption following heat treatment of the milk[1,37]. The Norwegian paediatrician Axel Johannessen mentioned the existence of lipase and other enzymes in breast milk as early as in 1902[17].

Trace elements are more completely absorbed from fresh human milk than from cow's milk[35,36]. The mechanism may be that artificial feeding is assoc-iated with higher concentrations of unabsorbed fatty acids in the intestinal lumen where they form complexes with calcium and trace metal ions.

The ample volume of water provided by natural breast feeding covers the high water requirement in hot climates and counteracts hypernatraemia during febrile illness[29].

BREAST FEEDING AND GROWTH

Breast-fed infants tend to grow without accretion of unnecessary fat. Frequent and brief suckling on demand establishes mechanisms which adjust the mother's milk production to the particular needs of her infant[15,16]. Yet, underfeeding at the breast does occur and can even lead to hypernatraemia[26]. Possibly these infants are more dependent than others on upright posture and close contact with the mother's moving body to stay alert and signal their needs. This is precisely the way orangutans[12] and foraging human bands 'in the wild' carry their young[20].

Furthermore, breast milk contains epidermal growth factor (EGF). This is a mitogenic polypeptide which is resistant to acid and trypsin[10]. EGF may stimulate intestinal growth. However, it is also absorbed and exerts systemic effects, notably growth and maturation of pulmonary epithelium[10].

IMMUNOLOGY AND MICROBIOLOGY

The secretory immunoglobulin A (s-IgA) of breast milk adheres to the infant's intestinal mucosa and repels pathogenic micro-organisms and possibly allergenic macromolecules[15,31]. The neonate's own production of s-IgA is low[31]. The plasma cells in breast milk stem from lymphocytes in the mother's intestinal mucosa where they are exposed to ingested antigens. In the milk they manufacture s-IgA with specificity against these antigens. This 'entero-mammary plasma-cell cycle' protects the infant against bacteria and viruses which have previously entered the maternal gut. Many other mechanisms contribute to the selective bacteriostatic effect of breast milk, e.g. the binding of iron to lactoferrin which deprives *E. coli* of this trace metal indispensable for growth[16]. Breast feeding ensures dominance of *Lactobacillus bifidus*[15] which produces lactic and acetic acids in the infant's intestinal lumen. The pH of the stools is about 5.5[16]. This low pH is hostile to gram-negative pathogens. One of the clinical benefits appears to be that colostrum and breast milk protect neonates against necrotizing enterocolitis[2,5]. The low morbidity associated with breast feeding is a combined effect of good nutrition, low exposure to pathogens and specific protection against infection[15-18,31].

BREAST MILK AND SHORT GUT SYNDROME

In these infants enteral feeding often results in diarrhoea necessitating resumption of intravenous infusions. Alternating periods of intravenous and attempted enteral feeding used to be the rule for 3–6 months. Prior to 1979 the length of hospital stay among our patients was comparable to the range reported by Tepas *et al.*[30]. When enteral feeding ultimately becomes better tolerated it often happens suddenly and for unknown reasons. However, the use of fresh human milk instead of artificial formulas promises to alleviate diarrhoea and permit earlier discharge from hospital. Børresen and Knutrud reported their first case at the ESPEN meeting in Stockholm in 1979. Feeding at the mother's breast was begun 2.5 weeks after intestinal resection due to midgut volvulus. The remaining 10 cm jejunum had been anastomosed to 30 cm terminal ileum leaving the ileocaecal valve intact. No diarrhoea occurred and weight increased normally within one week. The patient was sent home when 4 weeks old.

Breast milk probably prevents overgrowth of bacteria which deconjugate bile acids. The resultant amelioration of diarrhoea may be analogous to the effect of non-absorbable antibiotics given by mouth[33]. Furthermore, unconjugated bile acids are precipitated at the low colonic pH associated with breast feeding[32], and their toxic affect on mucosal cells[33] may therefore be attenuated.

HUMAN MILK AND PREMATURE INFANTS

The preterm infant's need for the nutritional, growth-promoting and bacteriostatic benefits of (unheated) breast milk is unquestionable. However, human milk cannot provide enough calcium, magnesium and phosphate for mineralization of bone corresponding to intrauterine growth. McCance and Widdowson called attention in the early sixties to the paucity of calcium, magnesium and phosphate in breast milk[22]. Phosphate was shown to be the factor limiting skeletal mineralization and possibly the growth of soft tissues[34]. Taking advantage of these data as reviewed by Shaw in 1973[27], Børresen and Knutrud reported in 1976 that supplementation of pasteurized human milk with phosphate and calcium had caused weight gain in a 1300 g premature infant and the continuation of the intrauterine growth rate without clinical evidence of rickets[4]. Guided by daily analyses of electrolytes in urine it was found desirable to add potassium, sodium and chloride also[4,6,7]. Zinc and other trace elements were given as well. Small premature infants should not be given unsupplemented breast milk since the predictable outcome is rickets which may even culminate in disastrous hypophosphat-aemia[25].

BREAST FEEDING AND FERTILITY

The analysis by Knodel and van de Walle of data from Germany in the 19th century suggests that breast feeding had little independent effect on marital fertility[18]. The probable reason was that the average duration of lactation did not exceed 6 months. Recent studies of populations not using contraception show that prolonged breast feeding can maintain postpartum amenorrhoea for more than 2 years with a corresponding increase of child spacing[3,23].

In the nomadic foraging !Kung people in the Kalahari-region of Africa the contraceptive effect of breast feeding lasts longer and is strikingly more reliable[19,20] than in urban and agricultural societies. The nomadic !Kung women carry their infants during daily life. Such physical contact allows frequent nursing bouts during the day and several times each night. This pattern continues with gradual reduction of suckling frequency, for almost 4 years on the average. Weaning occurs when the next pregnancy is evident. Having only one small child at a time, the !Kung mother enjoys a relaxed life with physical closeness and a warm emotional relationship to her children. Interviews with 47 old Kung women disclosed that their average number of live births was only 4.87[20]. Such low natural fertility may be characteristic of primates 'in the wild'. In Kalimantan (Borneo) infant orangutans cling to their mothers for 4 years, and are not weaned until the age of seven[12]. Child spacing among these animals is probably about 8 years. The fact that most orangutans reach mature age testifies to the adaptive value of these child rearing practices among primates. When the !Kung bands settle down to a more 'advanced' sedentary life, the tragedies occur in a very orderly sequence. The availability to infants of easily digestible cow's milk or food made from milled grains leads to earlier weaning. Postpartum amenorrhoea shortens and so does the child spacing. The result is baby boom, exhausted mothers and unhappy children[20]. Our own ancestors went through this transition in the millennia following the last ice age. Since infant mortality did not keep pace with the increasing fertility the present population explosion was inevitable.

THE COST OF CIVILIZATION

The shortened child spacing has tied women to reproduction, isolated them from production activities and severely limited their mobility. Industrialization and urban life have drawn the males away from the home and thus reinforced the polarization of sex roles. Ideological superstructures have been adopted to reinforce this life pattern under the pretext that the mentality of women is such that their desires are fulfilled and their capabilities fully challenged within the boundaries of the 'core family'. Likewise the rapid succession of childbirths is responsible not only for maternal strain, but causes nutritional and emotional deprivation of children. History provides evidence of appalling abuse, morbidity and mortality among children due to neglect and ignorance[24].

These cultural problems can be traced to the abandonment of natural prolonged and frequent breast feeding with ensuing failure of prolactin-mediated infertility[19,23]. Low fertility appears to be a fundamental characteristic of primates in 'the state of nature'. To the extent that reinstatement of adequate lactation infertility in the human female is impractical, the only possible remedy is artificial contraception. This perspective is conspicuously absent from the 1968 Papal encyclical letter *'Humanae vitae'* denouncing contraception by methods 'alien to nature'.

References

1 Atkinson, S. A., Bryan, M. H. and Anderson, G. H. (1981). Human milk feeding in premature infants: protein, fat and carbohydrate balances in the first two weeks of life. *Pediatrics*, **99**, 617

2 Barlow, B., Santulli, T. V., Heird, W. C., Pitt, J., Blanc, W. A. and Schullinger, J. N. (1974). An experimental study of acute neonatal enterocolitis − the importance of breast milk. *J. Pediatr. Surg.*, **9**, 587

3 Bongaarts, J. (1980). Does malnutrition affect fecundity? A summary of evidence. *Science*, **208**, 564

4 Borresen, H. C. and Knutrud, O. (1976). Intravenous feeding of low birthweight infants. *Nutr. Metabol.*, **20**, (Suppl. 1), 88

5 Borresen, H. C. (1979). Necrotizing enterocolitis, formula feeding and colostrum. (Letter to the Editor) *J. Am. Med. Assoc.*, **242**, 713

6 Borresen, H. C. (1979). Urine electrolytes and body weight changes in the routine monitoring of total intravenous feeding. (Editorial) *Scand. J. Clin. Lab. Invest.*, **39**, 591

7 Borresen, H. C. (1982). Aldosterone, angiotensin II, and renal salt sparing. (Letter to the Editor) *Scand. J. Clin. Lab. Invest.*, **42**, 93

8 Borum, P. R., York, C. M. and Broquist, H. P. (1979). Carnitine content of liquid formulas and special diets. *Am. J. Clin. Nutr.*, **32**, 2272

9 Borum, P. R. (1981). Possible carnitine requirement of the newborn and the effect of genetic disease on the carnitine requirement. *Nutr. Rev.*, **39**, 385

10 Carpenter, C. (1980). Epidermal growth factor as a major growth-promoting agent in human milk. *Science*, **210**, 198

11 Crawford, M. A., Hassam, A. G. and Rivers, J. P. W. (1978). Essential fatty acid requirements in infancy. *Am. J. Clin. Nutr.*, **31**, 2181 (See also pp. 267–287 in ref. 38)
12 Galdikas, B. M. F. (1980). Indonesia's orangutans. Living with the great orange apes. *National Geographic*, June
13 Hayes, K. C. and Sturman, J. A. (1981). Taurine in metabolism. *Ann. Rev. Nutr.*, **1**, 401
14 Hernell, O. (1975). Human milk lipases. *Eur. J. Clin. Invest.*, **5**, 267
15 Jelliffe, D. B. and Jelliffe, E. F. P. (1977). 'Breast is best': modern meanings. *N. Engl. J. Med.*, **297**, 912
16 Jelliffe, D. B. and Jelliffe, E. F. P. (1978). *Human Milk in the Modern World.* (Oxford University Press)
17 Johannessen, A. (1902). Spaedbarnets ernaering og pleie (in Norwegian). In A. Arstal (ed.). *Foraeldre og born.* pp. 19–48. (Kristiania: H. Aschehoug)
18 Knodel, J. and van de Walle, E. (1967). Breast feeding, fertility and infant mortality: an analysis of some early German data. *Pop. Stud.*, **21**, 109
19 Konner, M. and Worthman, C. (1980). Nursing frequency, gonadal function, and birth spacing among !Kung hunter-gatherers. *Science*, **207**, 788
20 Lee, R. B. (1979). *The !Kung San. Men, Women and Work in a Foraging Society.* (Cambridge University Press)
21 Lie, S. O. and Motzfeldt, K. (1982). Pers. comm.
22 McCance, R. A. and Widdowson, E. M. (1961). Mineral metabolism of the foetus and newborn. *Br. Med. Bull.*, **17**, 132
23 McNeilly, A. S. (1979). Effects of lactation on fertility. *Br. Med. Bull.*, **35**, 151
24 Meuse, L. de. (1976). The evolution of childhood. In L. de Meuse (ed.). *The History of Childhood. The Evolution of Parent–Child Relationships as a Factor in History.* pp. 1–73. (London: Souvenir Press)
25 Rowe, J. C., Wood, D. H., Rowe, D. W. and Raisz, L. G. (1979). Nutritional hypophosphatemic rickets in a premature infant fed breast milk. *N. Engl. J. Med.*, **300**, 293
26 Rowland, T. W., Zori, R. T., Lafleur, W. R. and Reiter, E. O. (1982). Malnutrition and hypernatremic dehydration in breast-fed infants. *J. Am. Med. Assoc.*, **247**, 1016
27 Shaw, J. C. L. (1973). Parenteral nutrition in the management of sick low birthweight infants. *Pediatr. Clin. N. Am.*, **20**, (No 2), 333
28 Slonim, A. E., Borum, P. R., Tanaka, K., Stanley, C. A., Kasselberg, A. G., Greene, H. L. and Burr, I. M. (1981). Dietary-dependent carnitine deficiency as a cause of nonketotic hypoglycemia in an infant. *Pediatrics*, **99**, 551
29 Sunderland, R. and Emery, J. L. (1979). Apparent disappearance of hypernatraemic dehydration from infant deaths in Sheffield. *Br. Med. J.*, **2**, 575
30 Tepas, III, J. J., MacLean, W. C., Kolbach, S. and Shermata, D. H. (1978). Total management of short gut secondary to midgut volvulus without prolonged total parenteral alimentation. *J. Pediatr. Surg.*, **13**, 622
31 Walker, W. A. and Isselbacher, K. J. (1977). Intestinal antibodies. *N. Engl. J. Med.*, **297**, 767
32 Watkins, J. B. (1974). Bile acid metabolism and fat absorption in newborn infants. *Pediatr. Clin. N. Am.*, **21**, No 2, 501
33 Watkins, J. B. (1975). Mechanisms of fat absorption and the development of gastrointestinal function. *Pediatr. Clin. N. Am.*, **22**, No 4, 721
34 Widdowson, E. M. and McCance, R. A. (1963). Effect of giving phosphate supplements to breast-fed babies on absorption and excretion of calcium, strontium, magnesium and phosphorus. *Lancet*, **ii**, 1250
35 Widdowson, E. M. (1965). Absorption and excretion of fat, nitrogen, and minerals from 'filled' milks by babies 1 week old. *Lancet*, **ii**, 1099
36 Widdowson, E. M., Dauncey, J. and Shaw, J. C. L. (1974). Trace elements in foetal and early postnatal development. *Proc. Nutr. Soc.*, **33**, 275
37 Williamson, S., Finucane, E., Ellis, H. and Gamsu, H. R. (1978). Effect of heat treatment of human milk on absorption of nitrogen, fat, sodium, calcium, and phosphorus by preterm infants. *Arch. Dis. Child.*, **53**, 555
38 Winick, M., Rosso, P. and Brasel, J. A. (1972). Malnutrition and cellular growth in the brain: existence of critical periods. In *Lipids, Malnutrition and the Developing Brain. A Ciba Foundation Symposium.* pp. 199–206. (Amsterdam: Elsevier)

Section 4
Current Perspectives in the Use of Lipid Emulsion

15
Intravenous fat emulsions: a neonatologist's point of view

D. H. ADAMKIN

INTRODUCTION

Survival of premature neonates has increased markedly over the past 10 years. This is related to advances in knowledge as well as an increase of sophisticated equipment improving both the diagnosis and treatment of neonatal problems. The advances have improved care of respiratory failure, thermoregulation, fluid and electrolyte balance, and maintenance of energy needs. A shift in focus has occurred toward other problems of clinical concern during the neonatal period such as infections, necrotizing entero-colitis (NEC) and the consideration of how to best nourish the rapidly growing premature neonate. Last year our neonate survival for very-low-birth-weight (VLBW) infants (<1500 g) was 80%.

The average weight of a fetus reaches 1000 g around the 27th week of gestation[1] and will double in the subsequent 5 weeks. During that period there will be an increase of 120 g of protein, 10 g of mineral, and 120 g of fat[2]. The remaining weight is due to water accretion. If a neonate is born prematurely during this period of rapid growth, many argue that it should be nutritionally supported to continue this intrauterine rate of growth. Often the neonato-logist is confronted with a VLBW infant who, because of extreme immaturity and serious medical problems, cannot be adequately nourished according to these standards during the first days or weeks of life. Attempts to provide nutritional support under conditions where enteral feedings may not be possible or adequate, have led to the use of either total parenteral nutrition (TPN) or parenteral supplementation in those neonates who are able to take some enteral feedings but not enough to satisfy their total requirements.

This paper will review our clinical experience with the recently released group of fat emulsions as a supplement to parenteral feedings. A recent national survey of neonatal intravenous alimentation practices revealed that

87% or 211 of 242 responding hospitals were using fat emulsions in their neonatal nurseries[3].

The risks of fluid overload and problems with infusion sites limit the quantity of solution that can be infused safely into the peripheral veins of these VLBW infants when using glucose-amino acid mixtures. A further limitation is that the VLBW infant may be intolerant of glucose[4,5]. The intravenous fat preparations have the highest caloric density of any nutrient and are also isotonic, exerting negligible osmotic effects. The possibility of their addition to peripheral vein regimens has made them an attractive alternative energy source for the neonate.

Despite the extensive use of these lipid emulsions over the past decade, many questions about their efficacy and potential hazards remain unanswered and led to a recent statement by the American Academy of Pediatrics about their use in paediatric patients[6].

A review of the available fat preparations and the metabolism of parenterally administered lipid is covered in other chapters. This paper will review specific considerations of fat emulsions as they apply to the neonate and include our own clinical experience with these emulsions.

PREVENTION AND CORRECTION OF ESSENTIAL FATTY ACID DEFICIENCY (EFAD)

There are two functions for fat as part of a total parenteral nutrition (TPN) regimen. The first is its use in small amounts as a source of linoleic acid to prevent or treat essential fatty acid deficiency (EFAD). The second is when fat is given in relatively large amounts as a partial replacement for glucose as a major source of calories.

Four decades ago Holt and co-workers demonstrated that intravenous fat emulsions had therapeutic value[7]. They observed a weight gain of 20–37 g/day in a series of malnourished infants receiving 1–2 g of lipid per kilogram bodyweight by vein. Hansen first reported human EFAD in full-term infants fed a skim milk formula containing less than 0.1% of energy as linoleic acid for 6 weeks[8]. The infants developed a dry, scaly skin condition while their plasma fatty acid analysis showed a triene/tetraene (T/T) ratio of 1.55 compared to a ratio of 0.06 for control infants fed human milk. Soderhjelm and co-workers subsequently reported that when a diet was devoid of linoleic acid, the T/T ratio attained a value as high as 5.0 in 2–4-month-old infants[9]. Friedman and co-workers studied the effect of fat-free alimentation on five sick newborns, four of whom were less than 32 weeks in gestation. All of these neonates developed biochemical evidence of EFAD during the first week of life – the smallest neonate did so by the second day[10].

Figure 15.1 is from a study we performed in which seven neonates were studied for a minimum of 2 weeks while they were maintained exclusively on

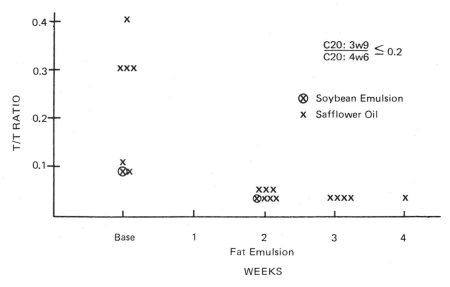

Figure 15.1   Effect upon triene/tetraene ratio in neonates of supplementing TPN formula with lipid emulsion. Four of the seven neonates showed evidence of EFAD prior to therapy

TPN. Six of the neonates were receiving 10% safflower oil and one received 10% soybean emulsion. Five of the seven were VLBW infants ranging in gestation from 26 to 28 weeks. Supplementation with intravenous fat began at 3–11 days of age. Fat contributed 8–30% of their total caloric intake. Four of the seven neonates demonstrated biochemical evidence of EFAD prior to the start of intravenous fat therapy. By the end of the second week of TPN with fat, all the infants had T/T ratios of <0.2 and these remained normal throughout the study period which lasted 15–30 days.

The human fetus depends entirely on placental transfer of essential fatty acids (EFA). The proportions of linoleic acid and arachidonic acids in cord plasma lipids are influenced by the length of gestation and show an abrupt rise at about 37 weeks[11]. The concentration of linoleic acid in fetal tissue is less than in the adult[12]. Furthermore, the proportion of linoleic acid in muscle phospholipids also increases with gestational age[13]. The VLBW infant with limited non-protein caloric reserve must mobilize fatty acids for caloric needs when receiving intravenous nutrition devoid of fat. The borderline stores of EFA in VLBW infants contributes to this rapid onset of EFAD as demonstrated in the two previous studies when these infants were receiving fat-free parenteral alimentation.

The specific clinical syndrome of EFAD (dermatitis) usually does not develop unless there is a prolonged period of fat-free nutrition typically greater than 3 weeks and sometimes as long as 3 months[8].

The full clinical implications of EFAD in the first week of life are not known. Friedman and co-workers demonstrated abnormal platelet aggregation in sick neonates with EFAD[14]. The tendency for sick newborns to develop coagulopathy is a relatively common problem in neonatal intensive care. If a relationship can be found between platelet function and EFAD, then this form of nutritional deprivation takes on greater significance earlier in life. The potential impact of EFAD on neuronal development was demonstrated by Heird and associates. They maintained a group of newborn beagle puppies on fat-free alimentation and showed identical cerebral DNA and the same levels of total brain lipids as controls fed a normal fat diet; however, they found the levels of two essential fatty acids, arachidonic and linoleic, decreased in the animals on fat-free diets. The dogs not receiving fat had abnormally high levels of eicosatrienoic acid – an abnormal metabolic product that occurs in states of EFAD[15]. The implications of these alterations in lipid composition during periods of neuron development are not currently understood but do necessitate further investigation in humans.

PROVISION OF FAT AS CALORIC SUBSTRATE

As mentioned earlier, because of anatomical, biochemical or functional limitation, it may be necessary to feed VLBW infants and other compromised neonates parenterally. Our preference is to avoid central venous catheters whenever possible and to use peripheral veins for TPN. High complication rates have been reported in association with central catheters[16-18].

Table 15.1 Normal basal energy expenditure of a non-stressed premature neonate

Category of energy expenditure	Energy expended (cal/kg day)
Resting caloric	50
Intermittent activity	15
Stress from body temperature decrease (absence of term-infant fat)	10
Calories lost in stool	12
Specific dynamic action of food	8
Growth allowance	25
Total	120

Table 15.1 illustrates the normal basal energy expenditure of a typical non-stressed premature neonate and serves as a guide to the role of various exogenous caloric substrates.

With peripheral venous nutrition, the concentration of glucose which can be administered safely is limited by the local tissue necrosis following extravasation. Other potential problems with large amounts of glucose in the

VLBW infant include hyperglycaemia and osmotic diuresis with dehydration. We try never to exceed 15% dextrose in peripheral vein infusions. Many of the VLBW infants may be intolerant to glucose even in modest concentrations[4,5].

Of particular importance to the neonate and especially the VLBW infant is the risk of fluid overload if large volumes of fluid are given intravenously in an effort to boost caloric intake. In particular, cardiopulmonary problems may be exaggerated or created in the VLBW infant[19]. For example, a peripheral TPN regime of 10% dextrose and amino acids at 150–200 ml/kg day would provide a 1 kg neonate with only 51–68 cal/kg day. To achieve 90 cal/kg day such a VLBW infant would require more than 250 ml/kg day of such an infusate. Hyaline membrane disease complicated by patent ductus arteriosus is common among these neonates. A retrospective study of 62 newborns weighing less than 2 kg showed that 31 went on to develop a functionally significant patent ductus arteriosus. The infants that developed a patent ductus had significantly higher fluid intakes immediately before its appearance. As the infants' fluid intake increased, they developed an increased shunt through the patent ductus. As fluid intake decreased, the shunting decreased accordingly[20]. We attempt to maintain the fluid volume below 140 ml/kg day whenever possible in the VLBW infants with hyaline membrane disease.

To avoid the complications of fluid overload and hyperglycaemia, and to be able to provide adequate energy intake to VLBW infants, the selective and careful use of fat supplementation during TPN has some advantages.

Figures 15.2–15.4 demonstrate study data for six neonates we evaluated ranging in weight from 0.8 to 2.5 kg, and in postnatal age from 1 to 31 days, in which the infants received TPN with 10% safflower oil in the first week and TPN with 20% safflower oil in the second week. The diagnoses in these neonates which precluded enteral feedings included hyaline membrane disease and patent ductus arteriosus in three, and four had surgical conditions including two with necrotizing enterocolitis and two with gastroschisis. The neonates received adequate calories and four of the infants gained weight in the first week during the induction phase of TPN. By the second week all of the neonates were gaining weight with four gaining nearly 20 g/kg day. Figure 15.4 demonstrates that even while receiving increased calories, the use of 20% fat emulsion allowed total fluid volumes to remain in an acceptable range and actually decrease from the first week in four of the neonates including all of those with cardiopulmonary problems.

Table 15.2 presents data from a series of eight neonates who were maintained for a minimum of 7 days on a 20% soybean emulsion. The group was divided with equal numbers receiving one of the two available soybean emulsions. Five of the neonates were VLBW infants with hyaline membrane disease and/or patent ductus arteriosus and three had surgical conditions including malrotation, gastroschisis and jejunal atresia, respectively. Illust-

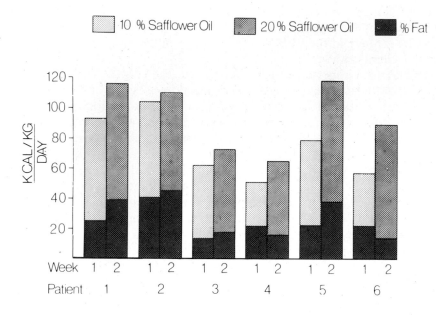

Figure 15.2 Concentration of lipid emulsion supplement given to six neonates, ages ranging between 1 and 31 days, maintained on TPN

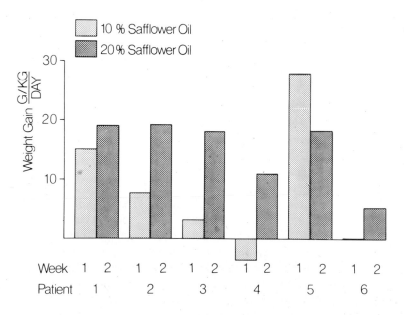

Figure 15.3 Change in weight of six neonates on TPN including lipid emulsion supplements as illustrated in Figure 15.2

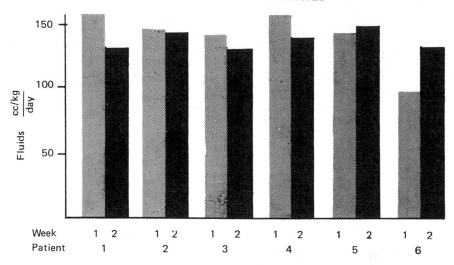

Figure 15.4 Fluid intake in six neonates on TPN given lipid emulsion supplements as illustrated in Figure 15.2. Use of 20% fat emulsion enables fluid volumes to be more readily controlled

rated are data for the first 8 days with initiation of fat in the TPN at a mean postnatal age of 7 days. No untoward side-effects were observed and the infusions were tolerated without difficulty for the first 8 days allowing more calories to be provided by the 20% emulsion while fluid volumes remained acceptable.

The potential usefulness of these 20% lipid emulsions is to provide a highly concentrated (2.0 kcal/ml) source of calories to VLBW neonates isotonically thus permitting relative fluid restriction and the use of peripheral veins.

Table 15.2 Data from a study of eight neonates maintained on a 20% soybean emulsion for 8 days

Fluid volume (ml/kg day)	% Calories fat	Total calories (kcal/kg day)	Weight gain (g/kg day)
146	45	99	7.8

SPECIAL NEONATAL CONSIDERATIONS

Vitamin E and lipid emulsions

Vitamin E (α-tocopherol) has been credited with a number of beneficial and protective effects in the VLBW infant. A deficiency of vitamin E in premature neonates has been linked to a haemolytic anaemia which occurs 4–6 weeks after premature birth[21]. Large parenteral doses of vitamin E have recently been shown to help prevent and/or limit the degree of retrolental fibroplasia[22]. The mechanism for this protective function is the action of vitamin E as a free-radical scavenger and natural antioxidant that protects cell membranes against lipid peroxidation[23,24].

The amount of vitamin E required to prevent lipid peroxidation depends upon the dietary intake of fat, particularly the polyunsaturated fatty acids (PUFA)[25,26]. The greater the intake of PUFA, the more vitamin E is required to protect vital membranes from oxidative injury.

Table 15.3 Vitamin E and polyunsaturated fatty acid (PUFA) content of soybean and safflower oil emulsions

	Soybean 10%	Safflower 10%
α-Tocopherol (mg/l)*	<1	20
γ-Tocopherol (mg/l)*	12	0
Vitamin E activity †	4	20
PUFA (g/l) ‡	63	78
Vitamin E:PUFA (mg d-α-tocopherol/g)	0.06	0.26

*Average data from analyses by R. J. Roberts (unpublished results).
†mg d-α-tocopherol equivalent per litre, (assuming γ-tocopherol has 30% of the activity of α-tocopherol (ref. 3).
‡From manufacturers' data (Cutter Laboratories and Abbott Laboratories, respectively)

The use of TPN with fat emulsions influences the intake and requirement of vitamin E by the VLBW infant. Table 15.3 from Bell and Filer illustrates the vitamin E and PUFA contents of the soybean and safflower oil emulsions[27]. They conclude that a 1 kg VLBW infant receiving 150 ml of TPN including 1 ml of M.V.I. Concentrate (USV Laboratories, Tuckahoe, NY) per day and 10% soy will receive adequate ratios of vitamin E to PUFA (>0.6 mg/g) as long as the lipid intake is 2.0 g/kg day or less. The same regimen with 10% safflower oil would yield vitamin E:PUFA ratio adequate for up to 2.5 g lipid/kg day. The vitamin E:PUFA ratio diminishes with increasing lipid dose, unless additional vitamin E is given orally or parenterally.

Fat emulsions and albumin – bilirubin binding
The potential problem of the infused lipid emulsions resulting in an increase in free fatty acids which will compete with bilirubin for binding to albumin is possible but preventable[28]. *In vitro*, displacement of bound bilirubin by free fatty acids depends on the relative concentrations of bilirubin, free fatty acids and albumin[29,30]. *In vivo*, studies[31] have shown no free bilirubin generated if the molar ratio of free fatty acids to albumin is less than 6. There are also no reports indicating a higher incidence of kernicterus in infants receiving intravenous fat emulsions. There is variability in the free fatty acid concentrations that may be generated. Other factors such as the degree of prematurity and severity and nature of illness also play a role in individual susceptibility to bilirubin encephalopathy. Therefore, we do not provide more fat than that

required to prevent EFAD (0.5–1.0 g/kg day) when the bilirubin concentrations are greater than 8 mg/100 ml.

Fat and the neonatal lung

Fat globules have been identified at autopsy in alveolar macrophages and capillaries of newborn infants[32]. Fox and colleagues studied seven preterm neonates all less than 1500 g to evaluate direct pulmonary effects while a 10% soybean emulsion was being infused[33]. Each neonate received 1 g/kg over 4 h. Their results indicated no significant changes in any pulmonary function tests, and no significant changes in pH, PCO_2, bicarbonate, or base excess levels. The PaO_2 levels fell from a control level of 66.4 mmHg to a low of 54.1 mmHg at the same time that triglyceride and free fatty acid levels peaked. 4 h after infusion, the PaO_2 was 62.6 mmHg.

Recently a total of 11 neonates have been described with fat accumulation in the lungs at the time of autopsy[34,35]. The administration of fat to VLBW infants with respiratory distress syndrome (RDS) must be assessed carefully and the infant's ability to eliminate infused fat from the circulation monitored carefully (triglyceride and/or free fatty acid plasma levels).

SUMMARY

The availability of both soybean and safflower oil emulsions with their relatively high content of linoleic acid simplifies the task of providing this essential nutrient to neonates requiring TPN. These preparations also allow for a more balanced caloric intake and are particularly attractive in VLBW infants as a source of concentrated isotonic calories that can be delivered by peripheral vein infusion circumventing the many hazards of indwelling central vein catheters.

However, uncertainties still exist concerning the neonates' ability to clear these fat emulsions and the adverse effects of serum lipaemia with specific reference to pulmonary fat accumulation in VLBW infants with RDS.

Based on current experience, the use of small amounts of fat to prevent EFAD is not likely to be associated with any harmful effects in the neonate. When using fat as an energy source every attempt should be made to monitor fat metabolism.

References

1 Freeman, M. G., Graves, W. L. and Thompson, R. L. (1970). Indigent Negro and Caucasian birth-weight-gestational age tables. *Pediatrics*, **46**, 9

2 Brans, Y. W. and Cassady, G. (1975). Fetal nutrition and body composition. In Ghadimi, H. (ed.). *Total Parenteral Nutrition: Premises and Promises.* pp. 301–333. (New York: John Wiley)

3 Malky, J. M. and Haspela, N. A. (1980). The national survey of neonatal intravenous alimentation services. *Am. J. Intravenous Ther. Clin. Nutr.*, Dec/Jan

4 Brans, Y. W., Sumners, J. E., Dweck, H. S. *et al.* (1974). Feeding the low-birth-weight infant: orally or parenterally? Preliminary results of a comparative study. *Pediatrics*, **54**, 15

5 Bryan, M. H., Wei, P., Hamilton, J. R. *et al.* (1973). Supplemental intravenous alimentation in low-birth-weight infants. *J. Pediatr.*, **82**, 940

6 Committee on Nutrition, AAP. (1981). Use of intravenous fat emulsions. *Pediatrics*, **68**, 738

7 Holt, L. E., Todwell, H. C. and Scott, T. F. M. (1935). The intravenous administration of fat. *J. Pediatr.*, **6**, 151

8 Hansen, A. E., Stewart, R. A., Hughes, G. and Soderhjelm, L. (1962). The relation of linoleic acid to infant feeding. *Acta Paediatr.*, **51**, (Suppl. 137), 1

9 Soderhjelm, L., Wiese, H. F. and Holman, R. T. (1968). The role of polyunsaturated acid in human nutrition and metabolism. In Holman, R. T. (ed.) *Progress in the Chemistry of Fats and Other Lipids. Vol. IX*, p. 555. (Elmsford, N.Y.: Pergamon)

10 Friedman, A., Danon, A., Stahlman, M. T. and Oates, J. A. (1976). Rapid onset of essential fatty acid deficiency in the newborn. *Pediatrics*, **58**, 640

11 Fosbrooke, A. S. and Wharton, B. A. (1973). Plasma lipids in umbilical cord blood from infants of normal and low birth weight. *Biol. Neonate*, **23**, 330

12 Roux, J. F., Takeda, Y. and Grigorian, A. (1971). Lipid concentration and composition in human fetal tissue during development. *Pediatrics*, **48**, 540

13 Bruce, A. and Svennerholm, L. (1971). Skeletal muscle lipids: 1. Changes in fatty acid composition of lecithin in man during growth. *Biochim. Biophys. Acta*, **239**, 393

14 Friedman, Z., Lamberth, E. L., Stahlman, M. T. *et al.* (1975). Platelet aggregation in infants with essential fatty acid deficiency, abstracted. *International Congress of Prostaglandins*, May, Florence, Italy

15 Heird, W. C., Bieber, M. A., Bassi, J. *et al.* (1976). Effects of total parenteral nutrition on brain growth of beagle puppies. *Pediatr. Res.*, **10**, 323

16 Bends, G. I. M. and Babson, S. G. (1971). Peripheral intravenous alimentation of the small premature infant. *J. Pediatr.*, **79**, 494

17 Bryan, M. H., Wei, P., Hamilton, J. R., Chance, G. W. and Swyer, P. R. (1973). Supplemental intravenous alimentation in low-birth-weight infants. *J. Pediatr.*, **82**, 940

18 Pildes, R. S., Ramamurthy, R. S. Cordero, G. V. and Wong, P. W. K. (1973). Intravenous supplementation of L-amino acids and dextrose in low-birth-weight infants. *J. Pediatr.*, **82**, 945

19 Bell, E. F., Warburton, D., Stonestreet, B. S. and Oh, W. (1979). Randomized trial comparing high and low-volume maintenance fluid administration in low-birth-weight infants. *Pediatr. Res.*, **13**, 489

20 Stevenson, J. G. (1977). Fluid administration in the association of patent ductus arteriosus complicating respiratory distress syndrome. *J. Pediatr.*, **90**, 257

21 Oski, F. A. and Barness, L. A. (1967). Vitamin E deficiency: a previously unrecognized cause of hemolytic anemia in the premature infant. *J. Pediatr.*, **70**, 211

22 Hittner, H. M., Godio, L. B., Rudolph, A. J. *et al.* (1981). Retrolental fibroplasia: efficacy of vitamin E in a double-blind clinical study of preterm infants. *N. Engl. J. Med.*, **305**, 1365

23 Phelps, D. L. (1979). Vitamin E: where do we stand? *Pediatrics*, **63**, 933

24 Farrell, P. M. (1979). Vitamin E deficiency in premature infants. *J. Pediatr.*, **95**, 869

25 Mason, K. E. and Filer, L. J., Jr. (1947). Interrelationships of dietary fat and tocopherols. *J. Am. Oil. Chem. Soc.*, **24**, 240

26 Hashim, S. A. and Asfour, R. H. (1968). Tocopherol in infants fed diets rich in polyunsaturated fatty acids. *Am. J. Clin. Nutr.*, **21**, 7

27 Bell, E. F. and Filer, L. J. (1981). The role of vitamin E in the nutrition of premature infants. *Am. J. Clin. Nutr.*, **34**, 414

28 Thiessen, H., Jacobsen, J. and Brodersen, R. (1972). Displacement of albumin-bound by fatty acids. *Acta Pediatr. Scand.*, **61**, 285

29 Jacobsen, J. (1969). Binding of bilirubin to human serum albumin. *FEBS Lett.*, **5**, 112

30 Starinsky, R. and Shafrir, E. (1970). Displacement of albumin-bound bilirubin by free fatty acids: implications for neonatal hyperbilirubinemia. *Clin. Chim. Acta*, **29**, 311

31 Andrew, G., Chan, G. and Schiff, D. (1976). Lipid metabolism in the neonate. II, The effect of intralipid on bilirubin binding *in vitro* and *in vivo*. *J. Pediatr.*, **88**, 279

32 Freedman, Z., Marks, K. H., Maisels, J. *et al.* (1978). Effect of parenteral fat emulsion on the pulmonary and reticuloendothelial systems in the newborn infant. *Pediatrics*, **61**, 694

33 Fox, W. W., Pererra, G. and Schwartz, J. G. (1978). Effects of intralipid infusions on arterial blood gases and pulmonary function in small premature neonates. *Pediatr. Res.*, **10**, 524

34 Levene, M. I., Wigglesworth, J. S. and Desai, R. (1980). Pulmonary fat accumulation after 'intralipid' infusion in the preterm infant. *Lancet*, **2**, 815

35 Dahms, B. B. and Halpin, T. C. (1980). Pulmonary arterial lipid deposit in newborn infants receiving hyperalimentation. *J. Pediatr.*, **97**, 800

16
Clinical applications of 20% fat emulsions

J. D. HIRSCH AND D. L. BRADLEY

In 1976 fat emulsion for intravenous therapy was reintroduced in the United States. This new formulation, based on soybean oil, emulsified with egg yolk phospholipid and glycerol had a long history of safety in the European market. Although only the 10% emulsion was initially released in the United States, 20% emulsions were the preparation used predominantly in both Europe and Canada[1]. During the late 1960s and early 1970s parenteral nutrition was developing in the United States based on hypertonic dextrose/amino acid solutions. The initial use of fat emulsions was for the treatment and prevention of essential fatty acid deficiencies. The need for a fat emulsion in a total parenteral nutrition regime had been demonstrated by the occurrence of essential fatty acid deficiency. This was expressed clinically by dermatitis, increased capillary permeability, alopecia, liver and kidney damage and thrombocytopenia. Biochemically it was manifested by depressed levels of linoleic and arachidonic acid in the phospholipid fraction along with increased levels of Δ 5, 8, 11 eicosatrienoic acid. Although reports varied it appeared that between 2.5 and 20 g of linoleic acid per day are necessary to prevent an essential fatty acid deficiency. Prevention required the administration of between 4 and 8% of the total weekly calories as fat emulsions. Although a relatively large volume of literature exists to document the safety and efficacy of 10% fat emulsion, scant information is available on the utilization of 20% fat emulsion in the American literature. Both 10% and 20% fat emulsion are equally effective in treating or preventing essential fatty acid deficiency.

The use of lipids as an energy source has continued to expand in this country. It is in this area that a choice between 10% and 20% fat emulsion may be significant. Dextrose calories are significantly cheaper than lipid calories. Furthermore, the lack of substantial evidence that fat as a caloric source is significantly better than dextrose has reinforced this prejudice. Other problems have plagued the utilization of fat emulsions. Information regarding exogenous energy utilization is at best incomplete. Hyper-triglyceridaemia is frequently seen in stressed and septic patients as is hyperglycaemia. Holliday et al.[2] suggested that fat emulsions may be detrimental in the septic patient. Furthermore, the appropriate ratio of dextrose/lipid calories in a dual energy system is unclear and the formulation of such a system is most difficult based on the present method of supplying fat emulsions. Few would consider discarding a portion of the product because of the expense nor would they 'share' portions of solutions among patients. There appears to be an increasing interest in many centres in basing protein and non-protein calories on metabolic need with solutions tailored to meet each patient's requirement. Some modification in the method of supplying fat emulsions would be a significant benefit in this area.

In spite of potential problems, lipids offer significant advantages in a nutritional support system. Fat emulsions' principal advantages are:

(1) High caloric content, from 1.1 to 2.0 kcal/ml based on the emulsion used.

(2) Isotonicity – osmolalities from 280 to 330 mOsmol based on the preparation supplied.

(3) The availability of essential fatty acids necessary for the treatment and prevention of essential fatty acid deficiency.

20% fat emulsion's principal value is that it remains isotonic while delivering a richer caloric load per ml than 10% fat emulsions. 20% fat emulsion should allow the prescribing physician greater flexibility in formulating a dual energy nutrition support system while maintaining isotonicity and essential fatty acid requirements. The main chemical difference between the solutions is in the amount of emulsifying agent, egg yolk phospholipid. Essentially, per unit dose, 20% has twice the concentration of triglycerides and half the concentration of phospholipid as compared to 10%. The clearance and metabolism of the 20% solution is essentially the same as with the 10% solution. In adults, the methods for administration are similar with the same initial precautions. Adverse reactions are rare and are no different than the reported side-effects from the 10% solution. This difference is most apparent in the fluid-restricted patient, where fluid volume is critical. The paediatric or neonatal patient, the patient with either renal or cardiovascular compromise, or the critically ill patient with multiple system failure are the most obvious examples. In each of these areas successful therapy with fat emulsion has been reported. However, for any given nutritional support service to switch to 20% fat emulsions as the primary fat emulsion must involve cost comparison. It would be difficult to justify, in this time of cost containment, the substitution of 1 unit of 20% lipid for 2 units of 10% lipid unless the price of the 20% emulsion was equal to or less than 2 units of 10% emulsion.

A common use of fat emulsions is for peripheral parenteral nutrition (PPN). The obvious benefit of administering calories through a peripheral vein is that central vein catheterization is not necessary. To safely administer nutrients through a peripheral vein, osmolalities should, in general, be less than $600\,mOsmol^3$. Lipid emulsions constitute a low osmolality, high caloric density source of non-protein calories. In theory, lipids should serve as a protective coating to the peripheral vein. Whether this works in practice, is still debatable. Peripheral parenteral nutrition using a lipid-based system has been popularized by the work of Jeejeebhoy, Silberman, and others[4-6]. They have reported successful treatment of protein calorie malnutrition by peripheral administration. Others have been less impressed[7]. Peripheral parenteral nutrition has been plagued in many centres by high rates of phlebitis, fluid overload, relatively low caloric content and inconsistent results. The addition of 20% fat emulsion allows the physician greater flexibility in setting up a PPN system and may alleviate some of the problems with PPN, but not all of them.

The amount of CO_2 produced during the combustion of dextrose is significantly greater than in lipid combustion. This appears to have an important impact on the nutritional support of the patient with chronic obstructive pulmonary disease or acute respiratory failure. Several investigators have demonstrated dramatic improvements in P_{CO_2} and minute ventilation when patients have been switched from a dextrose-based system to a lipid-based TPN system[8]. For many this has meant less difficulty in weaning from mechanical ventilatory assistance. Certainly many of those patients are fluid-restricted and the use of a 20% fat emulsion would potentially improve their management.

The use of a centrally administered dual energy system has gained acceptance. Using fat emulsion as an energy source as well as a source of essential fatty acids appears to alleviate glucose intolerance in selected patients. Hyperglycaemia may be associated with an osmotic diuresis that often complicates fluid management. In the centrally administered dual energy system 20% fat emulsion would appear to have a significant advantage only in the severely fluid-restricted patient. Fluid overload is a significant problem in the care of the critically ill, especially in the elderly population with concurrent cardiovascular and renal disease. Unfortunately, in many centres when fluid overload does occur, nutritional support appears to be the usual first choice to be deleted or drastically reduced. 20% fat emulsion may help maintain appropriate caloric load in those critically ill patients.

Recently, clinical trials of a new 20% fat emulsion derived from soybean oil and egg yolk phospholipid were performed. 20% Travamulsion Injection (Baxter-Travenol Laboratories) was compared to 20% Intralipid (Cutter Laboratories) in a randomized non-blinded study of adult patients requiring nutritional support for at least 7 days. Patients did not receive lipids for at least 3 days prior to the study. Pretreatment values for serum triglycerides, cholesterol, coagulation screen, bilirubin, SGOT and SGPT were in the normal range or, if abnormal, not felt to be clinically significant. All patients were free of severe acute pancreatitis, severe chronic obstructive pulmonary disease, severe renal or hepatic disease, severe diabetes mellitus, severe hyperlipidaemia, and severe anaemia of any cause. Patients received nutritional support according to the hospital protocol and received approximately 40% of their non-protein calories as lipids. The two study groups were demographically similar. Statistical analysis of the two groups revealed no significant difference between the treatment groups (p $\alpha=0.05$). The variables analysed included albumin, triglycerides, glucose, total protein, bilirubin, SGOT, SGPT, LDH, cholesterol, weight, free fatty acids and phospholipids. Although the total number of patients included in the study was small, it appeared that both 20% Travamulsion Injection and 20% Intralipid were safe and effective during the period of study.

References

1 Wretland, A. (1981). Development of fat emulsions. *J. Parenter. Enter. Nutr.*, **5**, 230
2 Holliday, R. L., Viidik, T. and Jennings, B. (1978). Lipid metabolism in stress. In I. D. Johnston (ed.). *Advances in Parenteral Nutrition* (Baltimore University Park Press)
3 Blackburn, G. L., Gazitun, R., Wilson, K. and Bistrian, B. (1979). Factors determining peripheral vein tolerance to amino acid infusions. *Arch. Surg.*, **114**, 897
4 Jeejeebhoy, K. N., Anderson, G. H., Nakhooda, A. F., Greenberg, G. R., Sanderson, I. and Marliss, E. B. (1976). Metabolic studies in total parenteral nutrition with lipid in man. *J. Clin. Invest.*, **57**, 125
5 Silberman, H., Freehaut, M., Fong, G. and Rosenblatt, N. (1977). Parenteral nutrition with lipids. *J. Am. Med. Assoc.*, **238**, 1380
6 Hansen, L., Hardie, R. W. and Hidalgo, J. (1976). Fat emulsions for intravenous administration. *Ann. Surg.*, **184**, 80
7 Long, J. M., Wilmorc, D. W., Mason, A. and Pruitt, B. A. (1977). Effect of carbohydrate and fat intake on nitrogen excretion during total intravenous feedings. *Ann. Surg.*, **185**, 417
8 Askanazi, J., Weissman, C., Rosenbaum, S. H., Hyman, A. I., Milio-Emili, J. and Kinney, J. M. (1982). Nutrition and the respiratory system. *Critical Care Med.*, **10**, 163

17
Changes in binding of bilirubin due to intravenous lipid nutrition

P. F. WHITINGTON AND G. J. BURCKART

INTRODUCTION

Intravenous lipid products are a potentially important source for nutritional support of sick neonates because they have high caloric density, they include fatty acids which are essential for the rapidly developing CNS, and they have a low osmolality. One of the major constraints to using them is hyperbilirubinaemia because the products of free fatty acids, appearing in blood after the hydrolysis of triglycerides, possibly increase the risk of bilirubin toxicity.

Bilirubin encephalopathy (kernicterus) results when toxic quantities of non-conjugated bilirubin enter brain tissue. Death often results, and survivors can suffer sequelae such as choreoathetotic cerebral palsy, deafness, and mental retardation. It may also cause dyslexia, hyperactivity and impaired learning. The mechanisms by which bilirubin produces the changes in the brain which result in kernicterus are unknown, but most investigation has been focused on the binding of bilirubin to albumin[1-3]. Simply stated, when binding capacity is exceeded, free bilirubin can enter the central nervous system and produce kernicterus.

Pharmacological agents may reduce bilirubin binding by competing for binding sites on albumin. Over 150 drugs have been investigated with regards to their effect on bilirubin binding[4], but only one nutritional compound, the fatty acid, has shared the same degree of interest. From *in vitro* studies, several authors have reported that fatty acids displace bilirubin from its albumin binding sites. Starinsky and Shafrir[5] used the Sephadex gel filtration system to demonstrate that high concentrations of linoleic acid displaced unconjugated bilirubin from albumin. Chan, Schiff, and Stern[6] used the HBABA dye technique and obtained similar results with oleic acid. Other reports confirmed these results and stressed that caution should be used when prescribing therapy that might enhance serum fatty acid concentration in hyperbilirubinaemic infants. This recommendation has effectively precluded the use of intravenous lipid therapy in such infants.

However, several reports in the late 1970s have suggested that fatty acids do not alter bilirubin binding. Brodersen used the peroxidase technique to demonstrate a reduction in free bilirubin at a free fatty acid to albumin molar ratio of 2, and Soltys and Hsia[7] demonstrated enhancement of binding of a mono-anionic spin label by laurate and palmitate. Because of our interest in intravenous nutrition in infants and because there appeared to be some question regarding the effects of intravenous lipid administration on bilirubin binding, we studied lipid emulsions and free fatty acids with regard to changes in binding of bilirubin to albumin *in vitro* and *in vivo*.

METHODOLOGY

The difference spectroscopy assay first described by Ash[8,9] was modified to study alterations of bilirubin binding due to the administration of intravenous lipid. This assay quantitates the shift in the absorption spectrum of bilirubin in aqueous solution when it is bound to albumin (Figure 17.1). The magnitude of the positive change in the difference spectrum at 482 nm is proportional to the number of binding sites available when a sample containing albumin is added to a solution containing excess bilirubin. The procedure is as follows.

All of the measurements are made in semi-micro cuvettes (1 cm light path and 4 mm width) in a Zeiss DM-4 recording dual beam spectrophotometer with wavelength at 482 nm and band width at 0.5 nm:

A_B = the absorbance of buffer plus bilirubin.

A_{SB} = the absorbance of buffer plus bilirubin plus sample.

A_S = the absorbance of buffer plus sample.

$\Delta A_{482} = A_{SB} - (A_S + A_B)$

The cuvettes contain, and recordings are made, as follows:

Cuvette I: 500 μl buffer
+ 10 μl bilirubin solution (records A_B)
+ 10 μl sample (records A_{SB})

Cuvette II: + 500 μl buffer
+ 10 μl sample (records A_S)

The buffers used in these experiments are isotonic (280–300 mOsm/kg), containing 5 mmol/l EDTA and are as follows: borax, pH 10.5 and 9.5 and sodium-potassium phosphate, pH 7.4. The bilirubin solution is 3.0 mmol/l in 100 mmol/l NaOH and contains 5.0 mmol/l EDTA.

We found that under the conditions prescribed by the method each mole of

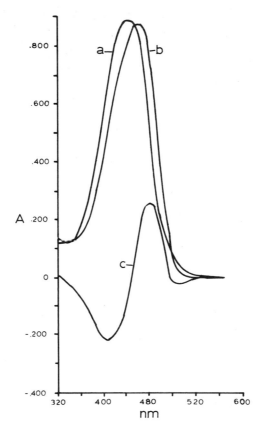

Figure 17.1 The spectral absorption curves of free (a) and albumin-bound (b) bilirubin: Curve (a) was of 20 μmol/l bilirubin in isotonic borax buffer, pH 10.5, and curve (b) was of the same solution to which was added an equimolar amount of human serum albumin. Curve (c), the difference spectrum (b − a), peaks at 482 nm and the height of this peak correlates with the concentration of albumin in the presence of excess bilirubin

albumin binds two moles of bilirubin. We attempted to measure each of these binding sites by including salicylate at a concentration of 15 mmol/l in the buffer. This concentration of salicylate produces roughly a 25:1 salicylate: albumin molar ratio after the sample is added.

The design of the assay uses bilirubin as the test ligand, making questions of specificity irrelevant. During the performance of the assay, a sample containing binding sites is added to excess bilirubin. Any free binding sites and any sites occupied by a ligand with lesser affinity than bilirubin will be occupied, resulting in a change in the absorption spectrum of bilirubin. The magnitude of the change is proportional to the concentration of available binding sites. The absorbance is measured at 482 nm for each of the test parameters: A_{SB}, A_B and A_S. ΔA_{482} is calculated. Any interfering absorbance at 482 nm contributed by pigments in serum is negated by subtracting A_S. Likewise, contribution by the excess bilirubin is subtracted as A_B. The result is the measurement of absorbance resulting from the binding of bilirubin to previously unoccupied sites.

The relationship between ΔA_{482} and the concentration of binding sites in the sample was established by performing the assay using as the sample a series of albumin solutions varying from 0 to 0.70 mmol/l. The data from such an experiment are shown in Figure 17.2. From such a plot a binding capacity can be graphically derived from ΔA_{482} for an unknown sample.

Inclusion of 15 mmol/l salicylate in the borate buffer reduced by approximately one-half the slope of the regression line of ΔA_{482} versus albumin concentration (mmol/l). These data indicate that salicylate interferes with binding at approximately half of the available sites in the sample: generally, low affinity binding sites.

In vitro studies using lipid emulsion[10]

Freshly prepared solutions of lipid emulsion in 0.48 mmol/l albumin were assayed by difference spectroscopy. Using dilutions of both safflower and soybean oil emulsions, concentrations of 50, 100, 200, 300, 500 and 1000 mg/100 ml were prepared. The lipid products did not scatter sufficient

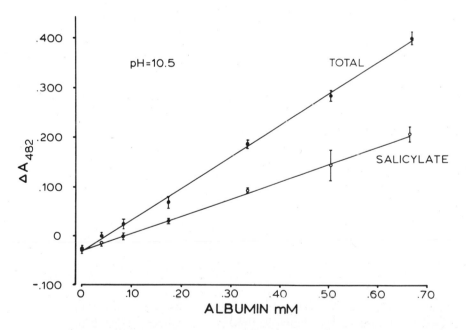

Figure 17.2 The relationship between ΔA_{482} and the concentration (mmol/l) of FFA-free human serum albumin in the sample. The test performed in borate buffer, pH 10.5, (●) yielded the regression line, $y = 0.640x - 0.032$; $r = 0.998$. In the same buffer containing 15 mmol/l sodium salicylate (O), it yielded the line $y = 0.346x - 0.028$; $r = 0.988$ ($n = 6$ at each point)

light at the concentrations tested to result in measurable absorbance at 482 nm nor did they alter the absorbance of the albumin or the bilirubin solutions. Therefore, no correction for the lipid other than subtraction of the serum blank was needed. Two components of the lipid emulsion, egg phosphatides and glycerol, were also assayed. Concentrations of each component similar to that contained in the lipid emulsions were prepared in 0.1 mmol/l potassium phosphate buffer, pH 7.4. The solutions were added to human albumin to yield dilutions equivalent to 100, 500 and 1000 mg/100 ml lipid emulsion for glycerol and 50, 100, 200, 300, 500 and 1000 mg/100 ml lipid emulsion for egg phosphatides. Five observations were recorded in borate buffer and in borate–salicylate buffer for each of the specimens.

The effects of the lipid products on bilirubin binding are shown in Figure 17.3. Binding was significantly increased from baseline at concentrations of the soybean oil emulsion equal to 200, 300 and 500 mg/100 ml and concentrations of the safflower oil emulsion equal to 300 and 500 mg/100 ml. Both products produced about 15% enhancement. Greater enhancement was seen when salicylate buffer was used indicating that the lipid products had a differential effect on primary binding sites. Egg phosphatides and glycerol produced minor though significant alterations in binding. Phosphatide produced enhancement of total and primary binding capacity whereas glycerol reduced both to a minor degree.

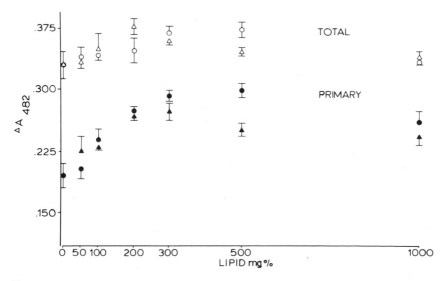

Figure 17.3 The relationship between ΔA_{482} and the concentration of lipid product in the sample. Soybean oil emulsion $\triangle$ and safflower oil emulsion (O) were added in various concentrations to an albumin solution, and binding capacity was assayed by difference spectroscopy. Total binding capacity (open symbols) and binding capacity at primary sites (test performed in borate–salicylate buffer, closed symbols) are first seen to increase and then decrease as lipid concentration increases ($n = 6$ at each point)

In vitro studies using fatty acids[11]

Studies of the alterations of binding produced by fatty acids were performed by preparing solutions of linoleic and octanoic acid in human serum albumin and by analysing them by difference spectroscopy. Fatty acids were dissolved in chloroform and quantities of these solutions were added to tubes and dried under nitrogen gas. A solution of albumin in isotonic phosphate buffered saline, pH 7.4, was added, and the tube was vortexed until the solution cleared. Solutions with free fatty acid to albumin molar ratios from 0 to 10 were prepared. The solutions exhibited no light scattering, indicating that micelles of fatty acid salts were not being formed. The solutions were used immediately after preparation. Measured free fatty acid concentrations[12] agreed very closely with the calculated concentrations, indicating all free fatty acid added to the tubes was in solution.

The data plotted in Figure 17.4 demonstrate that increasing the linoleic acid to albumin molar ratio from 0 to 3.5 resulted in 37% enhancement of total binding capacity. Further increasing the linoleic acid concentration resulted in a decline in binding capacity, and there was 35% reduction when the molar ratio was 8.7. Similar results were obtained when this experiment was performed in salicylate–borate buffer. A 90% enhancement in binding at primary sites occurred at a molar ratio of 3.8, and a 23% reduction occurred when the ratio was 10.9. The same experiments performed at pH 7.4 demonstrated that enhancement occurs at this pH as well.

To further investigate the site of enhancement of binding, bilirubin was added to the albumin solution at a molar ratio of 1.0 and linoleic acid was added as previously described. Figure 17.5 illustrates the results of this experiment. The plot in salicylate–borate buffer demonstrates that at a linoleic acid to albumin molar ratio of 0 there are essentially no reserve binding sites. This is predicted because bilirubin would occupy the primary binding site and salicylate the secondary site. As the molar ratio increases, there is enhancement of the binding capacity with a peak at a ratio of 4.0. From this point a gradual decline in binding occurs, but even at a ratio of 10.0 it remains enhanced as compared to that at 0 ($p<0.01$), indicating no depletion of binding capacity at primary sites. In buffer alone there was no enhancement of binding, but fatty acid reduced binding at higher molar ratios.

Adding octanoic acid to solutions of albumin did not enhance the ΔA_{482}; the binding capacity was stable until the fatty acid to albumin molar ratio was 2.5, after which a gradual decline occurred. At a ratio of 10.0 there was a 12.6% reduction in total binding. In salicylate–borate buffer no changes in binding were noted.

In separate experiments, linoleic acid alone did not alter the spectral characteristics of bilirubin in the absence of the albumin, indicating that the observations noted were due to changes in the binding of bilirubin.

These data indicate that free fatty acids in moderate concentrations

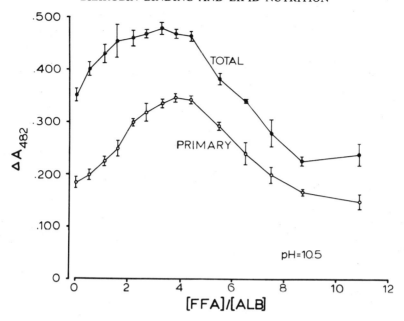

Figure 17.4 The relationship between ΔA_{482} and FFA/Alb in borate buffer, pH 10.5, (●) and borate buffer containing 15 mmol/l sodium salicylate, pH 10.5, (O). The albumin concentration remained constant at 0.63 mmol/l ($n=6$ at each point)

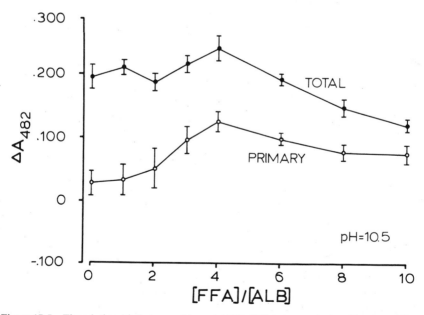

Figure 17.5 The relationship between ΔA_{482} and FFA/Alb in borate (●) and borate−salicylate (O) buffer, pH 10.5, when the sample also contained one molar equivalent of bilirubin. The albumin and bilirubin concentrations were held constant at 0.63 mmol/l ($n=6$ at each point)

enhance rather than reduce the bilirubin binding capacity of the albumin *in vitro*. Linoleic acid was chosen for study because it is a major component in intravenous lipid products. It enhances binding primarily at the high affinity binding site. Reasons for this effect are not understood completely. Free fatty acid binding to albumin is very complex and is not produced by any simple model[13]. However, free fatty acid probably causes steric alterations in the high affinity binding sites[14]. Our data are consistent with the overall theory that binding of the first few molar equivalents of free fatty acid to albumin causes allosteric enhancement of binding of a second ligand[13], in this case bilirubin.

Free fatty acids produce both increases and decreases in bilirubin albumin binding depending on their concentration. Increases in the linoleic acid to albumin molar ratio above 4 caused a reduction in binding. Our data suggest that the reduction occurred mainly as a result of interference with binding at the secondary or low affinity binding site.

Fatty acid chain length also determines the effect on bilirubin binding. Octanoic acid caused no enhancement of binding capacity, and when the free fatty acid to albumin molar ratio exceeded 2.5, a gradual reduction was demonstrated. This effect appeared to be entirely at secondary binding sites, supporting the conclusion of Soltys and Hsia that free fatty acids of chain lengths of less than 10 compete for secondary binding sites only[15].

In vivo **studies**
These studies were performed in a group of neonates requiring parenteral nutrition. Patients who had received feedings of fat, either enterally or parenterally, within 96 h or who were receiving xenobiotics known to interfere with bilirubin binding to albumin were excluded. Informed parental consent was obtained for each study patient.

A 1.0 g/kg test dose of 10% intravenous safflower oil emulsion was administered over 4 h. Blood samples were obtained in a consistent manner for each study patient prior to the study and at 2, 4, 6 and 8 h after the initiation of the infusion. The serum was separated from blood within 30 min and refrigerated at 4 °C. The samples were analysed by the difference spectroscopy assay within 8 h of obtaining them. Free fatty acid was measured[12] in serum samples frozen at -20 °C for less than 7 days. Total bilirubin[16] and albumin[17] were measured in fresh serum from the baseline sample (Table 17.1).

The table lists the general characteristics of the 15 neonates studied. Drug therapy during the study period included ampicillin and gentamicin in six patients, phenobarbital in three patients and theophylline in one patient. The data from the difference spectroscopy assay of the patients' samples are given in Figure 17.6 where they are computed as a percentage change in ΔA_{482} from the baseline sample and are plotted against the corresponding serum free fatty acid concentrations expressed as free fatty acid to albumin molar ratios.

Table 17.1 Characteristics of the study population

Patient	Age (days)	Weight (g)	Gestational age (weeks)	Gestational size*	Primary diagnosis	Total bilirubin (mg/dl)
1	24	1760	33	SGA	Tracheal stenosis	1.6
2	2	3000	41	AGA	Meconium aspiration	9.2
3	3	3620	40	AGA	Perinatal asphyxia	1.2
4	11	2360	36	AGA	Necrotizing enterocolitis	0.4
5	8	670	31	SGA	Respiratory insufficiency of prematurity	6.4
6	9	1880	34	AGA	Pneumothorax	5.5
7	2	2400	34	AGA	Hyaline membrane disease	6.5
8	4	1630	31	AGA	Perinatal asphyxia	6.9
9	4	2950	38	AGA	Pneumothorax	3.8
10	1	1700	35	SGA	Meconium aspiration	5.3
11	5	1300	33	SGA	Respiratory insufficiency of prematurity	8.8
12	4	680	29	SGA	Respiratory insufficiency of prematurity	5.2
13	30	1170	29	AGA	Necrotizing enterocolitis	3.2
14	2	1660	32	AGA	Pneumonia	5.7
15	19	3200	40	AGA	Multiple congenital anomalies	29.4

*Small for gestational age (SGA) or appropriate for gestational age (AGA)

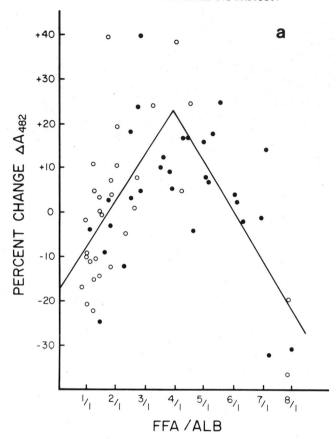

Figure 17.6a Relationship between the percentage change in ΔA_{482} determined in borate buffer, pH 9.5, and the FFA/Alb for each of the 60 neonatal serum samples. Closed circles represent samples obtained during and at the completion of the fat infusion. Open circles represent samples obtained after the administration of the lipid emulsion. The best fit lines for the points with FFA/Alb<4/1 ($n=41$, $y=10.5x-17.7$; $r=0.601$) and FFA/Alb>4/1 ($n=19y$, $=-11.0x+65.3$; $r=0.760$) intersect at a 3.9/1 FFA/Alb. An FFA/Alb of 4/1 was chosen for use in the regression analysis after examining *in vitro* data

Total binding capacity increased to above 30% over baseline in the range of free fatty acid to albumin molar ratios of 1.6 to 4.0. As the molar ratio increased, binding capacity fell. When the molar ratio exceeded 7.0 a decrease of 30% below baseline was observed. The test performed in buffer containing salicylate demonstrates that binding at primary sites was increased to 30% above the baseline at a fatty acid to albumin molar ratio of 4.0 and thereafter fell gradually, never falling to below baseline. It is interesting to note that binding capacity of samples obtained after the infusion were often reduced 10% or more despite free fatty acid values remaining at or

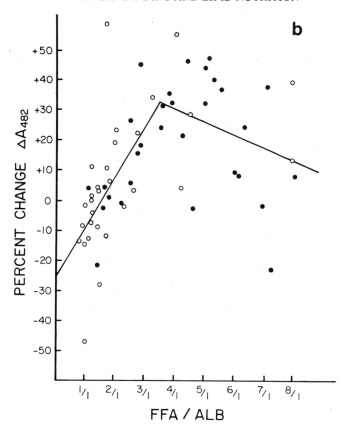

Figure 17.6b Relationship between the percentage change in ΔA_{482} determined in borate buffer containing 15 mmol/l sodium salicylate and the FFA/Alb ratio for each of the 60 samples obtained during and at the completion of the fat infusion. Open circles represent samples obtained after the administration of the lipid emulsion. The best fit lines for the points with FFA/Alb<4/1 ($n=41$, $y=17.1x-25.2$, $r-0.718$) and FFA/Alb>4/1 ($n=19$, $y=4.2x+46.5$, $r=0.280$) intersect at a 3.4/1 FFA/Alb

above baseline. This finding suggests that some component of the emulsion other than free fatty acid negatively affects ΔA_{482}.

These *in vivo* data in general confirm our *in vitro* work and demonstrate that lipid infusion in human neonates tends to enhance rather than reduce bilirubin binding capacity of serum.

CONCLUSIONS

The clinical implications of these studies are against withholding intravenous fat from icteric neonates when proper monitoring techniques, such as difference spectroscopy, are available and when calorie supplementation is critical for the patient's recovery. The *in vitro* data indicate that neither soybean nor safflower oil emulsions reduce bilirubin binding capacity. Components of liquid emulsions, specifically egg phosphatides and glycerol, produce minimal effects on bilirubin binding, phosphatides enhance binding and glycerol reduces it. The data indicate that the concentration and molecular species of fatty acids which would result from the hydrolysis of infused lipid in human neonates would probably not increase the risk of bilirubin toxicity. Rather, in the usual molar ratios of free fatty acid to albumin, there appears to be an enhancing effect. The *in vivo* data confirm this observation. We conclude that judicious use of intravenous lipid in infants needing this therapy is justified when measurements of reserve binding capacity and free fatty acid to albumin molar ratios are possible. Extra caution should be exercised with sick or small-for-gestational-age infants in whom free fatty acid concentrations which exceed safe limits are observed occasionally. The general application of these findings to practical newborn medicine should be reserved until such time that confirmation of the findings are made because the bilirubin binding determination used in these studies is limited, as are all others[18], and because factors other than binding may play a role in the development of kernicterus[19].

ACKNOWLEDGEMENTS

We gratefully acknowledge the technical assistance of Sharon R. Gross and the secretarial assistance of Karen Willard. This work was supported by University of Tennessee New Faculty Grant H00039 (PFW) and by a grant from Abbott Laboratories, Chicago.

References

1 Wennberg, R. P., Ahlfors, C. E. and Rasmussen, L. F. (1979). The pathochemistry of kernicterus. *Early Hum. Dev.*, 3 and 4, 353
2 Brodersen, R. (1980). Bilirubin transport in the newborn infant, reviewed with relation to kernicterus. *J. Pediatr.*, 96, 349
3 Odell, G. B. (1973). Influence of binding on the toxicity of bilirubin. *Ann. New York Acad. Sci.*, 226, 225
4 Brodersen, R. (1978). Free bilirubin in blood plasma of the newborn: effects of albumin, fatty acids, pH, displacing drugs and phototherapy. In L. Stern, W. Oh and B. Friis-Hansen (eds.). *Intensive care of the newborn.* Vol. II, pp. 331–345. (New York: Masson)
5 Starinsky, R. and Shafrir, E. (1970). Displacement of albumin-bound bilirubin by free fatty acids. *Clin. Chim. Acta*, 29, 311
6 Chan, G., Schiff, D. and Stern, L. (1971). Competitive binding of free fatty acids and bilirubin to albumin; differences in HBABA dye versus Sephadex G-25 interpretation of results. *Clin. Biochem.*, 4, 208
7 Soltys, B. J. and Hsia, J. C. (1978). Human serum albumin. I. On the relationship of fatty acid and bilirubin binding sites and the nature of fatty acid allosteric effects – a monoanionic spin label study. *J. Biol. Chem.*, 253, 3023
8 Ash, K. O., Holmer, M. and Johnson, C. S. (1978). Bilirubin–protein interactions monitored by difference spectroscopy. *Clin. Chem.*, 24, 1491
9 Burckart, G. J., Whitington, P. F. and Gross, S. R. (1982). Reserve bilirubin binding capacity determined by difference spectroscopy: modifications and performance of the difference spectroscopy assay. *Pediatr. Gastr. Nutr.* (In press)

10 Burckart, G. J., Whitington, P. F. and Helms, R. A. (1982). The effect of two intravenous fat emulsions and their components on bilirubin binding to albumin. *Am. J. Clin. Nutr.* (In press)
11 Whitington, P. F., Burckart, G. J., Gross, S. R., Korones, S. B. and Helms, R. A. (1982). Alterations in reserve bilirubin binding capacity of albumin by free fatty acids: *In vitro* and *in vivo* studies using difference spectroscopy. *Pediatr. Gastr. Nutr.* (In press)
12 Nixon, M. and Chan, S. H. P. (1979). A simple and sensitive colorimetric method for the determination of long-chain free fatty acids in subcellular organelles. *Anal. Biochem.*, **97**, 403
13 Spector, A. A. (1975). Fatty acid binding to plasma albumin. *J. Lipid Res.*, **16**, 165
14 Soetewey, F., Rosseneu-Motereff, M., Lamote, R. and Peeters, H. (1972). Size and shape determination of native and defatted bovine serum albumin monomers. *J. Biochem.*, **71**, 705
15 Soltys, B. J. and Hsia, J. C. (1978). Human serum albumin. II. Binding specificity and mechanisms − a dianionic spin label study. *J. Biol. Chem.*, **253**, 3029
16 Malloy, H. and Evelyn, K. (1937). The determination of bilirubin with the photoelectric colorimeter. *J. Biol. Chem.*, **119**, 481
17 Doumas, B. T., Watson, W. A. and Biggs, H. G. (1971). Albumin standards and the measurement of serum albumin with bromocresol green. *Clin. Chim. Acta*, **31**, 87
18 Cashore, W. J., Gartner, L. M., Oh, W. and Stern, L. (1978). Clinical application of neonatal bilirubin-binding determinations: current status. *J. Pediatr.*, **93**, 827
19 Levine, R. L. (1979). Bilirubin: worked out years ago? *Pediatrics*, **64**, 380

18
Clinical experience with fat-containing TPN solutions

A. SITGES-SERRA, E. JAURRIETA, R. PALLARES, L. LORENTE
AND A. SITGES-CREUS

The administration of fat to patients receiving TPN is now gaining popularity in the United States but it has been routine in some European countries since a safe fat emulsion became available in the early sixties. Fat has been given to lower the osmolarity of TPN mixtures making possible their administration through peripheral veins[1,2] or, in weekly infusions, to prevent essential fatty acid deficiency. Solassol and Joyeux[3] from Montpellier introduced the concept of 'mélange nutritif' (nutritional mixture) in 1972. They advanced the idea that TPN solutions should be as complete and physiological as possible and suggested that all the basic nutrients could be delivered simultaneously after being mixed in a single container. They showed convincingly that fat emulsions could be incorporated into solutions containing glucose, amino acids, electrolytes and vitamins and that the resultant mixture was stable at 4 °C. A silicone U-shaped bag was designed by these authors in which all the nutrients were mixed and then infused through a central venous catheter over 12–24 h[3-5]. These bags have now been replaced by plastic disposable containers.

In July, 1974 we began to use Intralipid-containing TPN mixtures and a report is presented on the clinical experience gained in the treatment of 395 surgical patients for a total of 9100 days of TPN.

METABOLIC AND TECHNICAL CONSIDERATIONS

A survey of the literature showing the safety and efficacy of fat infusions is beyond the scope of the present paper. The subject has been recently reviewed by Wretlind[1] and Pelham[6]. Hallberg[7] and Carpentier[8] have shown that fat is well metabolized in fasting as well as in stressed and septic patients.

Routine administration of a part of the non-protein calories as fat has several metabolic advantages. It avoids hyperinsulinism[9], fatty liver, rebound hypoglycaemia, depressed visceral protein synthesis and acute hypophosphataemia. In patients receiving 50% of the calories as glucose and 50% as fat, insulin levels are in the postprandial range of orally fed man[10]. The use of fat emulsions reduces the amount of glucose necessary to meet daily caloric requirements and thus facilitates the control of glucose metabolism. Hyperglycaemia and glycosuria are minimized and exogenous insulin is rarely needed except in diabetic patients or in those with marked glucose intolerance. Even in these cases the infusion of fat will reduce the risk of major metabolic complications resulting from disturbed glucose homeostasis. Lipid emulsions have a high energy/volume ratio and prevent the essential fatty acid deficiency that may occur early in the course of TPN[11,12]. Fat seems to be the fuel of choice for the hypermetabolic patient with compromised respiratory function in whom glucose overload increases CO_2 production[13,14]. A metabolic stress secondary to glucose overfeeding has been described by Kinney and co-workers[13,15] in hypermetabolic patients. They showed that loading these patients with glucose resulted in an increase of the BMR associated with a high urinary excretion of norepinephrine and increased plasma T_3.

The addition of lipids to the conventional glucose–amino acid mixtures is desirable as it makes TPN delivery easier and safer by reducing the number of manipulations of the i.v. line. This will reduce the nurse's workload and the risk of sepsis. The single bag system makes unnecessary the peripheral infusion of fat and lessens the risk of glycosuria in intermittent infusion schedules. All these advantages apply to patients in hospital as well as at home.

STABILITY OF LIPID CONTAINING TPN MIXTURES

The stability of complete TPN mixtures has been studied by Solassol's group[3,5] who found no change in lipid particle size after periods of storage ranging from 7 days to 2 months at 4 °C. At an ambient temperature (24 °C) there were no changes in stability of the mixture at 24 h. Composition studies did not reveal qualitative alterations after prolonged storage at 4 °C. After 6 months a slight decline in triglycerides with a parallel increase in fatty acid levels was noted. Stability of TPN mixtures with lipids has also been studied by Hardy and Klim[16]. They mixed Synthamin * 8, 5%, with and without electrolytes, with glucose and Intralipid and showed that the solution remained stable after storage at room temperature for 24–48 h. In our experience, breaking of the fat emulsion has only happened in prepared mixtures inadvertently left at high ambient temperatures for a week or more. If this takes place, oil droplets coalesce forming an oil–water interphase which can be seen on the top of the mixture. The solutions should be discarded.

*Travasol[R]

Table 18.1 Standard formulations at the 'Principes de España' hospital

Glucose (g)	Fat (g)	Nitrogen (g)	Energy (cal)	N/cal ratio
200	100	9	1800	1/200
300	100	12	2200	1/183
400	100	15	2600	1/173
400	150	18	3100	1/180

All formulations contained the following components

NaCl	0–250 mEq	Zn	3 mg
KCl	0–200 mEq	Mn	0.5 mg
K_2HPO_4	20–40 mEq	Cr	10 μg
Mg	13 mEq	Cu	1 mg

Vitamins, not including folic acid, biotin, vitamins K and B_{12} which are given intramuscularly

COMPOUNDING AND DELIVERY

The composition of our basic nutritional solutions is shown in Table 18.1. Water and electrolytes are given as required taking account that some patients receive fluids, ions and drugs through a second i.v. line. As a rule 2–3 l of water are given daily to deliver 35–45 kcal/kg day and 1.5 g/kg day of crystalline amino acids. Non-protein calories are given as glucose (60–40% of the total) and fat (40–60%). Zinc, chromium, manganese and copper are routinely added. Vitamins, excepting vitamin K, biotin, folic acid and B_{12}, are also incorporated in the mixture.

The TPN solutions are prepared in the pharmacy, in an aseptic area equipped with a laminar flow hood. The electrolytes and vitamins are added first to the glucose solutions. Then, the different components of the mixture are pooled by gravity into a 3 l plastic bag. The fat emulsions are then added. After compounding, the solutions are stored at 4 °C and administered within the next 24 h. The TPN bags are supplied to the wards where a specialized nurse changes the bags. The flow rate is adjusted to provide a regular infusion over 24 h. We use no pumps but careful control of the flow rate is necessary to avoid fluid overload or hyperkalaemia due to rapid infusion. Intermittent infusions are carried out in suitable cases. The urine is checked for glucose twice a day while in most centres using glucose as the only energy source this control is carried out every 6 h.

CLINICAL EXPERIENCE

Between July 1974 and March 1982, 395 adult surgical patients received fat-containing TPN mixtures. The main indications for TPN were post-operative enterocutaneous fistulas (30%), severe acute pancreatitis (20%), septic post-operative complications (20%), oesophageal and gastric cancer (15%) and inflammatory bowel disease, chemotherapy and other disorders (15%). The only contraindication for fat administration has been essential or secondary hyperlipidaemia. The mean treatment time has been 23 days, with a fifth of our patients receiving TPN for 30–120 days. About 25 000 l of lipid-containing TPN mixtures were prepared and administered. The amount of Intralipid given varied between 1.4 and 2.9 g/kg day, with a mean of 2.15 g/kg day.

Table 18.2 Substrate-related metabolic complications in 395 patients treated with fat-containing TPN solutions

Complications	No. of patients	% of total
Hyperosmolar dehydration	3	0.75
Hyperglycaemia needing insulin	32	8
Rebound hypoglycaemia	2	0.5
Essential fatty acid deficiency	0	0

Very few metabolic complications related to substrate infusion were seen (Table 18.2). Three patients had hyperosmolar hyperglycaemic dehydration with impairment of consciousness. All of them were over 60 years of age and were receiving less than 300 g of glucose per day. One died in septic shock and the other two recovered after discontinuing TPN and starting treatment with insulin. Less than 10% of the patients required insulin because of previous diabetes or marked glucose intolerance due to their primary illness (sepsis in old people, acute necrotizing pancreatitis, pancreatic resections, etc.). In these cases fat administration was increased to 2.5–2.9 g/kg day accounting for 60–70% of the total caloric requirement. A better control of the blood glucose was achieved with this regimen combined with 6-hourly subcutaneous injections of insulin (8–18 units). As the final glucose concentration of these mixtures rarely exceeds 15%, rebound hypoglycaemia is exceptional even with intermittent infusions. The administration of complete TPN mixtures can be stopped at any time and there is no need to start an infusion of 5% glucose. Obviously, there was no essential fatty acid deficiency.

COMPLICATIONS

No complications attributable to the infusion of fat were observed. Acute reactions secondary to rapid infusions such as fever, shivering and pain in the neck were not seen when lipids were mixed with other TPN components and delivered over 12–24 h. In no case did the fat emulsion break before or during its administration. No coagulation disorders or respiratory problems were identified that could be linked with the infusion of lipids. Contamination of the bags has been kept under 0.2%. In a survey of catheter-related sepsis in 173 consecutive patients on TPN, we identified two cases of *Staphylococcus epidermidis* bacteraemia secondary to contaminated solutions[17]. No case of bacteraemia due to contamination of the bags has been seen during the last 6 months in which 1340 TPN bags have been prepared under the care of a specialized pharmacist. Of our TPN patients, 75% develop minor abnormalities of liver function and 1.5% had severe clinical symptoms due to intrahepatic cholestasis[18]. Minor hepatic dysfunction is as frequent in patients receiving fat as in those receiving only glucose[19,20]. Fatty infiltration of the liver is often seen in liver biopsies of patients not given lipids[19,20] while only mild periportal inflammatory changes have been noted in our cases (unpublished observations). In longterm TPN patients, hyperbilirubinaemia was not related to lipid dosage. Of 32 recent patients treated for 21–118 days (mean 46 days) 16 developed mild hyperbilirubinaemia (less than 5 mg/dl or 85 mmol/l) and 16 ended the TPN treatment with normal bilirubin values. The mean dosage of Intralipid in the first group was 2.17 ± 0.2 g/kg day while the second group received 2.11 ± 0.5 g/kg day. This slight difference is not significant. It is important to notice that no patient received 3 g of lipid or more per day. Higher dosages resulted in a 60% incidence of frank hyperbilirubinaemia and hypercholesterolaemia in 18 patients treated for a mean of 39 days by Allardyce (personal communication). At this time we feel that fat, given at dosages below 3 g/kg day, is not an aetiological factor in the pathogenesis of liver abnormalities during TPN.

References

1 Wretlind, A. (1977). Lipid emulsions and technique of peripheral administration in parenteral nutrition. In J. M. Greep *et al.* (eds.). *Current Concepts in Parenteral Nutrition.* pp. 273–297. (The Hague: Martinus Nijhoff Medical Division)

2 Deitel, M. and Kaminski, V. (1974). Total parenteral nutrition by peripheral vein – the lipid system. *CMA Journal*, **111**, 152

3 Solassol, C. and Joyeux, H. (1974). Nouvelles techniques pour nutrition parentérale chronique. *Ann. Anesth. Fran. Spécial*, **2**, 75

4 Astruc, B. (1973). L'intestin artificiel experimental. Nutrition parentérale exclusive par voie portale chez le chien. *Thèse de Doctorat*, Faculty of Medicine, Montpellier

5 Solassol, Cl., Joyeux, H., Astruc, B., Fourtillan, J. B., Hazane, C., Saubion, J. L. and Jalabert, M. (1980). Complete nutrient mixture with lipids for total parenteral nutrition in cancer patients. *Acta Chir. Scand.*, **498** (Suppl.), 151

6 Pelham, L. D. (1981). Rational use of intravenous fat emulsions. *Am. J. Hosp. Pharm.*, **38**, 198

7 Hallberg, D., Holm, I., Obel, A. L., Schuberth, O. and Wretlind, A. (1967). Fat emulsions for complete intravenous nutrition. *Postgrad. Med. J.*, **43**, 307

8 Carpentier, Y. A., Nordenstrom, J., Robin, A. and Kinney, J. M. (1981). Glycerol turnover and kinetics of exogenous fat in surgical patients. *Acta Chir. Scand.*, **507** (Suppl.), 226

9 Jeejeebhoy, K. N., Anderson, G. H., Nakhooda, A. F., Greenberg, G. R., Sanderson, I. and Marliss, E. B. (1976). Metabolic studies in total parenteral nutrition with lipid in man. *J. Clin. Invest.*, **57**, 125

10 Jeejeebhoy, K. N. (1977). Relationship of energy and protein input to nitrogen retention and substrate hormone profile. In J. M. Greep *et al.* (eds.). *Current Concepts in Parenteral Nutrition*, pp. 313–322. (The Hague: Martinus Nijhoff Medical Division)

11 Wilmore, D. W., Moylan, J. A., Helmkamp, G. M. and Prueitt, B. A. Jr. (1973). Clinical evaluation of a 10% intravenous fat emulsion for parenteral nutrition in thermally injured patients. *Ann. Surg.*, **178**, 503

12 O'Neill, J. A., Caldwell, M. D. and Meng, H. C. (1977). Essential fatty acid deficiency in surgical patients. *Ann. Surg.*, **185**, 535
13 Askanazi, J., Carpentier, Y. A., Elwyn, D. H., Nordenstrom, J., Jeevanandam, M., Rosenbaum, S. H., Gump, F. E. and Kinney, J. M. (1980). Influence of total parenteral nutrition on fuel utilization in injury and sepsis. *Ann. Surg.*, **191**, 40
14 Elwyn, D. H., Askanazi, J., Kinney, J. M. and Gump, F. E. (1981). Kinetics of energy substrates. *Acta Chir. Scand.*, **507** (Suppl.), 209
15 Elwyn, D. H., Kinney, J. M. and Askanazi, J. (1981). Energy expenditure in surgical patients. *Surg. Clin. N. Am.*, **61**, 545
16 Hardy, G. and Klim, R. A. (1981). The physical stability of parenteral nutrition mixtures with lipids. Presented at the *3rd European Congress on Parenteral and Enteral Nutrition.* September 27–30, Maastricht, The Netherlands
17 Sitges Serra, A., Puig, P., Jaurrieta, E., Garau, J., Alastrue, A. and Sitges Creus, A. (1980). Catheter sepsis due to *Staphylococcus epidermidis* during parenteral nutrition. *Surg. Gynecol. Obstet.*, **151**, 481
18 Sitges Creus, A., Cañadas, E. and Vilar, L. (1978). Cholestatic jaundice during parenteral alimentation in adults. In I. D. A. Johnston (ed.). *Advances in Parenteral Nutrition.* pp. 461–466. (Lancaster: MTP Press)
19 Grant, J. P., Cox, Ch. E., Kleinman, L. M., Maher, M. M., Pittman, M. A., Tangrea, J. A., Brown, J. H., Gross, E., Beazley, R. M. and Scott Jones, R. (1977). Serum hepatic enzyme and bilirubin elevations during parenteral nutrition. *Surg. Gynecol. Obstet.*, **145**, 573
20 Lindor, K. D., Fleming, C., Abrams, A. and Hirschkorn, M. A. (1979). Liver function values in adults receiving total parenteral nutrition. *J. Am. Med. Assoc.*, **241**, 2398

19
The stability of fat emulsions for intravenous administration

S. S. DAVIS

INTRODUCTION

An emulsion is the dispersion of one immiscible liquid in another in the form of small droplets. Such systems are thermodynamically unstable and will revert to two separate phases by the coalescence of droplets unless a barrier to coalescence can be created using a third component, the emulsifying agent. The barrier produced by the emulsifying agent can be mechanical; in the form of a thick rigid interfacial film, and/or electrical in the form of electrostatic repulsion.

Table 19.1 Lipid composition of lecithin from two natural sources

Lipid	Composition of lecithins (%)	
	Egg yolk	Soybean
Phosphatidylcholine (PC)	73	40
Lysophosphatidylcholine (LPC)	5.8	–
Phosphatidylethanolamine (PE)	15.0	35
Lysophosphatidylethanolamine (LPE)	2.1	–
Phosphatidylinositol (PI)	0.6	20
Phosphatidic acid (PA)	–	–
Sphingomyelin (SP)	2.5	5

Data from references 2 and 3

Fat emulsions
Fat emulsions employed in parenteral nutrition are stabilized by the natural emulsifying agent lecithin obtained from animal (egg) or vegetable (soybean) sources. Strictly speaking, lecithin is the name for phosphatidylcholine (PC), the most common of the naturally occurring phospholipids[1] but the name is

often used to embrace all the phospholipids found in emulsifiers used for intravenous emulsions. Typical compositions of egg and soy lecithin are given in Table 19.1[2,3].

Phospholipids are good emulsifying agents because they possess within the same molecule regions of opposing affinities for oil and water[4], i.e. hydrophilic and hydrophobic regions. The basic structure of a phospholipid is shown in Table 19.2. The group X is the polar or hydrophilic function while

Table 19.2 Phospholipid structure and functional groupings

Basic structure

$$H_2COOCR'$$
$$HCOOCR''$$
$$H_2CO\!-\!\!-\!\!P(\!=\!\!O)\!-\!\!OX$$
$$OH$$

	R'	R''	X
PC	Fatty acid residues (see Table 19.3)		$-CH_2CH_2N(CH_3)_3$
PE	,,	,,	$-CH_2CH_2NH_3$
PS	,,	,,	$-CH_2CH\!<^{NH_3}_{COO^-}$
PI	,,	,,	inositol ring
PA	,,	,,	$-H^+$
LPC	,,	OH	$-CH_2CH_2N(CH_3)_3$

PS = phosphatidylserine. For other abbreviations see Table 19.1

Table 19.3 Fatty acid residues in egg lecithin

Residue	%
Myristic acid	0.09
Palmitic acid	32.5
Palmitoleic acid	0.4
Stearic acid	15.7
Oleic acid	32.0
Linoleic acid	11.3
Linolenic acid	0.3
Arachidic acid	0.1
Arachidonic acid	0.2
Behenic acid	3.4
Unidentified acids	0.2

Data from reference 5

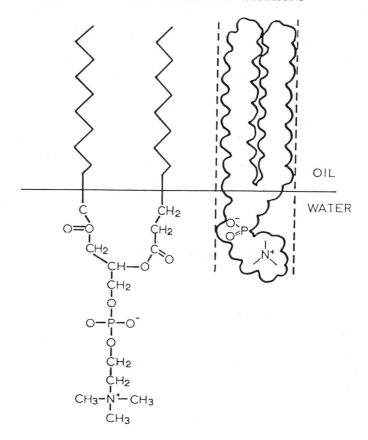

Figure 19.1 The adsorption of phospholipid (PC) at the oil–water interface. (Space-filling Corey–Pauling–Koltun molecular model)

the groups R_1 and R_2 are not specific even for one type of phospholipid but in general they are of about equal proportions of saturated and unsaturated species. A typical breakdown for egg yolk phospholipids is given in Table 19.3[5].

The dual characteristics of the phospholipid molecule cause it to orientate itself at an oil–water interface in order to satisfy both affinities[6] (Figure 19.1).

Emulsion stabilization

The presence of the emulsifier at the oil–water interface creates a mechanical barrier to coalescence and if the emulsifier molecules can pack into a coherent film then good stability results. Close-packed films can be produced by the use of mixtures of simple emulsifiers or the use of co-emulsifiers that can

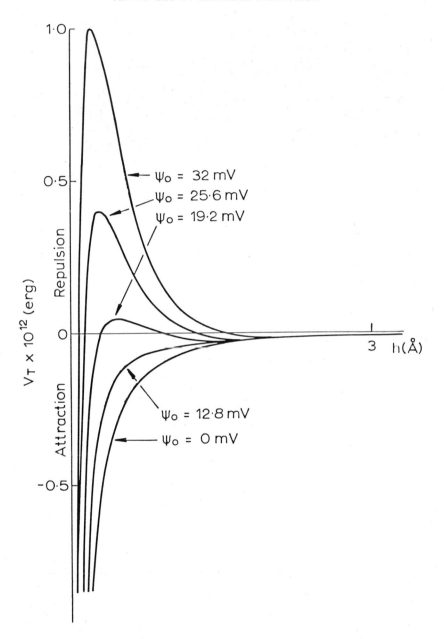

Figure 19.2 Potential energy–separation curves, showing the influence of surface potential on the total potential energy of interaction of two spherical particles (from reference 13). Particle radius (a) = 100 nm; Hamaker constant = 10^{-19} J; h = R/a where R is the distance between centres of spheres

complex with the emulsifier at the interface to form a solid-like (liquid crystalline) barrier[7]. Other emulsifiers such as proteins and other macro-molecules produce thick, interfacial films either by a process of denaturation or molecular rearrangement and the building-up of multiple layers of stabilizing agent.

Some emulsifiers confer stability not just through a mechanical barrier but through the creation of an electrical (electrostatic) barrier; for example by the ionization of the hydrophilic part of the emulsifier molecule. The charged surface of the emulsion droplet will then act to repel the close approach of the other similarly charged emulsion droplets. Emulsifying agents that can produce mechanical as well as electrical barriers to coalescence have obvious advantages in pharmaceutical formulations.

Electrostatic repulsion

The importance of electrostatic forces in colloid and thus emulsion stability has been analysed quantitatively. The theory so developed has been called the DLVO theory after the names of the two groups who developed it[8,9]. Full details of this theory are given in textbooks of colloid science[10,11] and only the relevant aspects will be considered here.

The theory considers the two different forces that will operate in a dispersion of small colloidal particles; those of repulsion (electrostatic effect) and attraction. Small particles will attract each other through what are known as van der Waals forces that arise from interactions between pairs of atoms or molecules on neighbouring particles. Thus, the particles in an emulsion will tend to come together and coalesce. The charge on the droplets arising from the ionization of groups on the surface will provide a repulsion if the two interacting particles have the same surface charges and potentials. The DLVO theory combines the electrostatic repulsive energy V_R with the attractive potential energy V_A to give the total potential energy of interaction

$$V_{total} = V_A + V_R$$

V_{total} will change as the particles approach each other. The influence of surface potential on the total potential energy for interaction is shown in Figure 19.2 for a model system.

At small separations, the V_A term dominates and a strong attraction will result in particle aggregation and coalescence.

At intermediate separations and surface potentials greater than about 15 mV, V_R is dominant and there is a maximum in the potential energy diagram. This can be considered to be an energy barrier that will prevent coalescence. The higher the energy barrier the greater the stability. Emulsion droplets with high surface potentials will have correspondingly high energy barriers.

The surface potential on an emulsion droplet will be dependent upon the state of ionization of the emulsifying agent. Many emulsifiers are ionized at all pH values (e.g. sodium lauryl sulphate) or used at pH values when the emulsifying molecule is effectively 100% ionized (e.g. the soaps). However, with natural materials the ionization state can depend markedly upon the pH value. Zwitterionic emulsifiers such as proteins will have a pH value for zero charge, the isoelectric point, which may coincide with aggregation of the emulsion, and a decrease in stability[12]. However, reduction in ionization can also lead to a corresponding ability for closer packing of emulsifier molecules at the oil—water interface (less mutual repulsion from adjacent molecules) as well as a decreased solubility of the emulsifier in the aqueous phase. These factors will lead to an enhanced mechanical stability.

Effect of added electrolyte
The addition of electrolytes will have a profound effect by reducing the

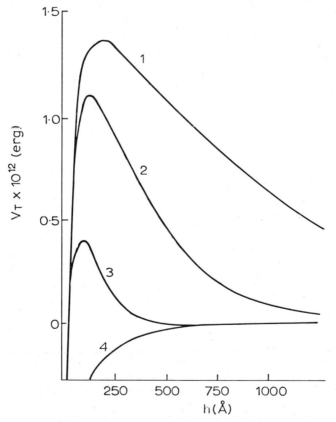

Electrolyte content (1:1 electrolyte) (mol l^{-1}): curve $1 - 10^{-5}$, curve $2 - 10^{-4}$, curve $3 - 10^{-3}$, curve $4 - 10^{-2}$

Figure 19.3 The effect of electrolyte content on the total potential energy of interaction of two spherical particles (from reference 13). Physical parameters as in Figure 19.2. Surface potential in absence of electrolyte = 25.6 mV

surface potential and thereby leading to aggregation of the emulsion (Figure 19.3)[13]. The DLVO theory will allow estimation of the concentration of electrolyte that will cause aggregation for a system with a given surface potential[13]. The theory predicts that the aggregation concentrations of electrolytes containing ions of opposite charge to the charge on the surface of the droplet will depend markedly on the charge numbers, so that for the addition of NaCl, CaCl$_2$ and AlCl$_3$ to a negatively charged emulsion, the concentrations required to produce aggregation will be in the ratio

$$\frac{1}{1^6} : \frac{1}{2^6} : \frac{1}{3^6}$$

or   100 : 1.6 : 0.14

This is a quantitative statement of the well known Schutze Hardy rule of colloid stability that defines the effectiveness of monovalent, divalent and trivalent salts in causing flocculation (aggregation)[14].

These theoretical considerations will be very helpful to us later in the chapter when we consider the possible consequences of adding electrolytes to fat emulsions or changing the pH by the addition of an acidic solution such as dextrose.

STABILITY OF FAT EMULSIONS

The properties of phospholipids

Commercial fat emulsions are stabilized by egg or soy lecithins (Table 19.4). As we have seen earlier these lecithins are not pure chemical entities but are complex mixtures of phospholipids. In the development of suitable fat emulsions for parenteral nutrition pure phospholipids (PC and phosphatidyl-ethanolamine (PE)) were employed but the emulsions produced had poor stability[15]. This was because some pure phospholipids, especially PC, form neither good mechanical nor electrical barriers to droplet coalescence.

Rydhag[16] has shown the necessity of incorporating ionic lipids into lecithin emulsifiers in order to produce stable emulsions. These ionic lipids (materials

Table 19.4 Some commercial fat emulsions

Trade name	Oil phase	Emulsifier	Other components
Intralipid (Kabi-Vitrum)	Soybean 10% or 20%	Egg lecithin 1.2%	Glycerol 2.5%
Lipofundin (S) (Braun)	Soybean 10% or 20%	Soybean lecithin 0.75% or 1.2%	Xylitol 5.0%
Lipofundin (Braun)	Cotton-seed 10%	Soybean lecithin 0.75%	Sorbitol 5.0%
Liposyn (Abbott)	Safflower 10% (and 20%)	Egg lecithin 1.2%	Glycerol 2.5%
Travemulsion (Travenol)	Soybean 10% or 20%	Egg lecithin 1.2%	Glycerol 2.5%

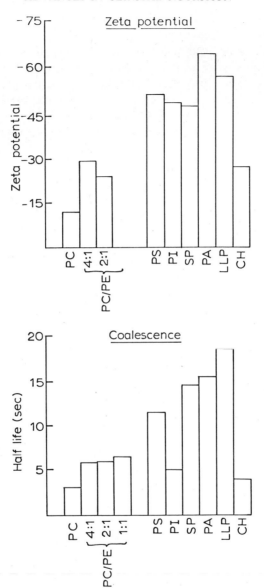

PS = Phosphatidylserine
PI = Phosphatidylinositol
SP = Sphingomyelin

PA = Phosphatidic acid
LLP = Lysophosphatidylcholine
CH = Cholesterol

Figure 19.4 The effect of minor phospholipid components on the coalescence behaviour and surface potential of soybean oil droplets stabilized by mixtures of phosphatidylcholine (PC) and phosphatidylethanolamine (PE). All minor impurities examined at a concentration of 0.25% total phospholipid added to a 4:1 mixture of PC:PE (10^{-3} mol l^{-1} for coalescence studies) (1.2% for emulsion microelectrophoresis)

such as phosphatidic acid (PA) and phosphatidylserine (PS) not only increased the surface potential but also produced complex interfacial films with PC and PE. Liquid crystalline gel structures were formed at the interface. The formation of a double bilayer structure of about 80 Å in thickness at the oil–water interface has been proposed[17]. Thus acceptable emulsion stability can be achieved by choosing a commercially available lecithin that contains a suitable amount of negatively charged phospholipid[16]. The exact nature of the charged phospholipids is not critical provided of course that they are acceptable from toxicological considerations.

Our own work at Nottingham supports this view. Hansrani[18] has studied the mechanical strength of phospholipid films by examining coalescence of individual droplets of soybean oil at the plane oil–water interface. The measurement of the times for coalescence for a large number of separate droplets allowed her to derive a rate constant ($T_{1/2}$) (being the time in which 50% of the drops would coalesce). Figure 19.4 shows how the value of this parameter can be altered by the addition of charged phospholipids as well as the hydrolysis product of lecithin, lysolecithin (lysophosphatidylcholine). Cholesterol has a destabilizing effect.

Surface potential
The various additives also bring about an increase in the surface potential on the droplets thereby enhancing stability by increased electrostatic repulsion. The surface potential data in Figure 19.4 are referred to at a pH value of 7.00. Figure 19.5 shows the complete pH–potential diagram for a commercial fat

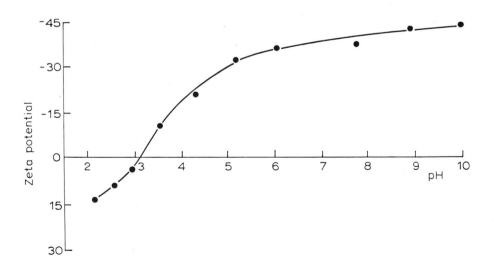

Figure 19.5 Surface (zeta) potential versus pH for Intralipid (10% soybean oil emulsified with 1.2% egg lecithin) (from reference 19)

Table 19.5 Ionization characteristics of phospholipids[20]

Phosphatide	Ionic species	Ionization characteristics
Phosphatidic acid (PA)	Primary phosphate PO_4^{2-}	Strong acid (pK_a 3.8, 8.6)
Phosphatidylcholine (PC) ⎫ Lysophosphatidylcholine (LPC) ⎬	Secondary phosphate−choline $PO_4^- - NMe_3^+$	Isoelectric over wide range of pH
Phosphatidylethanolamine (PE)	Secondary phosphate−amine $PO_4^- - NH_3^+$	Negative at pH 7 (pK_a 4.1, 7.8)
Phosphatidylserine (PS)	Secondary phosphate−carboxyl−amine $PO_4^- - COO^- - NH_3^+$	Negative at pH 7.5 (pK_a 4.2, 9.4)
Phosphatidylinositol (PI)	Secondary phosphate−sugar $PO_4^- - sugar$	Negative above pH 4 (pK_a 4.1)

emulsion (Intralipid)[19]. At low pH value the emulsion carries a positive charge but passes through zero charge around a pH value of 2.5 to become negative with a plateau region above a pH value of 5.5. The surface (zeta) potential for this plateau region is above −30 mV. The negative charge arises from the ionization of the various minor components. Ionization characteristics are given in Table 19.5[20]. Figure 19.6 shows the zeta potential−pH curves for various phospholipid liquid crystalline particles[21]. The neutrality of pure PC and the properties of the minor components are well demonstrated.

The effect of additives
Intravenous fat emulsions normally have a pH value around neutrality. However, the addition of certain additives (e.g. dextrose) can reduce the pH value considerably[22] so that the zeta potential is reduced and the systems will become unstable. The zeta potential can also be reduced by electrolyte as is

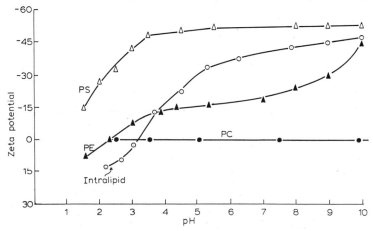

Figure 19.6 Surface (zeta) potential−pH curves for various phospholipid liquid crystalline particles (from reference 20). ■ Intralipid, ▲ PE, ▼ PS, ● PC

shown in Figure 19.7[18]. The DLVO theory suggests that the zeta potential (as an approximation of the surface potential) should fall with the square root of concentration[13]. Dawes and Groves[23] investigated the effect of added electrolyte on the properties of Intralipid. Even small amounts of electrolyte were able to bring about a change in the surface potential. Monovalent cations

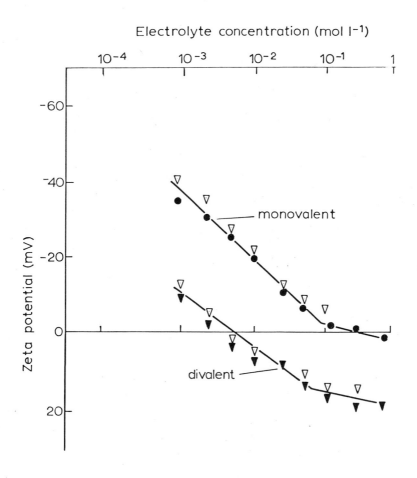

Figure 19.7 The effect of electrolyte concentration and type on the zeta potential of Intralipid (from reference 23). ∇ NaCl, ● KCl, ▼ CaCl₂, ∇ MgCl₂

(K^+, Na^+) both produced similar effects, as did the divalent ions Ca^{2+} and Mg^{2+}. For the divalent ions a reversal of charge occurred with the fat droplets becoming positively charged. The emulsions would therefore be expected to undergo aggregation followed by deaggregation as the concentration of divalent cation was increased. Aluminium ions (Al^{3+}) were extremely effective in providing a very rapid reversal of zeta potential. The number of charges on the anion was of little significance for all cations studied.

The continued stability of emulsions in the complete or virtual absence of charge has been interpreted in terms of steric stabilization arising from the mere presence of the adsorbed chains of emulsifier as well as an osmotic repulsion which occurs as the chains approach and increase the local concentration of high molecular weight material[24].

ASSESSMENT OF EMULSION STABILITY

Mechanisms of instability
The physical stability of an emulsion system can be manifested by a number of processes (Figure 19.8). The coalescence of droplets to form larger droplets and eventually free oil is one form of instability that will have important clinical consequences. However, emulsions can demonstrate more subtle forms of instability that may increase the likelihood of coalescence.

If the electrostatic repulsive forces keeping droplets apart are suppressed the droplets can undergo aggregation. These aggregates have a much larger size than the individual droplets and consequently will rise to the surface of the emulsion to form a cream layer. The particles in this cream layer will be close-packed thereby encouraging coalescence.

Aggregation, although reversible, may not be desirable from a different standpoint, clinical acceptability. Opinion is divided as to the break-up or not

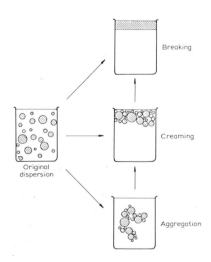

Figure 19.8 Mechanisms of emulsion instability

of aggregated fat droplets inadvertently administered to patients[25]. Some would argue that the shear forces in the circulation would result in the rapid break-up of aggregates into individual droplets, whereas other opinion considers that the electrolyte content of the blood (especially the free calcium content) and the adsorption of blood components to the surface of the aggregate could enhance the aggregation state.

Measurement of stability
As a consequence of the factors discussed above the assessment of emulsion stability will need to include the (quantitative) measurement of the following.

(a) Particle size and particle size distribution.

(b) Surface (zeta) potential.

(c) Aggregation state.

Also important are other parameters such as pH value, free fatty acid content etc. This chapter will consider methods for (a), (b) and (c).

Particle size analysis
The sizes of droplet in a commercial fat emulsion range from less than 10 nm to about 1 μm. Larger sizes are found in fat emulsion systems that have been mixed with carbohydrates and electrolytes. Unstable and clinically unacceptable systems may have particles of 20 μm or more; the appearance of droplets of free oil on the surface of the emulsion may also occur. Clearly no one method of particle size analysis can cover such a wide range of particle size. In our studies we use not one method but three. Each has its advantages and disadvantages and provides different information.

Figure 19.9 shows the particle size versus frequency distribution for a typical fat emulsion and the applicability of the methods. It should be noted that the lower limit of the Coulter Counter is substantially lower than that shown in the figure, but the measurement of particles less than 800 nm is a difficult procedure and the alternative method of laser light scattering (photon correlation spectroscopy) is to be preferred. Full description of the principles of measurement for photon correlation spectroscopy and the Coulter Counter can be found elsewhere[26,27] and only a brief description of the methods will be given here.

Each method will provide a different answer so far as particle size is concerned depending on its level of resolution. The light microscope will not see particles below 1 μm and effectively these will be ignored. Similarly the Coulter Counter will see only those particles within its measurement range. Consequently the quoted size parameters will be method-dependent.

Laser light scattering   If one wishes to obtain a true estimation of the mean diameter of the particles in a fat emulsion then a method that will size

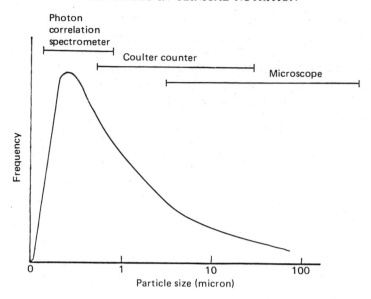

Figure 19.9 The particle size distribution of an intravenous fat emulsion and methods of particle size determination

particles in the range 10–1000 nm is essential. Methods based upon laser light scattering (photon correlation spectroscopy (Malvern) Nanosizer (Coulter))[28] are ideal for this purpose. When laser light is passed through a sample of colloidal particles undergoing Brownian motion the number of photons scattered in a given direction per unit time is a function of the number, size and relative positions of the particles. Changes in the position of the particles because of their Brownian motion alter the rate of photon arrival in the detection system of the instrument which results in a departure from the expected randomness of arrival. Advanced digital processing techniques are used to follow this departure from randomness (statistically this is handled by an autocorrelation function). The change in the autocorrelation function with time can be used to obtain a value for mean particle diameter and the polydispersity of the distribution.

Some particle size values obtained using the Malvern photon correlation spectrometer with commercial fat emulsions are given in Table 19.6 together with their polydispersity index. The values for fat emulsions agree well with determinations using alternative methods, e.g. electron microscopy, centrifuge[29,30].

It will be noted that the mean sizes of the fat emulsion are well below 1 μm.

Laser light scattering can be used to follow changes of particle size with changes in formulation, storage etc., however the method is not able to sample particles above 3 μm both on theoretical grounds and for statistical reasons. In clinical practice adverse reactions and side-effects result from

Table 19.6 Particle size analysis of commercial fat emulsions using laser light scattering

Emulsion	Mean diameter (nm)	Polydispersity factor
Intralipid (10%)	250	2.5
Intralipid (20%)	410	2.3
Liposyn (10%)	190	1.1
Lipofundin (S) (20%)	340	3.6
Travemulsion (10%)	210	2.0
Travemulsion (20%)	240	1.4

particles in the 5–20 μm size range or from large free oil droplets and not from those below 1 μm. Thus it is useful in some cases to ignore the smaller particles and to create a particle size analysis window to see those particles that matter clinically.

The Coulter Counter The Coulter Counter is an electronic particle size counter developed originally for the counting of red blood cells. It is ideally suited to examine particles in the 0.8–20 μm size range. It works on the principle of measuring the change in resistance of a charged orifice caused by the presence of a non-conducting particle. The Coulter Counter will give an apparent mean diameter for the particle it sees but since the lower limit in our work is set at 0.8 μm it effectively ignores particles smaller than this lower limit. Thus the quoted diameter is an apparent diameter. To avoid confusion we prefer to quote instead the percentage of particles below an arbitrary size (in our case chosen as 1.5 μm)[25].

Microscopy The Coulter Counter, like laser light scattering, will not see a small number of very large particles outside its window of measurement. These particles may be totally insignificant so far as the statistics of size analysis are concerned but could be very relevant clinically. Hence the conventional light microscope is used to measure particles of 20 μm or more. Visual inspection of the emulsion is also undertaken to detect the presence of free oil droplets on the surface as well as the presence of a cream layer. The importance of using the three different methods of size analysis as well as visual observation was made clear by work undertaken by Hansrani[18] who found for some experimental emulsion systems that the mean particle diameter decreased slightly upon autoclaving even though visible free oil droplets could be seen on the surface of the emulsion.

Particle size limits
It is difficult to set an upper limit for particles in a fat emulsion. Some guidance can be obtained from the literature on natural fat particles, the chylomicra, that range in size from 100 nm or less to 3 μm in diameter[31]. The British Pharmacopoeia states that intravenous fat emulsions should not have

any particles greater than 5 μm in diameter. Certainly it has been established that emulsion containing large oil droplets can be hazardous[32-34] and emulsion droplets exceeding 6 μm in diameter are known to cause adverse reactions, particularly emboli in the lungs[32,35].

Zeta potential

The surface potential on emulsion droplets can be measured using the technique of microelectrophoresis. The methodology has been described by Bangham *et al.*[36]. A sample of the emulsion diluted using the appropriate continuous phase is placed in a capillary tube between two electrodes. A potential difference is applied and the particles move to the electrode of opposite charge. The rate of movement (electrophoretic mobility) is related to the zeta potential. Special electrode systems have to be employed when dealing with systems of high ionic strength (electrolyte content). The commercial apparatus (Rank Mark II) can be fitted with laser light illumination to enable dark ground visualization of particles below 1 μm.

Aggregation

The aggregation of a fat emulsion (for instance in the presence of electrolyte) can usually be detected by visual examination. However, in order to determine the critical aggregation concentration a quantitative procedure is required. A nephelometer or spectrophotometer can be used to measure the change in light transmission when aggregation occurs but such techniques normally require dilution of the emulsion sample. Instead we have adapted a method used to study the aggregation of red-blood cells[37]. A modified microscope with photo cell attachment is used to measure the light transmission through a thin film of emulsion. The kinetics of aggregation can be followed for different electrolyte concentrations. An alternative procedure is to measure the change in emulsion viscosity that occurs upon aggregation using a simple U-tube viscometer[38].

Accelerated stability testing

The commercially available fat emulsions all have excellent longterm stability when stored at 4 °C (for some room temperature also)[39]. The chemical and physical changes that occur during the normal shelf life of 2 years are minimal.

When testing new fat emulsion systems, or mixtures of fat with amino acids, carbohydrates and electrolyte, it can be an advantage not only to conduct the relevant in-use storage tests but also to accelerate the processes of instability. We have shown that conventional fat emulsions can be destabilized by two simple processes[18].

(a) Shaking at 200 shakes/min at 25 °C for periods of up to 24 h.

(b) Freeze/thaw cycles that comprise rapid deep freeze (−20 °C) followed by thawing at room temperature.

Even so-called stable systems such as Intralipid succumb to these accelerated tests (Figure 19.10). As described below ('Combined systems') these accelerated tests can be used to assess different formulations and admixtures.

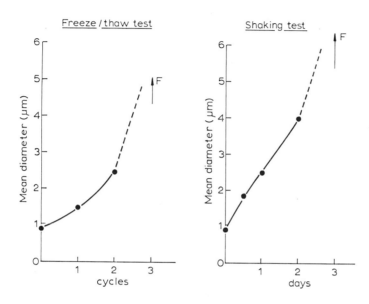

Figure 19.10 The effect of freeze/thaw cycles and shaking on emulsion stability; F indicates separation of free oil (from reference 18)

THE MIXING OF FAT EMULSIONS WITH CARBOHYDRATE, ELECTROLYTES AND AMINO ACIDS

Mixed systems
The use of fat emulsions in parenteral nutrition is now well accepted practice but up until recently it has been believed (certainly in the United Kingdom as well as by manufacturers of fat emulsion) that the fat should be kept separate from other nutrients to avoid problems of instability. Sequential or simultaneous infusion via administration sets is then necessary[40]. This is a more complicated procedure than the single 3 l bag commonly used when dextrose is employed as the calorie source. A stable feeding mixture comprising fat emulsion, amino acids, carbohydrate and electrolyte would greatly facilitate nutritional support. Reports from France have indicated that such mixed systems can be produced and that these are acceptable clinically. Solassol and colleagues[41-43] have advocated the use of mixed systems contained in special silicone rubber bags as a means of allowing patients on longterm parenteral nutrition greater independence. However, no data on the physicochemical properties of these systems have been presented. The literature contains

conflicting reports about the mixing of fat emulsions with other nutrients although Black and Popovich have attempted to give guidelines and recommendations[22].

Chapter 20 by Hardy *et al.*[44] describes the *in vitro* and *in vivo* evaluation of some mixed lipid systems and consequently this chapter will concentrate on the general principles of mixed systems and provide some guidelines. The mixed parenteral nutritional system is a sensible approach to therapy provided that the clinical team has a full understanding of the properties of fat emulsions and their susceptibilities to certain additives such as electrolytes. Commercial products are available from France and Germany that are mixed systems containing fat, amino acid and carbohydrate (Table 19.7) (electrolytes are omitted for stability reasons).

Table 19.7   Mixed parenteral feeding systems

Trade name	Oil phase	Amino acids	Emulsifier	Other components
Trive 1000 (Egic)	Soybean 3.8%	6.0%	Soybean lecithin 0.7%	Sorbitol
Nutrifundin (Braun)	Soybean 3.8%	6.0%	Soybean lecithin 0.38%	Xylitol

In the following discussion it is assumed that the mixed systems will be prepared aseptically by the hospital pharmacy department and used within a period of a few days. Some systems clearly have much longer shelf lives than this but until a definite limit for maximum permissible particle size has been set it is difficult to define a realistic shelf life.

Any modification to a commercial fat emulsion will bring about a reduction in its stability. Nevertheless, in many cases the resultant systems are quite acceptable for clinical use and offer advantages to those carrying out parenteral nutrition programmes[25].

Carbohydrates

Carbohydrates can be mixed with fat emulsion systems without giving rise to unacceptable stability problems. One exception is dextrose which has been reported to reduce the pH value of Intralipid from 7.00 to 3.5 over a 48 h storage period[22]. A cream layer of aggregated droplets was observed at the top of the emulsion.

Such changes are to be expected if we refer to Figure 19.5 showing the pH-zeta potential profile for Intralipid. The high (stabilizing) zeta potential of more than -30 mV is found in the pH value range of 5 to 10 whereas at pH 3.5 the zeta potential is below -15 mV. The electrostatic contribution to stability will be reduced as a consequence and the emulsion will be able to aggregate.

Amino acids

Mixtures of fat emulsion with amino acids (and certain carbohydrates) have good longterm stability. Indeed there is evidence that amino acids can provide a protective effect when fat emulsions are mixed with electrolytes.

Electrolytes

As discussed above the presence of electrolyte is expected to have an adverse effect on emulsion stability. The electrolyte will reduce zeta potential which in turn results in a reduction to the electrostatic barrier to repulsion. The cumulative effect of different electrolytes can be predicted using an expression of the form derived from the predictions of the DLVO theory.

$$x < a + 64b + 729c$$

where a, b and c are the concentrations of mono-, di- and trivalent cations in mmol l^{-1}. x is defined as the critical aggregation number and is the concentration of sodium chloride (in mmol l^{-1}) required to aggregate the emulsion system. The value of x will change somewhat from system to system depend-

Table 19.8 Formulations for mixed emulsion systems

System 1	10% Intralipid	500 ml
	Glucose 50%	500 ml
	Synthamin 14 with electrolytes	500 ml
		(Mg 1.7 mmol/l)
System 2	1000 ml system 1 + 2 ml Ca^{2+}/Mg^{2+} ampoule	
	(Ca^{2+} = 0.9 mmol/l and Mg^{2+} = 2.9 mmol/l)	
System 3	10% Intralipid	500 ml
	Glucose 50%	500 ml
	Synthamin 14	500 ml
System 4	10% Intralipid	500 ml
	Glucose 20%	500 ml
	Synthamin 9	500 ml
System 5	10% Intralipid	500 ml
	Glucose 20%	500 ml
	Synthamin 9 with electrolytes	500 ml
System 6	1000 ml system 5 + 2 ml Ca^{2+}/Mg^{2+} ampoule	
	(Ca^{2+} = 0.9 mmol/l and Mg^{2+} = 2.9 mmol/l)	
System 7	20% Intralipid	1 litre
	Vamin glucose	1.5 litre
	10% dextrose	0.5 litre
	51 mmol Sodium chloride (as 30% solution)	
	54 mmol Potassium chloride	
	30 mmol Potassium dihydrogen phosphate	
	(Total monovalent electrolyte = 80 mmol l^{-1})	
System 8	System 7 + Addamel (10 ml)	
	(Total divalent electrolyte = 4.24 mmol l^{-1})	
System 9	System 8 + Solvito, one vial + Vitlipid, one vial	
System 10	System 7 + Addamel (5 ml)	
System 11	System 10 + Heparin 200 i.u.	

ing on the pH value and the presence of protective substances such as amino acids. Burnham *et al.*[25] have reported recently a value of 130 for a mixed system based on Intralipid.

Combined systems

Systems comprising fat, electrolyte, amino acid and carbohydrate have been examined at Nottingham University using the various experimental techniques discussed above (Table 19.8).

The electrophoretic mobility data (expressed in terms of zeta potential) are given in Table 19.9. The effect of electrolyte on the zeta potential is readily apparent. Studies were also conducted on the particles dispersed in distilled water (effectively washing the particles free from electrolyte, carbohydrate and amino acid). These 'washed' droplets have the same zeta potential as the original particles in Intralipid indicating that no permanent changes have been induced in the stabilizing layer of phospholipid.

Table 19.9 Zeta potential of emulsion particles in some mixed nutritional systems after 12 h storage at 4 °C

| System | Zeta potential (mV) | |
	In feeding mixture as continuous phase	Distilled water
Intralipid	–	– 33
7	– 6.0	– 31
8	– 4.5	– 31
9	– 4.0	– 31

Mean of 20 measurements on individual droplets. Standard error of mean no greater than 15%. Data from reference 25

Particle size analysis data obtained using the Coulter Counter are shown in figures in histogram form. Figure 19.11 shows the change in the percentage of particles less than 1.5 μm for mixed systems (some containing monovalent and divalent cations).

All systems have good stability, even for storage up to 9 days. The stability tests can be accelerated by one freeze thaw cycle which discriminates between systems with or without added electrolytes. The effects of divalent electrolyte are shown in Figure 19.12. There is no statistical difference in the stability parameters for 1 day of storage but after 2 days the destabilizing effect of Ca^{2+} and Mg^{2+} are more evident. Freeze thaw cycles and shaking were used to accelerate the destabilization processes (Figure 19.13)[25]. Mixed systems containing divalent and trivalent cations should be prepared with some thought to the total electrolyte content as defined by the expression on page 000. If possible the prescription should be adjusted to keep well below the critical value of x. The clinical team and especially the pharmacist should be

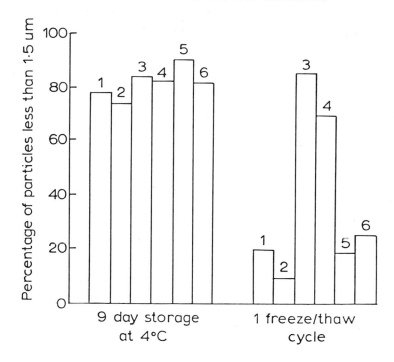

Figure 19.11 The stability of mixed parenteral nutritional systems. (Formulation details for the various systems are given in Table 19.8)

aware of the electrolyte content of the various added supplements and the important role of valency of the cations. For example 10 ml of Addamel in 3 l of a nutritional system gives the following

$a = 0$
$b = 2.19 \, \text{mmol} \, (Ca^{2+}, Mg^{2+}, Zn^{2+}, Mn^{2+}, Cu^{2+})$
$c = 0.017 \, \text{mmol} \, (Fe^{3+})$

Thus total 'electrolye content' = 152 mmol l^{-1} which is just above the critical aggregation number of 130. Any added monovalent cations (K^+, Na^+) present in the feeding mixture will make the situation worse.

In 1979, Gove et al.[46] reported a parenteral feeding regimen that comprised:

Vamin N (a pure crystalline L-amino acid mixture with added electrolyte)	1000 ml
Intralipid 20%	500 ml
20% glucose solution	1000 ml
Addamel	10 ml
Potassium phosphate injection	10 ml

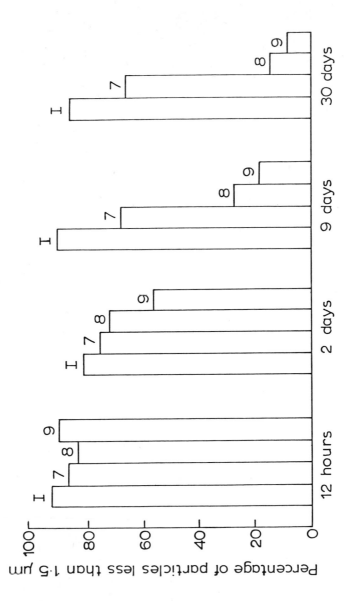

Figure 19.12 The stability of mixed parenteral nutritional systems. The effect of storage and the presence of divalent ions. Formulation details for the various systems are given in Table 19.8). I = Intralipid control, no additives

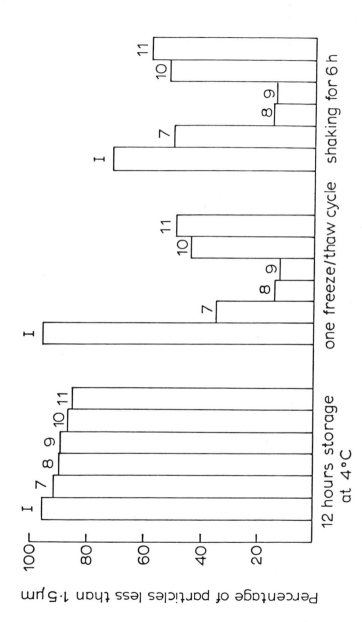

Figure 19.13 The stability of mixed parenteral nutrition systems. The use of accelerated testing procedures. (Formulation details are given in Table 19.8). I = Intralipid control, no additives

The resulting mixture after 12 h storage had a distinct 2 cm layer of cream on the upper surface and microscopic examination revealed aggregates. The authors concluded that the mixing of parenteral feeding solutions with fat emulsions should not be undertaken until further work on stability and safety had been undertaken.

We can evaluate the above regimen for total electrolyte content. $Na^+ = 50$ mmol, $K^+ = 40$ mmol, $Ca^{2+} = 7.5$, $Mg^{2+} = 3$, $Fe^{2+} = 0.05$. This gives a total electrolyte content of 798 mmol in a total volume of 2520 ml, equivalent to 317 mmol in 1 l. This is over twice the critical aggregation number of 130 mmol l^{-1}. It is therefore not surprising that the system was aggregated and unstable.

Table 19.10 Critical aggregation concentrations for electrolytes added to a diluted Intralipid system (oil content 10%, egg lecithin concentration 0.6%)

Electrolyte	Critical aggregation concentration (mmol l^{-1})	
	Intralipid	Silver iodide
NaCl	110	–
NaNO$_3$	–	140
KCl	150	–
KNO$_3$	–	136
CaCl$_2$	2.4	–
Ca(NO$_3$)$_2$	–	2.4
MgCl$_2$	2.6	–
Mg(NO$_3$)$_2$	–	2.6

Data from references 13 and 25

Emulsion aggregation

More basic knowledge about the properties of fat emulsions and their susceptibility to aggregation by added electrolyte has been gained by the microscope–photocell system described on page 228. Some critical aggregation concentrations are given in Table 19.10 for diluted Intralipid together with values for a model colloid – silver iodide. The agreement between the two different systems and the predictions of the DLVO theory is good, indicating that the behaviour of fat emulsion particles can be treated using the established theories of colloid stability[13].

Aggregation experiments have also been conducted in the presence of various amino acids[47] (Table 19.11). It is clear that certain amino acid mixtures give rise to a protective effect. The nature of this protective effect has yet to be elucidated. Black and Popovich[22] have reported a similar effect in their studies on mixed systems. They suggested that the amino acid could be adsorbed at the oil–water interface by analogy to the known surface active properties of proteins. We believe that the attachment of the amino acids at

Table 19.11 Effect of added amino acids on the aggregation of Intralipid by calcium ions. (10 ml 20% Intralipid, 5 ml calcium chloride solution and 5 ml of amino acid)

System	Concentration of CaCl₂ required for aggregation (mmol l⁻¹)	Total electrolyte expressed in terms of divalent cation (64 mmol monovalent cation equivalent to 1 mmol divalent cation) (mmol l⁻¹)
Intralipid + water	1.8	1.8
Synthamin 9	0.8	2.0
14	1.0	2.3
17	1.8	3.1
Synthamin 14 (no electrolyte)	2.5	2.5
Freamine	2.0	2.0

the interface is unlikely and that the protective effect is related to the known action of hydrophilic colloids on states of aggregation[48] and a weak interaction between the amino acid and phospholipid layer. (The effect reported by Black and Popovich[22] could also be explained in part by the dilution of the electrolyte by the added amino acid and change in pH.)

Biological considerations

The data shown above indicate that mixed emulsion systems can be prepared using fat, amino acid, carbohydrate and even electrolytes and that these have acceptable (short-term) physical stability.

The biological implications of using mixed systems have yet to be evaluated. Wretlind[49] has commented on the fact that two apparently similar fat emulsions, Intralipid and Lipofundin S, although both made from soybean oil and lecithin (Table 19.4) are cleared by the body in different ways. Subtle differences in the nature of the layer of emulsifying agent (probably the higher content of phosphatidylinositol) were suggested as a cause.

The nature of the emulsifying layer of phospholipid on fat emulsions in mixed systems is certainly modified as shown by electrophoretic mobility measurements and the protective effect of amino acids against aggregation by electrolyte.

As soon as an emulsion (or another foreign colloidal particle) is administered to the blood, plasma components are adsorbed to the surface extremely rapidly; the nature of the adsorbed material changing with the surface properties of the particle[50-52]. These adsorbed components will influence the fate of the particles in the body, especially the interaction of lack of it with the reticuloendothelial system. Thus there is the possibility that administered fat droplets from a mixed system may be coated in a different way to the same type of droplets given directly using an unmixed system, and are thus handled differently by the body.

References

1 Hauser, H. and Phillips, M. C. (1979). Interactions of the polar group of phospholipid bilayer membranes. *Prog. Surface Membr. Sci.*, **13**, 297

2 Rhodes, D. N. and Lea, C. H. (1957). Phospholipids. 4. On the composition of hen's egg phospholipids. *Biochem. J.*, **65**, 526

3 Sartoretto, P. (1964). Lecithin. In *Kirk-Othmer Encyclopaedia of Chemical Technology*. Vol. 12, p. 343. (Wiley–Interscience: New York)

4 Maggio, B. and Lucy, J. A. (1976). Polar group behaviour in mixed monolayers of phospholipids and fusogenic lipids. *Biochem. J.*, **155**, 353

5 Hanahan, D. J., Brockerhoff, H. and Barron, E. J. (1960). Attack of phospholipose A on lecithin − reevaluation − position of fatty acids on lecithins and triglycerides. *J. Biol. Chem.*, **235**, 1917

6 Ries, H. E., Matsumoto, M., Uyeda, N. and Suito, E. (1975). Studies of monolayers of lecithin. In Goddard, E. D. (ed.). *Monolayers. Advanced Chemistry Series*. Vol. 144, p. 286. (Washington: American Chemical Society)

7 Becher, P. (1965). *Emulsions Theory and Practice*. p. 1. (New York: Reinhold)

8 Deryagin, B. V. (1940). Repulsive forces between charged colloidal particles and the theory of slow coagulation and stability of lyophobic sols. *Trans. Faraday Soc.*, **36**, 203

9 Verwey, E. J. W. and Overbeek, J. Th. G. (1948). *Theory of the Stability of Lyophobic Colloids*. (Amsterdam: Elsevier)

10 Davies, J. T. and Rideal, E. K. (1963). *Interfacial Phenomena*. (New York: Academic Press)

11 Shaw, D. J. (1970). *Introduction to Colloid Chemistry*. (London: Butterworths)

12 Kitchener, J. A. and Mussellwhite, P. R. (1968). The theory of stability of emulsions. In Sherman, P. (ed.). *Emulsion Science*. p. 77. (London: Academic Press)

13 Overbeek, J. Th. G. (1952). Stability of hydrophobic colloids and emulsions. In Kruyt, H. R. (ed.). *Colloid Science*. Vol. 1. Irreversible systems, pp. 302−341. (Amsterdam: Elsevier)

14 Alexander, A. E. and Johnson, P. (1949). *Colloid Science*. p. 115. (Oxford: Clarendon Press)

15 Davis, S. S. (1974). Pharmaceutical aspects of intravenous fat emulsions. *J. Hosp. Pharm. (Supplement)* 149−160, 165−170

16 Rydhag, L. (1979). The importance of the phase behaviour of phospholipids for emulsion stability. *Fette Surf. Anstrich*, **81**, 168

17 Rydhag, L. and Wilton, I. (1981). The function of phospholipids of soybean lecithin in emulsions. *J. Am. Oil Chem. Soc.*, **58**, 830

18 Hansrani, P. K. (1980). Studies on intravenous fat emulsions, *PhD Thesis*, Nottingham University

19 Davis, S. S. and Galloway, M. (1981). Effects of blood plasma components on the properties of an intravenous fat emulsion. *J. Pharm. Pharmacol.*, **33**, (Suppl.), 88P

20 Bangham, A. D. (1968). Membrane models with phospholipids. *Prog. Biophys. Mol. Biol.*, **18**, 29

21 Papahadjopoulos, D. (1968). Surface properties of acidic phospholipids: interaction of monolayers and hydrated liquid crystals with uni- and bivalent metal ions. *Biochim. Biophys. Acta*, **163**, 240

22 Black, C. D. and Popovich, N. G. (1981). A study of intravenous emulsion compatibility. Effects of dextrose, amino acids and selected electrolytes. *Drug Intell. Clin. Pharm.*, **15**, 184

23 Dawes, W. H. and Groves, M. J. (1978). The effect of electrolytes on phospholipid-stabilised soybean oil emulsions. *Int. J. Pharm.*, **1**, 141

24 Becher, P., Trifiletti, S. E. and Machida, Y. (1976). Emulsions stabilized by non-ionic surface-active agents: effects of electrolytes. In Smith, A. L. (ed.). *Theory and Practice of Emulsion Technology*. p. 271. (London: Academic Press)

25 Burnham, W. R., Hansrani, P. K., Knott, C., Cook, J. and Davis, S. S. (1982). Stability of a fat emulsion-based intravenous feeding mixture. *Intern. J. Pharmaceut.*, (In press)

26 Berne, B. J. and Pecora, R. (1977). *Dynamic Light Scattering with Applications to Chemistry, Biology and Physics*. (New York: Plenum Press)

27 Groves, M. J. (1980). Particle size characterisation in dispersions. *J. Dispersion Sci. Technol.*, **1**, 97

28 Randle, K. J. (1980). *Chemistry and Industry*, **19**, 74

29 Jeppsson, R. I., Groves, M. J. and Yalabik, H. S. (1976). The particle size distribution of emulsions containing diazepam for intravenous use. *J. Clin. Pharmacol.*, **1**, 123

30 Jeppsson, R. I. and Schoefl, G. I. (1974). The ultrastructure of lipid particles prepared with various emulsifiers. *Aust. J. Expl. Biol. Med. Sci.*, **52**, 697

31 Pinter, G. G. and Zilversmit, D. B. (1962). A gradient centrifugation method for the determination of particle size distribution of chylomicrons and of fat droplets in artificial fat emulsions. *Biochim. Biophys. Acta*, **59**, 116

32 Geyer, R. P., Watkin, D. M., Matthews, L. W. and Store, J. (1951). Parenteral nutrition. XI. Studies with stable and unstable fat emulsions administered intravenously. *Proc. Exp. Biol.*, **77**, 872

33 Atik, M., Marrero, R., Isla, F. and Mariale, B. (1965). Hemodynamic changes following infusion of intravenous fat emulsions, *Am. J. Clin. Nutr.*, **16**, 68

34 Le Veen, H. H., Giordano, P. and Johnson, A. (1965). Flocculation of intravenous fat emulsions. *Am. J. Clin. Nutr.*, **16**, 129

35 Fujita, T., Sumaya, T. and Yokohama, K. (1971). Fluorocarbon emulsion as a candidate for artificial blood. Correlation between particle size of the emulsion and acute toxicity. *Europ. Surg. Res.*, **3**, 436

36 Bangham, A. D., Flemans, R., Heard, D. H. and Seaman, G. V. F. (1958). Apparatus for microelectrophoresis of small particles. *Nature*, **182**, 642

37 Klose, H. J., Volger, E., Brechtelsbauer, H., Heinich, L. and Schmid-Schonbein, H. (1972). Microrheology and light transmission of blood. I. The photometric effect of red cell aggregation and red cell orientation. *Pflugers Arch. Gen. Physiol.*, **333**, 126

38 Hassan, M. A. (1982). Studies on the aggregation of emulsions stabilized by anionic emulsifiers. *PhD Thesis*, Nottingham University

39 Wells, P. A. (1981). Comment on intravenous emulsion compatibility. *Drug Intell. Clin. Pharmacol.*, **15**, 908

40 Borresen, H. C., Coran, A. G. and Knutrud, O. (1970). Metabolic results of parenteral feeding in neonatal surgery: a balanced parenteral feeding programme based on a synthetic L-amino acid solution and a commercial fat emulsion. *Ann. Surg.*, **172**, 291

41 Solassol, C. and Joyeux, H. (1974). Nouvelles techniques pour nutrition parenterale chronique. *Ann. Anaesth. Franc. Special*, **2**, 75

42 Solassol, C., Joyeux, H., Etco, L., Pujol, H. and Romieu, C. (1974). New technique for long term intravenous feeding: an artificial gut in 75 patients. *Ann. Surg.*, **179**, 519

43 Solassol, C., Joyeux, H., Serrou, R., Pujol, H. and Romieu, C. (1973). Nouvelles techniques de nutrition parenterale a long terme pour suppleance intestinale propositions d'un intestin artificial. *J. Chir.*, **105**, 15

44 Hardy, G., Cotter, R. and Dawe, E. (1982). This volume, Chapter 20

45 Pourriat, J. L., Garnier, M., Rathat, C., Hoang, P. and Cupa, M. (1979). Modifications des lipides erythrocytaires au cours de l'alimentation parenterale avec emulsion lipidique. *Medico. Chir. Dig.*, **8**, 183

46 Gove, L., Wallis, A. D. F. and Scott, W. (1979). Mixing parenteral nutrition products. *Pharm. J.*, Dec. 8th, p. 587

47 Davis, S. S. and Galloway, M. (1982). Studies on mixed parenteral feeding systems. (In press)

48 Zsigmondy, R. and Thiessen, P. A. (1925). *Das Kolloide Gold.* p. 173. (Leipzig: Akademische Verlagsgesellschaft m.t.h.)

49 Wretlind, A. (1976). Current status of Intralipid and other fat emulsions. In Meng, H. C. and Wilmore, D. W. (eds.). *Fat Emulsions in Parenteral Nutrition.* p. 109. (Chicago: American Medical Association)

50 Hoekstra, D. and Scherphof, G. (1979). Effect of fetal calf serum and serum protein fractions on the uptake of liposomal phosphatidylcholine by rat hepatocytes in primary monolayer culture. *Biochim. Biophys. Acta*, **551**, 109

51 Carlson, L. A. (1980). Studies on the fat emulsion Intralipid. I. Association of serum proteins to Intralipid triglyceride particles. *Scand. J. Clin. Invest.*, **40**, 139

52 Allen, C., Sata, T. M. and Molnar, J. (1973). Isolation, purification and characterisation of opsonic protein. *J. Reticuloend. Soc.*, **13**, 410

20
The stability and comparative clearance of TPN mixtures with lipid

G. HARDY, R. COTTER AND R. DAWE

INTRODUCTION

The optimal utilization of nutrients with a minimum of side-effects is a basic requirement for parenteral nutrition. It has been known since 1937[1] that orally administered nutrients are better utilized when given simultaneously and, when protein, carbohydrates and fats are taken as a mixture, nitrogen balance is better than that observed when the same nutrients[2] are taken consecutively. This is also true for parenteral nutrients, as demonstrated by Elman[3] and others[4-6]. However, creation of a pre-mixed total parenteral nutrition (TPN) system has posed certain problems. Addition of lipid has sometimes resulted in unstable mixtures, particularly when certain electrolytes are also added. In addition, there are questions concerning storage of such mixtures.

In practice, throughout Northern Europe and North America, a double or triple bottle system is usually used to administer parenteral nutrients. Amino acids and carbohydrates are sometimes admixed before infusion, but in many cases are administered in sequence. Minerals and vitamins are often added to different bottles and infused at different times. Fat is most commonly administered separately or not at all.

Such sequential or parallel infusion means changing each container every 4–6 h. Thus, it is not uncommon to have up to six bottle changes and set reconnections each day. Unless pumps or controllers are used, considerable time and attention are required to monitor the lines, make the various additions, control glucose homeostasis and manage the set changes safely and efficiently.

In 1971, Dudrick and Rhoads[7] proposed a single plastic container with separate compartments for each TPN component. Creation of such a container still remains elusive. However, the search for a more simple, nutritionally complete system, that would especially facilitate mobility of longterm TPN patients has continued.

TPN MIXTURES WITHOUT LIPIDS

One approach utilizes the 3 litre Viaflex[R] container in which most of the nutrients (excluding fat) can be mixed under aseptic conditions. This system, developed and described by Powell-Tuck[8] and Fielding et al.[9], reduces the number of bottle changes and the number of manipulations of the infusion system in bacteriologically uncontrolled environments[10]. Moreover, a reduced incidence of glycosuria and a threefold reduction in insulin requirements were reported as compared to the results with a multibottle regimen[11,12]. Reduced growth of many bacterial species has been reported with some mixtures prepared in these containers[13].

Although the containers are usually filled prior to immediate use, bags can be stored for 72 h at 4 °C to accommodate weekend requirements and for up to 1 month at this temperature for home patients[14]. Nedich reported[15] that 2.75% and 4.25% amino acid solutions (Travasol[R]*) in 25% dextrose are stable when stored for up to 7 days at room temperature or for 30 days at 4–5 °C in Viaflex containers. More recently, he has shown that similar TPN solutions are chemically stable for 28 days at room temperature or 112 days (16 weeks) under refrigeration[16]. These preparations can also be frozen for at least 60 days and thawed using microwave[17,18].

TPN MIXTURES WITH LIPIDS

In hospitals throughout Europe, it is becoming increasingly common to use this 3 litre plastic bag system[19,20]. It has also proved particularly appropriate for home parenteral nutrition[21–23] when it is not necessary to provide fat each day. The inclusion of fat emulsion is an obvious next step which could further simplify clinical procedures.

Clinical experience

The use of complete TPN mixtures with lipid was pioneered by Solassol and Joyeux[5,24,25]. These authors described the clinical use of mixtures previously stored for up to 3 months at 4 °C in silicone rubber containers. However, their publications did not include detailed stability data. In addition, it was necessary to reuse the containers after washing and resterilization, which was a time-consuming disadvantage.

In spite of the apparent success of the so-called Montpelier system, there have been conflicting reports over the stability of TPN mixtures with lipid. Wretlind confirmed that Intralipid[R] could withstand such mixing, but did not recommend it[26], and occasional reports from other workers of incompatible mixtures[27] have deterred widespread adoption of this approach. Neverthe-

*Synthamin[R]

less, such mixtures are increasingly accepted in France, Austria[28], and Spain[29]. There was renewed interest in 1980 with the publication of clinical results obtained at Montpelier during the period 1972–1978[30]. A total of 64 095 mixtures with lipid were given to 2669 patients during this time. Metabolic complications occurred in only 1.4% of the cases and none could be linked to the mixtures. An additional 395 patients given similar mixtures for a total of 9100 days with no mixture-related complications is reported by Sitges-Serra[29]. Clinical experience with various TPN–lipid mixtures has now been reported by Kaminski and co-workers[31], Burnham *et al.*[32], and Pamperl and Kleinberger[33]. Refrigerated storage is universally recommended. However, storage times have varied from 24 h and 7 days to 1 month. No adverse clinical effects have been reported.

Effect of electrolyte additions
Several workers have investigated the compatibility of various electrolyte and drug additives with fat emulsions, but often at concentrations unlikely to be seen in practice. Studies by LeVeen[34] and others[35] found that emulsions cream and flocculate when admixed with sodium, potassium, or calcium ions. Franck reported no apparent instability when calcium chloride was added to Intralipid. In contrast, calcium gluconate at the same concentration (2 mg/ml) produced a 'cracked' emulsion[36].

Black and Popovich[37] also added monovalent (Na^+, K^+) and divalent (Ca^{2+}, Mg^{2+}) ions directly to emulsions. Four hours after addition of 50 mmol NaCl and/or KCl to 0.5 l of Intralipid, noticeable surface creaming occurred. After 48 h, significant globule coalescence was apparent. Addition of divalent cations, for example, 3.4 mmol $CaCl_2$, to Intralipid produced widespread flocculation within 4 h.

Effect of glucose and amino acids
Addition of hypertonic glucose solutions (10–25%) to lipid emulsions also produces significant changes in particle size distribution (PSD), with markedly increased globule coalescence 48 h after mixing. In contrast, Black and Popovich showed that addition of 8.5% amino acid solution produces no significant difference in PSD after 72 h. Furthermore, if the amino acid and glucose solutions are combined before mixing them with the emulsion, the combination is stable. The amino acids apparently buffer the system, inhibiting the pH lowering effect of glucose solution. Similarly, when a 3.4 mmol/l mixture of $CaCl_2$, amino acids, and Intralipid was tested, changes in PSD were delayed. Although some globule growth occurred over the 72 h observation period, flocculation was prevented in every such dispersion tested. Amino acid solutions thus appear to have a protective effect and a dilution effect that delays coalescence and enhances stability.

A report by Pamperl and Kleinberger[33] confirms that mixing *all* TPN components in one bottle does not lead to rapid instability providing the fat emulsion is added last, followed by storage at 4 °C. These authors measured a decrease in the diameter of emulsion liposomes from 0.8 to 0.6 μm and a pH drop from 7.5 to 6.2–6.8 immediately after mixing but there were no further signs of instability over a 1 month period.

Combined effects of additives

There is a lack of published stability data on amino acid–glucose–electrolyte–emulsion combinations. Moreover, a 3 litre solution may have a dilution and stabilization effect that has not previously been recognized. In 1980, we were fortunate that a new nonplasticized container material (ethyl vinylacetate) became available. We therefore began to investigate the compatibility and changes in particle size distribution of some typical 3 litre TPN mixtures at concentrations consistent with current clinical practice.

The results of our first experiments with amino acids (Synthamin[R]), glucose, and lipid emulsion (Intralipid[R] or Lipofundin[R]) confirmed that, in the absence of electrolytes, room temperature storage for 24 h did not destabilize the systems[17]. Coulter Counter analysis showed that more than 80% of the emulsion particles were less than 1.5 μm. Further studies in our laboratories[39] and with the Department of Pharmacy, University of Nottingham showed that 3 litres mixtures of 5.5 or 8.5% amino acids (Synthamin), 20 or 50% glucose and 10% Intralipid with mono- and divalent electrolytes did not flocculate. Mean particle size (1.2 μm) was maintained for 6 days at room temperature (Table 20.1). (The value for unmixed Intralipid in these experiments was 0.95 μm using a Coulter Counter with 30 μm orifice.)[†]

Table 20.1 Particle size distribution of 3 l TPN mixtures with time

TPN mixtures* stored at room temperature	Percentage of particles less than 1.5 μm				Mean particle size (μm)
	0 h	24 h	48 h	6 days	
10% lipid emulsion[†], 20% dextrose, 5.5% amino acids[‡]					
Without electrolytes	85.0	83.0	81.0	82.0	1.2
With electrolytes**	88.0	80.0	84.0	89.5	1.2
With extra electrolytes[††]	83.0	82.5	84.0	81.5	1.2
10% lipid emulsion[†], 50% dextrose, 8.5% amino acids[‡]					
Without electrolytes	86.5	90.5	83.0	82.5	1.2
With electrolytes**	87.5	82.0	84.5	77.5	1.2
With extra electrolytes[††]	86.5	79.0	75.0	73.0	1.2

*Equal volumes (1 l) lipid emulsion, dextrose and amino acids in a 3 l bag
[†] Intralipid[R]
[‡] Synthamin[R]
**Synthamin with electrolytes contains Na^+ 70, K^+ 60, Mg^{2+} 5, P 30 mmol per litre
[††] Synthamin with electrolytes plus extra Mg^{2+} 4 mmol and Ca^{2+} 3 mmol per litre

[†] For a more detailed discussion of mean particle size determination with the Coulter Counter see previous chapter

The results confirmed that divalent cations such as calcium, previously shown to have a destabilizing effect on fat emulsion alone, can be incorporated in stable mixtures at a concentration of 1 mmol/l if in combination with the other nutrient components in 3 litres of solution. Several different factors are therefore influencing the stability of emulsion mixtures. Amino acid content, relative ionic strength of cations and anions, glucose concentration, pH, and container surface could all play a role. We were therefore encouraged to investigate in more depth the stability of a range of typical TPN solutions to create a body of information for future reference. A concurrent study was also instituted to assess the metabolism and elimination kinetics of the mixtures in a dog model.

STABILITY OF TPN MIXTURES

Experimental design

In the first part of the study, four different admixtures were prepared, as listed in Table 20.2. Mixtures I and II contained high and low amino-acid concentrations, and Mixtures II and III represented the extremes in glucose concentration. Mixture IV is a 'worst case' for emulsion stability given the high glucose concentration, expected low pH, and high electrolyte content.

Replicates of each mixture were prepared once a day on 4 days. To avoid direct contact between the glucose solution and the emulsion, the amino acid solution and then the glucose were aseptically transferred to a sterile 3 litre plastic bag containing lipid emulsion. The solutions were mixed by inverting the bag ten times.

Mixtures were examined at regular intervals under a high-intensity lamp and were compared to a freshly prepared reference solution consisting of 1 litre of Travamulsion and 2 litres of sterile water for injection. Units were also inspected for oil separation near the meniscus, the presence of a 'watery' layer, loss of homogeneity, and appearance of oil droplets at the fluid/bag interface.

The solutions were sampled when mixed and after 72 h of storage (48 under refrigeration at 5–9 °C and 24 at room temperature) by using a syringe to withdraw a sample from the injection port. The four amino acids most susceptible to oxidation and therefore deemed most likely to reflect any change in stability (methionine, phenylalanine, tryptophan, and tyrosine) were analysed using high-performance liquid chromatography (HPLC). Soybean oil and glucose concentrations and duplicate pH determinations were also made. Particle size distribution (PSD) was determined on a Coulter Counter Model TA II, with a 10 μm orifice.

The effect of adding a trace element preparation (MTE-4™, containing zinc 10 mg, copper 4 mg, manganese 1 mg and chromium 40 μg) was investigated in a second separate study using similar mixtures and storage conditions.

Table 20.2 Combinations of lipid emulsion, amino acids, and dextrose used to assess stability of lipid in TPN solutions

Mixture	10% Lipid emulsion* (ml)	10% Dextrose† (ml)	70% Dextrose† (ml)	5.5% Amino acids without electrolytes‡ (ml)	10% Amino acids without electrolytes‡ (ml)	10% Amino acids with electrolytes‡ (ml)
I	1000	1000	–	1000	–	–
II	1000	1000	–	–	1000	–
III	1000	–	1000	–	1000	–
IV	1000	–	1000	–	–	1000

*10% Travamulsion™ injection
†USP dextrose injection
‡Travasol® injection

Table 20.3 Changes in physical and chemical stability* of TPN mixtures shown in Table 20.2

	Mean particle size (μm)	Tyrosine (mg/100 ml)	Tryptophen (mg/100 ml)	Methionine (mg/100 ml)	Phenylalanine (mg/100 ml)	Glucose (%)	Soybean oil (%)
Mixture I							
0 h	0.28	7.55	29.8	110	129	3.22	3.47
72 h	0.28	7.53	29.1	112	128	3.10	3.39
% Change	0	0.3	2.3	1.8	0.8	3.7	2.3
Mixture							
0 h	0.28	13.7	56.8	204	222	3.28	3.36
72 h	0.28	14.0	56.5	209	221	3.12	3.36
% Change	0	2.2	0.5	2.4	0.5	4.9	0
Mixture III							
0 h	0.28	13.4	56.5	197	217	21.7	3.31
72 h	0.29	13.7	56.5	206	219	21.4	3.31
% Change	3.5	2.2	0.	4.6	0.9	1.4	0
Mixture IV							
0 h	0.28	13.9	56.8	210	216	21.7	3.37
72 h	0.29	14.0	56.6	217	218	21.6	3.38
% Change	3.5	0.7	0.4	3.3	0.9	0.5	0.3

*After storage for 48 h under refrigeration and 24 h at room temperature

Results

There were no visible signs of creaming, flocculation, free oil droplets, or other indications of emulsion instability (i.e. colour changes, mould, or bacterial growth). No statistically significant changes with storage time were found in the glucose, soybean oil, or amino acid concentrations. pH of all mixtures remained within the range 5.5–5.8.

The changes in mean particle size and chemical parameters of the mixtures with time are presented in Table 20.3. The PSD for the high glucose Mixtures III and IV did show a small but statistically significant increase of 3.5% in mean particle size (from 0.28 to 0.29 μm) after 72 h of storage. However, PSD was still within acceptable limits.

Table 20.4 Particle size distribution for TPN mixture with electrolytes and trace elements *after storage for 72 h (48 under refrigeration and 24 at room temperature)

Particle size	Percentage distribution by volume			
(μm)	0 h	24 h	48 h	72 h
0–1.2	87.2	83.7	77.5	68
1.2–1.5	10.1	11.2	15.7	18.6
1.5–1.8	1.8	2.9	3.9	6.9
1.8–2.4	0.9	1.6	2.5	5.1
2.4–3.0	–	0.6	0.3	1.4

*1 l of 10% amino acids with electrolytes,
  1 l of 50% glucose,
  1 l of 10% lipid emulsion, and 10 ml of trace elements (10 mg of zinc, 4 mg of copper, 1 mg of manganese, 40 μg of chromium)

Changes in PSD for a typical mixture with trace elements are presented in Table 20.4. Mean particle size for each sample over ten determinations, made using photon correlation spectroscopy, changed from 0.22 μm to 0.24 μm over the 72 h period. 86.6% of emulsion particles remained below 1.5 μm and there were no signs of instability.

Discussion

Two predominant factors affect the stability of TPN mixtures containing lipid. One is the presence of electrolytes which may induce coalescence of the emulsion particles. The second factor is pH which, below a certain level, will cause separation of the emulsion. Our studies have demonstrated that under certain conditions emulsion mixtures are stable. The relative ionic strengths of the monovalent and divalent cations are important to stability and influence the degree to which coalescence is induced. Dawes and Groves[35] and others have shown that these ions can change the negative potential at the surface of the emulsion particles, which normally maintains globule separation. An excess of cations can therefore reduce the zeta potential and electrophoretic mobility of the emulsion particles. Furthermore, divalent cations such as calcium may form a bridge between the negatively charged phosphate groups of the lecithin molecules, again inducing aggregation.

However, previous studies have not taken into account the very important protective action of amino acids demonstrated by Black and Popovich[37,38]. Amino acids appear to adsorb at the oil–water interface, thereby reducing the tendency for particles to associate and coalesce. In every dispersion tested by these workers and ourselves, there was no flocculation when amino acids were present.

Amino acids also minimize the adverse effects of low pH hypertonic dextrose solutions. Dextrose alone causes denaturation of the lecithin surfactant of the emulsion with subsequent release of fatty acids into the aqueous phase. The findings of Zeringue et al.[40] confirm that lecithin-stabilized emulsions lose significant stability when the pH falls below 5.0.

In this context, the additional buffering capacity of phosphate and acetate ions could be expected to maintain pH at an acceptable level and enhance stability. The amino acid solutions used in our studies contained up to 150 mmol/l acetate and, when electrolytes were present, contained 30 mmol/l phosphate.

Kawilarang et al.[41] concluded that any interaction of lipid emulsion with amino acids is pH dependent. At a pH below their isoelectric point, amino acids are predominantly positively charged. They can thus react more rapidly with the negatively charged emulsion particles to exert their protective effect. Conversely, at a pH higher than the isoelectric point amino acids will be predominantly anionic. They could thus repulse the emulsion particles and perhaps afford access to the electrolyte cations which reduce stability.

The pH of all the Synthamin–glucose–lipid mixtures remained within the range 5.5–5.8 during the entire test period. In this pH range, most individual amino acids will be electrically neutral since their isoelectric point (pI) is between 5.5 and 6.3 (Table 20.5). The three basic amino acids (arginine, lysine and histidine) present in most commercially available solutions have pI greater than 7 and presumably help to neutralize the pH lowering effect of glucose solutions and maintain stability. In contrast, the three 'acidic' amino acids with low pI values (cystine 5.02, aspartic acid 2.98 and glutamic acid 3.08) which were not present in the mixtures tested exist in the anionic form at a pH above 5.5 and could have had a deleterious effect on emulsion stability, especially in the presence of electrolytes. Koida et al.[43] have shown that aspartic and glutamic acids accelerate coalescence of fat particles and this may, in part, account for some of the literature reports of unstable mixtures with different amino acid solutions[27].

METABOLIC STUDIES

The metabolic stability of an amino acid–glucose–lipid mixture was assessed by studying the metabolism of the emulsion to triglyceride, free fatty acids,

Table 20.5 pH value at the isoelectric points of amino acids in water at 25 °C (pI)

Amino acid	pI
L-Arginine	10.76
L-Lysine	9.47
L-Histidine	7.64
DL-Alanine	6.11
Glycine	6.06
DL-Isoleucine	6.04
DL-Leucine	6.04
DL-Methionine	5.74
DL-Phenylalanine	5.91
L-Proline	6.30
DL-Serine	5.68
L-Tryptophan	5.88
L-Tyrosine	5.63
DL-Valine	6.00
L-Cystine	5.02
L-Aspartic acid	2.98
L-Glutamic acid	3.08

Taken from the *Handbook of Chemistry and Physics*. 57th Edn., Reference 42

and phospholipids in dogs. The mixture was compared to an identical TPN regimen administered separately. In addition, the elimination kinetics of lipid emulsion administered in each form were assessed. This is a quantitative indication of lipid metabolism reflecting removal from the bloodstream via the lipoprotein lipase enzyme system and conversion from exogenous to endogenous lipid.

Material and methods

The test mixture consisted of 10% lipid emulsion (10% Travamulsion™), amino acids with electrolytes (5.5% Travasol[R] (Amino Acids) Injection), and 10% dextrose injection, premixed approximately 24 h before use in a 3 litre plastic bag and stored at room temperature. As a control, the 10% lipid emulsion, amino acids and dextrose solutions were infused separately.

The test and control solutions were administered to six healthy male beagles from 9 to 18 months old weighing between 8 and 15 kg. Beagles were selected for this study because of our experience with this animal in similar TPN studies. Three of the dogs were given the premixed solution, and three the separate components. The test and control solutions were administered daily over a 7 h period for 5 days, via peripheral intravenous infusion.

The basic formula was selected on the basis of canine nutritional requirements[44] and the osmolarity limitations of peripheral injections. Each dog received the TPN formula at a dosage of approximately 74 kcal/kg/day (metabolic energy). 60% of the calories were provided by the lipid emulsion, 20% from the glucose and 20% from the amino acids.

The dose provided approximately 4.9 g of lipid, 3.7 g of protein, and 3.7 g of carbohydrate per kg every 7 h. Calculated osmolarity of this formulation is 584 mOsm/l.

To establish the metabolism and elimination of the lipid, on days 1, 3 and 5 venous blood samples (approximately 5 ml) were collected at 0, 15, 20, 60, 90, 120, 180, 240, 300, 360 and 420 min after the start of infusion. These samples were analysed for triglyceride content. Phospholipid and free fatty acid determinations were made on samples (approximately 5 ml of blood from the saphenous or cephalic vein) drawn at 0, 60, 120, 180, 240, 300, 360 and 420 min.

The data were analysed in two phases. Phase one, metabolism, consisted of examining the actual blood concentration–time data for triglycerides, free fatty acids and phospholipids. The second phase of analysis involved kinetic evaluation of elimination of the lipid emulsion from the bloodstream. The kinetic model used in this analysis was as follows:

$$\frac{dc}{dt} = R_0 - \frac{a_1 c}{a_2 + c} \qquad \text{Equation 1}$$

where

c = triglyceride concentration at time t,

t = time in minutes,

$R_0 = \dfrac{\text{infusion in mg/ml emulsion per min}}{\text{distribution volume}}$,

a_1 = maximal elimination capacity (mg/ml blood per min),

a_2 = a parameter whose magnitude relative to concentration indicates the nature of the elimination process (mg/ml blood),

$\dfrac{a_1}{a_2}$ = fractional elimination rate (% min)

Specific elimination processes may be identified in terms of the parameters derived using the model. These can be one of three types: first order (i.e. fractional, elimination is exponential), zero order (i.e. linear, indicating saturation of the elimination mechanism), and mixed order (i.e. both processes are operating simultaneously). Thus, maximal eliminating capacity (a_1) and the fractional elimination rate (a_1/a_2) are applicable when the process is operating in a zero-order or mixed-order region. For fractional elimination, the ratio a_1/a_2 is applicable.

The actual nature (first, mixed, or zero order) of the elimination process is determined by the magnitude of the a_2 parameter in relation to blood triglyceride concentrations attained during infusion. If concentrations are predominantly ten times or more lower than the a_2 value, the process is first

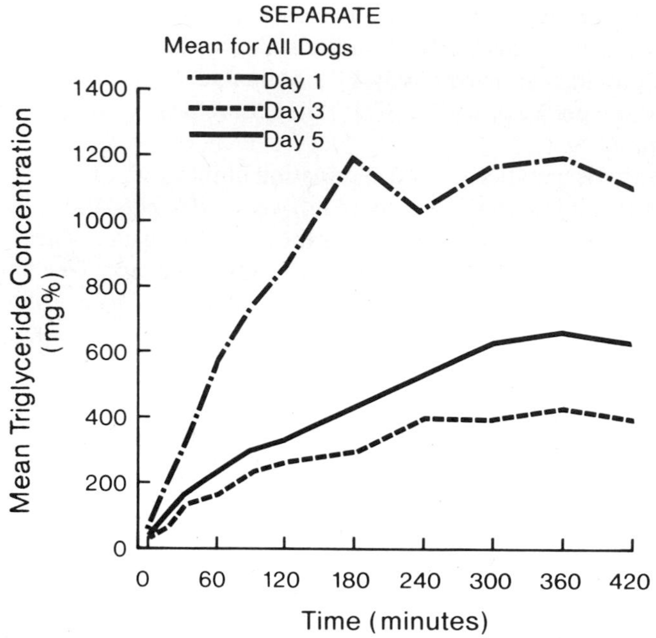

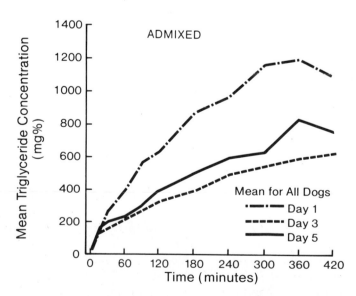

Figure 20.1 Mean triglyceride concentration data obtained during infusion of 10% lipid emulsion to beagles on days 1, 3 and 5. The emulsion was administered as a separate entity in a TPN regimen with amino acids and dextrose

Figure 20.2 Mean triglyceride concentration data obtained during infusion of 10% lipid emulsion to beagles on days 1, 3 and 5. The emulsion was administered admixed with amino acids and dextrose in a TPN regimen

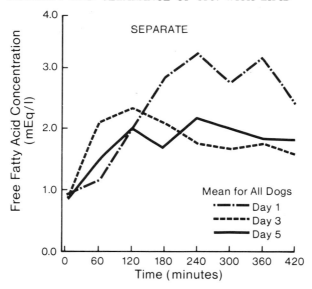

Figure 20.3 Mean free fatty acid concentration data obtained during infusion of 10% lipid emulsion to beagles on days 1, 3 and 5. The emulsion was administered as a separate entity in a TPN regimen with amino acids and dextrose

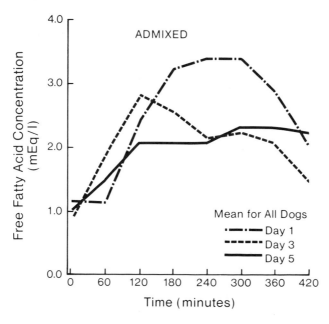

Figure 20.4 Mean free fatty acid concentration data obtained during infusion of 10% lipid emulsion to beagles on days 1, 3 and 5. The emulsion was administered admixed with amino acids and dextrose in a TPN regimen

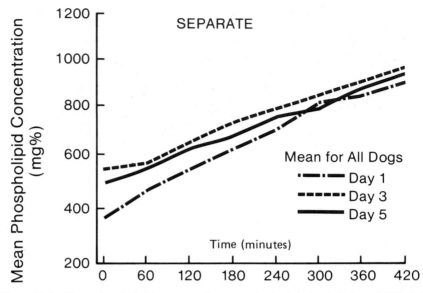

Figure 20.5 Mean phospholipid concentration data obtained during infusion of 10% lipid emulsion to beagles on days 1, 3 and 5. The emulsion was administered as a separate entity in a TPN regimen with amino acids and dextrose

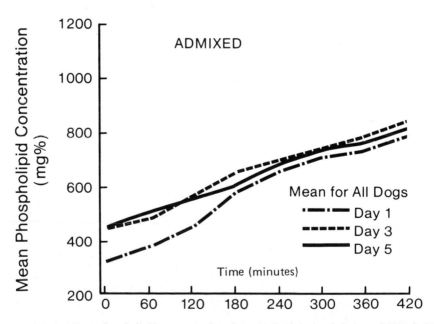

Figure 20.6 Mean phospholipid concentration data obtained during infusion of 10% lipid emulsion to beagles on days 1, 3 and 5. The emulsion was administered admixed with amino acids and dextrose in a TPN regimen

order. Concentrations in a range ten times higher or lower than a_2 indicate a mixed-order process is operating, while concentrations predominantly more than ten times greater than a_2 indicate zero-order elimination.

Results

Serum triglyceride and free fatty acid concentrations increased during infusion for both groups (Figures 20.1–4). However, the increase was greatest on the first day of infusion. Phospholipid concentrations (Figures 20.5 and 6), on the other hand, increased steadily during infusion, with starting concentrations also increasing. Paralleling the triglyceride and free fatty acid data, the overall increase was also greatest on day 1. The above data for all 18 infusion episodes were analysed for differences between the test and control groups in these lipid classes. No statistically significant differences were observed.

The second phase of analysis involved the elimination kinetics of the lipid emulsion from the bloodstream in the test and control groups. The kinetic parameters (Table 20.6) obtained using the elimination model (Equation 1) indicate that the elimination process for the test and control groups is predominantly mixed order over the infusion period. However, there was a difference in elimination reflected in the parameter estimates for day 1 as compared to days 3 and 5.

Statistical evaluation of the elimination parameters for all 18 infusion episodes showed no significant differences between the test and control groups. A simulation of the elimination profile for the test and control groups using mean parameter estimates for day 1 and days 3 and 5 combined is shown in Figure 20.7.

Discussion

The metabolic phase of this study clearly showed that a 10% soybean oil lipid emulsion (10% Travamulsion™ injection) is metabolized in a similar fashion when administered as a separate entity or admixed with amino acids and dextrose in a TPN regimen.

Adaptation to lipid infusion was observed in both groups, with higher levels of triglycerides and fatty acids being observed on day 1 when compared to days 3 and 5. Although there was an increase in starting concentrations of phospholipid over the test period, the greatest increase during infusion was on day 1. This adaptation is most probably due to the induction of the lipoprotein lipase enzyme system in response to lipid infusions.

The adaptability of the test system to lipid TPN is indicated by a shift in elimination. While a mixed-order process is observed throughout the test period, on day 1 the zero-order aspect predominates, while first-order kinetics predominate on days 3 and 5. The shift is also reflected in the difference in the magnitude of the a_2 parameter on day 1 as opposed to days 3 and 5. It should also be pointed out that the relatively constant parameters obtained on days 3 and 5 make the data for these days more useful for comparing

Table 20.6 Estimates of the parameters for the elimination kinetics of 10% lipid emulsion admixed with amino acids and dextrose

Group	Dog	Day	a_1*	a_2†	a_1/a_2 (× 100)‡
Admixed	178	1	0.207	2.834	7
	190	1	0.178	0.643	28
	195	1	0.184	2.233	8
	178	3	0.190	0.651	29
		5	0.200	0.957	21
	190	3	0.201	0.533	38
		5	0.183	0.760	24
	195	3	0.165	0.170	97
		5	0.159	0.000	00
Separate	181	1	0.170	0.292	58
	182	1	0.520	29.709	2
	183	1	12.805	871.031	1
	181	3	0.183	0.098	87
		5	0.182	0.163	12
	182	3	0.186	0.391	48
		5	0.170	0.683	25
	183	3	0.181	0.225	80
		5	0.179	0.181	99

*mg/ml blood per minute
†mg/ml blood
‡%/minute

elimination in the test and control groups. In contrast, there is wide variation in the day 1 parameters.

CONCLUSIONS

In summary, we have shown that TPN mixtures of Travasol (Synthamin) with or without electrolytes, trace elements, glucose and 10% soybean oil lipid emulsions are stable when prepared aseptically in 3 litre plastic containers and stored at refrigerated temperature for 72 h. Four representative mixtures with different proportions of these nutrients are shown in Table 20.7.

The pH buffering property of the amino-acid solution is an important contributor to mixture stability. The relative proportions of individual amino

Table 20.7 Components for four stable TPN mixtures

Solutes and physical characteristics	Travasol (Synthamin)			Ca/Mg Additive	Dextrose				MTE-4 trace element vial	Fat emulsion 10%	Mixed solutions			
	5.5%	8.5%	10.0%		10%	20%	50%	70%			High N High Cal*	Med N High Cal†	Med N Low Cal‡	Low N Low Cal**
Amino acid (g)	55	85	100								100	85	85	55
Nitrogen (g)	9.3	14.3	16.9								16.9	14.3	14.3	9.3
Carbohydrate (g)					100	200	500	700			700	500	200	100
Fat (g)										100	100	100	100	100
Kilocalories (kcal)					400	800	2000	2800		1100	3900	3100	1900	1500
Kcal:N ratio											230	216	133	161
Electrolytes														
Sodium (mmol)	70	70	70								70	70	70	–
Potassium (mmol)	60	60	60								60	60	60	–
Chloride (mmol)	70	70	70								70	70	70	22
Magnesium (mmol)	5	5	5	4							5	5	9	–
Calcium (mmol)				3							–	–	–	–
Phosphorus (mmol)	30	30	30								30	30	30	–
Acetate (mmol)	100	135	150								150	135	135	35
Trace minerals														
Zinc (mg)									10		10	–	10	–
Copper (mg)									4		4	–	4	–
Manganese (mg)									1		1	–	1	–
Chromium (µg)									40		40	–	40	–
Volume (ml)	1000	1000	1000	5	1000	1000	1000	1000	10	1000	3010	3005	3010	3000
Mean particle size (µm)										0.25	0.29	0.23	0.23	0.28
pH	6.0	6.0	6.0	6.0			3.5 to 6.0			5.5 to 9.0	5.6	5.8	5.9	5.7
MOsm/litre	520	1160	1300	14	555	1100	2775	3885		270	1818	1406	846	448

*10% amino acids with electrolytes, 70% dextrose, 10% lipid emulsion   ‡8.5% amino acids with electrolytes, 20% dextrose, 10% lipid emulsion
†8.5% amino acids with electrolytes, 50% dextrose, 10% lipid emulsion   **5.5% amino acids without electrolytes, 10% dextrose, 10% lipid emulsion

257

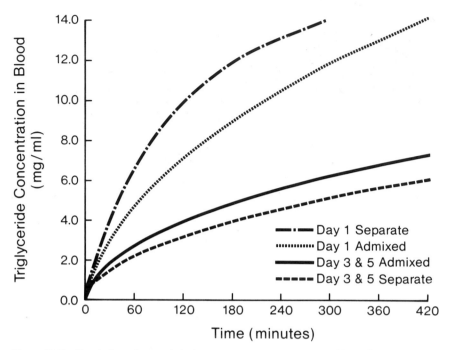

Figure 20.7 Simulation of mean infusion profiles for 10% lipid emulsion administered as a separate entity (day 1 (——·——)) and days 3 and 5 combined (— — — —)) and admixed in a TPN regimen (day 1 (------)) and days 3 and 5 (————)). Simulation obtained using mean parameter estimates obtained for the triglyceride data from the test and control groups (one outlier for day 1, dog 183, rejected)

acids and the additional buffering capacities of phosphate and acetate ions in certain mixtures may also be significant.

The lipid emulsion, admixed with amino acids and glucose for parenteral administration to dogs, is eliminated and metabolized to triglyceride, free fatty acids and phospholipids in a similar fashion to that observed during separate administration of the same nutrients.

ACKNOWLEDGEMENTS

The authors wish to thank R. Klim, L. Martis, J. Hedstrom, S. Hall and T. Staff for their technical assistance and Professor S. S. Davis for scientific advice and encouragement.

References

1 Larson, P. S. and Chaikoff, I. L. (1937). The influence of carbohydrate on nitrogen metabolism in the nutritional state. *J. Nutr.*, **13**, 287

2 Munro, H. N. (1949). The relationship of carbohydrate to protein metabolism. *J. Nutr.*, **39**, 375

3 Elman, P. (1953). Time factors in the utilisation of a mixture of amino acids (protein hydrolysate) and dextrose given intravenously. *Am. J. Clin. Nutr.*, **1**, 287

4 Voss, E. U. and Schnell, J. (1970). Parenterale Ernährung mit einer aminosäurenhaltigen Fettemulsion in Tierversuch. *Med. Ernähr.*, **11**, 71

5 Solassol, C., Joyeux, H., Serrou, B., Pujol, H. and Romieu, C. (1973). Nouvelles techniques de nutrition parentérale à long terme pour suppléance intestinale. Proposition d'un intestin artificiel appliqué à 54 cs. *J. Chir.*, **105**, 15

6 Nube, M., Bos, L. P. and Winkelman, A. (1979). Simultaneous and consecutive administration of nutrients in parenteral nutrition. *Am. J. Clin. Nutr.*, **32**, 1505

7 Dudrick, S. J. and Rhoads, J. (1971). New horizons for intravenous feeding. *J. Am. Med. Assoc.*, **215**, 939

8 Powell-Tuck, J., Farwell, J. A., Nielsen, T. and Lennard-Jones, J. E. (1978). Team approach to long term intravenous feeding in patients with gastrointestinal disorders. *Lancet*, **2**, 825

9 Fielding, L. P., Humfress, A., Mouchizadeh, J., Dudley, H. and Gilmour, M. (1981). Experience with three-litre bags. *Pharm. J.*, 590

10 Powell-Tuck, J., Lennard-Jones, J. E., Lowes, J. E., Twum-Danso, K. and Shaw, E. (1979). Intravenous feeding in a gastroenterological unit. *J. Clin. Pathol.*, **32**, 549

11 McGeehan, D., Radcliffe, A. G. and Fielding, L. P. (1978). Comparison of glucose homeostasis in TPN using small and large capacity containers. *Br. J. Surg.*, **65**, 818

12 Fielding, L. P. and Ellis, B. (1981). Intravenous nutrition − technical aspects. In Hill, G. L. (ed.) *Nutrition and the Surgical Patient*. p. 150. (London: Churchill Livingstone)

13 Duffett-Smith, T. and Allwood, M. C. (1979). Further studies on the growth of microorganisms in TPN solutions. *J. Clin. Pharm.*, 219

14 Farwell, J. (1980). Pharmaceutical factors in long term parenteral nutrition. *Proc. Guild Hosp. Pharm.*, **8**, 3

15 Nedich, R. L. (1978). The compatibility of extemporaneously added drug additives with Travasol (amino acid) injection. In Johnston, I. D. A. (ed.) *Advances in Parenteral Nutrition*. p. 415. (Lancaster: MTP Press)

16 Nedich, R. L. and Needham, T. (1982). Private communication

17 Klim, R. A. and Hardy, G. (1980). The stability of parenteral nutrition mixtures after freezing and microwave thawing in 3 litre plastic containers. Presented at the *2nd European Congress on Parenteral and Enteral Nutrition*, September 7−10, Newcastle, UK

18 Ausman, R. K., Kerkhof, K., Holmes, C. J., Cantwell, R., Kundsin, R. B. and Walter, C. W. (1981). Frozen storage and microwave thawing of parenteral nutrition solutions in plastic containers. *Drug Intell. Clin. Pharm.*, **15**, 440

19 De Keersmaecker, L. (1981). Pharmacy role in total parenteral nutrition therapy − methods and equipment. *Acta Chir. Scand.*, **507** (Suppl.), 375

20 Allwood, M. C. (1981). Microbial contamination of parenteral and enteral nutrition solutions. *Acta Chir. Scand.*, **507** (Suppl.), 382

21 Irving, M. and Tresadern, J. (1981). The benefits of central vein feeding and long term access to the circulation. *Acta Chir. Scand.*, **507** (Suppl.), 394

22 Hardy, G. and Kiepert, D. R. (1981). The role of the pharmaceutical industry in home parenteral nutrition. *Acta Chir. Scand.*, **507** (Suppl.), 120

23 Milewski, P. J., Gross, E., Holbrook, I., Clarke, C., Turnberg, L. A. and Irving, M. (1980). Parenteral nutrition at home in the management of intestinal failure. *Br. Med. J.*, **280**, 1356

24 Solassol, C., Joyeux, H., Etco, L., Pujol, H. and Romieu, C. (1974). New techniques for long-term intravenous feeding: an artificial gut in 75 patients. *Ann. Surg.*, **179**, 519

25 Solassol, C. *et al.* (1976). Long term parenteral nutrition: an artificial gut. *Int. Surg.*, **61**, 266

26 Wretlind, A. (1976). Current status of Intralipid and other fat emulsions. In Meng, H. C. and Wilmore, D. W. (eds.) *Fat Emulsions in Parenteral Nutrition*. (Chicago: American Medical)

27 Gove, L., Walls, A. D. F. and Scott, W. (1979). Mixing parenteral nutrition products. *Pharm. J.*, Dec. 8th, 587

28 Kleinberger, G. (1980). Parenterale Ernährung bei schwerer Leberinsuffizienz. *Z. Gastroent.*, **16** (Suppl.), 99

29 Sitges-Serra, A., Jaurrieta, E., Pallares, R., Lorente, L. and Sitges-Creus, A. (1982). Clinical experience with fat containing TPN solutions. This volume, Ch. 18.

30 Solassol, C. *et al.* (1980). Complete nutrient mixtures with lipids for TPN in cancer patients. *Acta Chir. Scand.*, **498** (Suppl.), 151

31 Knutsen, C., Miller, P. and Kaminski, M. V. (1981). Compatibility, stability, and effect of mixing a 10% fat emulsion in TPN solutions: a case report. *J. Parent. Enteral. Nutr.* **5**, 579

32 Burnham, W. R., Hansrani, P., Knott, C. E., Cook, J. A., Davis, S. S. and Longman, M. J. S. (1981). Simplified intravenous nutrition using Intralipid based mixtures in a 3-litre bag system. *Gut*, **22** A, 898

33 Pamperl, H. and Kleinberger, G. (1982). Morphologic changes of Intralipid 20% liposomes in all-in-one solutions during prolonged storage. *Infusiontherapie*, **9**, 86

34 LeVeen, H. H., Giordano, P. and Johnson, A. (1965). Flocculation of intravenous fat emulsions. *Am. J. Clin. Nutr.*, **16**, 129

35 Dawes, W. H. and Groves, M. J. (1978). The effect of electrolytes on phospholipid-stabilized soybean oil emulsions. *Int. J. Pharm.*, **1**, 141

36 Franck, J. T. (1973). Intralipid compatibility study. *Drug. Intell. Clin. Pharm.*, **7**, 351

37 Black, C. D. and Popovich, N. G. (1981). Stability of intravenous fat emulsions. *Arch. Surg.*, **115**, 891

38 Black, C. D. and Popovich, N. G. (1981). A study of intravenous emulsion compatibility. Effects of dextrose, amino acids and selected electrolytes. *Drug Intell. Clin. Pharm.*, **15**, 184

39 Hardy, G. and Klim, R. A. (1981). The physical stability of parenteral nutrition mixtures with lipids. *J. Parent. Enteral Nutr.*, **5**, 363 and 569

40 Zeringue, H. J., Brown, M. L. and Singleton, W. S. (1964). Chromatographically homogeneous egg lecithin as stabilizer of emulsions for intravenous nutrition. *J. Am. Oil Chem. Soc.*, **41**, 688

41 Kawilarang, C., Georghiou, K. and Groves, M. J. (1980). The effect of additives on the physical properties of a phospholipid stabilized soybean oil emulsion. *J. Clin. Hosp. Pharm.*, **5**, 151

42 Weast, R. C. (ed.). *Handbook of Chemistry and Physics.* 57th Edn. (1976–1977). C-767. (Cleveland, Ohio: CRC Press)

43 Koida, Y., Matsuda, S. and Miura, H. (1976). Preparation of infusion for total parenteral nutrition. *Acta Chir. Scand.*, **466**, 116

44 *Basic Guide to Canine Nutrition* (1977). White Plaines NY: Gaines Professional Services

21
Problems and prospects in lipid metabolism during parenteral nutrition

T. P. STEIN

INTRODUCTION

Parenteral nutrition differs from enteral nutrition in two important aspects — other than the obvious one of delivery route. They are (1) the patient has little choice in the amount and type of nutrients he receives and (2) current formulations are biochemically devised diets. They have been designed to approximate to normal human dietary requirements as far as is compatible with limitations in the manufacturing process. Examples of the limitations are (1) the differential solubilities of amino acids seriously limit the amino-acid patterns that can be used; (2) the lipid emulsions used are pseudo-chylomicrons derived from semi-purified vegetable oils and (3) glucose is usually the major calorie source rather than the complex polysaccharides found in natural foodstuffs.

Bypassing normal food intake regulatory mechanisms can lead to fat accumulation in the liver if the caloric intake is excessive[1-4]. The reason for concern with the TPN induced hepatic steatosis is eventual damage to the liver as fat continues to accumulate as long as the TPN given is in excess of metabolic capacity.

The accumulation of fat in the liver is not necessarily due to the body's nutrient requirements being exceeded, but rather because the liver's metabolic capacity to process either the quantity or type of nutrients given has been exceeded[4,5]. The response of the liver to excessive demands, be they nutritional or stress related, is to stretch liver functions to the limit and nutrients whose processing can be completed at a later time are converted temporarily to fat. The export of liver fat has a low priority because fat is not toxic.

In contrast amino acids must be processed rapidly and any excess detoxified promptly because amino acids are very toxic[6]. Impaired detoxification of

261

excess amino acids is widely suspected as being a major contributing factor to hepatic encephalopathy[7]. Thus, when calorie input is excessive, the excess is (temporarily) stored in the liver mainly as fat, with a little as glycogen.

In orally fed subjects problems secondary to excess nutrient intake can usually be avoided by stopping eating. Furthermore, liver damage from excess nutrients is less likely to occur because the gut also plays an important role in the metabolism of foodstuffs, decreasing the workload on the liver. With TPN, the role of the gut is virtually eliminated. And, with TPN, where the input of nutrients is continuous the opportunity for clearing a damaged liver may not occur. Furthermore, the problems are compounded by the natural human tendency to overfeed, the philosophy being that if food is good, more is better. Consequently, damage can easily be induced in seriously ill patients with unhealthy livers who have little metabolic reserve capacity.

As fat continues to accumulate in the liver, it eventually leads to liver damage. Rupture of hepatocytes occurs and the plasma liver enzymes (SGOT, SGPT, etc.) become elevated and other liver functions, such as secretory protein synthesis, are impaired. We recently documented this sequence of events in a study on the TPN induced deposition of fat in the liver and its relationship to albumin synthesis. A series of malnourished patients with tumours of the gastrointestinal tract were studied[8]. These patients subsequently underwent a surgical procedure during which liver biopsies were obtained.

The patients were divided into two groups and both groups were nutritionally depleted. Half the subjects were given fat-free TPN for 10 days prior to surgery and the other half continued on their inadequate oral intake for the 10 days before operation. Eighteen hours before surgery a continuous infusion of [^{15}N]glycine was given for the estimation of protein synthesis rates. These patients did not receive any fat during their period of parenteral nutrition, and consequently their livers had lost most of their linoleate (Figure 21.1). Even though there must have been a more than adequate supply of linoleate in the periphery of the body, the high glucose levels blocked the transfer of linoleate to the liver.

About half of the TPN treated patients developed fatty livers, whereas none of the orally nourished group did. On further investigation it turned out that those patients with fatty livers were the ones whose plasma albumin levels did not improve during the 10 days of TPN (Figure 21.2). In contrast, the plasma albumin levels in the TPN treated patients who did not develop fatty livers were increased significantly.

The patients who developed fatty livers were the most nutritionally depleted and presumably had less metabolic reserve for coping with any excess or deficiency in their nutrient regimen. They had lower initial plasma protein levels, iron binding capacity and albumin levels (Figure 21.2). There is also a good correlation between the development of a fatty liver and

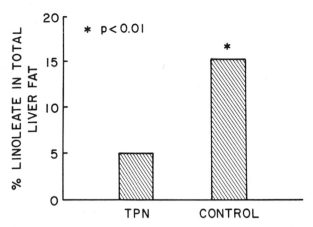

Figure 21.1 Depletion of linoleate in the liver after 1 week of fat free TPN. The study population consisted of adult males undergoing TPN for 7–10 days prior to surgery

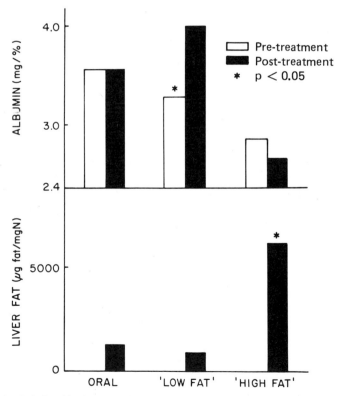

Figure 21.2 Relationship between fat accumulation in the liver and decreased albumin synthesis. The patients were the same as those in Figure 21.1. 'High' and 'low' fat refers to whether the patient accumulated fat ('high') or did not ('low') during the TPN programme. The controls were a group of similar patients who did not receive TPN

impaired albumin synthesis.

Fatty livers were not found in the semi-starved orally fed patients but only in the malnourished patients given TPN. Apparently the metabolic capacity of the liver (and the intestine) was adequate for a reduced oral intake but was inadequate for the increased amount of nutrients entering the circulation when the gut was bypassed.

In this study, the hepatomegaly was due to excess glucose, but a series of rat studies showed that it is even easier to cause hepatomegaly by giving parenteral emulsion containing long chain fatty acids.

Parenterally nourished rats do well on a caloric regimen when glucose is the non-protein calorie source, but, when the glucose is replaced by an isocaloric long chain fatty acid emulsion (Intralipid) hepatomegaly occurs (Figure 21.3). Of particular interest is the tendency of linoleate and linolenate to accumulate in the liver rather than oleate or palmitate which are also major constituents of Intralipid (Figure 21.4).

The preferential accumulation of the polyunsaturated fatty acids raises the question as to whether an excess of parenteral polyunsaturated fatty acids with their potential of preferential deposition in the liver and spleen may not have some adverse effects. Possibly an optimal mix of fatty acid triglycerides would provide enough linoleate and linolenate to prevent deficiencies from developing, and enough energy such as palmitate, stearate, and oleate.

If the metabolic capacity of the liver has been compromised previously, as for instance by prior protein depletion, then glucose too will cause hepatomegaly (Figure 21.5). But, by giving a mixed calorie source, i.e. a mixture of glucose and fat, hepatomegaly is avoided. This is probably because the liver has a finite capacity to metabolize different substrates and by using mixed fuel substrates alternate energy pathways can be utilized, thereby permitting more energy to be given to the rest of the body than if a single fuel substrate was used[9-11]. By varying the ratio of carbohydrate to lipids, it was shown that the optimal ratio is 75% glucose to 25% lipid[4,5].

Fat that accumulates in the liver is not only damaging to the liver, but is not available to the tissues for energy. Hence the reports of 'glucose calories being more effectively utilized than fat calories'[12-14]. When fat from Intralipid accumulates in the liver, both the whole body (Figure 21.6) and plasma albumin synthesis rates (Figure 21.7) are depressed because the tissues are 'semi-starved' (Figure 21.3).

In the rat experiments, the animals paradoxically accumulated Intralipid in the liver (Figure 21.3) and depleted their endogenous fat stores in an attempt to compensate for the shortfall in calories (Figure 21.8). Apparently, the amount of energy obtained from the depot fat was not enough to maintain a 'fed state' level of whole body and albumin protein synthesis (Figures 21.6 and 21.7). The situation is likely to be similar in man with the difference that a non-depleted human has proportionately more fat calories stored than the rather lean rat.

MEDIUM CHAIN FATTY ACIDS IN PARENTERAL NUTRITION

One of the great potential advantages of TPN is the prospect of varying the substrate and calorie sources given to patients based on the hypotheses

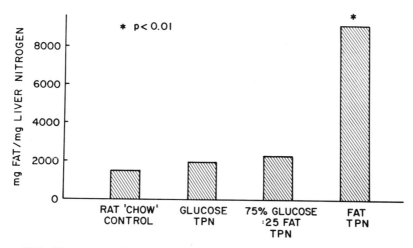

Figure 21.3 Hepatomegaly induced by giving Intralipid as the sole non-protein calorie source to parenterally fed rats. Unlike the rats in Figure 21.5, these rats had not been previously protein depleted

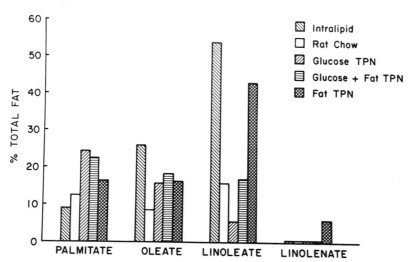

Figure 21.4 Pattern of fatty acids accumulated in the liver on various TPN regimens. The first set of bars (Intralipid) is the chemical composition of Intralipid. The second set, the fat distribution pattern from rats given standard laboratory rat chow and the last three bars are for the TPN regimens indicated

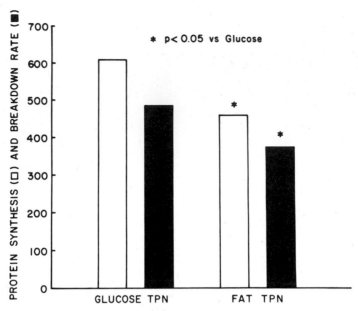

Figure 21.5 Fat deposition in the liver induced by both glucose and Intralipid in rats which had been previously compromised by prior protein depletion. The rats were studied during repletion with the dietary regimens indicated in the figure. The significant point is that although the three parenteral regimens were isocaloric, the mixed caloric source was clearly superior because it did not cause hepatomegaly

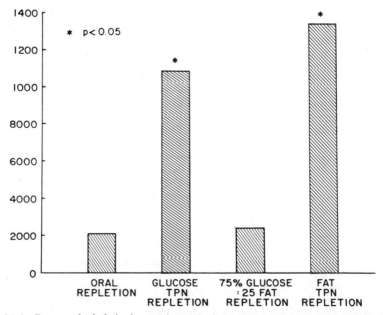

Figure 21.6 Decreased whole body protein synthesis in parenterally fed rats with Intralipid accumulation in their liver (Figure 21.3) while depleting their endogenous fat (Figure 21.8)

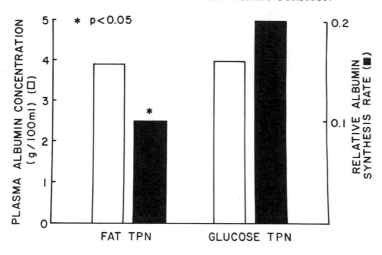

Figure 21.7   Decreased albumin synthesis in parenterally fed rats with Intralipid accumulation in their liver (Figure 21.3) while depleting their endogenous fat (Figure 21.8)

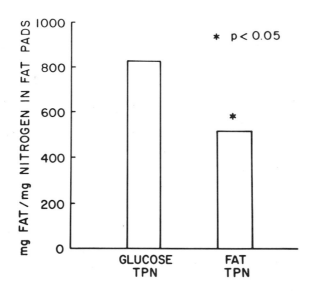

Figure 21.8   Depletion of endogenous fat in rats while Intralipid is accumulating in the livers of the rats given Intralipid as their sole non-protein calorie source. To meet energy demands the rats are forced to use their fat reserves because fat deposited in the liver is unavailable. The rats given parenteral glucose in this experiment did not develop hepatic steatosis, hence their fat stores were not depleted

derived from acute *in vitro* biochemical experiments. Two such substrates have now been proposed which are specifically designed to selectively nourish certain tissues in stressed patients. They are the branched-chain amino acids, which are discussed elsewhere in this volume and medium-chain length fatty acids (usually abbreviated MCT or MCFA). A medium-chain length fatty acid is defined as a saturated fatty acid with either 8, 10 or 12 carbon atoms in the chain. MCTs are of interest for four reasons.

(1) By providing mixed fuel substrates there is the possibility of using other potentially available pathways for fuel utilization. It has been suggested that in healthy male adults the capacity for glucose may be as little as 1800 kcal/day which may be less than daily energy needs[9,10,15]. The rat experiments described above document the advantages of mixed fuel substrates, a 75:25 glucose:fat regimen was superior to either fat or glucose alone or any other combination of fat and glucose.

(2) The idea of giving a substantial proportion of the daily non-protein requirement as fat is attractive because lipid emulsions have low osmolarities and so can be given through a peripheral vein[16]. There may be less risk to the patient from peripheral as opposed to central vein nutrient administration and there are more access sites.

(3) The suggestion has been made that MCFAs might be better utilized than LCFAs in the septic state particularly if malnutrition and/or multisystems failure is also present. The argument is that LCFA oxidation is impaired due to either decreased carnitine levels in the muscle or increased competition for available carnitine. Carnitine is required for translocating LCFAs into the cells. MCFAs can also enter cells via a carnitine independent mechanism thereby circumventing the potential blockage in fat oxidation[10,17-22].

(4) As pointed out above, an excess of parenteral calories, either as glucose or lipid, leads to the deposition of fat in the liver. Conceivably alternate fuel sources may be able to overcome this limitation.

The literature on medium-chain fatty acid metabolism is conflicting. There are reports that they are well tolerated when given as part of the non-protein calories but as the proportion of MCTs is increased there is a failure to promote weight gain in rats[23-25]. For the above reasons we decided to investigate the parenteral metabolism of an MCT-containing emulsion in parenterally nourished rats with the objective of testing them as a supplementary caloric source in the septic state.

An MCT-containing emulsion was prepared and its effectiveness as a caloric source on lipid and protein metabolism was compared with rats given either an isocaloric glucose or long-chain fatty acid-containing emulsion (Intralipid). The MCT-containing emulsion was designed to match Intralipid

(20%) in calorie and glycerol content and lecithin concentration.

The animals appeared to tolerate the three parenteral regimens well, but there was blood in the urine of the glucose treated rats and ketone bodies in the urine of the Intralipid animals (Table 21.1). The uteri and fallopian tubes of the Intralipid treated animals were filled with a milky white fluid and their spleens were enlarged and bloody.

Table 21.1 Survival data, urine and plasma analyses results for parenterally nourished rats given (1) glucose, (2) MCTs, and (3) Intralipid for 5 days

| | Nutritional regimen | | |
	Glucose	MCT + LCT	Intralipid
Total rats	12	11	12
Died	1*	2*	1*
Excluded	1	3	1
URINE			
Protein	5	4	3
Blood	8	2	3
Ketones	0	0	8
PLASMA			
β-hydroxy-butyrate (mmol/ml plasma)	0.01 (0.02)	0.17 (0.05)*	1.04 (0.38)* †

The rats were also given a balanced mixture of parenteral amino acids, trace minerals, vitamins etc. They were allowed 2 days to recover from the surgery and then given the 5 days of TPN. 'Total rats' is the number of rats cannulated within each group, 'excluded' rats are animals that were alive at the end of the study but appeared to be unhealthy to the investigators, and 'died' means that the rat died during the study. * denotes death from fluid overload on days 1–4, and a catheter problem. The urine analyses were done on the 7th day using a 'Dipstick' (Miles Laboratories, Elkart, Indiana). The number of rats with positive measurements for the 3 parameters listed above are tabulated

The rats were able to metabolize the infused MCT. However, there were differences between the MCT and Intralipid treated rats. The MCT emulsion did not cause ketosis as evidenced by plasma β-hydroxy butyrate levels and the excretion of ketone bodies in the urine and neither did it produce hepato-megaly (Table 21.1).

We found no evidence of substantial MCT deposition in either the liver or fat pads, or of MCT induced lipogenesis in the MCT treated animals. The plasma lipid analyses showed no significant difference in the levels of the total free fatty acids between Intralipid and the MCT-containing emulsion. More importantly, the plasma concentration of medium-chain length fatty acids was less than 1% of the total plasma free fatty acids. These findings are in accord with earlier studies that show that MCFAs are preferentially oxidized[26,27].

CONCLUSION

In summary, TPN not only offers an alternate route for delivery of nutrients to the tissues but may allow selected targeting. The parenteral route differs from the enteral route because the gut and liver combine to select the nutrient mix from that presented to the periphery. In TPN this selection has to be done by the nutritionist. Numerous studies have documented that TPN regimens can be improved by the addition of EFA, the use of the RQ as an early warning for fat deposition in the liver and the use of specific amino-acid combinations for renal disease and hepatic insufficiency. The next few years will show if MCTs have a role in this further fine tuning of parenteral nutrition.

References

1 Sheldon, G. F., Petersen, S. R. and Sanders, R. (1978). Hepatic dysfunction during hyperalimentation. *Arch. Surg.*, **113**, 504
2 Ota, D. M., Imbembo, A. L. and Zuidema, G. D. (1980). Total parenteral nutrition. *Surgery*, **83**, 503
3 Rogers, B. M., Hollenbeck, J. I., Donnelly, W. H. *et al.* (1976). Intrahepatic cholestasis with parenteral alimentation. *Am. J. Surg.*, **131**, 149
4 Stein, T. P., Buzby, G. P., Leskiw, M. J. *et al.* (1981). Protein and fat metabolism during repletion with parenteral nutrition. *J. Nutr.*, **111**, 154
5 Buzby, G. P., Mullen, J. L., Stein, T. P. *et al.* (1981). Manipulation of TPN caloric substrate and fatty infiltration of the liver. *J. Surg. Res.*, **31**, 46
6 Owen, O. E., Reichard, G. A., Boden, G. *et al.* (1978). Interrelationships among key tissues in the metabolic utilization of substrates. In Katzen, H. M. and Mahler, R. I. (eds.) *Advances in Modern Nutrition.* pp. 517−550. (New York: Wiley)
7 Fischer, J. F. (1981). The etiology of hepatic encephalopathy − nutritional implications. *Acta Chir. Scand. Suppl.*, **507**, 50
8 Stein, T. P., Buzby, G. P., Gertner, M. H. *et al.* (1980). Effect of parenteral nutrition on protein synthesis and liver fat metabolism in man. *Am. J. Physiol.*, **239**, G280
9 Matzkies, F. (1975). Untersuchungen zur Pharmacodynamik von Kohlenhydraten als Grundlage ihrer Anwendung zur parenteralen Ernahrung. *Z. Ernaehrungswiss.*, **14**, 184
10 Birkhahn, R. H., Long, C. L. and Blakemore, W. S. (1981). New substrates for parenteral feeding. *J. Parent. Enterol. Nutr.*, **3**, 346
11 Cahill, G. F. Jr. (1981). Ketosis. *J. Parent. Enterol. Nutr.*, **5**, 281
12 Freund, H. R., Yoshimura, N. and Fischer, J. E. (1980). Does intravenous fat spare nitrogen in the injured rat? *Am. J. Surg.*, **140**, 377
13 Long, J. M., Wilmore, D. W., Mason, A. D. *et al.* (1977). Effect of carbohydrate and fat on nitrogen excretion during total intravenous feeding. *Ann. Surg.*, **185**, 417
14 Allison, S. P. (1977). Modifying nitrogen loss after injury. In Richards, J. R. and Kinney, J. M. (eds.) *Nutritional Aspects of Care in the Critically Ill.* pp. 389−342. (Edinburgh: Churchill Livingstone)
15 Birkhahn, R. H. and Border, J. R. (1981). Alternate or supplemental energy sources. *J. Parent. Enteral Nutr.*, **5**, 24
16 Jeejeebhoy, K. N., Anderson, G. H., Nakooda, F. *et al.* (1976). Metabolic studies in total parenteral nutrition with lipid in man. *J. Clin. Invest.*, **57**, 125
17 Border, J. R., Burns, G. P., Rumph, C. *et al.* (1970). Carnitine levels in severe infection and starvation. A possible key to the prolonged catabolic state. *Surgery*, **68**, 175
18 Border, J. R., Chernier, R., McMenamy, R. H. *et al.* (1976). Multiple systems organ failure. Muscle fuel deficit with visceral protein depletion. *Surg. Clin. N. Am.*, **56**, 1147
19 Cerra, F. B., Siegel, J. H., Coleman, B. *et al.* (1980). Septic autocannibalism, a failure of exogenous nutrition support. *Ann. Surg.*, **192**, 570
20 Moyer, E. D., Border, J. R., McMenamy, R. H. *et al.* (1981). Multiple systems organ failure.

IV. Alterations in the acute phase plasma profile in septic trauma: effects of intravenous amino acids. *J. Trauma*, **21**, 645

21 O'Donnell, T. F., Clowes, G. H. A., Blackburn, G. L. *et al.* (1976). Proteolysis associated with a deficit of peripheral energy substrates in septic man. *Ann. Surg.*, **80**, 192

22 Ryan, N. T. (1976). Metabolic adaptations for energy production during trauma and sepsis. *Surg. Clin. N. Am.*, **56**, 1073

23 Baba, N., Bracco, E. P. and Hashim, S. A. (1982). Enhanced thermogenesis and diminished deposition of fat in response to overfeeding with diet containing medium chain triglyceride. *Am. J. Clin. Nutr.*, **35**, 678

24 Wiley, G. H. and Leveille, G. A. (1973). Metabolic consequences of dietary medium chain triglycerides in the rat. *J. Nutr.*, **103**, 829

25 Lavau, M. M. and Hashim, S. A. (1978). Effect of medium chain triglyceride on lipogenesis and body fat in the rat. *J. Nutr.*, **108**, 613

26 Scheig, R. (1968). Hepatic metabolism of medium chain triglycerides. In Senior, J. R. (ed.) *Medium Chain Triglycerides*. pp. 39–50. (Philadelphia: University of Pennsylvania Press)

27 Greenberger, N. J. and Skillman, T. G. (1969). Medium triglycerides: physiological considerations and clinical implications. *N. Engl. J. Med.*, **280**, 1045

Section 5
Energy requirements in stressed patients

22
Alterations in fuel metabolism in stress

P. R. BLACK AND D. W. WILMORE

The hyperglycaemia and glucose intolerance observed in patients following severe trauma appear to be central to the metabolic response to injury. Oral and intravenous glucose tolerance tests following injury[1,2], burn shock[2,3] and systemic infection[5,6] demonstrated delayed disposal of glucose from the plasma into body tissues. Hyperglycaemia and the increased urinary nitrogen losses in these patients were consistent with an insulin-deficient state and terms such as 'stress diabetes' or the 'diabetes of injury' were used to describe this metabolic response to injury. Although it has been shown that during the early or 'ebb' phase of injury hypoinsulinaemia does exist[1,3,7], following resuscitation, as the patient enters the 'flow' phase of injury, the beta cell response to glucose load is normal or even higher than that observed in normals[3,7]. Hyperglycaemia and glucose intolerance in the presence of a normal or increased insulin response would suggest that certain target tissues of the injured patient are relatively resistant to circulating insulin. The aetiology of the resistance to circulating insulin in the injured patient is not presently known. However, recent advances made in insulin physiology have produced a clearer understanding of the insulin resistance following injury.

Insulin resistance can be defined as a condition in which normal concentrations of insulin produce less than a normal biological response. Moreover, the biological response to insulin can be characterized by two distinct terms: the maximal response that can be achieved with any hormonal concentration, and the sensitivity of the response to the hormone[8]. Sensitivity to insulin or any hormone can be defined in terms of the concentration at which a half-maximal response is elicited; the higher the concentration required, the lower the sensitivity. Thus, insulin resistance can be the result of decreased responsiveness, decreased sensitivity, or a combination of the two. The construction of dose-response curves allows the investigator to distinguish between these possible aetiologies of insulin resistance.

At the cellular level insulin resistance may result from alterations prior to the interaction of insulin with its receptor, a defect in the interaction of insulin with its receptor, or changes in steps distal to the formation of the insulin-receptor complex. Those effects causing a reduction in free insulin concentrations (e.g. increased insulin degradation or increased insulin binding) are examples of insulin resistance occurring as the result of alterations prior to the level of the receptor. Changes in receptor concentration on the cell membrane or alterations in the affinity of the receptor to insulin represent causes of insulin resistance resulting from defects in the interaction of insulin with its receptor.

It is obvious that the biological response observed in patients manifesting insulin resistance secondary to either of these two aetiologies would be dependent upon the concentration of circulating insulin. Higher levels of insulin would be required in these subjects to obtain a normal response. Thus, the dose-response curves in these subjects would be shifted to the right indicating decreased sensitivity to plasma insulin concentrations. By appropriately increasing circulating insulin levels, maximal responses similar to those observed in normals could be achieved in these patients. Since these two causes of insulin resistance have the same biological implications, they are frequently referred to as prereceptor defects, a term generally indicating decreased sensitivity to insulin as the aetiology of insulin resistance.

Insulin resistance resulting from alterations in the interaction between insulin and its target tissues distal to the receptor level is frequently referred to as a postreceptor defect. This pathophysiological condition usually results in a decrease in the maximal response that can be elicited by circulating insulin regardless of its concentration and, therefore, represents an example of insulin resistance secondary to decreased responsiveness. The distinction between prereceptor and postreceptor defects as the aetiology of insulin resistance has important clinical implications especially in those critically ill patients who must be given nutritional support. If the patient manifests glucose intolerance because of insulin resistance secondary to a prereceptor defect, infusing exogenous insulin at sufficiently high doses will return the insulin response to normal in these patients. In contrast, if the insulin resistance is secondary to a postreceptor defect, exogenous insulin will not alter maximal responsiveness and will not fully correct the glucose intolerance.

Over the past year we have used the euglycaemic insulin clamp technique and the hyperglycaemic glucose clamp technique[9] to characterize glucose–insulin interactions following trauma. In the euglycaemic insulin clamp, exogenous insulin is infused at a fixed rate to increase plasma insulin concentration and maintain it at a constant level. A variable speed infusion pump is frequently adjusted to deliver glucose at a rate which maintains euglycaemia and prevents hypoglycaemia. Under the steady-state conditions of our experimental model, the rate of infusion of glucose approximates to glucose disposal by the tissues. By infusing insulin at varying rates, dose-response curves can be constructed defining the relationship between glucose disposal and plasma insulin concentration. These curves can then be used to characterize the nature of the insulin resistance following injury. Moreover, because the rate of infusion of insulin is constant in these studies, the metabolic clearance rate of insulin could be measured.

The hyperglycaemic glucose clamp acutely elevates plasma glucose concentration 1250 mg/l above basal levels and examines the individual's beta cell response to fixed hyperglycaemia. Furthermore, disposal of glucose by target tissues can be quantitated concurrently since the rate of glucose infusion necessary to maintain fixed hyperglycaemia approximates to tissue glucose disposal under the study conditions. This technique mimics the clinical situation during glucose infusion and characterizes glucose–insulin interactions during hyperglycaemia and hyperinsulinaemia.

Nine previously healthy patients were studied using the euglycaemic insulin clamp following acute trauma. Five had multiple injuries resulting from motor vehicle accidents and four had body surface burns. These patients were compared to 14 age-matched controls at infusion rates of 1.0, 2.0 and 5.0 mU/kg min of crystalline zinc insulin. At each infusion rate the amount of glucose infused to maintain euglycaemia was significantly less in the patients than in the controls. Moreover, the plasma insulin concentration achieved for each infusion rate was less in the patients than in the controls. Calculating the metabolic clearance rate of insulin for both groups revealed that the insulin clearance rate of 734 ± 52 mg/ m^2 for the patients was almost 50% greater than the clearance rate of 497 ± 41 observed in the controls.

Analysis of the dose-response curves constructed from these data demonstrated that the maximal glucose disposal rate in the patients was 9.17 ± 0.78 mg/kg min. This value was significantly less than the maximal disposal rate of 14.3 ± 0.87 noted in the controls. However, the insulin concentration at which a half-maximal response occurred in both groups was virtually the same (86 and 85 mU/l in the patients and controls, respectively). These data would suggest that the insulin resistance following injury is manifest by decreased responsiveness and normal sensitivity to insulin. These findings are consistent with the concept that the insulin resistance following injury is secondary to a postreceptor defect.

Table 22.1 Glucose clamp − basal values (Mean ± SEM)

	Glucose (mg/l)	Insulin (mU/l)	Glucagon (ng/l)
PATIENTS	1020 ± 50	22 ± 5	200 ± 50
CONTROLS	950 ± 30	10 ± 1	108 ± 22
	n.s.	$p<0.005$	n.s.

Six additional acutely injured patients were studied using the hyper-glycaemic glucose clamp and compared to 12 age-matched controls. Four of these patients had multiple injuries following motor vehicle accidents and two had large body surface burns. Although basal plasma glucose levels were similar in both groups, both basal plasma insulin and glucagon concentrations were elevated in the patients (Table 22.1). In response to fixed hyperglycaemia, plasma insulin concentrations increased in both study groups. However, during the 2 h study period the mean plasma insulin concentration in the patients (145 ± 61 mU/l) was almost four times that observed in the normal controls (43 ± 7). Furthermore, in spite of the marked hyperinsulinaemia, the patients required an infusion rate of glucose which was significantly less than that of the controls to maintain fixed hyperglycaemia 1250 mg/l above basal (6.23 ± 0.87 versus 9.46 ± 0.79 mg/kg min). Both groups demonstrated glucagon suppression with fixed hyperglycaemia. However, despite prolonged elevation of plasma glucose concentrations, glucagon concentrations in the patients did not decline to the levels observed in the controls suggesting insufficient suppression of the alpha cell to glucose infusion following injury.

The hyperglycaemic glucose clamp also demonstrated that over the 2 h of the study the controls required increasing amounts of exogenous glucose to maintain fixed hyperglycaemia. In contrast, the glucose infusion rates in the patients remained relatively constant throughout the study period (Figure 22.1). The controls also demonstrated a strong correlation between plasma insulin concentration and glucose disposal. A similar correlation did not exist in the patient group. In fact, over a wide range of plasma insulin concentrations, there was little change in the glucose disposal rate among the patients (Figure 22.2). These data would suggest that with fixed hyperglycaemia the rate of maximal glucose uptake is rapidly attained by the injured patient and remains constant thereafter. Increasing plasma insulin concentration by endogenous secretion appears to be ineffective in augmenting glucose disposal in the traumatized patient with this level of fixed hyperglycaemia. In contrast, normal controls respond to fixed hyperglycaemia by increasing glucose disposal to correct the hyperglycaemia. Thus, the results of the hyperglycaemic glucose clamp also indicate that a significant difference in the maximal glucose uptake exists between the two groups and confirm the findings of the euglycaemic insulin clamp studies lending support to the concept that insulin resistance in the injured patient results from a postreceptor defect.

What are the clinical implications of these findings? Many critically ill patients receive constant infusions of hypertonic glucose as a common component of most intravenous feeding regimens. The results of the hyperglycaemic insulin clamp suggests that approximately 6–7 mg/kg min (30–35 kcal/kg day) approaches the maximal glucose disposal in the critically ill patient given intravenous glucose when exogenous insulin is not

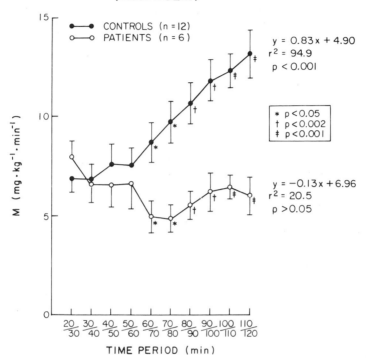

GLUCOSE CLEARANCE WITH TIME
(Mean ± SEM)

Figure 22.1 With time the rate of glucose disposal (M) in the control subjects progressively increased during the hyperglycaemic glucose clamp. In contrast, glucose uptake was relatively constant in the patients over the 2 h of the study. In the patients the infusion rate plotted versus time was similar to a horizontal line. Between group responses were significantly different; differences between individual time points are noted

administered concurrently. Wolfe and associates[10] studied post-operative patients using varying rates of infusion of standard central parenteral nutrition solutions. Utilizing tracer techniques to measure glucose clearance, they concluded that glucose infusion rates of 6–7 mg/kg min were optimal in these patients and that infusion of glucose above these rates resulted in the deposition of triglycerides. Studies by Long and co-workers[11] suggested that the maximal protein-sparing effect of glucose infusion alone in critically ill patients occurred at infusion rates which approximated to resting metabolic expenditures, or an average of about 35 kcal/kg day. These studies would suggest that optimal glucose infusion in the injured patient is about 30–35 kcal/kg day or approximately 2400 glucose kcal/day in a 70 kg patient. This represents an amount equal to 2 l of 'standard' central vein solution per day.

EFFECT OF PLASMA INSULIN ON GLUCOSE CLEARANCE

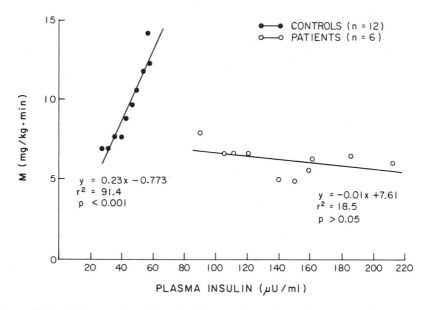

Figure 22.2   In the controls, as plasma insulin rose, glucose disposal increased during the hyperglycaemic glucose clamp studies. In contrast, over a wide range of plasma insulin concentrations glucose disposal did not change in the injured patients. Points shown are group means

Several investigators[11,12] have suggested that the decrease in nitrogen wasting elicited by glucose infusion can be augmented by the administration of exogenous insulin to the stressed patient. The results of the euglycaemic insulin clamp would imply that the infusion of exogenous insulin will increase the maximal disposal rate of glucose in the injured patient to about 9 mg/kg min. If this increase in maximal disposal rate represents an increase in the utilization of glucose for energy, protein catabolism to meet the body's energetic needs would be lessened and nitrogen wasting would be decreased. Thus, the simultaneous administration of insulin during glucose infusion in the injured patient would increase the maximal glucose disposal rate in the injured patient to approximately 3000 kcal/day in a 70 kg patient or about 3 l of 'standard' central vein solution. However, because of the complications recently associated with large quantities of glucose infused into hypermetabolic patients[13,14], these data would suggest that most critically injured patients should maintain energetic balance by receiving only enough glucose calories to meet their resting metabolic requirements (30–35 kcal/kg day). If additional calories are required, fat emulsion should be administered.

   In summary, we have used the insulin and glucose clamp techniques to assess the interaction between glucose disposal and plasma insulin concentration in an attempt to characterize the nature of the insulin resistance and glucose intolerance following surgical illness. Both techniques reconfirmed that insulin resistance was indeed a sequela of injury and the hyperglycaemic glucose clamp technique demonstrated that glucose disposal was diminished in the injured patient despite normal or increased pancreatic beta cell responsiveness. Moreover, dose-response curves constructed from data derived from the euglycaemic insulin clamp suggested that the insulin resistance in these patients could be attributed to a decrease in the responsiveness of the target tissues to circulating insulin, a so-called postreceptor defect. Finally, these studies were in agreement with other studies in the literature which would suggest that optimal glucose utilization and the maximal protein-sparing effects of exogenous glucose infusion occurred at an infusion rate of approximately 30–35 kcal/kg day. The simultaneous administration of exogenous insulin increases maximal glucose disposal in the traumatized patient and may augment the protein-sparing effects of glucose administration following injury.

References

1 Carey, L. C., Lowery, B. D. and Cloutier, C. T. (1970). Blood sugar and insulin response in human shock. *Ann. Surg.*, **172**, 342

2 Thomsen, V. (1938). Studies in trauma and carbohydrate metabolism with special reference to the existence of traumatic diabetes. *Acta Med. Scand. Suppl.*, **91**, 1

3 Allison, S. P., Hinton, P. and Chamberlain, M. J. (1968). Intravenous glucose tolerance, insulin and free fatty acid levels in burn patients. *Lancet*, **2**, 1113

4 Taylor, F. H. L., Levenson, S. M. and Adams, M. A. (1944). Abnormal carbohydrate metabolism in human thermal burns. *N. Engl. J. Med.*, **231**, 437

5 Rayfield, E. J., Curnow, R. T., George, D. T. and Beisel, W. R. (1973). Impaired carbohydrate metabolism during mild viral illness. *N. Engl. J. Med.*, **289**, 618

6 Williams, J. L. and Dick, G. F. (1932). Decreased dextrose tolerance in acute infectious disease. *Arch. Int. Med.*, **50**, 801

7 Wilmore, D. W., Mason, A. D. Jr. and Pruitt, B. A. (1976). Insulin response to glucose in hypermetabolic burn patients. *Ann. Surg.*, **183**, 314

8 Kahn, C. R. (1978). Insulin resistance, insulin sensitivity, and insulin unresponsiveness: a necessary distinction. *Metabolism*, **27**, 1893

9 DeFronzo, R. A., Tobin, J. D. and Andres, R. (1979). Glucose clamp technique: a method for quantifying insulin secretion and resistance. *Am. J. Physiol.*, **237(3)**, E214

10 Wolfe, R. R., O'Donnell, T. F. Jr., Stone, M. D., Richmand, D. A. and Burke, J. F. (1980). Investigation of factors determining the optimal glucose infusion rate in total parenteral nutrition. *Metabolism*, **29**, 892

11 Long, J. M., Wilmore, D. W., Mason, A. D. and Pruitt, B. A. (1977). Effects of carbohydrate and fat intake on nitrogen excretion during total intravenous feeding. *Ann. Surg.*, **185**, 417

12 Woolfson, A. J. M., Heatley, R. V. and Allison, S. P. (1979). Insulin to inhibit protein catabolism after injury. *N. Engl. J. Med.*, **300**, 14

13 Askanazi, J., Elwyn, D. H., Silverberg, P. A., Rosenbaum, S. H. and Kinney, J. M. (1980). Respiratory distress secondary to a high carbohydrate load: a case report. *Surgery*, **87**, 596

14 Askanazi, J., Rosenbaum, S. H., Hyman, A. I., Silverberg, P. A., Milic-Emili, J. and Kinney, J. M. (1980). Respiratory changes induced by the large glucose loads of total parenteral nutrition. *J. Am. Med. Assoc.*, **243**, 1444

23
The carbohydrate content of parenteral nutrition

J. M. KINNEY

Carbohydrate has been the predominant source of energy in the human diet since the beginning of history. The largest source of dietary carbohydrate has traditionally been the polysaccharide, starch, derived from cereals. Although cereals have been cultivated for over 10 000 years, primitive man obtained glucose, fructose and sucrose from fruits, berries and honey. During the 19th and 20th centuries, large amounts of various sugars have been obtained from agricultural products, such as sucrose from sugar canes and sugar beets and glucose−fructose mixtures from cornstarch. These purified mono- and disaccharides contribute a significant amount to the total energy consumption of contemporary western man, particularly in the United States. The consumption of sucrose and glucose has risen as a percentage of total carbohydrate in the diet from around 23% in 1900 to over 50% by 1965. A parallel decrease has occurred in the amount of carbohydrate consumed as starch[1]. The rising dietary intake of refined simple sugars has prompted criticism from those who feel that other nutrients which are present with starch in natural foods may be deficient in a modern diet.

The uncertainties which relate to the optimum content of carbohydrate in parenteral nutrition also involve questions concerning the proper chemical form of the carbohydrate and the total amount of carbohydrate that should be given. Some European investigators have recommended that parenteral solutions include fructose, sorbitol and xylitol and there are scattered references to maltose[2]. Since each of these sources of carbohydrate have been shown to be either impractical, or to have undesirable side-effects, the use of glucose has been the common form of carbohydrate for the majority of parenteral nutrition in Europe and essentially all of parenteral nutrition in the United States. All body cells have the capacity to oxidize glucose, either aerobically via the Embden−Meierhoff pathway to pyruvate, or anaerobically to lactate. Pyruvate and lactate can then be further oxidized via the tricarboxylic acid cycle to yield energy. The metabolic utilization of glucose, including oxidation, is dependent upon the presence of insulin to facilitate its entrance into body cells. When insulin synthesis and output is inhibited during periods of increased catecholamine activity, the uptake of glucose into cells is reduced with an associated tendency toward hyperglycaemia.

GLUCOSE METABOLISM – THE DEPLETED PATIENT

The infusion of excessive quantities of glucose above that required to meet the resting energy expenditure (REE) results in lipogenesis. This conversion of glucose to fatty acids is associated with a rise in the non-protein respiratory quotient, which is largely the result of an increased CO_2 production. The magnitude of these changes is a function of the patient's clinical state and the amount of the glucose load. In the depleted patient, lipogenesis occurs readily and the non-protein respiratory quotient commonly rises to 1.1 or 1.2, at a time when there is essentially no increase in the resting energy expenditure as represented by the O_2 consumption despite the large increase in CO_2 production[3]. In a study with graduated glucose intakes given to depleted patients, Elwyn et al.[4] showed that there was no increase in resting energy expenditure with increasing energy intake below energy equilibrium. However, when glucose intake achieved a positive energy balance, the REE was found to increase by 1 kcal for each 5 kcal of intake. At zero energy balance, nitrogen balance in these depleted patients was only slightly positive at an intake of 173 mg N per kg. This is about twice the intake of nitrogen required to maintain zero nitrogen balance in normal adults. These authors concluded that improving the nitrogen balance by increasing glucose intake above energy expenditure restored mainly the portion of the lean body mass associated with fat deposition.

GLUCOSE METABOLISM – THE HYPERMETABOLIC PATIENT

The administration of large amounts of glucose in lipid-free total parenteral nutrition to the hypermetabolic patient, who is injured or septic, reveals a different metabolic response from the depleted patient. The hypermetabolic patient shows major increases in not only the resting CO_2 production, but also the resting O_2 consumption, so that the non-protein respiratory quotient remains below 1.0[3] (Figure 23.1). This indicates that either lipogenesis has been inhibited or that some fat oxidation persists in the face of the large carbohydrate intake which would normally abolish fat oxidation and produce net lipogenesis. These acute patients have an elevated urinary excretion of norepinephrine, while receiving 5% dextrose and water, and the

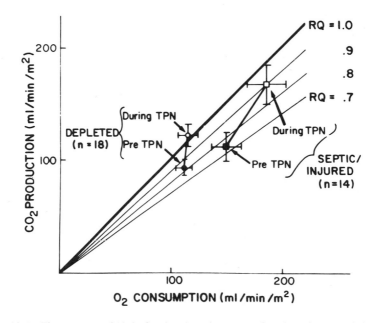

Figure 23.1 The response of 18 depleted patients is compared to that of 14 acutely injured or septic patients, when the O_2 consumption and CO_2 production is compared before and after the administration of high glucose, lipid-free total parenteral nutrition in amounts to provide approximately 150% of the resting energy expenditure. Note that the excess glucose intake did not achieve a respiratory quotient (RQ) above 1.0 in the acutely ill patients, indicating persistence of some net fat oxidation. (from Askanazi *et al.*, reference 3)

glucose loads with TPN cause the urinary norepinephrine to more than double in amount. Nordenström *et al.*[5] examined the urinary norepinephrine excretion in injured and septic patients who were given two kinds of TPN, either with all of the non-protein calories as glucose, or as 50% glucose + 50% intravenous fat. TPN with the all glucose system significantly increased the norepinephrine excretion, while TPN with the glucose + lipid system showed only a tendency to increase which was not clinically significant.

Elwyn *et al.*[6] have summarized the effects of increasing glucose intake on glucose metabolism in acutely injured patients, as follows:

(1) Plasma glucose concentrations are maintained at the already high level associated with injury, but are not significantly increased.

(2) Rates of glucose oxidation and also non-oxidative metabolism increase proportionately with intake.

(3) There is an abnormally high rate of glycogen deposition which is proportional to glucose intake.

(4) Gluconeogenesis from protein is abnormally high, but can be completely suppressed with an intake of approximately 600 g of

glucose per day (over four times the amount needed for suppression in normal subjects).

(5) Glucose recycling with glycogen, glycerol, or both, is increased with increasing intake.

Fatty infiltration of the liver and the development of liver function abnormalities have been thought to be correlated with the infusion of excess glucose intake, both in malnourished and hypermetabolic patients[7-10]. Investigations have been conducted to determine the maximum rate of glucose infusion beyond which physiologically significant increases in N balance and the oxidation of glucose to CO_2 do not occur. Burke et al.[8] have reported that there is little increase when the infusion rate exceeds 5 mg of glucose/kg min in burn patients, and Wolfe et al.[11] have reported no oxidation of glucose in general surgical patients when glucose is administered in excess of 6–7 mg/kg min.

FAT METABOLISM IN THE TPN PATIENT

The proper perspective for considering the amount of glucose to be given to patients requiring TPN must include some consideration of the fat metabolism of such a patient, both in terms of endogenous fat metabolism in the post-absorptive state and the metabolism of exogenous fat given instead of glucose. Net fat oxidation is the result of two opposing factors, lipogenesis in the liver and adipose tissue versus fat oxidation by peripheral tissues. The altered patterns of gas exchange when acutely ill patients are given glucose loads as part of TPN could be due to increased utilization of fatty acids, or a relative inhibition of lipogenesis. Mobilization of triglyceride from adipose tissue stores is increased in the injured or infected patient, as judged by glycerol turnover studies which have revealed consistently higher levels than normally occur in relation to the plasma glycerol concentration[12,13]. This increased mobilization of triglyceride is not inhibited by glucose intake to the degree that is seen in the nutritionally depleted patient. Therefore, more circulating free fatty acids are potentially available for oxidation providing a possible mechanism for the relative increase seen in the net fat oxidation of the acutely ill patients. There is evidence to indicate that the major regulatory enzymes involved in fatty acid synthesis and oxidation are acetyl CoA carboxylase and carnitine acetyltransferase[14,15]. The former enzyme catalyses the rate limiting step in lipogenesis from two carbon fragments, while the latter is necessary for the entry of fatty acids into mitochondria which may be the rate limiting step in fat oxidation. Malonyl CoA exerts an inhibitory effect upon the carnitine acyltransferase reaction and, thus, upon fat oxidation[16,17]. The two processes appear to be regulated in a reciprocal manner; when one is accelerated, the other is necessarily slow. Therefore, one may hypothesize that the altered response of the acutely ill patient to carbohydrate loading above energy equilibrium can be accounted for, not by an isolated change in one process, but rather by a combination of continued fat oxidation and a diminished capacity for lipogenesis. Since the acutely ill patient demonstrates an inhibition of lipogenesis, excess glucose intake must be either oxidized or converted to glycogen. Carbohydrate loading in acutely ill patients results in only small changes in the turnover rate of plasma free fatty acids, although a marked increase occurs in urinary norepinephrine[6]. This large increase in norepinephrine excretion is not seen in depleted patients. Although the idea of a calorigenic effect of free fatty acids is not new[18], Kjekshus et al.[19] have recently shown that a rise in plasma free fatty acids is associated with an increase in oxygen consumption when in the presence of catecholamine stimulation. These workers suggest that the activation of cyclic AMP may facilitate a waste oxidation or a 'futile cycling' of free fatty acids. In acutely ill patients, given excessive glucose, the observed marked rise in O_2 consumption, continued fat oxidation, and increased triglyceride turnover rate concomitant with a rise in norepinephrine excretion are consistent with this type of 'free fatty acid hypothesis'.

NITROGEN RETENTION – GLUCOSE VS FAT

It is common practice to provide parenteral nutrients with sufficient calories and nitrogen to achieve a positive balance for both calories and nitrogen. The nutrient mixture for TPN in the United States is most often composed of amino acids and hypertonic glucose, without the addition of intravenous lipid. Fat emulsions are generally administered once, or twice, a week to prevent a deficiency of essential fatty acids. The nitrogen sparing effect of fat emulsions has been compared both favourably and unfavourably in relation to that of glucose[20-25]. However, it is most important to keep in mind that nitrogen balance, while important, should not be the only factor to be considered in determining the substrate composition to be used for TPN.

The effect of TPN on the utilization of an intravenous fat emulsion was studied by Nordenström et al.[26] in patients with injury, infection, and nutritional depletion, using [1-^{14}C]trioleate labelled Intralipid[R]. The plasma fractional removal rate and oxidation rate were 55% and 25% higher in patients receiving 5% dextrose, following trauma and sepsis, when compared with healthy, controlled subjects. TPN was then administered to patients with all of the non-protein calories given as glucose, or with equal proportions of glucose and intravenous fat emulsions. Studies in patients receiving only glucose resulted in higher plasma clearance rates and lower oxidation rates of a tracer dose of lipid in both acutely ill and depleted patients. There was no correlation between the rates of plasma removal and oxidation of the labelled fat, indicating that the removal rate of exogenous fat from the plasma cannot be used to predict the rate of oxidation. Glucose intakes exceeding energy expenditure did not abolish oxidation of the fat emulsion. The oxidation rate of the labelled Intralipid was linearly related to net whole body fat oxidation calculated using indirect calorimetry, suggesting that the fat emulsion was oxidized in a similar manner to endogenous fat.

GLUCOSE INTAKE AND VENTILATION

Is has been noted that the metabolic response of the injured or septic patient to a high glucose load seems to differ from the response of the nutritionally depleted patient. A study was performed to examine the influence of TPN on the ventilation of both acutely ill and depleted patients[27]. Spirometry and gas exchange measurements were performed prior to and during lipid-free TPN, which was provided in amounts to produce a daily calorie intake of approximately 50% above the measured energy expenditure. As energy sources shift from being predominantly fat, when receiving 5% dextrose, with increasing glucose intake, there is a rise in CO_2 production concurrent with the increase in the respiratory quotient. Conversion of carbohydrate to fat is associated with a respiratory quotient of approximately 8.0 and, hence, a large increase in CO_2 production occurs as glucose is given above energy equilibrium (Figure 23.2). Since there is a small energy expenditure associated with lipogenesis, O_2 consumption increases by a much smaller amount when a positive calorie balance is produced by glucose administration. Minute ventilation at rest was found to increase by 32% after the administration of the high glucose TPN to the depleted patients; there was 121% increase in other patients with a normal metabolic rate before TPN was administered, while hypermetabolic patients had an elevated resting ventilation before TPN, which was further increased by 71% during the administration of TPN. The majority of the increase in minute ventilation of these three groups was due to an increase in tidal volume, with little change in frequency (see Figure 23.3).

The CO_2 load, induced by high glucose TPN, would be expected to cause a compensatory increase in minute ventilation. However, in patients with decreased sensitivity to CO_2, or in patients with compromised lung function, who are already hyperventilating as the result of some increase in alveolar ventilation together with some increase in dead space ventilation, the added ventilatory stimulus of high glucose TPN may aggravate the pre-existing pulmonary situation. The ventilatory workload imposed by TPN in this group of patients may occasionally precipitate respiratory failure. Weaning a patient from mechanical ventilation may be more difficult than usual in the face of the extra CO_2 production, associated with high glucose TPN[28]. During the brief period for weaning from mechanical ventilation CO_2 production should be reduced from that seen with high glucose TPN, either by limiting total calorie intake or substituting fat for carbohydrate to reduce the respiratory quotient.

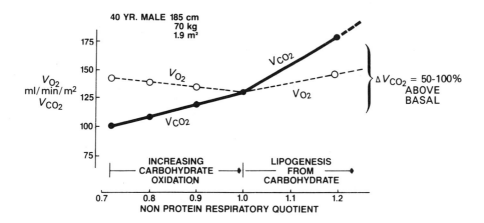

Figure 23.2 Average values of O_2 consumption and CO_2 production for a normal middle-aged male, as the non-protein respiratory quotient is raised by increasing carbohydrate administration. Note that the CO_2 production is raised by 50–100% above normal basal levels, as high glucose parenteral nutrition is used to produce a non-protein respiratory quotient between 1.1 and 1.2

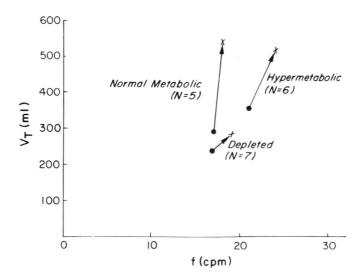

Figure 23.3 Changes in the tidal volume and frequency of ventilation are shown for three groups of patients given high glucose parenteral nutrition. The depleted and hypometabolic patients showed little response, while the patients with a metabolic rate in the normal range had a sharp rise in the tidal volume and the hypermetabolic patients had a large rise in tidal volume and a small increase in frequency of ventilation (from Askanazi *et al.*, reference 27)

OBJECTIVES OF TOTAL PARENTERAL NUTRITION

The amount of glucose which should be administered to a patient as part of parenteral nutrition must depend in part upon the goals of the physician when he gives TPN at any particular stage of convalescence. It is clear that the metabolic response to acute injury or infection includes a period of catabolic tissue breakdown which is difficult to abolish by nutritional means. As convalescence progresses, a turning point is reached where the metabolic response of such patients enters an anabolic phase, where nutritional intake is used more efficiently for tissue restoration. Many of the undesirable side-effects of high glucose loads given with TPN are seen during the height of the catabolic phase. Therefore, our group at Columbia has become convinced that the goal of TPN during the acute catabolic phase should be to provide enough calories and nitrogen to achieve equilibrium, but to postpone higher intakes in order to obtain a positive balance. We believe that providing enough TPN to obtain a positive calorie and nitrogen balance is logical once the anabolic phase has been reached. Extra calories and nitrogen administered at this time are used most efficiently for tissue restoration and are associated with less undesirable side-effects, including hyperglycaemia, catecholamine stimulation, a thermogenic response and excessive ventilatory stimulation.

References

1 Connor, W. E. and Connor, S. L. (1976). Sucrose and carbohydrate. In Hegsted, D. M. (ed.) *Present Knowledge in Nutrition.* 4th Edn., p. 33. (New York, Washington: The Nutrition Foundation)

2 Phillips, G. D. and Odgers, C. L. (1982). *Parenteral and Enteral Nutrition.* 2nd Edn., p. 55. (Bedford Park, South Australia: The Flinders University of South Australia)

3 Askanazi, J., Carpentier, Y. A., Elwyn, D. H., Nordenström, J., Jeevanandam, M., Rosenbaum, S. H., Gump, F. E. and Kinney, J. M. (1980). Influence of total parenteral nutrition on fuel utilization in injury and sepsis. *Ann. Surg.*, **191**, 40

4 Elwyn, D. H., Gump, F. E., Munro, H. N., Iles, M. and Kinney, J. M. (1979). Changes in nitrogen balance of depleted patients with increasing infusions of glucose. *Am. J. Clin. Nutr.*, **32**, 1597

5 Nordenström, J., Jeevanandam, M., Elwyn, D. H., Carpentier, Y. A., Askanzi, J., Robin, A. and Kinney, J. M. (1981). Increasing glucose intake during total parenteral nutrition increases norepinephrine excretion in trauma and sepsis. *Clin. Physiol.*, **1**, 525

6 Elwyn, D. H., Kinney, J. M., Jeevanandam, M., Gump, F. E. and Broell, J. R. (1979). Influence of increasing carbohydrate intake on glucose kinetics in injured patients. *Ann. Surg.*, **190**, 117

7 Sheldon, G. F., Petersen, S. R. and Sanders, R. (1978). Hepatic dysfunction during hyper-alimentation. *Arch. Surg.*, **113**, 504

8 Burke, J. F., Wolfe, R. R., Mullany, C. J., Mathews, D. E. and Bier, D. M. (1979). Glucose requirements following burn injury. Parameters of optimal glucose infusion and possible hepatic and respiratory abnormalities following excessive glucose intake. *Ann. Surg.*, **190**, 274

9 Lowry, S. F. and Brennan, M. F. (1979). Abnormal liver function during parenteral nutrition: relation to infusion excess. *J. Surg. Res.*, **26**, 300

10 Skidmore, F. D., Tweedle, D. E. F., Gleave, E. N., Gowland, E. and Knass, D. A. (1979). Abnormal liver function during nutritional support in postoperative cancer patients. *Ann. R. Coll. Surg. Eng.*, **61**, 183

11 Wolfe, R. R., O'Donnell, T. F., Jr., Stone, M. D., Richmand, D. A. and Burke, J. F. (1980). Investigation of factors determining the optimal glucose infusion rate in total parenteral nutrition. *Metabolism*, **29**, 892

12 Carpentier, Y. A., Askanazi, J., Elwyn, D. H., Gump, F. E., Nordenström, J. and Kinney, J. M. (1980). The effect of carbohydrate intake on the lipolytic rate in depleted patients. *Metabolism*, **29**, 974

13 Carpentier, Y. A., Askanazi, J., Elwyn, D. H., Jeevanandam, M., Gump, F. E., Hyman, A. I., Burr, R. and Kinney, J. M. (1979). Effects of hypercaloric glucose infusion on lipid metabolism in injury and sepsis. *J. Trauma*, **19**, 649

14 Newsholme, E. A. and Start, C. (1973). Adipose tissue and the regulation of fat metabolism. In Newsholme, E. A. and Start, C. (eds.) *Regulation in Metabolism*. Chap. 5. (Chichester, New York, Brisbane, Toronto: Wiley-Interscience)

15 Newsholme, E. A. and Start, C. (1973). Regulation of fat metabolism in liver. In Newsholme, E. A. and Start, C. (eds.) *Regulation in Metabolism*. Chap. 7. (Chichester, New York, Brisbane, Toronto: Wiley-Interscience)

16 McGarry, J. D. and Foster, D. W. (1979). In support of the roles of malonyl-CoA and carnitine acyltransferase 1 in the regulation of hepatic fatty acid oxidation and ketogenesis. *J. Biol. Chem.*, **254**, 8163

17 McGarry, J. D., Takaboyashi, Y. and Foster, D. W. (1978). The role of malonyl-CoA in the coordination of fatty acid synthesis and oxidation in isolated rat hepatocytes. *J. Biol. Chem.*, **253**, 8294

18 Havel, R. J., Carlson, L. A., Ekelund, L. G. and Holmgren, A. (1964). Studies on the relation between mobilization of free fatty acids and energy metabolism in man, effects of norepinephrine and nicotinic acid. *Metabolism*, **13**, 1402

19 Kjekshus, J. K., Elekjair, E. and Rinds, P. (1980). The effect of free fatty acids on oxygen consumption in man: the free fatty acid hypothesis. *Scand. J. Clin. Lab. Invest.*, **40**, 63

20 Bark, S., Holm, I., Hakanson, I. and Wretlind, A. (1976). Nitrogen-sparing effect of fat emulsion compared with glucose in the postoperative period. *Acta Chir. Scand.*, **142**, 423

21 Brennan, M. F. and Moore, F. D. (1973). An intravenous fat emulsion as a nitrogen sparer: comparison with glucose. *J. Surg. Res.*, **14**, 501

22 Brennan, M. F., Fitzpatrick, G. F., Cohen, K. H. and Moore, F. D. (1975). Glycerol: major contributor to the short term protein sparing effect of fat emulsions in normal man. *Ann. Surg.*, **182**, 386

23 Jeejeebhoy, K. N., Anderson, G. H., Nakhooda, A. F., Greenberg, G. R., Sanderson, I. and Marliss, E. B. (1976). Metabolic studies in total parenteral nutrition with lipid in man. Comparison with glucose. *J. Clin. Invest.*, **57**, 125

24 Long, J. M., Wilmore, D. W., Mason, A. D. and Pruitt, B. A. (1977). Effect of carbo-hydrate and fat intake on nitrogen excretion during total intravenous feeding. *Ann. Surg.*, **185**, 417

25 Parodis, C., Spanier, A. H., Calder, M. and Shizgal, H. M. (1978). Total parenteral nutrition with lipid. *Am. J. Surg.*, **135**, 164

26 Nordenström, J., Carpentier, Y. A., Askanazi, J., Robin, A. P., Elwyn, D. H., Hensle, T. W. and Kinney, J. M. (1982). Metabolic utilization of intravenous fat emulsion during total parenteral nutrition. *Ann. Surg.*, **196**, 221

27 Askanazi, J., Rosenbaum, S. H., Hyman, A. I., Silverberg, P. A., Milic-Emili, J. and Kinney, J. M. (1980). Respiratory changes induced by the large glucose loads of total parenteral nutrition. *J. Am. Med. Assoc.*, **243**, 1444

28 Askanazi, J., Nordenström, J., Rosenbaum, S. H., Elwyn, D. H., Hyman, A. I., Carpentier, Y. A. and Kinney, J. M. (1981). Nutrition for the patient with respiratory failure: glucose vs fat. *Anesthesiology*, **54**, 373

24
Fat utilization in critically ill patients

Y. A. CARPENTIER

INTRODUCTION

Most tissues of the body are able to utilize either glucose or fatty acids for their energy requirements. However, some particular cells (central nervous system, blood cells, etc.) are not capable of oxidizing fatty acids. This can explain why glucose is still often considered as the sole energy substrate to be parenterally administered, exogenous fat being given only to prevent essential fatty acid deficiency.

Recent observations have shown that infusion of large glucose doses to critically ill patients could induce severe side-effects such as respiratory distress consecutive to high CO_2 production[1] and impaired hepatic function due to massive deposition of fat and/or glycogen[2,3]. Increased release of catecholamines[4] and glucagon[5] was also observed as if the glucose load was acting as a supplementary stress in these patients. Moreover, the respiratory quotient of these patients consistently remained below 1.0, indicating persistent net fat oxidation despite glucose intake in excess of their energy expenditure.

These recent findings generated an interest in studying the effect of injury and sepsis, as well as the influence of glucose infusion on the metabolic utilization of endogenous and exogenous fat.

METABOLISM OF ENDOGENOUS FAT

The major part of energy storage in the body is represented by the accumulation of triglycerides in the various kinds of adipose tissue. The hydrolysis of one molecule of triglyceride in the adipocyte produces one molecule of glycerol that is almost entirely released in the blood stream and three molecules of free fatty acids (FFA) that are re-esterified to a variable degree within the cell under the control of several hormonal and biochemical factors

(Figure 24.1). Therefore, the rate of glycerol turnover indicates the breakdown of triglycerides while the rate of FFA turnover represents the release of substrate that is available for utilization in the other tissues.

Glycerol turnover has been measured in normal subjects and in hypermetabolic patients using a staged infusion of cold glycerol[6]. As expected, a close correlation was found in normal subjects between the rate of turnover and the plasma concentration of glycerol. After an injury or during sepsis, the turnover of glycerol was increased which was not reflected in determinations of plasma concentration. Total parenteral nutrition providing 135–175% of energy expenditure and containing glucose as sole non-protein calorie source, induced a marked decrease in the glycerol turnover of noninjured depleted subjects[7] but had less effect in hypermetabolic patients[6].

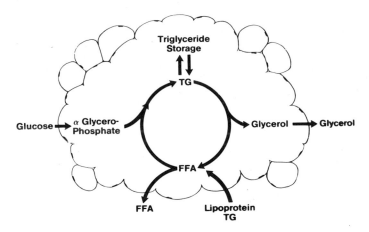

Figure 24.1 Adipose tissue metabolism

FFA turnover was determined in hypermetabolic patients using a primed-infusion of [1-^{14}C]palmitate during 70 min[8]. The values of FFA turnover were found to be greater in the hypermetabolic patients than would have been expected from the values of plasma FFA concentration by applying the correlation observed in normal subjects by Hagenfeldt[9]. This finding is very similar to the observation reported for glycerol turnover. The oxidation rate of plasma FFA, estimated by determining the excretion rate of $^{14}CO_2$ in the expired air during the 390 min following the initiation of [1-^{14}C]palmitate infusion, was studied in relation to carbohydrate intake in patients receiving TPN. The inhibition of fat oxidation was by far less marked in hypermetabolic patients than in depleted patients (Figure 24.2).

These studies show that:

(1) Dynamic measurements are needed for studying the mobilization of endogenous fat in hypermetabolic patients.

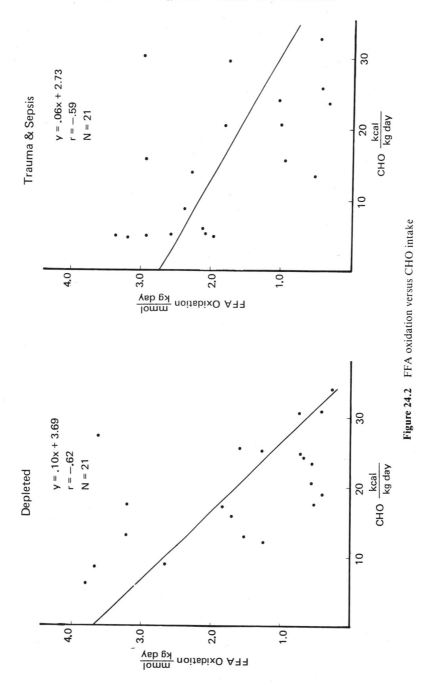

Figure 24.2 FFA oxidation versus CHO intake

(2) The rates of fat mobilization and oxidation are higher in hyper-metabolic patients than in normal or depleted subjects.

(3) The inhibitory effect of carbohydrate intake on fat oxidation is less pronounced in hypermetabolic than in depleted patients.

METABOLISM OF EXOGENOUS FAT

Since some resistance to the utilization of carbohydrates has been demon-strated in hypermetabolic surgical patients who, by contrast, seem to prefer-entially utilize endogenous fat as an energy source – at least in some tissues – we decided to investigate some metabolic steps in the utilization of exogenous lipids by these patients. A 10% Intralipid emulsion labelled with [1-^{14}C]trioleate (Kabivitrum, Stockholm) has been used for this study[10].

The physical structure and the size of Intralipid particles are comparable with endogenous chylomicra[11]. These particles differ from endogenous lipo-proteins in that:

(1) They are not covered with apoproteins.

(2) They contain no cholesteryl ester and little, if any, free cholesterol.

(3) The fatty acid composition of Intralipid triglycerides is very rich in linoleic and linolenic acids.

However, particles of exogenous fat can rapidly acquire apoproteins C (namely C-II and C-III) and E by exchange with high density lipoproteins in the plasma[12–14]. Apoprotein C-II allows these particles to be recognized by lipoprotein lipase and removed by the tissues containing this enzyme, namely adipose tissues and muscles.

The activity of lipoprotein lipase varies in opposite directions in both kinds of tissues, under the effect of hormonal and nutritional factors. This means that, according to the hormonal and nutritional milieu, exogenous lipids could be removed in varying proportions by fat tissues (where they are stored) and muscles (where they could presumably be oxidized). Therefore we measured not only the plasma clearance but also the oxidation rate of [^{14}C]Intralipid in various types of surgical patients receiving different nutritional regimens.

One ml/kg b.w. of 10% Intralipid (tagged with 36 μCi of [1-^{14}C]trioleate) was injected intravenously within 2 min. The fractional removal rate K2 was determined following the intravenous fat tolerance test method described by Rossner[15]. An estimation of $^{14}CO_2$ excretion in the expired air was obtained by repetitive gas exchange measurements over 450 min following the injection. The oxidation rate was expressed as the fraction of the injected radioactive dose that was recovered in the expired air during that period.

In comparison with the data obtained in four normal subjects studied after an overnight fast, significantly higher rates of plasma fractional removal and oxidation were observed during the hypermetabolic phase in eight severely traumatized patients, while no significant change was observed in three other patients having undergone minor surgical injury[10] (Table 24.1).

Table 24.1 Effect of injury on [^{14}C]Intralipid plasma release and oxidation (measurements carried out in the morning on fasting patients)

	Normal controls	Minor injury	Major injury
Number	4	3	8
Clearance rate (%/min)	8.5 ± 0.3	8.4 ± 1.1	13.4 ± 2.4
Oxidation rate (%/450 min)	29.5 ± 4.1	33.2 ± 1.5	38.3 ± 3.9

The same parameters were measured in hypermetabolic and depleted patients receiving total parenteral nutrition[16]. Two intravenous diets were compared. The nitrogen and the total energy intake (150% of the measured energy expenditure) were the same in both regimens but glucose was the sole non-protein caloric source in the first one (glucose system) while approximately equal proportions of exogenous fat and glucose represented the non-protein calorie intake in the second one (lipid system). The fractional removal rate of [^{14}C]Intralipid was higher in hypermetabolic and depleted patients receiving TPN with the glucose system than in normal subjects after an overnight fast; however, the values observed during TPN in hypermetabolic patients were not statistically different from the control data measured during infusion of 5% dextrose. The rate of [^{14}C]Intralipid oxidation was much lower in both groups of depleted and hypermetabolic patients receiving the glucose system than in the normal subjects studied after an overnight fast.

In both depleted and hypermetabolic patients receiving TPN with the lipid system, the rate of fractional removal did not differ from the values observed in normal subjects, but was significantly lower than in similar groups of patients studied during administration of the glucose system.

In these conditions, the rate of [^{14}C]Intralipid oxidation was significantly higher in both groups of patients than during infusion with the glucose system. These values were similar to those observed in the normal subjects studied after an overnight fast.

In the patients studied during TPN, a very close positive correlation was observed between the rate of [^{14}C]Intralipid oxidation and the net fat oxidation measured by indirect calorimetry; there was a negative correlation ($r = 0.92$; $p < 0.001$) between the rate of [^{14}]Intralipid oxidation and carbohydrate intake; no correlation could be found between oxidation and plasma clearance of exogenous fat ($r = 0.04$).

These results show that:

(1) Both the rates of plasma fractional removal and oxidation are increased in severely injured patients.

(2) The plasma clearance is increased while the oxidation of exogenous fat is decreased during TPN with the glucose system.

(3) Both the plasma clearance and oxidation of exogenous fat are in the normal range during TPN using the lipid system.

(4) The rate of fractional removal is not a suitable parameter for studying the metabolic utilization of exogenous lipids.

SOME METABOLIC EFFECTS OF TOTAL PARENTERAL NUTRITION WITH EXOGENOUS FAT

The above-mentioned studies have demonstrated that both endogenous and exogenous fat are important energy substrates in hypermetabolic surgical patients. In these patients, parenteral nutrition with high glucose intake was also reported to induce various side-effects such as respiratory distress due to high CO_2 production and impaired liver function due partly to excessive glycogen or fat deposition.

We have therefore investigated the effect of including exogenous fat in parenteral regimens on respiratory and hepatic functions.

Effect on ventilatory parameters

Two different intravenous regimens have been given to five depleted and 13 hypermetabolic surgical patients. Both diets contained the same amino-acid load and total calorie intake which was equal to 150% of the resting energy expenditure previously measured during 5% dextrose infusion. Glucose was the sole non-protein energy source in one regimen (glucose system) while glucose and fat provided approximately equal amounts of calories in the other regimen (lipid system). The depleted patients received each regimen for 1 week, following a random order. The hypermetabolic patients were randomly assigned for 1 week to one or the other intake. Ventilatory parameters were measured after 4 days of TPN. Oxygen consumption, CO_2 production and non-protein respiratory quotient were recorded for both types of patients; supplemental data concerning minute-ventilation, inspiratory flow and respiratory rate were obtained in depleted patients[17].

In both types of patients, the respiratory quotient was much lower during TPN with the lipid system compared to the glucose system (0.87 vs 1.00, $p<0.001$ for the depleted; 0.85 vs 0.94, $p<0.001$ for the hypermetabolic patients). Compared to the values measured during TPN with the glucose system, TPN with the lipid system resulted in the depleted patients in lower values for CO_2 production (-20%; $p<0.025$), minute-ventilation (-26%;

$p<0.01$) and inspiratory flow (-30%; $p<0.001$) while the oxygen consumption and the respiratory rate were not different. In hypermetabolic patients, the values of CO_2 production (147 vs 179 ml/min; $p<0.01$), O_2 consumption (172 vs 190 ml/min; $p<0.025$) and respiratory quotient (0.85 vs 0.94; $p<0.001$) were lower during TPN with the lipid system than during TPN with the glucose system.

These data show that the ventilatory load induced by TPN with high glucose intake can be reduced significantly when half of the non-protein calories is provided by exogenous fat.

Effect on the liver function

A high incidence of alterations of liver function has been reported in the past 5 years in relation to the administration of a hypercaloric parenteral nutrition[18-21]. In most of these studies, glucose was the main or the only non-protein calorie source. A correlation has been demonstrated between the incidence of these biological alterations and the glucose intake[2].

The purpose of a prospective study was to determine the incidence of liver dysfunction (judged on alterations of biological tests) in patients receiving a mid-term parenteral nutrition adjusted to their energy expenditure and providing isocaloric amounts of glucose and exogenous lipids[22]. 36 patients who required TPN in our unit (Hôpital Saint-Pierre, Brussels) in 1979 were included in the study. These patients had normal liver function tests prior to TPN; they did not develop severe sepsis and received no hepato-toxic agents. The mean duration of TPN was 22 days. The total caloric intake was equal to 125% of the estimated energy expenditure using the Harris–Benedict formula[23]. Amino-acid intake provided 1 g nitrogen for 150 calories. Hepatic function tests were performed prior to TPN and then once a week during TPN. A rise in alkaline phosphatase values above the normal range was observed in five patients (14%) after 2 weeks of TPN; determinations of γ-glutamyltranspeptidase confirmed hepatic dysfunction in these patients. No alteration in serum glutamic-pyruvic transaminase was observed in any patient while serum glutamic-oxaloacetic transaminase level was abnormally increased in one patient after 1 week of TPN. Serum bilirubin concentration rose (up to 20 mg/l) in one patient after 26 days of TPN. All these biological alterations of the liver function spontaneously resolved within 2 or 3 weeks after cessation of TPN.

A randomized study comparing two different TPN regimens would obviously be more appropriate for drawing definitive conclusions, but infusion of large glucose loads was not approved by the local Ethical Committee. Nevertheless, both the incidence and the severity of liver alterations appear to be very low in comparison with the results observed in patients receiving hyperalimentation with high glucose intake.

Beside excessive glucose intake, several other factors could be involved in the development of liver dysfunction during TPN. The functional bypass of

299

the gastrointestinal tract, by altering the composition of the bile, could rapidly induce the formation of lithiases. A direct toxic effect on the liver could result from the degradation products of some amino acids, from the addition of stabilizing agents in the amino-acid solutions, as well as from the resorption of toxic products by the remaining bowel.

CONCLUSIONS

The studies reported in the present paper demonstrate an increased utilization of endogenous and exogenous lipids by hypermetabolic surgical patients. The inclusion of exogenous fat parenteral nutrition regimens reduces the incidence and the severity of the complications observed during TPN with very high glucose loads.

The results of these studies do not deny the essential role of glucose in the nutrition of surgical patients but argue against excessive administration of carbohydrates. Caution should also be drawn against the infusion of excessive amounts of fat. Some side-effects (such as decreased leukocyte migration and chemotaxis) have already been reported during rapid infusion of Intralipid[24] and the effect of exogenous lipids on the metabolism of endogenous fat is still poorly understood.

It seems reasonable to avoid hyperalimentation in critically ill patients and adjust parenteral intake to the patient's energy expenditure. Nutritional support should therefore be started early enough to avoid severe nutritional depletion.

References

1 Askanazi, J., Elwyn, D. H., Silverberg, P. A. *et al.* (1980). Respiratory distress secondary to a high carbohydrate load of TPN: a case report. *Surgery*, **87**, 596
2 Lowry, S. F. and Brennan, M. F. (1979). Abnormal liver function during parenteral nutrition: relation to infusion excess. *J. Surg. Res.*, **26**, 300
3 Mashima, Y. (1979). Effect of calorie overload on puppy livers during parenteral nutrition. *J. Parent. Enter. Nutr.*, **3**, 139
4 Nordenstrom, J., Jeevanandam, M., Elwyn, D. H. *et al.* (1981). Glucose loading during total parenteral nutrition increases norepinephrine excretion in trauma and sepsis. *Clin. Physiol.*, **1**, 525
5 Elwyn, D. H., Kinney, J. M., Jeevanandam, M. *et al.* (1979). Influence of increasing carbohydrate intake on glucose kinetics in injured patients. *Ann. Surg.*, **190**, 117
6 Carpentier, Y. A., Askanazi, J., Elwyn, D. H. *et al.* (1979). Effect of hypercaloric glucose infusion on lipid metabolism in injury and sepsis. *J. Trauma*, **19**, 649
7 Carpentier, Y. A., Askanazi, J., Elwyn, D. H. *et al.* (1980). The effect of carbohydrate intake on the lipolytic rate in depleted patients. *Metabolism*, **29**, 974
8 Nordenstrom, J., Carpentier, Y. A., Askanazi, J. *et al.* (1982). Free fatty acid mobilization and oxidation during total parenteral nutrition in trauma and sepsis. (In preparation).
9 Hagenfeldt, L. (1975). Turnover of individual free fatty acids in man. *Fed. Proc.*, **34**, 2246
10 Carpentier, Y. A., Nordenstrom, J., Askanazi, J. *et al.* (1979). Relationship between rates of clearance and oxidation of [^{14}C]Intralipid in surgical patients. *Surg. Forum*, **30**, 72
11 Schoefl, G. I. (1968). The ultrastructure of chylomicron and of the particles in the artificial fat emulsion. *Proc. R. Soc. Lond.*, **169**, 147

12 Robinson, S. F. and Quarfordt, S. H. (1979). Apoproteins in association with Intralipid incubation in rat and human plasma. *Lipids*, **14**, 343

13 Weinberg, R. B. and Scanu, A. M. (1982). *In vitro* reciprocal exchange of apoproteins and non-polar lipids between human high density lipoproteins and an artificial tri-glyceride–phospholipid emulsion quoted by Rosenberg, I. H. and Weinberg, R. B. The composition and metabolism of IV fat emulsions. In *Balanced Parenteral Nutrition*. (North Chicago, Ill.: Abbott Laboratories 60064) (In press)

14 Erkelens, D. W., Brunzell, J. D. and Bierman, E. L. (1979). Availability of apolipoprotein CII in relation to the maximal removal capacity for an infused triglyceride emulsion in man. *Metabolism*, **28**, 495

15 Rossner, S. (1974). Studies of an intravenous fat tolerance test. Methodological, experimental and clinical experiences with Intralipid. *Acta Med. Scand.*, **564** (suppl.), 1

16 Nordenstrom, J., Carpentier, Y. A., Askanazi, J. *et al.* (1982). Metabolic utilization of intravenous fat emulsion during total parenteral nutrition. *Ann. Surg.* (In press)

17 Askanazi, J., Nordenstrom, J., Rosenbaum, S. H. *et al.* (1981). Nutrition for the patient with respiratory failure: glucose vs fat. *Anesthesiology*, **54**, 373

18 Messing, B., Bitoun, A., Galian, A. *et al.* (1977). La stéatose hépatique au cours de la nutrition parentérale dépend-elle de l'apport calorique glucidique? *Gastroenterol. Clin. Biol.*, **1**, 1015

19 Grant, J. P., Cok, C. E., Kleinman, L. M. *et al.* (1977). Serum hepatic enzyme and bilirubin elevations during parenteral nutrition. *Surg. Gynecol. Obstet.*, **147**, 573

20 Sheldon, G. P., Petersen, S. R. and Sanders, R. (1978). Hepatic dysfunction during hyper-alimentation. *Arch. Surg.*, **113**, 504

21 Lindor, K. D., Fleming, C. R., Abrams, A. *et al.* (1979). Liver function values in adults receiving total parenteral nutrition. *J. Am. Med. Assoc.*, **241**, 2398

22 Carpentier, Y. A. and Van Brandt, M. (1981). Effect of total parenteral nutrition on liver function. *Acta Chir. Belg.*, **2 and 3**, 141

23 Harris, J. A. and Benedict, F. G. (1919). *Biometric Studies of Basal Metabolism in Man*. (Carnegie Institution of Washington. Publication no 279)

24 Nordenstrom, J., Jarstrand, C. and Wiernik, A. (1979). Decreased chemotactic and random migration of leukocytes during Intralipid infusion. *Am. J. Clin. Nutr.*, **32**, 2416

25
Liver malfunction associated with parenteral nutrition

R. K. AUSMAN, E. J. QUEBBEMAN AND C. L. ALTMANN

Signs of liver malfunction appear in some patients after initiating continuous intravenous infusion of a high concentration glucose/amino-acid formulation[1]. An abnormal histologic picture develops which includes cholestasis and fatty infiltration[2,3] associated with measured biochemical abnormalities. Several causes for this pathophysiologic state have been suggested[4-7] including lack of enteral feeding, plasma amino-acid imbalance, use of parenteral lipid, excess dextrose in the parenteral nutrition formulation, and degradation products of preservatives in the amino-acid solutions used.

The practice of supplying carbohydrate calories to meet the physician's concept of enhanced energy requirements associated with a catabolic state is widespread[8-10]. Generally, this amount of calories is some multiple of the resting or basal energy expenditure. Recently our group evaluated the calorie needs of patients who received parenteral nutrition[1] and, contrary to widely held belief, found actual energy utilization to be lower than has been stated in other reports[9,10,12]. As a result it occurred to us there might be an association between non-protein calorie intake, blood glucose levels, and liver malfunction.

The purpose of this study was to determine if a relationship could be detected between the incidence and magnitude of apparent liver malfunction and differing levels of calorie and/or amino-acid support. We sought to find what role, if any, could be assigned to 'excess' carbohydrate, protein, or fat components as a cause of abnormal hepatic function and fat deposition in the liver during intravenous feeding.

MATERIALS AND METHODS

This study was retrospective and included the records of all patients seen at the Milwaukee County Medical Center between October, 1979 and September, 1981, who had been given parenteral nutrition for at least 3 weeks, supervised by a member of the Department of Surgery parenteral nutrition team. Excluded were individuals who demonstrated overt hepatic failure, conditions which would be likely to cause abnormal liver function, or an abnormal alkaline phosphatase, bilirubin and/or SGOT levels preceding the administration of parenteral nutrition.

Patients received parenteral nutrition through a central vein catheter when one or more of the usual indications were present. Crystalline amino-acid solution (Travasol*–Travenol Laboratories) was the nitrogen source, usually in a final concentration of 27.5 g amino acid per litre of solution. The carbohydrate was given at 250 g dextrose per litre. The solution also contained electrolytes, vitamins, and trace elements as indicated. Three patients were given intravenous lipid as a non-protein calorie source on several occasions during nutritional support. Patients received two 500 ml units of intravenous lipid per week to counter the development of essential fatty acid deficiency[13].

The volume of solution administered and the quantities of carbohydrate, fat, and protein varied among individuals who were evaluated. Usually 2000–3500 ml of fluid were infused each day.

Insulin was placed in the solution container with the parenteral nutrient formulation when indicated by the continuing presence of more than a trace of glucose in the urine.

Data obtained from patient records, parenteral nutrition records, and the pharmacy included age, sex, height, weight (at the beginning of nutritional support), total carbohydrate calories per week, total grams of nitrogen per week, blood glucose, and SGOT and alkaline phosphatase levels. In some patients an actual measurement of resting energy expenditure (REE) was made during parenteral nutrition support according to the technique described previously[11]. When this value was not available, a calculation of basal energy expenditure was made using the equation developed by Harris and Benedict[14] which we found to be a suitable approximation of REE.

Fever can alter REE, although no adjustments were made in energy utilization calculations since the duration of pyrexia was only a small proportion of the total days on parenteral nutrition. No patients with surface burns which cause a significant increase in REE[15] were included in this study.

Patients receiving less than 3 weeks of intravenous nutritional support were excluded because they provided insufficient opportunity to observe the development of liver abnormalities. When parenteral nutrition therapy was given for a prolonged period, only the initial 30 days of treatment were considered to maintain comparability in the study.

*Synthamin[R]

RESULTS

The age, sex, height, weight, disease process, and energy expenditure relating to each patient in this study are shown in Table 25.1.

Table 25.1 Results of study

Patient identity	(Age) (y)	Sex	Height (cm)	Weight (kg)	BEE	Disease, operation, etc.
Group I						
LP	24	F	162	81	1695*	Hyperemesis gravidarum
GE	20	M	145	32	1140	Crohn's disease
SS	60	F	163	35	900	Post-gastrectomy; GI bleeding
VK	57	M	174	47	1257	Carcinoma of oesophagus
WB	18	M	183	44	1382*	Malnutrition, neurological deficit, aspiration pneumonia
Group II						
KW	67	M	178	96	1816	Carcinoma of oesophagus; alcoholism
GS	68	M	175	72	1793*	Colon carcinoma; wound dehiscence
KD	60	F	155	44	1276	Gun shot wound to oesophagus
PL	58	M	183	54	1326	Carcinoma of oropharynx; malnutrition
JT	62	M	170	43	1084	Malnutrition, dumping syndrome, multiple injuries
SR	31	M	168	44	1228	Gastric outlet obstruction
ER	68	F	160	85	1450	Carcinoma of stomach; GI fistula
GK	48	M	170	66	1400	Pancreatitis; pseudocyst
Group III						
DE	35	M	178	58	1514	Pancreatic pseudocyst; alcoholism
GP	40	M	174	54	1400	Pancreatitis; pseudocyst; diabetes, malnutrition
RW	66	F	160	67	2000*	Small bowel obstruction; carcinomatosis
IM	77	F	160	79	1314	Small bowel obstruction
EK	62	M	179	109	2032	Ileal conduit leak; wound dehiscence
VL	71	F	157	67	1206*	Small cell carcinoma of lung
EM	70	F	170	60	1170	Vesico vaginal fistula
DT	65	F	157	54	1218*	Abdominal necrotizing fasciitis; sepsis; diabetes
LW	51	F	167	64	1314*	Carcinoma of pancreas; wound infection peritonitis

*REE (measured)

The patient population was divided into three groups on the basis of liver function studies (SGOT and alkaline phosphatase) as follows:

Group I   no liver function abnormality

Group II   *either* SGOT *or* alkaline phosphatase values were elevated during the 3 week period

Group III   *both* the SGOT and alkaline phosphatase values became abnormal

Mean basal energy expenditure and resting energy expenditure values for each group were not different (Table 25.2).

Table 25.2   Mean energy expenditure

	kcal/day	Number of patients
Group I	1275	5
II	1422	8
III	1463	9

A rise in bilirubin did not occur in these patients. The mean non-protein calories given, expressed as a multiple of energy expenditure (calories/calorie of energy expended), was comparable among the groups as shown below:

Group I   1.6 non-protein calories/calorie of energy expended
Group II   1.5
Group III 1.5

The quantities of nitrogen (via crystalline amino-acid solution) administered to patients in each group are shown in Table 25.3. Individual values differ to some extent, but these differences were not significant; Group I patients received the highest dose per kg of body weight.

Table 25.3   Nitrogen administered (g/kg day). Mean of daily values

	Weeks		
	1	2	3
Group I	0.24	0.26	0.25
II	0.16	0.22	0.21
III	0.13	0.19	0.21

No significant differences

Table 25.4   Liver function values

	Treatment (weeks)			
	1	2	3	4
Mean SGOT (IU)				
Group I	24	24	22	25
II	25	37	51	46
III	23	52	81	72*
Mean alkaline phosphatase (mg/dl)				
Group I	77	82	77	94
II	70	94	110	113*
III	80	138	210	246†

* $p = <0.05$, † $p = <0.01$ compared to week 1 analysis of variance

Liver function data are shown in Table 25.4. They demonstrate the logic by which the groups were formed. All values were normal when parenteral nutrition was started; subsequent abnormalities were not present in Group I but were seen in Groups II and III.

Table 25.5 includes summary data for blood glucose. The number of data points for each group was different because (a) there were more patients in Groups II and III than in Group I, and (b) the higher blood glucose levels in

Table 25.5   Blood sugar levels (mg/dl)

	Mean (3 weeks)	Range	Number of values	Number of patients
Group I	122	65–240	54	5
II	164	70–395	94	8
III	223	53–696	123	9

$p < 0.0001$, analysis of variance

Group III undoubtedly caused more clinical watchfulness which stimulated frequent measurements of blood glucose. Although the quantity of dextrose administered to the patients was comparable, there was a considerable difference in their blood glucose response to the infusion. There is a highly significant difference between each group (analysis of variance; $p < 0.0001$).

DISCUSSION

The appearance of altered liver function in parenteral nutrition patients has been an enigma to physicians. Sometimes the manifestation is limited to enzyme changes which are reversible. In these instances the value of parenteral nutrition to the patient is judged to outweigh the temporary chemical derangements which have an uncertain adverse effect. However, frequently in children and sometimes in adults the problem progresses to significant cholestasis[1,4] and clinical jaundice. Many physicians are unwilling to continue parenteral nutrition, even if the illness for which it was being administered has not resolved. Liver biopsies almost invariably demonstrate a fatty metamorphosis and intrahepatic cholestasis when liver malfunction is present[3,16,17].

Several authors have conjectured about the treatment of this phenomenon. One suggestion has been that cyclic parenteral nutrition may resolve the clinical problem[18]. This practice involves eliminating the carbohydrate component of the parenteral nutrition formula for approximately 6 h during each 24 h period. Patients with demonstrated fatty liver were reported to benefit by the disappearance of fat during cyclic parenteral nutrition. No patients with sepsis were included in this study.

Animal experiments have been conducted which showed that a 'balanced' formulation of 75% carbohydrate and 25% fat did not cause an abnormal increase in liver weight when compared to orally fed controls[16]. Total carbohydrate or total fat fed animals demonstrated gross liver changes as excess weight of the organ. These results suggested that a more 'balanced' formulation may be efficacious in avoiding liver changes. Other investigators have stated that excess glucose may be the cause of liver malfunction[16,19] or that glucose inhibits the normal breakdown of triglyceride[3].

Our data were examined for a possible influence of nitrogen intake on the genesis of liver function problems. We found that all patients received approximately the same amount of nitrogen during the second and third week; in the first week Groups II and III were given less nitrogen than Group I. To implicate nitrogen differences as the reason for liver malfunction would require a belief that lower nitrogen levels early in treatment are the cause. This concept seems unlikely. We conclude that nitrogen intake is an important variable.

As the initial step in analysing our patients, we divided them into groups according to the presence and degree of liver dysfunction. Approximately 25% (5/22) did not have adverse changes in the enzymes measured (Group I). Others showed varying degrees of change, nearly equally divided between moderate (Group II − 8/22) and significant (Group III − 9/22). SGOT and alkaline phosphatase parameters were used to measure liver changes because they are indicators of acute cell damage and cholestasis, respectively[20].

Data in this study do not support a causal relationship between the quantity of dextrose administered and the development of liver dysfunction. There was no difference between the ratio of calories of dextrose infused to calories of energy expended in the groups studied. Another reflection of the similarity of carbohydrate supply in the patient population is shown in Table

Table 25.6 Carbohydrate supply (kcal/week)

	Mean	Number of observations
Group I	13478	19
II	15420	29
III	15497	31

25.6. These values are not statistically different. The amount of carbohydrate does not appear to be related to the hepatic malfunction which occurs during parenteral nutrition.

A combination of nitrogen and glucose likewise seems improbable because these intakes are so similar among the groups. No combination of these constituents can be identified which differentiates the groups.

The findings for blood glucose are most interesting. Table 25.5 shows there are significant differences among the three groups, both with reference to the mean values for all blood glucose measurements and the degree of variation about the mean within each group. Figures 25.1, 25.2 and 25.3 are line

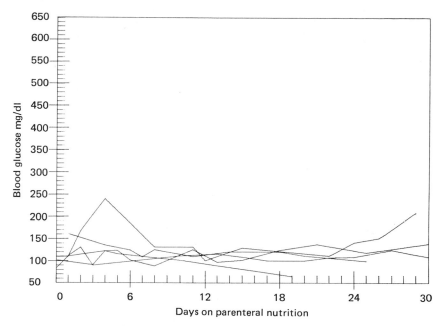

Figure 25.1   Glucose levels in Group I patients

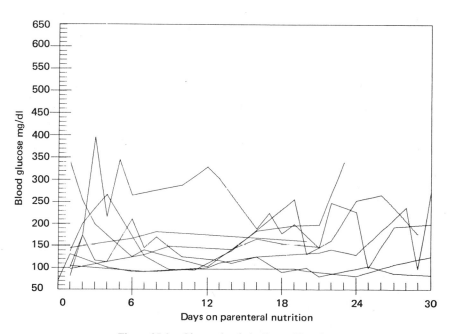

Figure 25.2   Glucose levels in Group II patients

309

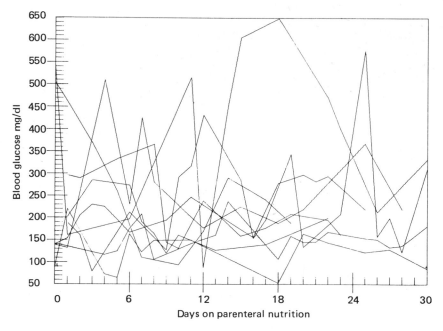

Figure 25.3 Glucose levels in Group III patients

delineations of blood glucose data for patients in their respective groups. The dispersion suggested by the range of each group can be verified visually.

Contrasting quantities of dextrose intake were not the cause of different blood glucose levels. Trauma and particularly sepsis increase the insulin resistance of peripheral tissues and cause difficulty in the utilization of glucose. For such patients one would expect to see a consistently higher than normal blood glucose level and a variation in levels as the 'internal milieu' is restored. Five patients in Group III and one in Group II had obvious sepsis. There was none in Group I. Therefore, disease processes that cause difficulty in glucose control[21] are concentrated in the group which has the most significant alterations in hepatic function.

Glucose is freely permeable to liver cells whereas in muscle and adipose tissue a specific transport system involving insulin is necessary for glucose utilization[22]. As the blood sugar rises above 150 mg/dl, release of glucose from the liver decreases and uptake increases because of the presence of insulin[23,24]. This phenomenon reverses as lower levels are attained. Glucokinase, an enzyme found only in the liver in man, catalyses the initial reaction of glucose within the liver cell. Insulin induces glucokinase and in the absence of insulin there is no glucokinase in the liver. The presence of insulin also increases lipogenesis in the liver and decreases triglyceride escape from that organ.

We did not obtain serum insulin values in the patients described here. However, with the exception of diabetics, we believe that the insulin response mechanism was intact, and we would expect to have seen high levels of insulin in some of the patients. Other authors have demonstrated already elevated insulin levels during parenteral nutrition[25].

In parenteral nutrition there seem to be two phenomena operating to cause the liver to increase lipogenesis and become a storage site for fat: (1) a peripheral insulin resistance typically associated with sepsis and major trauma, and (2) a related hyperglycaemia and associated insulin release, thereby converting the liver to a net glucose uptake organ. Under normal conditions and for short times, excluding diabetic patients, the insulin response induced by infusion of high dose glucose can moderate the blood glucose level rather rapidly as demonstrated by our data in Group I subjects. Such patients seem unlikely to manifest liver function abnormalities. However, when glucose utilization is impaired and more endogenous insulin is present, as in sepsis, hepatic glucose uptake rises followed by deposition of fat in the liver. We believe these physiological mechanisms explain adequately several reports of hepatic function abnormalities, especially those where sepsis and major trauma have been prominent.

Finally, there is the question about the detrimental effect of this overall phenomenon. Protein synthesis is impaired while liver steatosis is present. Albumin levels are depressed compared to normal controls[19,20]. Since the primary goal of parenteral nutrition is to restore or maintain protein, the presence of hepatic steatosis is harmful.

Physicians managing parenteral nutrition should be sensitive to the levels of blood glucose in their patients, particularly those with clinical problems known to produce peripheral insulin resistance. In elderly patients when the renal threshold for glucose may be higher than normal or in cases of renal impairment, the estimation of blood sugar levels from sugar excreted into the urine can be inappropriate. Based on data from our Group I patients, blood sugar levels above 125 mg/dl should be avoided. Prolonged high levels of glucose infusion with compensating exogenous insulin do not appear to be wise therapeutic alternatives. Quite often substantial carbohydrate calories are not needed, and the amount of glucose administered can be moderated. Estimates for energy replacement are frequently too high.

An effective alternative for controlling blood sugar levels in patients receiving parenteral nutrition may be to substitute intravenous lipid as a non-protein calorie source for some of the high concentration carbohydrate. A balance in these components of the parenteral nutrition prescription would seem a wise course to follow, particularly in patients who have diseases which predispose to variable glucose metabolism.

References
1 Peden, W. H., Witzgleben, C. L. and Skelton, M. A. (1971). Total parenteral nutrition. *J. Pediatr.*, **78**, 80
2 Ausman, R. K. and Hardy, G. (1978). Metabolic complications of parenteral nutrition. In Johnston, I. D. A. (ed.) *Advances in Parenteral Nutrition.* pp. 403–410. (Lancaster: MTP Press)
3 Carpentier, Y. A., Askanazi, J., Elwyn, D. *et al.* (1980). The effect of carbohydrate intake on the lipolytic rate in depleted patients. *Metabolism*, **29**, 974
4 Allardyce, D. B. (1978). Clinical experience of total parenteral nutrition. In Johnston, I. D. A. (ed.) *Advances in Parenteral Nutrition.* pp. 429–443. (Lancaster: MTP Press)
5 Burke, J. F. *et al.* (1979). Glucose requirements following burn injury. *Ann. Surg.*, **190**, 274
6 Grant, J. P., Cox, C. E., Kleinman, L. M. *et al.* (1977). Serum hepatic enzyme and bilirubin elevations during parenteral nutrition. *Surg. Gynecol. Obstet.*, **145**, 573
7 Seashore, J. H. (1980). Metabolic complications of parenteral nutrition in infants and children. *Surg. Clin. N. Am.*, **60**, 1239
8 Boraas, M. (1981). Hepatic complications of total parenteral nutrition. *Clinical Nutrition Newsletter*, University of Pennsylvania, October
9 Rutten, P., Blackburn, G. L., Flatt, J. P. *et al.* (1975). Determination of optimal hyperalimentation infusion rate. *J. Surg. Res.*, **18**, 477
10 Steffee, W. P. (1980). Malnutrition in the hospitalized patient. *J. Am. Med. Assoc.*, **244**, 2630
11 Quebbeman, E. J., Ausman, R. K. and Schneider, T. C. (1982). A reevaluation of energy expenditure during parenteral nutrition. *Ann. Surg.* (In Press)
12 Barot, L. R., Rombeau, J. L. *et al.* (1981). Energy expenditure in patients with inflammatory bowel disease. *Arch. Surg.*, **116**, 460
13 Barr, L. H., Dunn, G. D. and Brennan, M. F. (1981). Essential fatty acid deficiency during total parenteral nutrition. *Ann. Surg.*, **193**, 304
14 Harris, J. A. and Benedict, F. G. (1919). *A Biometric Study of Basal Metabolism in Man.* Publication No. 279 (Washington, DC: Carnegie Institute)
15 Long, C., Schaffel, N., Geiger, J. W. *et al.* (1979). Metabolic response to injury and illness. *J. Parent. Enteral Nutr.*, **3**, 452
16 Buzby, G. P., Mullen, J. L., Stein, T. P. and Rosato, E. F. (1981). Manipulation of TPN calorie substrate and fatty infiltration of the liver. *J. Surg. Res.*, **31**, 46
17 Sitges-Creus, A., Canadas, E. and Vilar, L. (1978). Cholestatic jaundice during parenteral alimentation in adults. In Johnston, I. D. A. (ed.) *Advances in Parenteral Nutrition.* pp. 461–469. (Lancaster: MTP Press)
18 Maini, B., Blackburn, G. L., Bistrian, B. R. *et al.* (1976). Cyclic hyperalimentation: an optimal technique for preservation of visceral protein. *J. Surg. Res.*, **20**, 515
19 Stein, T. P., Buzby, G. P., Gertner, M. H. *et al.* (1980). Effect of parenteral nutrition on protein synthesis and liver fat metabolism in man. *Am. J. Physiol.*, **239**, 6280
20 Winkel, P., Ramsoe, K., Lyngybye, J. and Tygstrup, N. (1975). Diagnostic value of routine liver tests. *Clin. Chem.*, **21**, 71
21 Robin, A. P., Greenwood, M. R. C., Askanazi, J. *et al.* (1981). Influence of total parenteral nutrition on tissue lipoprotein lipase activity during chronic and acute illness. *Ann. Surg.*, **194**, 681
22 Williams, R. H. (1968). *Textbook of Endocrinology.* 4th Edn., pp. 651–654. (Philadelphia: W. B. Saunders)
23 Craig, J. W., Drucker, W. R., Miller, M. and Woodward, H. (1961). A prompt effect of exogenous insulin on net hepatic glucose output in man. *Metabolism*, **10**, 212
24 Madison, L. L. (1969). Role of insulin in the hepatic handling of glucose. *Arch. Int. Med.*, **123**, 284
25 Wolfe, B. M., Culebras, J. M., Sim, A. J. W., *et al.* (1977). Substrate interaction in intravenous feeding. *Ann. Surg.*, **186**, 518
26 Cohen, C. and Olsen, M. (1981). Pediatric total parenteral nutrition. *Arch. Pathol. Lab. Med.*, **105**, 152

Section 6
Some new concepts in clinical nutrition

26
Role of carnitine supplementation in clinical nutrition

P. R. BORUM

INTRODUCTION

Carnitine is an absolute requirement for the transport of long-chain fatty acids into the matrix of the mitochondria which is the site of β-oxidation[1]. We frequently speak of carnitine stimulating fatty acid oxidation. However, the role of carnitine in long-chain fatty acid oxidation is much greater than that of mere stimulation. In the absence of carnitine, long-chain fatty acids cannot penetrate the inner membrane of the mitochondria; and since the cytosol does not contain any of the enzymes needed for β-oxidation of long-chain fatty acids, the cell cannot utilize the fatty acids for energy via β-oxidation. As clinical nutritionists, we must realize that when we provide long-chain fatty acids as an energy source, carnitine must be provided in adequate amounts by either endogenous or exogenous sources, or the patient will simply not be able to derive energy from the fatty acids. Nutritionists have traditionally ignored the possible need for exogenous carnitine because we assumed that endogenous synthesis was always adequate, and therefore exogenous sources were not required. We now know that endogenous synthesis of carnitine is not adequate in all individuals[1].

We also now realize that carnitine may function in other processes in addition to transport of long-chain fatty acids into the mitochondrial matrix. Thus a carnitine deficiency may lead to symptoms other than those which can be explained by impairment of fatty acid oxidation.

If endogenous synthesis of carnitine is not adequate in an individual, carnitine supplementation should be considered. Five practical questions need to be evaluated when considering carnitine supplementation. Each question will be discussed briefly.

INDIVIDUALS WHO WOULD BENEFIT FROM CARNITINE SUPPLEMENTATION

The first question is which individuals would benefit from carnitine supplementation and which individuals do not need extra carnitine. One stumbling block is that although we have now elucidated the biosynthetic pathway for carnitine[1], we have only a minimal understanding of the biochemical mechanisms regulating the pathway. Thus the knowledge we have available to try to predict which individuals have impaired carnitine biosynthesis is extremely limited. Direct measurement of carnitine biosynthetic capability is very difficult technically. Thus we must rely on tissue carnitine concentration measurements to identify individuals who may need carnitine supplementation. However, as diagrammed in Figure 26.1, tissue carnitine concentrations are the summation of several very different biochemical processes.

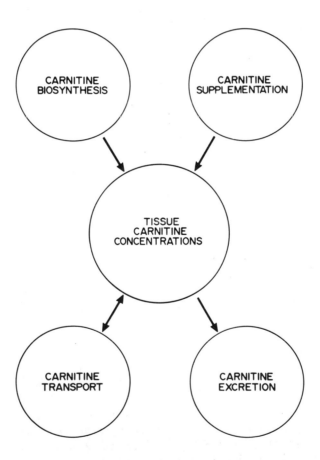

Figure 26.1 Tissue carnitine concentrations are the summation of the effects of carnitine biosynthesis, carnitine supplementation, carnitine transport and carnitine excretion

The carnitine concentration of a specific tissue reflects the different sources of carnitine, the transport into and out of that particular tissue as well as excretion of carnitine from the body. The liver and the kidney are the major sites of carnitine biosynthesis in the human. Although several patients with systemic carnitine deficiency have been identified, a defect in carnitine biosynthesis has never been documented in any of these patients. All of the biosynthetic enzymes have been shown to be present in normal amounts as measured *in vitro*[2] in many of these patients. Thus the low tissue carnitine concentrations in these patients may be the result of the impairment of one of the other processes influencing tissue carnitine concentrations. The only group of individuals who have been shown to have impaired carnitine biosynthetic capacity is the neonate. We discuss below the types of individuals who may benefit from carnitine supplementation.

Neonates

The neonate has a low level of activity of the last enzyme of the carnitine biosynthetic pathway known as γ-butyrobetaine hydroxylase[3]. At autopsy the γ-butyrobetaine hydroxylase activity in infant liver was approximately 12% of the adult level. By 1.5 y the activity had risen to 30% and by 15 y was within the standard deviation of the adult mean[3]. Although the neonate has a decreased ability to synthesize carnitine, the neonate's need for carnitine is great. Plasma and tissue carnitine concentrations of the newborn rat[4] and newborn human[5] are much lower than adult levels. The postnatal change from glucose to fatty acid oxidation is an extremely important postnatal adaptation which requires carnitine. In addition to its role in oxidation of long-chain fatty acids, carnitine is also known to stimulate lipolysis[6], ketogenesis[7], and thermogenesis[8] in the newborn. All of these processes are critical to the survival of the newborn. When endogenous carnitine biosynthesis is decreased, as is found in the newborn, exogenous sources such as dietary carnitine become very important. Carnitine has been measured in many different sources of milk. Human breast milk contains approximately 50 nmol of carnitine per millilitre[9]. Animal studies indicate that a significant portion of the infant's tissue carnitine is derived from milk carnitine[10]. Radioactively labelled γ-butyrobetaine injected into a lactating rat was converted into labelled carnitine and transferred via milk into the pups[10]. When labelled free carnitine was given to lactating rats, the carnitine was transferred orally to the suckling pup and preferentially accumulated in tissues oxidizing fatty acids[11]. Thus, dietary carnitine appears to be important to the survival of the newborn. However, some infant diets contain no exogenous carnitine. Commercial infant formula based on cow's milk has a similar or higher carnitine content than human milk, but infant formulae based on soybean protein have no detectable carnitine[9]. All intravenous solutions used in TPN support of newborns contain no carnitine. Intravenous fat, which contains predominantly long-chain fatty acids, is often supplied to the neonate as a caloric source because a large number of calories can be given in a small volume. Since TPN regimens which supply fat as a calorie source do not also supply the carnitine which is required for the utilization of that fat, the neonate must rely on endogenous carnitine to utilize the intravenous fat for energy production. There is evidence from different laboratories that endogenous carnitine is not adequate in the newborn. Unusually high levels of free fatty acids in plasma were found when small for gestational age infants were administered intravenous fat, and they remained significantly elevated during the postinfusion period. The authors suggested that the persisting higher levels of free fatty acids may reflect a slower rate of β-oxidation[12]. Significantly lower plasma total, free, and acyl-carnitine concentrations were found in infants maintained on TPN as compared to when they were fed orally with expressed milk or a proprietary formula which contains carnitine[13]. Plasma carnitine concentrations fell following 5 days of TPN[14] and liver and muscle carnitine concentrations[15] also fell following 15 days of TPN. Infants fed a soybean based carnitine-free formula had lower plasma carnitine concentrations and lower plasma ketone body concentrations than infants receiving exogenous carnitine[16].

Renal haemodialysis patients

The second group of patients who may benefit from carnitine supplementation are renal haemodialysis patients. Since the kidney is one of the major sites of carnitine biosynthesis[3], patients with an abnormally functioning kidney may have impaired carnitine biosynthesis. Renal patients treated with chronic haemodialysis lose carnitine via the dialysis membrane with resulting low plasma and tissue carnitine concentrations[17]. Oral carnitine supplementation[18], intravenous carnitine supplementation[19], and carnitine added to the dialysate fluid[20] have all been used to reverse the carnitine deficiency created by chronic haemodialysis.

Liver disease patients

The third group of patients who may benefit from carnitine supplementation are those with liver disease. The liver is the other major site of carnitine biosynthesis[3], and thus patients with an abnormally functioning liver may have impaired carnitine biosynthesis. Patients with liver cirrhosis[21] have been shown to have low levels of carnitine in plasma even when they are receiving adequate levels of lysine and methionine, which are the precursors for carnitine biosynthesis.

Muscle disease patients

The fourth group of patients who may benefit from carnitine supplementation are those with muscle disease. Since skeletal muscle cannot synthesize carnitine, all of the muscle carnitine must be transported into the muscle. The first identified carnitine deficient patient complained of slowly progressive muscle weakness and had lipid accumulation predominantly in Type I fibres[1]. Skeletal muscle carnitine concentrations were much lower than normal. More than 30 carnitine deficient patients with muscle weakness have now been identified. All of these patients had low muscle carnitine concentrations. Some of them had low plasma carnitine concentration while others were normal. The disease appears to be inherited in many cases, but the biochemical lesion has not been elucidated in any case[1]. However, defective transport into the tissue has been implicated in several cases. Carnitine supplementation has been attempted in many cases with dramatic positive results in some, but not in all cases.

Cardiac disease patients

The fifth group who may benefit from carnitine supplementation are cardiac disease patients. Since fatty acids are the preferred energy source for cardiac muscle, adequate carnitine is of the utmost importance in normal cardiac muscle. Carnitine must be transported into the cardiac muscle because the last enzyme of the biosynthetic pathway is not present in cardiac tissue. Some of the patients described above with muscle weakness and skeletal muscle carnitine deficiency died of cardiac failure, and cardiac carnitine deficiency was documented at autopsy[1]. Carnitine supplementation has improved the cardiac symptoms of some patients with skeletal muscle weakness. Carnitine administration has also been reported to result in both a lower exercise heart rate and pressure rate product with prolongation of exercise time, prior to the onset of angina pectoris[22]. Carnitine supplementation has also been shown to result in dramatic improvement in systemic carnitine deficiency presenting as familial endocardial fibroelastosis[23]. The authors of this report recommend that plasma and muscle carnitine concentration measurements be performed in all patients with familial cardiomyopathy or with a presumptive diagnosis of endocardial fibroelastosis.

Patients experiencing metabolic stress

The last group of patients who may benefit from carnitine supplementation are other patients experiencing metabolic stress. Pregnant and lactating women have greater demands for carnitine due to the carnitine needed for the fetus and for the production of milk. Most diets in the United States contain carnitine, and thus most women are receiving exogenous or supplemental carnitine. Our laboratory has assayed several hundred foods for carnitine[24]. Generally speaking, carnitine is found in foods from animal sources but is missing in foods from plant sources. Red meat is the best source of dietary carnitine. Strict vegetarians will receive little dietary carnitine. No one knows if the carnitine concentration in the milk of vegetarians is comparable to that of meat-eating populations[24]. Carnitine deficiency has also been documented in populations suffering from protein malnutrition[25].

319

FORM OF CARNITINE WHICH SHOULD BE USED FOR CARNITINE SUPPLEMENTATION

The second practical question concerns what form of carnitine should be used for carnitine supplementation. The L-isomer of carnitine is the natural form of carnitine. However, the racemic mixture is cheaper and more readily available than the pure L-isomer. At one time DL-carnitine was given to patients, assuming that the patient would utilize the L-isomer and simply excrete the D-isomer. However, recent evidence indicates that the presence of the D-isomer interferes with the utilization of the L-isomer. Some patients receiving intravenous DL-carnitine have developed symptoms of myasthenia gravis[26] which disappeared when the DL-carnitine was discontinued and did not return when intravenous L-carnitine was administered to the same patients[27]. All evidence presently available indicates that patients should be given the pure L-isomer rather than the racemic mixture.

ROUTE OF ADMINISTRATION USED FOR CARNITINE ADMINISTRATION

The third practical question concerns what route of administration should be used for carnitine supplementation. Use of the enteral or parenteral route of administration will be determined in large part by the presence or absence of a functioning gastrointestinal tract. Table 26.1 indicates the various physiological processes which must be operative in order for exogenous carnitine via the enteral or parenteral route of administration to be utilized by the body.

Table 26.1 Physiological processes required for use of different sources of carnitine

Physiological process	Endogenous carnitine via biosynthesis	Exogenous carnitine via enteral route	Exogenous carnitine via parenteral route
Carnitine uptake by gastrointestinal tract	No	Yes	No
Delivery of carnitine to blood stream	Yes	Yes	No
Transport of carnitine into and out of tissues	Yes	Yes	Yes
Carnitine excretion	Yes	Yes	Yes

Carnitine uptake by the gastrointestinal tract

If exogenous carnitine is given by the enteral route, carnitine absorption or uptake by the gastrointestinal tract is important. However, we know very little about this process. The results of one study indicate that in human infants the bioavailability of carnitine in breast milk is greater than that in commercial formulae. By 42 h of age the serum carnitine and the serum

320

ketone body concentrations of breast fed infants were significantly higher than those of infants fed commercial formula containing the same carnitine concentration as found in the human milk[28].

Delivery of carnitine to the blood stream

Delivery of carnitine to the blood stream is required for utilization of carnitine from biosynthesis or from the enteral route of administration. Our understanding of this process is severely limited. However, our laboratory has recently found that approximately 70–80% of the carnitine in cord blood is in the erythrocyte[29] and approximately 30–40% of the carnitine in blood from normal human adults is in the erythrocyte. We believe this finding to be very significant because it certainly indicates that the previous practice of only measuring serum or plasma carnitine must be altered to include erythrocyte carnitine concentration measurements when the carnitine status of an individual is being assessed.

Transport of carnitine into and out of tissues

Transport of carnitine into and out of the tissues is of critical importance in carnitine metabolism no matter what the source of carnitine. Investigations from several laboratories are consistent with a carrier-mediated transport process for carnitine in the liver[30], cardiac muscle[31] and skeletal muscle[32]. Detective carnitine transport into the tissues has been implicated in many patients, and thus an understanding of the normal transport mechanism for carnitine would be helpful in elucidating the biochemical lesion causing the defective transport of carnitine which results in carnitine deficiency. Our laboratory has recently documented the presence of tissue carnitine-binding proteins which may function in carnitine transport[33]. Our working hypothesis of the functions of these proteins is as follows. The liver carnitine-binding protein is involved in the transport of carnitine into and out of the liver. The red cell carnitine-binding protein functions in the transport of carnitine in the blood stream[34]. The muscle carnitine-binding protein functions in the transport of carnitine into the skeletal muscle. The cardiac carnitine binding-protein functions in the transport of carnitine into the cardiac muscle[33]. Carnitine deficiency could result from either an abnormal carnitine-binding protein or a decreased level of carnitine-binding protein.

Excretion of carnitine

Excretion of carnitine from the body greatly influences the utilization of carnitine no matter what the source of carnitine, but much more data is needed before we truly understand this process.

DOSAGE OF CARNITINE

The fourth practical question concerns what dosage of carnitine should be used for carnitine supplementation. Oral dosages of carnitine of 1–2 g per day or intravenous dosages of 1–2 g three times a week have been used for adult patients[1]. However, there has been little systematic investigation into establishing the optimal dosage of carnitine for each type of individual needing carnitine supplementation.

FACILITATION OF TRANSPORT OF SUPPLEMENTED CARNITINE INTO TISSUES

The last practical question concerns whether or not the transport of supplemented carnitine into the tissues can be facilitated. As discussed above, we have only limited knowledge concerning carnitine transport. Carnitine may be transported into tissues via carnitine-binding proteins and a carnitine-binding protein on the membrane of the erythrocyte may facilitate carnitine transport in the blood to the tissues. Provision of carnitine by the intravenous route in a bound form or in a form which is perhaps more readily available for binding are projects for future research. These proposed methods of treatment are very important for individuals who suffer from decreased carnitine-binding proteins or binding proteins with decreased binding activity.

SUMMARY

Carnitine supplementation has a definite role in clinical nutrition at the present time. If carnitine becomes readily available for clinical administration, its use will certainly increase. However, we need more research immediately on basic questions and on practical questions concerning the use of carnitine supplementation in clinical nutrition. Basic research needs to investigate carnitine biosynthesis, carnitine transport, carnitine excretion, use of exogenous carnitine, and the biochemical mechanisms regulating tissue carnitine concentrations. Clinical research needs to identify individuals who would benefit from carnitine supplementation, what form of carnitine should be used for carnitine supplementation, what route of administration should be used, what dosage of carnitine should be given, and whether or not carnitine can be supplied in such a manner that its transport into the tissues is facilitated.

References

1 Broquist, H. P. and Borum, P. R. (1982). Carnitine biosynthesis: nutritional implications. In Draper, H. H. (ed.) *Advances in Nutritional Research.* Vol. 4, pp. 181–204. (New York: Plenum Press)
2 Rebouche, C. J. and Engel, A. G. (1980). *In vitro* analysis of hepatic carnitine biosynthesis in human systemic carnitine deficiency. *Clin. Chim. Acta,* **106**, 295
3 Rebouche, C. J. and Engel, A. G. (1980). Tissue distribution of carnitine biosynthetic enzymes in man. *Biochim. Biophys. Acta,* **630**, 22
4 Borum, P. R. (1978). Variation in tissue carnitine concentrations with age and sex in the rat. *Biochem. J.,* **176**, 677

5 Battistella, P. A., Vergani, L. and Angelini, C. (1980). Tissue levels of carnitine in human growth. In Berra, B. and DiDonato, S. (eds.). *Fatty Acids and Triglycerides: Biosynthesis and Transport in Normal and Pathological Conditions.* pp. 151–162. (Melan: Edi Ermes)

6 Novak, M., Penn-Walker, D., Hann, P. and Monkus, E. F. (1975). Effect of carnitine on lipolysis in subcutaneous adipose tissue of newborns. *Biol. Neonate*, **25**, 85

7 McGarry, J. D. and Foster, D. W. (1980). Regulation of hepatic fatty acid oxidation and ketone body production. In Snell, E. (ed.). *Annual Reviews of Biochemistry.* Vol. 49, pp. 395–420. (Palo Alto: Annual Reviews Inc.)

8 Novak, M., Penn-Walker, D. and Monkus, E. F. (1975). Oxidation of fatty acids by mitochondria obtained from newborn subcutaneous (white) adipose tissue. *Biol. Neonate*, **25**, 95

9 Borum, P. R., York, C. M. and Broquist, H. P. (1979). Carnitine content of liquid formulas and special diets. *Am. J. Clin. Nutr.*, **32**, 2272

10 Robles-Valdes, C., McGarry, J. D. and Foster, D. W. (1976). Maternal-fetal carnitine relationships and neonatal ketosis in the rat. *J. Biol. Chem.*, **251**, 6007

11 Hahn, P. and Skala, J. P. (1975). The role of carnitine in brown adipose tissue of suckling rats. *Comp. Biochem. Physiol.*, **51**, 507

12 Schiff, D., Andrew, G. and Chan, G. (1978). Metabolism of intravenously administered lipid in the newborn. In Stern, L., Oh, W. and Fries-Hansen, B. (eds.) *Intensive Care in the Newborn.* Vol. II, pp. 267–273. (New York: Masson)

13 Schiff, D., Chan, G., Seccombe, D. and Hahn, P. (1978). Plasma carnitine levels during intravenous feeding of the neonate. *J. Pediatr.*, **95**, 1043

14 Penn, D., Schmidt-Sommerfeld, E. and Wolf, H. (1980). Carnitine deficiency in premature infants receiving total parenteral nutrition. *Early Hum. Dev.*, **4**, 23

15 Penn, D., Schmidt-Sommerfeld, E. and Pascu, F. (1981). Decreased tissue carnitine concentrations in newborn infants receiving total parenteral nutrition. *J. Pediatr.*, **98**, 976

16 Novak, M., Wieser, P. B., Buch, M. and Hahn, P. (1979). Acetylcarnitine and free carnitine in body fluids before and after birth. *Pediatr. Res.*, **13**, 10

17 DeFelice, S. L. and Klein, M. I. (1980). Carnitine and hemodialysis – a minireview. *Curr. Ther. Res.*, **28**, 195

18 Bizzi, A., Mingardi, G., Codegoni, A. M., Mecca, G. and Garattini, S. (1978). Accelerated recovery of post-dialysis plasma carnitine fall by oral carnitine. *Biomedicine*, **29**, 183

19 Gouarnieri, G. F., Ranieri, F., Toigo, G., Vasile, A., Cimon, M., Rizzoli, V., Moracchiello, M. and Campanacci, L. (1980). Lipid-lowering effect of carnitine in chronically uremic patients treated with maintenance hemodialysis. *Am. J. Clin. Nutr.*, **33**, 1489

20 Bizzi, A., Cini, M., Garattini, S., Mingardi, G., Licini, L. and Mecca, G. (1979). L-carnitine addition to hemodialysis fluid prevents plasma-carnitine deficiency during dialysis. *Lancet*, **1**, 882

21 Rudman, D., Sewell, C. W. and Ainsley, J. D. (1977). Deficiency of carnitine in cachectic cirrhotic patients. *J. Clin. Invest.*, **60**, 716

22 Kosalcharoen, P., Nappi, J., Peduzzi, P., Shug, A., Patel, A., Thomas, F. and Thomsen, J. H. (1981). Improved exercise tolerance after administration of carnitine. *Curr. Ther. Res.*, **30**, 753

23 Tripp, M. E., Katcher, M. L., Peters, H. A., Gilkert, E. F., Arya, S., Hedach, R. J. and Shug, A. (1981). Systemic carnitine deficiency presenting as familial endocardial fibroelastosis. *N. Engl. J. Med.*, **305**, 385

24 Borum, P. R., Vaughan, S. R., Graves, A. S. and Broquist, H. P. (19). Dietary carnitine. II. Carnitine content of foodstuffs in America. (In preparation)

25 Khan, L. and Banji, M. S. (1977). Plasma carnitine levels in children with protein-calorie malnutrition before and after rehabilitation. *Clin. Chim. Acta*, **75**, 163

26 Bazzato, G., Mezzina, C., Ciman, M. and Guarnieri, G. (1979). Myasthenia-like syndrome associated with carnitine in patients on long-term hemodialysis. *Lancet*, **1**, 1041

27 Bazzato, G., Coli, U., Landini, S., Mezzina, C. and Ciman, M. (1981). Myasthenia-like syndrome after D,L,- but not L-carnitine. *Lancet*, **1**, 1209

28 Warshaw, J. B. and Curry, E. (1980). Comparison of serum carnitine and ketone body concentrations in breast and in formula fed newborn infants. *J. Pediatr.*, **97**, 122

29 Shenai, J. P., Borum, P. R., Mohan, P. and Donlevy, S. C. (1982). Carnitine status at birth of newborn infants of varying gestation. *Pediat. Res.* (In press)

30 Christiansen, R. and Bremer, J. (1976). Active transport of butyrobetaine and carnitine into isolated liver cells. *Biochim. Biophys. Acta*, **448**, 562

31 Molstad, P., Bohmer, T. and Hovig, T. (1978). Carnitine-induced uptake of L-carnitine into cells from an established cell line from human heart (CCL 27). *Biochim. Biophys. Acta*, **512**, 557

32 Willner, J. H., Ginsburg, S. and Dimauro, S. (1978). Active transport of carnitine into skeletal muscle. *Neurology*, **28**, 721

33 Cantrell, C. R. and Borum, P. R. (1982). Identification of a cardiac carnitine binding protein. *J. Biol. Chem.* (In press)

34 Borum, P. R. and York, C. M. (1982). Red cell carnitine binding protein. *Fed. Proc.*, **41**, 1559

27
Experience with alternative fuels

R. H. BIRKHAHN

INTRODUCTION

Glucose is the predominant energetic substrate employed currently in clinical nutritional therapy although it may not produce the optimal benefits desired. This has led to the search for alternative fuel sources which has not been very successful. Perhaps there is no better energy source than glucose although such a conclusion has a weak foundation. The search for alternatives to glucose has been more haphazard and fortuitous than it is designed and premeditated. Natural compounds are all that are currently available. The failure of several synthetic substrates[1] stopped an alternative area of research that might have been valuable. It is only in the past 5 y that there has been a return to investigating synthetic compounds as alternative fuel sources. These investigations have been undertaken for quite different reasons than those of past studies.

Three basic reasons exist for seeking an alternative fuel source to replace or supplement glucose. The major reason, in the past, was to find a means of delivering more energy at lower osmolality. Under these conditions there is no real need to look further since lipid emulsions represent both a practical and theoretical ultimate nutrient. The second reason is simply to replace glucose with a compound or compounds equally effective as a nutrient but without the pathophysiological side-effects that can accompany intravenous glucose infusions[2,3]. The third reason, and the direction chosen by our laboratory, is as a specific treatment initially and a fuel source subsequently. The primary goal has been the control of the catabolic protein wasting that accompanies the critical illnesses of trauma and sepsis. It is possible that any nutrient effective under these conditions might also be beneficial in other clinical situations and hence prove a significantly useful substance. The remainder of this discussion will focus first on the problem caused by trauma, then on a theoretical solution to the problem, and finally on the practical work on identifying a new substrate.

THE PROBLEM

When a healthy person fasts for an extended period of time, there are characteristic changes in his metabolism which are designed to spare nitrogen. The reduction in urinary nitrogen excretion is accompanied by decreased plasma glucose and insulin[4]; elevated plasma free fatty acids and ketone bodies[4]; reduced resting energy expenditure[5] as shown in Figure 27.1, whole body protein turnover[6], and muscle protein breakdown[7]; and a switch from glucose to fat as the main energy source. When trauma or sepsis are superimposed on this fasting condition a quite different picture of changes presents itself. A patient who starves immediately after injury becomes hyperglycaemic, has unchanged to reduced free fatty acids and fails to develop hyperketonaemia[8,9]. This patient also responds with an increase in resting energy expenditure, as shown in Figure 27.2; and the classical excessive loss of urinary nitrogen. Other characterizing responses are increased oxidation of glucose[10], conversion of alanine to glucose that is unresponsive to intravenous glucose infusions[11], and the apparent failure to utilize endogenous fat efficiently as indicated by a non-protein RQ that remains about 0.8 even after a 72 h fast[8]. The protein component of metabolism responds to severe stress by increasing the rates of both whole body protein turnover[12], and skeletal muscle protein breakdown[13]. As shown in Figure 27.3, the muscle response is more exaggerated than the whole body response supporting the concept proposed below. Along with accelerated protein breakdown, the oxidation of leucine[14] is increased as shown in Figure 27.4 and our animal work has shown valine oxidation elevated[15].

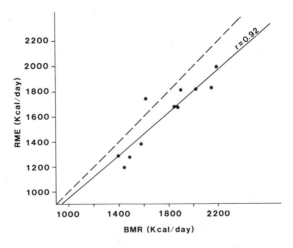

Figure 27.1 Comparison of energy expenditure predicted by the Harris−Benedict equation (BMR) with actual energy expenditure measured by indirect calorimetry (RME) for subjects on a 3 day fast. Measurements (star) were obtained on the third day. The dotted line is the theoretical perfect relationship while the solid line is the best fit of the data points

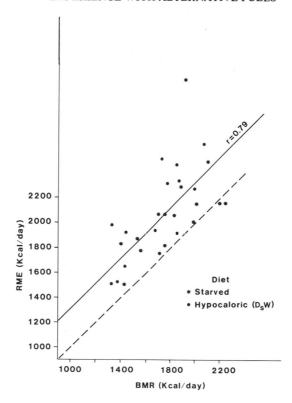

Figure 27.2 Effect of skeletal trauma on the resting energy expenditure (RME) measured by indirect calorimetry on the third day of a constant nutritional intake. Values were obtained on the third to fifth day post-injury. Measured RME are compared to predicted BMR (Harris–Benedict equation). The dotted line is the theoretical relationship and the solid line is the best fit of the data points

The summation of the data highlighted above is drawn up in a working hypothesis concerning the mechanism or chain of events which directly results in protein wasting, amino acid oxidation, and excretion of excessive urinary nitrogen following severe trauma or during sepsis. The hypothesis is that with the onset of severe stress skeletal muscle becomes insulin resistant, and, consequently, the glucose available for muscle metabolism is reduced. The loss of some fraction of glucose cannot be replaced by fat, either because the carnitine transport system limits fat oxidation or because fat mobilization is restricted by blood glucose and insulin levels. The net result is a reduction in carbon fuel from exogenous sources to maintain muscle metabolism leading to a stimulation of muscle to consume an alternate fuel. Branched-chain amino acids released by the breakdown of endogenous protein are the most likely candidates for this alternative fuel. The conclusion of this hypothesis is

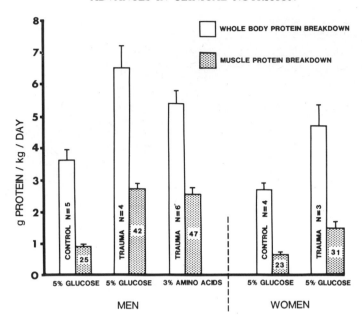

Figure 27.3 The contribution of skeletal muscle protein breakdown as estimated by 3-methyl-histidine to whole body protein breakdown as estimated by L-[15N]alanine is illustrated. Values for muscle and whole body were determined on the day for each individual and show the response to skeletal trauma

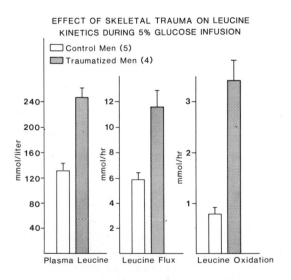

Figure 27.4 The kinetics of leucine metabolism were determined by a continuous intravenous infusion of L-[1-14C]leucine

that the wastage of protein may be preventable provided a carbon source for energy generation in muscle cells could be found. The obvious compounds, glucose and fatty acids, are at the centre of the problem, and our question was, what are the alternatives?

THE THEORETICAL SOLUTION

The concept arrived at was to use the ketone bodies (β-hydroxybutyrate and acetoacetate) as a parenteral nutrient. This concept is quite radical in clinical nutrition since a fundamental postulate has been to give glucose to prevent ketosis, while there is a significant literature which indicates that ketone bodies have an important metabolic role.

Ketosis has developed a tarnished reputation through association with diabetes where the appearance of ketone bodies is a danger signal. The beneficial effects observed in otherwise healthy, fasting man[16] have been overlooked. Fasting is not generally a disease whereas diabetes is.

The metabolic behaviour of the ketone bodies is both ubiquitous and enigmatic. The ketone bodies are fat-derived compounds that are water soluble and have unique metabolic characteristics more closely resembling glucose than fat. The free fatty acids, glucose and ketone bodies form a triumvirate of non-protein energy sources with each having a special role to play. The water solubility and fatty acid nature of the ketone bodies form a bridge between glucose and long-chain fatty acids when the metabolic economy switches from carbohydrate to fat as the primary fuel during the adaptation to fasting. In normal, healthy subjects, the ketone bodies provide energy in a wider variety of tissues than does either glucose or fat; they also can replace glucose in a number of metabolic regulatory functions.

The tissues of the body can be divided up according to the type of fuel source preferred. The divisions are carbohydrate preferring (brain, erythrocytes), long-chain fatty acid preferring (red muscle fibres) and scavenger tissues (liver, heart, lung, kidney) that use either fatty acids or glucose depending on availability. The ketone bodies enter this picture by way of being a more nearly universal energy source than either glucose or fatty acids and are the preferred fuel when a choice is available. This has been demonstrated by Windmueller and Spaeth[17] for the intestines, by Palaiologos and Felig[18] for diaphragm, by Williamson and Krebs[19] for heart, by Ruderman et al.[20] and Maizels et al.[21] for muscle and by Ruderman et al.[22], Palaiologos et al.[23], Gjedde and Crone[24], and Hawkins et al.[25] for brain. The strength of these observations is not just the oxidation of ketone bodies by the various tissues but that the ketone bodies can actually inhibit glucose or amino-acid oxidation. Randle et al.[26] and Maizel et al.[21] found a reduced tissue uptake of glucose in the presence of ketone bodies. Theoretically, the ketone bodies represent an energetic substrate equal to glucose which may even be preferred to glucose as an energy source.

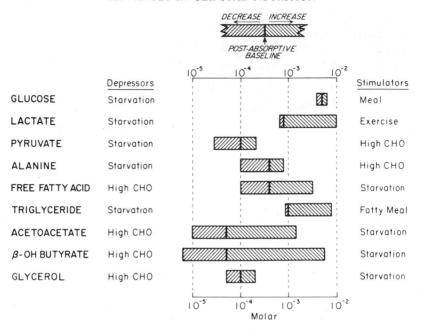

Figure 27.5 Physiological ranges of circulating substrates in normal man expressed on a logarithmic scale. Notice the narrow glucose range and the wide variation in acetoacetate and β-hydroxybutyrate. (Reprinted from Cahill, *J. Parent. Enteral. Nutr.*, **5**, 281, 1981 with permission)

The desirability of the ketones as energy sources does not stop there because they also play an influential role in total metabolic regulation. They can influence energy metabolism directly and also indirectly by modulating the neuroendocrine system. Ketone bodies inhibit glucose oxidation[21,23] and a number of workers[18,23,27,28] have shown that they can also inhibit the oxidation and deamination of essential amino acids. In addition, infused ketone bodies induce hypoglycaemia by inhibiting hepatic glucose production either at the site of glucose precursor release[28] or through the stimulation of insulin secretion[29-31]. They further have been shown to reduce free fatty acid and probably glycerol release and to restrict peripheral alanine release[28]. Ketones in the blood affect glucose metabolism to produce hypoglycaemia but the responses to hypoglycaemia do not occur. The ketone bodies can directly influence both the neuroendocrine centre and the secretory glands in the body to prevent this normal physiological response to hypoglycaemia. Johnson and Weiner[32] demonstrated that ketogenic diets protect rats from hypoglycaemia induced by insulin injections while Muller *et al.*[33], Flatt *et al.*[34] and Stricter *et al.*[35] showed that the ketones actually blocked the release of catabolic hormones which normally stimulate glucose production.

Consequently, the ketone bodies appear to replace glucose as a neuro-endocrine regulator. The ketones also apparently influence protein turnover as indicated by the longer survival of obese mice found by Cuendet et al.[36] and by the reduction of amino-acid oxidation.

The ketone bodies also have certain peculiarities of their own such as influencing neurocentres to reduce childhood epilepsy[37] and of inhibiting bacterial growth[38]. Several authors (see reference 38) have shown ketogenesis to inhibit infection by such bacteria as E. coli, Plasmodium berghei and Salmonella typhimurium.

There are several other reasons to believe that ketone bodies may be a better intravenous nutrient than glucose. As Krebs stated at the last Bermuda Symposium, 'when we are forced to do something artificial we must aim at remaining as close as possible to the natural situation'[39]. From examination of blood concentrations as illustrated in Figure 27.5 the ketones are certainly more suitable for intravenous infusion than any other substance. A second reason is that they avoid several of the regulatory problems confronting glucose and long-chain fatty acids. The ketones do not require either insulin or the carnitine transport system for oxidation nor any other known transport assistants. In fact, we have used the term 'carnitine independent' fat to describe the types of compounds desired[40]. Finally, many severely injured or septic patients are already hyperglycaemic and hypoketonaemic and have adequate energy stores to survive for a couple of months. What the patients need is to have protein and amino acids rather than be given kilo-calories only.

THE PRACTICAL SOLUTION

The identification of new fuel sources was subjected to several stringent conditions in addition to being ketogenic or 'carnitine independent': they should be water soluble, neutral, readily metabolized and relatively inexpensive to produce. The water solubility requirement was intended to avoid the problems of emulsification which have troubled fat nutrients in the past. Water soluble esters of short- and medium-chain fatty acids which include the ketone bodies was the line taken. It is probably worth noting that there are few guides to follow as to what can and cannot be safely given by intravenous infusion since the area is almost totally unexplored. Our efforts led us to try the monoglycerides of acetoacetate[41,42] and butyrate[43] and the ethyl ester of acetoacetate, the structures and properties of which are given in Table 27.1. Ethyl acetoacetate was given to several rats at a rate of 25 g/kg day for 7 days without noticeable problems, but further efforts with this compound have been abandoned because of concern for the effects of ethanol metabolism in the liver when hepatic metabolism may be compromised.

Table 27.1 Physical properties of ketogenic esters for parenteral feeding

	Ethyl acetoacetate	Monoacetoacetin	Monobutryin
Formula	$CH_3CH_2OCCH_2CCH_3$	$CH_2\text{-}OCCH_2CCH_3$ CHOH CH_2OH	$CH_2OCCH_2CH_2CH_3$ CHOH CH_2OH
Molecular weight (daltons)	130	176	162
Physical state at room temp.	liquid	liquid	liquid
Colour	clear, colourless	clear, yellow	clear, colourless
Energy density (kcal/g)	5.3	4.4	5.7
5% solution osmolality (mOsm)	405	299	325
Water solubility	10%	infinity	10%

The direction of the current work has focused on the monoglycerol ester of acetoacetic acid. The butyric acid ester was the first one tested[43] but it lacks sufficient water solubility to be considered as a nutrient which could deliver adequate energy or substrate for parenteral nutrition. This compound may have another role to play in clinical therapy but it probably will not assist much in protein-sparing nutrition. On the other hand, monoacetoacetin is totally miscible with water so that any desired concentration is possible, and being a liquid it could in theory be given intravenously in the absence of a water carrier.

Table 27.2 Maximal intravenous infusion rates of monoacetoacetin currently attempted in various species of animals and the associated degree of ketonaemia

Animal	Duration	Rate (g/kg day)	Total ketones (mmol/l)
Rat	7 days	50	1.16
Dog	30 min	72	0.41
Monkey	5 days	8.5	0.90

Monoacetoacetin infusion has been investigated in rats, dogs and monkeys under various conditions. Table 27.2 lists the maximal infusion rates used to date, the duration of that infusion, and the associated degree of hyperketon-aemia. The extent of ketosis appears variable for these animals but the differences in rates, duration and study conditions probably have a significant influence. It was initially of interest to determine if monoglycerides could be used at all and also used safely for energy. The early experience indicated energy generation from monoglycerides and relative safety for rats, at low doses, during extended infusions. There have not been any apparent contra-indications in any infusion. An article dating to 1896 by Hanroit[44], demonstrated that there are sufficient water soluble esterases and hydrolases in the body water spaces to rapidly degrade monoglycerides. Table 27.3 illustrates a dose relationship with ketonaemia during monoaceto-acetin infusion which supports the observations of Hanroit[44]. The ketone body formation, i.e. hyperketonaemia, for the 7 day infused rats was obtained together with free feeding of a carbohydrate diet. Rats on the same oral intake had ketones only at 0.03 mmol/l, some 1/40th of that found in monoacetoacetin infused rats, which indicates that the glycerol ester was the source of the ketosis.

Table 27.3 Dose responses by the rat and dog to increased infusions of monoacetoacetin

Rate (g/kg day)	Time	Total ketone bodies (mmol/l)
Dogs (n = 4)		
0	0	0.047
11.5	30 min	0.049
21.6	30 min	0.113
40.2	30 min	0.226
72.0	30 min	0.410
Rats (n = 5)		
0	0	0.032
27	7 days	0.839
50	7 days	1.158

Subsequently, monoacetoacetin given as a supplemental intravenous energy source to otherwise orally fed rats was found to support body weight gain similar to isocaloric glucose and to free oral feeding as shown in Figure 27.6[42]. While it should be noted that monoacetoacetin is, by weight, half carbohydrate precursor-glycerol and half ketone body, the evidence suggests that rats can grow in a hyperketonaemic state. A follow up investigation comparing monoacetoacetin, glucose and monobutyrin infused at 10 kcal/day for 7 days into otherwise starving rats gave the nitrogen excretion patterns shown in Figure 27.7. Glucose rapidly depressed urinary nitrogen followed by a rebound in urinary nitrogen losses. The monoglycerides took much longer to minimize urinary nitrogen and the effect appeared to endure longer than after glucose. In all cases, the extent of the depression was similar but longer studies are required to determine the effects on urinary nitrogen losses.

The basis for selecting ketones as an energy source is the extrapolation of the protein-sparing ability of ketones in healthy man to that of stressed man.

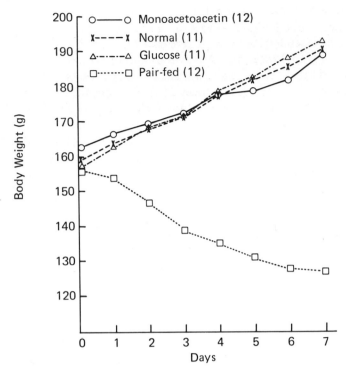

Figure 27.6 Comparison of *ad libitum* fed rats with animals fed an energy deficient diet supplemented with a continuous intravenous infusion of either glucose or monoacetoacetin. (Reprinted from Birkhahn *et al.*, *J. Nutr.*, **109**, 1168, 1979 with permission)

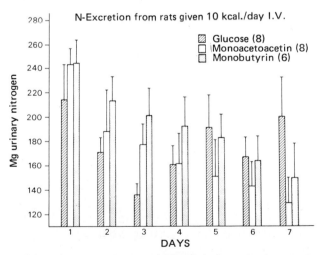

Figure 27.7 Urinary nitrogen response to hypocaloric intravenous support with glucose, monoacetoacetin, or monobutyrin in otherwise fasting rats. (Reprinted from Birkhahn *et al.*, *J. Parent. Enteral. Nutr.*, **3**, 346, 1976 with permission)

Turinsky and Shangraw[45] reported recently that acetoacetate altered glucose uptake and alanine formation in the burned limb of the rat. Wannemacher and co-workers (unpublished observations) have compared glucose, lipids and monoacetoacetin in conjunction with amino acids as hypernutritional support for infected monkeys. All three nutrients produced similar nitrogen-sparing over 5 days of study, but as indicated in Table 27.4 glucose infused monkeys were hyperglycaemic, lipid infused animals were hypercholesterol-aemic, and monoacetoacetin infused animals were neither although their ketones were elevated. These preliminary investigations using septic monkeys suggest that monoglycerides may be effective substrates in parenteral nutritional support of the critically ill.

Table 27.4 Changes in plasma substrates for septic monkeys receiving 48% BCAA plus a hypercaloric energy source of glucose, lipid or monoacetoacetin

	Glucose*	Lipid	Monoacetoacetin
Plasma glucose (mg/l)	+ 350	0	0
Plasma ketones (mol/l)	0	0	+ 0.40
Plasma free fatty acids (mEq/l)	− 1.6	+ 1.5	0
Plasma triglycerides (mg/l)	+ 0	+ 1250	0
Plasma cholesterol (mg/l)	0	+ 600	0

*Values are maximal levels obtained − pre-infusion levels. A zero (0) indicates no change while a plus (+) indicates a rise in blood levels

An advantage of using monoglycerides of ketogenic acids compared to glucose is that not only ketone bodies but also carbohydrate is provided for nutritional support. Since the carbohydrate is given as glycerol, the blood sugar level is allowed to remain under neuroendocrine and hepatic control. This is the most natural mechanism and in the rat and monkey studies monoacetoacetin did not increase the blood glucose. It appears from the limited data that synthetic substrates have the potential to play an important role in future parenteral nutritional therapy.

ACKNOWLEDGEMENTS

The work presented in this paper was supported by National Institutes of Health Grants GM-15768, GM-23065, and GM-26872.

References

1 LeVeen, H. H., Papps, G. and Restuccia, M. (1950). Problems in the intravenous administration of synthetic and natural fats for nutritional purposes. *Am. J. Dig. Dis.*, **17**, 20
2 Chang, S. and Silvis, S. E. (1974). Fatty liver produced by hyperalimentation of rats. *Am. J. Gastroenterol.*, **62**, 410
3 Stein, T. P., Buzby, G. P., Gertner, M. H., Hargrove, W. C., Leskiw, M. J. and Mullen, J. L. (1980). Effect of parenteral nutrition on protein synthesis and liver fat metabolism in man. *Am. J. Physiol.*, **239**, G280

4 Owen, O. E. and Reichard, G. A., Jr. (1975). Ketone body metabolism in normal, obese and diabetic subjects. *Isr. J. Med. Sci.*, **11**, 560

5 Long, C. L., Schaffel, N., Geiger, J. W., Schiller, W. R. and Blakemore, W. S. (1979). Metabolic response to injury and illness: estimation of energy and protein needs from indirect calorimetry and nitrogen balance. *J. Parent. Enteral Nutr.*, **3**, 452

6 Waterlow, J. C. and Stephen, J. M. L. (1966). Adaptation of the rat to a low-protein diet: the effect of reduced intake on the pattern of incorporation of L-[^{14}C]lysine. *Br. J. Nutr.*, **20**, 461

7 Young, V. R., Haverberg, L. N., Bilmazes, C. and Munro, H. N. (1973). Potential use of 3-methylhistidine excretion as an index of progressive reduction in muscle protein catabolism during starvation. *Metabolism*, **22**, 1429

8 Birkhahn, R. H., Long, C. L., Fitkin, D. L., Busnardo, A. C., Geiger, J. W. and Blakemore, W. S. (1981). A comparison of the effects of skeletal trauma and surgery on the ketosis of starvation in man. *J. Trauma*, **21**, 513

9 Miller, J. D. B., Blackburn, G. L., Bistrian, B. R., Rienhoff, H. Y. and Trerice, M. (1977). Effect of deep surgical sepsis on protein-sparing therapies and nitrogen balance. *Am. J. Clin. Nutr.*, **30**, 1528

10 Long, C. L., Spencer, J. L., Kinney, J. M. and Geiger, J. W. (1971). Carbohydrate metabolism in man: effect of elective operations and major injury. *J. Appl. Physiol.*, **31**, 110

11 Long, C. L., Kinney, J. M. and Geiger, J. W. (1976). A nonsuppressibility of gluconeogenesis by glucose in the septic patient. *Metabolism*, **25**, 193

12 Birkhahn, R. H., Long, C. L., Fitkin, D., Jeevanandam, M. and Blakemore, W. S. (1981). Whole body protein metabolism due to trauma in man as estimated by L-[^{15}N] alanine. *Am. J. Physiol.*, **241**, E64

13 Long, C. L., Birkhahn, R. H., Geiger, J. W. and Blakemore, W. S. (1981). Contribution of skeletal muscle protein in elevated rates of whole body protein catabolism in trauma patients. *Am. J. Clin. Nutr.*, **34**, 1087

14 Birkhahn, R. H., Long, C. L., Fitkin, D., Geiger, J. W. and Blakemore, W. S. (1980). Effects of major skeletal trauma on whole body protein turnover in man measured by L-[1,^{14}C]leucine. *Surgery*, **88**, 294

15 Birkhahn, R. H. and Long, C. L. (1982). Variation in valine oxidation with time following skeletal trauma to the rat. *Fed. Proc.*, **41**, 349

16 Cahill, G. F., Jr. (1981). Ketosis. *J. Parent. Enteral. Nutr.*, **5**, 281

17 Windmueller, H. G. and Spaeth, A. E. (1978). Identification of ketone bodies and glutamine as the major respiratory fuels *in vivo* for post-absorptive rat small intestine. *J. Biol. Chem.*, **253**, 69

18 Palaiologos, G. and Felig, P. (1976). Effects of ketone bodies on amino acid metabolism in isolated rat diaphragm. *Biochem. J.*, **80**, 709

19 Williamson, J. R. and Krebs, H. A. (1961). Acetoacetate as fuel of respiration in the perfused rat heart. *Biochem. J.*, **80**, 540

20 Ruderman, N. B., Houghton, C. R. S. and Hems, R. (1971). Evaluation of the isolated perfused rat hindquarter of the study of muscle metabolism. *Biochem. J.*, **124**, 639

21 Maizels, E. Z., Ruderman, N. B., Goodman, M. N. and Lau, D. (1977). Effect of acetoacetate on glucose metabolism in the soleus and extensor digitorum longus muscles of the rat. *Biochem. J.*, **162**, 557

22 Ruderman, N. B., Ross, P. S., Berger, M. and Goodman, M. N. (1974). Regulation of glucose and ketone-body metabolism in brain of anaesthetized rats. *Biochem. J.*, **138**, 1

23 Palaiologos, G., Koivisto, V. A. and Felig, P. (1979). Interaction of leucine, glucose and ketone metabolism in rat brain *in vitro*. *J. Neurochem.*, **32**, 67

24 Giedde, A. and Crone, C. (1975). Induction processes in blood−brain transfer of ketone bodies during starvation. *Am. J. Physiol.*, **225**, 1165

25 Hawkins, R. A., Williamson, D. H. and Krebs, H. A. (1971). Ketone-body utilization by adult and suckling rat brain *in vivo*. *Biochem. J.*, **122**, 13

26 Randle, P. J., Newsholme, E. A. and Garland, P. B. (1964). Regulation of glucose uptake by muscle. Eight effects of fatty acids, ketone bodies and pyruvate and of alloxan diabetes and starvation on the uptake and metabolic fate of glucose in rat heart and diaphragm muscles. *Biochem. J.*, **93**, 652

27 Landaas, S. (1977). Inhibition of branched-chain amino acid degradation by ketone bodies. *Scand. J. Clin. Lab. Invest.*, **37**, 411

28 Sherwin, R. S., Hendler, R. G. and Felig, P. (1975). Effect of ketone infusions on amino acid and nitrogen metabolism in man. *J. Clin. Invest.*, **55**, 1382

29 Fajans, S. S., Floyd, J. C., Jr., Knopf, R. F. and Conn, J. W. (1964). A comparison of leucine- and acetoacetate-induced hypoglycemia in man. *J. Clin. Invest.*, **43**, 2003

30 Mebane, D. and Madison, L. L. (1964). Hypoglycemic action of ketones. I. Effects of ketones on hepatic glucose output and peripheral glucose utilization. *J. Lab. Clin. Med.*, **63**, 177

31 Miles, J. M., Haymond, M. W. and Gerick, J. E. (1981). Suppression of glucose production and stimulation of insulin secretion by physiological concentrations of ketone bodies in man. *J. Clin. Endocrinol. Metab.*, **52**, 34

32 Johnson, W. A. and Weiner, M. W. (1978). Protective effects of ketogenic diets on signs of hypoglycemia. *Diabetes*, **27**, 1087

33 Muller, W. A., Aoki, T. T., Flatt, J. P., Blackburn, G. L., Egdahl, R. H. and Cahill, G. F., Jr. (1976). Effects of β-hydroxybutyrate, glycerol and free fatty acid infusions on glucagon and epinephrine secretion in dogs during acute hypoglycemia. *Metabolism*, **25**, 1077

34 Flatt, J. P., Blackburn, G. L., Randers, G. and Stanbury, J. B. (1974). Effects of ketone body infusions on hypoglycemic reaction in postabsorptive dogs. *Metabolism*, **23**, 151

35 Stricker, E. M., Rowland, N., Saller, C. E. and Friedman, M. I. (1977). Homeostasis during hypoglycemia: central control of adrenal secretion and peripheral control of feeding. *Science*, **196**, 79

36 Cuendet, G. S., Loten, E. G., Cameron, D. P., Renald, A. E. and Marliss, E. B. (1975). Hormone-substrate responses to total fasting in lean and obese mice. *Am. J. Physiol.*, **228**, 276

37 Livingston, S., Pauli, L. L. and Pruce, I. (1977). Ketogenic diet in the treatment of childhood epilepsy. *Dev. Med. Clin. Neurol.*, **19**, 833

38 Potezny, N., Atkinson, E. R., Rofe, A. M. and Conyers, R. A. J. (1981). The inhibition of bacterial cell growth by ketone bodies. *Aust. J. Exp. Biol. Med. Sci.*, **59**, 639

39 Krebs, H. A. (1977). Some general considerations concerning the use of carbohydrates in parenteral nutrition. In. Johnson, I. D. A. (ed.) *Advances in Parenteral Nutrition.* pp. 23−28. (Lancaster: MTP Press)

40 Birkhahn, R. H. and Border, J. R. (1981). Alternate or supplemental energy sources. *J. Parent. Enteral. Nutr.*, **5**, 24

41 Birkhahn, R. H. and Border, J. R. (1978). Intravenous feeding of the rat with short chain fatty acid esters. II. Monoacetoacetin. *Am. J. Clin. Nutr.*, **31**, 436

42 Birkhahn, R. H., McMenamy, R. H. and Border, J. R. (1979). Monoglyceryl acetoacetate: a ketone body-carbohydrate substrate for parenteral feeding of the rat. *J. Nutr.*, **109**, 1168

43 Birkhahn, R. H., McMenamy, R. H. and Border, J. R. (1977). Intravenous feeding of the rat with short chain fatty acid esters. I. Glycerol monobutyrate. *Am. J. Clin. Nutr.*, **30**, 2078

44 Hanroit, M. (1896). Sur un nouveau ferment du lang. *Compt. Rend. Soc. Biol.*, **48**, 925

45 Turinsky, J. and Shangraw, R. E. (1981). Augmented glucose uptake and amino acid release by muscle underlying the burn wound and their moderation by ketone bodies. *Exp. Molec. Pathol.*, **35**, 338

28
Topical nutrition – a preliminary report

A. W. GOODE, I. A. I. WILSON, C. J. C. KIRK AND P. J. SCOTT

INTRODUCTION

Methods of treating open wounds are many and varied, having been developed and passed on by succeeding generations of physicians and nurses. While most wounds heal satisfactorily with minimal intervention, healing may be delayed or almost non-existent and measures are instituted in an attempt to promote healing and restore skin continuity. Over the years many strange wound applications have been derived, including such diverse substances as cobwebs, faeces, milk, honey, vinegar, zinc, mercury and arsenic[1]. In current clinical practice some effective treatments have emerged but none is outstanding; most advances have been, in general, surgical techniques and related fields such as wound debridement and antibiotic therapy rather than in specific wound healing factors. A fundamental problem has been adequate assessment and valid comparison of specific treatments.

MODES OF HEALING

Primary wound healing occurs when clean fresh wound edges in healthy tissue are opposed. However, when wound edges are not brought together, when there is skin loss, when a wound has been laid open or with an ulcer, healing is by secondary intention. Granulation tissue, predominantly collagen, is laid down in the wound and the defect is bridged as the epithelial edge grows in.

Delayed wound healing is a multifactorial problem, the principal causes being given in Table 28.1.

WOUND NUTRITION

Viljanto and Raekallio[2] have suggested an amino acid/glucose/vitamin solution encouraged synthetic and proliferating functions of granulation tissue and implied direct cellular nutrition was possible. Ford *et al.*[3] have reported local amino-acid application improved healing in an abdominal wound, whilst Silvetti[4] and Calver and Stanley[5] have shown improved wound healing in a variety of situations using topical amino acids. Edwardson and Murphy[6] have described local tissue nutrition contributing to the healing of a leg ulcer and Herszage *et al.*[7] have treated 120 suppurating wounds with success using granulated sugar.

Table 28.1 Factors delaying wound healing

General	
Age	Vitamin deficiency
Malnutrition	Trace metal deficiency
Anaemia	Jaundice and uraemia
Neoplasia	Diabetes
Cortisone therapy	Cytotoxic therapy
Systemic infection	
Local	
Poor blood supply	Recurrent trauma
Haematoma formation	Wound tension
Necrotic tissue	Irradiation
Foreign body implantation	Infection

INITIAL STUDY

Forty patients with a variety of wounds have been treated using direct application of an amino acid/glucose solution. These included chronic lesions such as varicose ulcers and bed sores which had failed to respond to conventional therapy. Also included were acute lesions such as burns, dehisced abdominal wounds and pilonidal excision defects.

The wounds were infected when first seen and were initially cleaned with an application of either half strength Eusol or diluted iodine for 2–3 days to eradicate superficial infection. Debrisan and foam silicone dressings have been used in a minority of cases as recent reports suggest they may be more efficient in cleansing wounds compared to conventional methods[8,9]. Systemic antibiotics were only used for marked cellulitis. Once clean, the wounds were dressed with plain gauze moistened with the amino acid/glucose solution. Where practical, the patients were requested to keep the gauze moist by frequent (2 hourly) application of solution. In some patients this was achieved by placing a small cannula over the gauze and a pad over the whole dressing to allow continuous moistening without removing the dressing. In larger, less accessible sacral pressure sores continuous slow infusion (500 ml over 3 days) was delivered to the gauze using a fine urinary catheter. More recently, we have combined the amino-acid application with the use of individual wound contoured silastic foam dressing, especially in deeper wounds, pilonidal sinus excision defects and infected abdominal incisions, by inserting a cannula through the preformed silicone bung (Figure 28.1). The combination appears satisfactory and may even be complementary. Examples of the clinical results are shown in Figures 28.2–5 with 'before' and 'after' photographs. The chronic ulcers which had not healed despite multiple treatment regimens became rapidly pain free and showed marked improvement over a 2–3 week period. Most healed or provided an excellent surface for skin grafting. The quality of the skin appeared resilient and the wounds have remained healed.

METHODS OF QUANTITATING SKIN DEFICITS

This initial study was uncontrolled and while the photographs are adequate for clinical records, they are unsuitable for scientific assessment, as a deficit will heal both from the edge and base.

Three-dimensional measurement of the edge length, surface area and volume of skin ulcers and wounds at different times during treatment would give an objective method of clinical evaluation. Since the measurement is relatively slow, it is unreasonable to immobilize the patient during the complete process. To overcome this problem, a cast of the lesion may be made using a suitable moulding material such as silicone rubber. Pories *et al.*[10] described a method to measure the volume of moulding material used to make the impression. The casts are also eminently suited for measurement by three-dimensional measuring instrument, the Reflex Metrograph or, in the case of smaller casts, the Reflex Microscope[11]. Both instruments measure from a reflection of the object and are thus non-contacting. The measuring mark is carried on a digitized slide system, giving three-dimensional co-ordinates (x, y and z; Figure 28.6), which is on line to a microcomputer. These instruments are simple to use and the results are accurate and quickly obtained but a foreign substance has to be introduced into the lesion with the danger of infection or a reaction to the material used.

STEREOPHOTOGRAMMETRY

Williams[12] has reviewed the five ways in which photography may be used for clinical measurement. Monophotogrammetry, light sectioning, Moiré topography and holography are not, for various reasons, suitable for measuring surface lesions; however, stereophotogrammetry has proved to be of value.

In stereophotogrammetry the object to be measured is photographed from two positions (Figure 28.7). Differences in parallactic angles are thus recorded of all the different depths within the object. The resulting stereo pair when projected into space creates a three-dimensional image, the scale of which is dependent upon the magnification of the photograph. If this is done in a stereoplotting instrument, the operator sees the stereoscopic model and

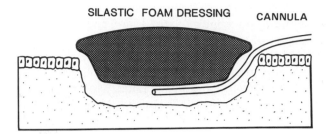

Figure 28.1 Silastic foam dressing contoured to the individual wound

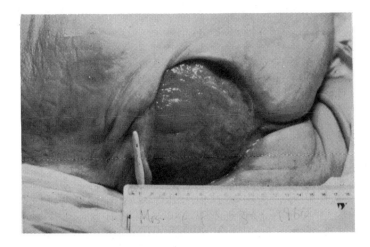

Figure 28.2 An intractable pressure sore, resistant to all previous therapy, before the topical application of an amino-acid solution

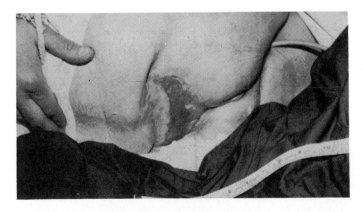

Figure 28.3 Pressure sore healing after 4 weeks treatment with the topical application of an amino-acid solution

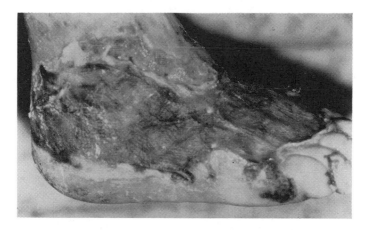

Figure 28.4 Ulcer on foot before topical application of an amino-acid solution

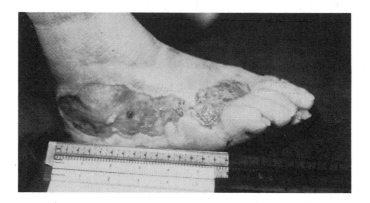

Figure 28.5 Ulcer on foot healing after 2 weeks topical application of an amino-acid solution

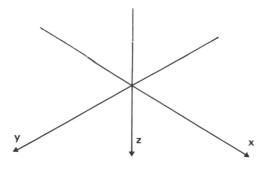

Figure 28.6 *x, y* (horizontal) and *z* (vertical) co-ordinates in isometric projection

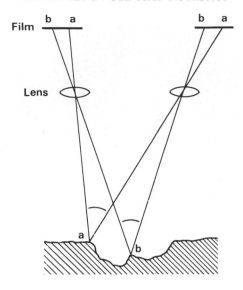

Figure 28.7   Camera positions for the stereophotogrammetry technique

by driving a mark around it can observe the x, y and z co-ordinates as required. Ulcer evaluation using this process has been reported by Eriksson *et al.*[13]. A parallel publication[14] explains the photogrammetric process, including the extensive calibration of the stereometric cameras, and shows how the length, area and volume were obtained. Such stereoscopic plotting instruments are expensive. Moreover, the reorientation of the photographs in the instrument is a very involved procedure. Several days are needed for an unskilled operator to learn the reorientation process.

An alternative, more complex, treatment of the stereopair involves measuring the surface x and y co-ordinates of the two pictures and using projective transformation formulae to reconstruct x, y and z co-ordinates for the object from the two sets of x, y co-ordinates. This process is lengthy and the surface co-ordinates must be measured very accurately since the photographic image is usually smaller than the ulcer. Errors in x, y co-ordinates are then magnified in reconstruction.

Measurement of small, less than 300 mm, non-stationary objects is greatly simplified by photogrammetry using the reprojection cameras as described by Beard and Burke[15]. The fixed-focus cameras are mounted rigidly in a jig which has a metal frame around the object being photographed (Figure 28.8). The frame has a series of control marks which will image on the photographs. The jig is portable and it can therefore be taken to the patient. The negatives are developed and replaced in the cameras and a projecting lamp is placed behind them. Each photograph is moved about until the control mark images

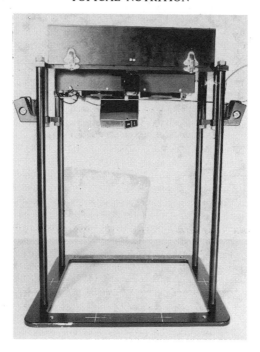

Figure 28.8   A fixed focus stereo camera with metal frame showing reprojection marks

reproject onto the frame marks themselves. It can now be assumed that any deformation has been reversed on projection and thus neutralized. The two pictures can be polarized and viewed stereoscopically. The model is theoretically free of distension and at a scale of 1.1. The setting up procedure is very simple, allowing an unskilled operator to learn quickly. The first system of this kind allowed only graphical plotting of contours and surface details but Beard and Tee[16] have described an ulcer evaluation using the contour and detail plot. The publication, however, also describes a further development of the camera system where the measuring mark is carried on a computerized slide system giving x, y and z co-ordinates which could allow direct calculation of lengths, areas and volumes.

A similar system is being used by the authors. The camera and jig are available for photographic recording and the x, y, z co-ordinates system of a Reflex Metrograph is used for measurement. A microcomputer program using the same method as Eriksson *et al.*[13] is being used.

Stereophotogrammetry provides an accurate reproducible method of assessing healing of irregular wounds. Initial studies using the described method of wound irrigation are to assess the effect of varying the osmolarity of Synthamin* Amino Acid solution.

*Travasol in U.S.A.

DISCUSSION

The mechanism whereby the direct application of an amino-acid solution may improve wound healing is uncertain and indeed a number of factors may be involved, as listed in Table 28.2.

Table 28.2 Possible mechanisms in the topical application of amino-acid solutions to wounds

Control of infection	Direct bacterial toxicity
	Alteration of pH
	Osmotic effect
Vascularization	Local irritation
	Oxygen availability
	Angioblast stimulation
Direct cellular nutrition	General
	Specific – histidine
	Vitamin and trace elements

Control of infection

Gross infection is uncommon but some degree of infection is usually present and control of this may be a factor. This could be achieved by a direct toxic effect on surface bacteria or possibly by alteration of local pH. The pH of an infected wound is in the range of 6.8–7.4[7]. But the use of an amino-acid solution may alter the local pH to acidity and this may both prevent the harmful effects of ammonia produced by urea splitting organisms and improve the quality of collagen laid down[17].

The water requirements for micro-organisms may be defined in terms of the water activity of the substrate[18,19]. This is the ratio of its water vapour pressure to that of pure water at the same temperature. Bacteria will not grow below a limiting water activity, with a critical ratio of 0.86–0.91 for most human pathogens. Certainly, the instillation of granulated sugar into a wound creates a low water activity environment[20] inhibiting bacterial growth.

Vascularization

An injury which initiates the repair process also results in injury to the local blood vessels. Those injured will thrombose whilst adjacent vessels dilate and a leukocyte emigration into the wound occurs. Macrophages are plentiful at the wound edge and within a zone 50–100 μm from the wound edge while the most distal capilliary forms endothelial buds to produce new capilliaries. The ingress of new vessels proceeds towards an area of hypoxia and acidosis; with each new capilliary loop oxygen and nutrients become available to nearby cells[12]. Local wound irritation probably mediated by histamine would further increase local cell oxygenation, nutrition and waste product removal.

Increased oxygen tension in wounds has extensive effects. It enhances leukocyte function and phagocytosis and results in a major increase in oxidative metabolism in the leukocyte[23], whilst hypoxia depresses these functions.

Increased oxygen tension in wounds will stimulate healing in rats, with 70% by volume oxygen administration, the tensile strength in 10 day wounds increased by 30%[24]. Collagen synthesis depends upon molecular oxygen for incorporation into the peptide chain to form the hydroxyprolyl and hydroxylysyl residues in collagen, there is augmented cross-linkage of collagen and an increased differentiation of wound cells shown by a rise in their RNA/DNA ratio. Recent work has demonstrated that collagen accumulation is impaired below a local pO_2 of 20 mmHg[25] and maximum synthesis and cross-linking occurs with a local pO_2 between 20 and 30 mmHg[26].

Cellular nutrition

The fibroblast is a metabolically complex cell. It synthesizes not only collagen but cholesterol, elastin, fibronectin and proteoglycans. The metabolic needs of the cell are sugar, fats, amino acids, the B vitamins, ascorbic acid and trace metals together with oxygen. Glucose is utilized in the early phase of healing via the pentose phosphate shunt while glycolysis is predominant during collagen synthesis[27]. A progressive increase in aminopeptidase activity in the wound during healing is a sensitive cytochemical indicator of connective tissue cell proliferation during repair[28].

Histamine is known to modify the rate of wound healing and carnosine and histidine availability may be of importance, although the exact relationship is unclear[29]. Certainly histidine deficient rats had impaired wound healing. Similarly other amino acids may have an as yet undetermined specific role.

Vitamins and trace elements have long been known to influence healing. Vitamin A is required for normal development and maturation of epithelium, vitamin B for fibroblast function and vitamin C has a direct influence on collagen synthesis and indirectly enhances the effect of copper in facilitating cross-link formation in collagen and elastin. Zinc acts both as a cofactor for nucleic acid and protein synthesis[30] and in immune function[31].

Conclusions

A firm clinical impression derived from the first part of this study was that topical nutrition was of benefit, healing otherwise intractable cutaneous deficiencies. The development of the technique of stereophotogrammetry allows quantitation of the efficacy of different solutions, and the comparison of the efficiency of this method with the use of porcine dermis where good results have been reported[32,33].

Although there are a number of methods by which a topically applied amino-acid solution may work, there is no evidence to indicate what should

347

be the composition of the ideal solution. Further studies into the use of topical nutrition are indicated. Certainly such a solution may be of clinical benefit in healing cutaneous ulcers or preparing a possible donor site for skin grafting.

A recent survey at Charing Cross Hospital, London suggested that annually at least 72 varicose ulcers were admitted, resulting in 3100 days of inpatient care with varying degrees of success, while between 4 and 8% of all nursed patients may develop pressure sores[34], some of which may not heal with conventional treatment. There is thus scope for improvement in the treatment of skin ulcers and granulating wounds.

References

1 Forrest, R. D. (1982). Development of wound therapy from the Dark Ages to the present. *J. Roy. Soc. Med.*, **75**, 268
2 Viljanto, J. and Raekallio, J. (1976). Local hyperalimentation of open wounds. *Br. J. Surg.*, **63**, 427
3 Ford, M., Burroughs, A. K. and Littler, J. (1977). A case report of wound healing by local tissue nutrition. In Baxter, D. H. and Jackson, G. M. (eds.) *Clinical Parenteral Nutrition.* pp. 248–253. (Chester: S. G. Mason)
4 Silvetti, A. N. (1981). An effective method of treating long-enduring wounds and ulcers by topical applications of solutions of nutrients. *J. Dermatol. Surg. Oncol.*, **7**, 501
5 Calver, R. F. and Stanley, J. K. (1980). Enhancements of secondary wound healing by local tissue nutrition. *Clin. Trial J.*, **17**, 144
6 Edwardson, K. F. and Murphy, S. (1977). A case report on local tissue nutrition – a contributing factor in leg ulcer healing. In Baxter, D. H. and Jackson, G. M. (eds.). *Clinical Parenteral Nutrition.* pp. 254–257. (Chester: S. G. Mason)
7 Herszage, L., Montengo, J. R. and Joseph, A. L. (1980). Treatment of suppurating wounds with commercial granulated sugar. *Argentina Soc. Surg. Bull.*, **315**, 21
8 Goode, A. W., Glazer, G. and Ellis, B. W. (1979). The cost effectiveness of dextranomer in the treatment of infected surgical wounds. *Br. J. Clin. Pract.*, **33**, 325 and 328
9 Young, H. L. and Wheeler, M. H. (1982). Report of a prospective trial of dextranomer beads (Debrisan) and silastic foam elastomer (silastic) dressings in surgical wounds. *Br. J. Surg.*, **69**, 33
10 Pories, V. J., Henzel, J. H., Rob, C. G. and Strain, W. (1967). Acceleration of wound healing in man with zinc sulphate given by mouth. *Lancet*, **1**, 121
11 Scott, P. J. (1981). The reflex plotters: measurement without photographs. *Photogrammetric Record*, **10**(58), 435
12 Williams, A. R. (1978). Medical biostereometrics. The measurement of biological form in three dimensions. *Photographic J.*, **118**, 104
13 Eriksson, G., Eklund, A. E., Torlegard, K. and Dauphin, E. (1979). Evaluation of leg ulcer treatment with stereophotogrammetry; a pilot study. *Br. J. Dermatol.*, **101**, 123
14 Ostmann, A. (1981). A review of stereophotogrammetric measurements of leg ulcers. *Fotogrammetriska Meddelanden*, **2**, 45
15 Beard, L. F. H. and Burke, P. H. (1967). The evolution of a system of stereophotogrammetry for the study of facial morphology. *Med. Biol. Illustrated*, **17**, 20
16 Beard, L. F. H. and Tee, J. E. (1980). An approach to the introduction of stereophotogrammetry as an alternative to traditional methods of measurement. *Int. Arch. Photogrammetry*, **23** (B5), 62
17 Wilson, I. A. I., Henry, M., Quill, R. D. and Byrne, P. J. (1979). The pH of varicose ulcer surfaces and its relationship to healing. *J. Vasc. Dis.*, **84**, 339
18 Scott, W. J. (1953). Water relations of S aureus at 30 °C. *Aust. J. Biol. Sci.*, **6**, 549
19 Christian, J. H. B. and Waltho, J. A. (1962). The water relations of staphylococci and micrococci. *J. Appl. Bacteriol.*, **25**, 369

20 Chirife, J., Scarmato, G. and Herszage, L. (1982). Scientific basis for use of granulated sugar in treatment of infected wounds. *Lancet*, **1**, 560

21 Niinikoski, J. (1980). Cellular and nutritional interactions in healing wounds. *Med. Biol.*, **58**, 303

22 Babior, B. M. (1978). Oxygen dependent microbicidal killing of phagocytes. *N. Engl. J. Med.*, **298**, 659

23 Stossel, T. P. (1974). Phagocytosis. *N. Engl. J. Med.*, **290**, 717

24 Niinikoski, J. (1969). Effect of oxygen supply on wound healing and formation of experimental granulation tissue. *Acta Physiol. Scand. Suppl.*, **334**, 1

25 Niinikoski, J. (1980). The effect of blood and oxygen supply on the biochemistry of repair. In Hunt, T. K. (ed.) *Wound Healing and Wound Infection: Theory and Surgical Practice.* (New York: Appleton-Century-Crofts)

26 Silver, I. A. (1980). The physiology of wound healing. In Hunt T. K. (ed.) *Wound Healing and Wound Infection: Theory and Surgical Practice.* (New York: Appleton-Century-Crofts)

27 Lampiaho, K. and Kulonen, E. (1967). Metabolic phases during the development of granulation tissue. *Biochem. J.*, **105**, 333

28 Monis, B. (1963). Variations in aminopeptidase activity in granulation tissue and in the serum of rats during wound healing – a correlated histochemical and quantitative study. *Am. J. Pathol.*, **42**, 301

29 Fitzpatrick, D. W. and Fisher, H. (1982). Carnosine histidine and wound healing. *Surgery*, **91**, 56

30 Fell, G. S. and Burns, R. R. (1978). Zinc and other trace elements. In Johnston, I. D. A. (ed.) *Advances in Parenteral Nutrition.* pp. 241–261. (Lancaster: MTP Press)

31 Zinc and immuno competence (1980). *Nutr. Rev.*, **38**, 288

32 Rundle, J. S. H., Cameron, S. H. and Ruckley, C. V. (1976). New porcine dermis dressing for varicose and traumatic leg ulcers. *Br. Med. J.*, **1**, 216

33 Kaisary, A. V. (1977). A temporary biological dressing in the treatment of varicose ulcers and skin defects. *Postgrad. Med. J.*, **53**, 672

34 Notes and News – The Tissue Viability Society (1982). *Lancet*, **1**, 1423

29
Nutritional and immunological parameters for identification of the high risk patient

R. DIONIGI, L. DOMINIONI, P. DIONIGI, AND S. NAZARI

INTRODUCTION

Malnutrition is a common problem in hospital patients; in surgical patients if not diagnosed and treated, malnutrition is often associated with an increased incidence of post-operative sepsis.

The physiological response to malnutrition is the result of complicated metabolic modifications which interfere with organ functions. Many studies in animals and man[1,2] have clarified most of the compensatory mechanisms in response to fasting. A constant energy supply is essential for those tissues, such as brain, which have continuous demands but minimal fuel reserves. This process is usually accomplished by endogenous calories and proteins. Malnutrition may lead to a marasmic picture, with depletion of fat and muscle protein but preservation of serum albumin; or it may involve visceral attrition (adult kwashiorkor), with near-normal anthropometry but severe hypoalbuminaemia[3]. Recently, various nutritional parameters have been used in an attempt to classify the type and degree of the hospital malnutrition[4]. These indicators lack sensitivity and/or specificity. Moreover, the metabolic response to starvation cannot be standardized because of many associated factors such as neoplasms, sepsis, malabsorption, etc., which further modify the pathophysiology.

In the attempt to quantify the degree of malnutrition using anthropometric, biohumoral and immunological parameters, Blackburn and colleagues[4] suggested standards that allow classification of patients as normal (90%–110% of standard), moderately malnourished (60%–90% of standard) or severely malnourished (<60% of standard). This approach to nutritional assessment is oversimplified since it does not characterize the specific role of each parameter in evaluating the degree of malnutrition. In fact, it is known that values of albumin and transferrin below 60% of normal, have different clinical and prognostic significance than similar low values of standard skinfold triceps and arm circumference[5]. Moreover, it is difficult to classify patients in one of the indicated types of malnutrition because of the prevalence of border line situations[6].

The prognostic significance of Nutritional Assessment is another important aspect in the evaluation of nutritional status. In fact, many studies have documented the relationships between malnutrition, morbidity and mortality[7-10]. Nevertheless standardized nutritional parameters are not yet available in clinical practice for identification of high risk patients.

New approaches have been recently proposed[11-13], the most recent being the method developed by Buzby and colleagues[12], which is a computer-based stepwise regression procedure. The comparison between the mean values of Nutritional Assessment determinations at admission in complicated and uncomplicated surgical patients, indicates that four parameters (albumin, total proteins, transferrin and delayed hypersensitivity response to skin antigens) have predictive significance and allow the calculation of a 'prognostic nutritional index' using a linear predictive model. This index correlated the post-operative morbidity and mortality risk with the baseline values of these variables. However, this and other types of assessments based only on a few nutritional parameters do not give information about the type of nutritional abnormalities and their possible correction by means of nutritional repletion.

CLUSTER ANALYSIS IN NUTRITIONAL ASSESSMENT

In order to collect more information concerning the predictive value of each of the many nutritional parameters used to identify the high risk surgical patient, we have used a mathematical approach applying cluster analysis technique[14]. Cluster analysis has been used in surgical research mainly by Siegel, who studied the haemodynamic and metabolic modifications of patients in stress, sepsis and cardiac failure[15-19]. In his studies cluster analysis represented an excellent tool to synthesize the results obtained with multiple complicated metabolic and haemodynamic determinations. This approach can also establish at any moment, the exact pathophysiological condition of the patient and can be used to rationally plan the therapy and evaluate its effectiveness. In brief, this statistical method analyses a computer data bank of clinical, physiological and therapeutic variables and by ordering all of these multivariable data sets in the high-dimensional space represented by the number of data directions, the computer can also permit the projection of these data-inferred relationships onto a hyperplane so that relevant patterns can be evaluated by the physician.

The results obtained in the first phase of our study characterized the entire spectrum of modifications of nutritional indicators, into four different nutritional states.

The nutritional condition of cluster 1 is characterized by minor variations of the indicators. Retinol binding protein and ceruloplasmin are higher than normal probably because of a discrepancy between the normal range indicated by the laboratory and that of our normal population. This nutritional pattern is show by most (96%) of the controls and some cancer patients. Patients belonging to this cluster presented the lowest incidence of post-operative infection. Moreover, all the neoplastic patients underwent radical operations and showed the lowest mortality rate. We concluded that at our institution this condition can be considered as representing a normal nutritional state.

The nutritional condition of cluster 2 is similar to an adult kwashiorkor-like syndrome[3,4]. In fact, the usual body weight (UBW) is unaffected, delayed hypersensitivity is impaired and the parameters of the visceral protein compartment are reduced with the exception of ceruloplasmin, which is abnormally high. The highest mortality rate, the highest incidence of palliative procedures as well as a high incidence of post-operative sepsis were recorded in patients with this nutritional pattern, indicating that this is a high risk situation.

The nutritional state of cluster 3 suggests a type of malnutrition which is similar to a mixed marasmus-kwashiorkor-like syndrome. This group has slight abnormalities of the visceral protein compartment, modifications of anthropometric measurements, and a decrease of some immunological parameters (C3c and lymphocytes). This nutritional state is associated with a high mortality rate and an increased risk of post-operative infections.

The nutritional condition of cluster 4 showed important alterations of most of the indicators. A pronounced increase of ceruloplasmin, C3c and transferrin as well as binding protein was noted; these modifications could be explained on the basis of an 'acute phase reaction' which is often observed in the progressive phase of neoplastic disease[20-25].

These four identified nutritional states when depicted in a multi-dimensional graph (Figure 29.1) can be considered as reference groups allowing the classification of patients depending upon their nutritional conditions. In this multidimensional model the position of a new patient can be obtained by measuring the euclidean distance between each determination of the reference states and each determination of the patient. Thus, the nature and degree of nutritional imbalance can be quantified by its absolute distance from a given reference state. However, since no patient is exactly similar to the patient represented in the reference chart, for practical purposes, a patient is considered to belong to a group to which he is found to be closest, on the basis of the euclidean distance.

The second phase of the study showed that a higher susceptibility to sepsis

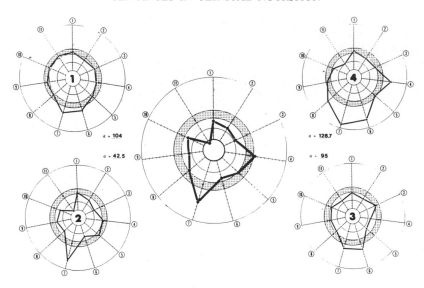

Figure 29.1 Circle diagrams of nutritional status. Small circles show the four reference nutritional states identified in phase 1 of the study. Large circle at centre shows nutritional pattern of patient with gastric cancer. Values are expressed as % of normal, shaded area represents normal range, and inner circles indicate 60% and 30% of normal values

can be identified by means of cluster analysis; when a patient is classified in one of the four reference states, he has the same risk of post-operative septic complications recorded for that state (Figure 29.2). These data confirm previous observations that sepsis is a disease of disordered metabolic control in which nutritional abnormalities play a major and preliminary role[10]. The progressive deterioration of nutrition and metabolism produces profound physiological changes which can be characterized and quantified by cluster analysis.

In the present study the distribution in the four nutritional states of GI tract cancer and neoplasms of other organs is different. In fact in both series of patients neoplasms of the GI tract are more frequent in state 2, 3 and 4, where the abnormalities of nutritional indicators are more marked. On the contrary, neoplasms of other organs are almost equally distributed in the four clusters. This finding confirms the observation that GI tract malignancy is more frequently associated with undernutrition.

Moreover, these results are in agreement with a recent study[26] on cancer patients which showed that individuals with cachexia and major variations of nutritional indicators all had GI cancer. The same study showed that in the patient group with no nutritional abnormalities, the incidence of GI cancer was only 25%. More recently Bozzetti *et al.*[27] in a perspective study observed more pronounced nutritional variations of nutritional indicators in GI cancer

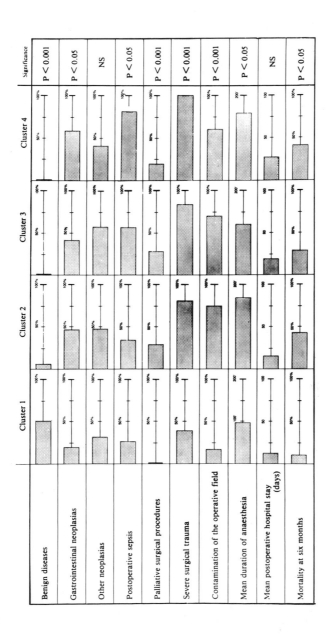

Figure 29.2 Incidence of different clinical variables in four reference clusters. Overall evaluation shows that clusters 2–4 represent more severe clinical conditions associated with poorer prognosis

355

patients than in patients with cancer of other organs. These reports support the hypothesis that cachexia in cancer patients is caused mainly by nutritional deficiencies which are more frequent in GI cancer patients.

CERULOPLASMIN

Ceruloplasmin is a blue, copper-containing glycoprotein (α_2-globulin) of serum that was isolated in 1948 by Holmberg and Laurell[28]. Its copper content is approximately 0.32% and its molecular weight is about 151 000, with eight atoms of copper per molecule[29]. Ceruloplasmin is synthesized in the liver and is present in the circulation at concentrations of 20–40 mg/100 ml[30,31]. In the plasma, 96% of the copper is bound to ceruloplasmin; the remaining 4% is bound unstably to albumin[32], and part is conveyed to a β_1-globulin identified by Gubler et al.[33]. The ratio between plasma and albumin-bound ceruloplasmin is constant in both normal individuals and patients with different diseases[34,35]. Ceruloplasmin is also an acute-phase protein, since its serum concentration increases after trauma or stress[36]. Moreover, a preliminary study has shown that a pre-operative increase of ceruloplasmin (CP) levels could be predictive of higher suscepti-bility to sepsis[37] and more recently it has been shown that ceruloplasmin is significantly increased in the sera of patients with solid tumours and lymphomas[35,37]. Moreover, a strict correlation between the stage of the disease and the serum levels of this protein has been recorded. This latter observation seems to be in contrast with the results of other authors[35], who found that the presence of metastatic disease does not affect the plasma levels of copper and ceruloplasmin. Our results suggest a possible relationship between tumour mass and CP levels in the serum. From this point of view, CP could be considered an important clinical determinant of tumour stage. The biochemical mechanism that determines these variations in the course of tumour growth is still not known. Our results have also shown that CP increases in the post-operative period in patients with benign lesions, indicat-ing that this protein is a positive acute-phase reactant. This increase is not evident after surgery for lymphomas or melanomas; whereas, after opera-tions for cancer of the GI tract, there is a negative acute-phase response. A possible explanation is the relationship between nutritional status and acute-phase response; GI cancer patients present more often than other patients in fact severe nutritional impairment that could be the cause of the negative acute-phase response. Pre-operative concentration of the CP is always significantly higher in patients with benign lesions who develop post-opera-tive infections (Figure 29.3).

DELAYED HYPERSENSITIVITY RESPONSE (DHR)

In recent years the measurement of delayed hypersensitivity response (DHR) to primary and recall cutaneous antigens has been extensively used for

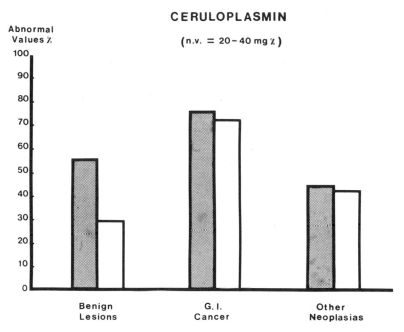

CERULOPLASMIN

Abnormal
Values %

(n.v. = 20 – 40 mg %)

Figure 29.3 Percentage distribution of abnormal pre-operative serum ceruloplasmin levels in septic (shaded area) and non-septic patients with benign lesions, GI cancer, and other neoplasias

monitoring the immunological status of cancer patients[38-40] and sometimes to evaluate their response to antineoplastic therapy[41,42]. It is now recognized that some factors associated with the growth of human tumours do cause immunosuppression, but they have not yet been identified clearly. Several authors indicate that the impairment of cellular immunity in malignant disease may be caused by inhibitors released by the cancer cells[43-45]. However, other factors such as advanced age[46-48] and malnutrition[49,50] may play a significant immunosuppressive role in cancer patients.

Multiple synergistic factors are likely to be involved in the aetiology of depression of cellular immunity in malignant disease, including some forms of antineoplastic therapy such as radiation and chemotherapy[51,52]. Consequently, the biological meaning of the responses to skin tests in cancer patients are very difficult to interpret.

In spite of the fact that the aetiology of depression of cellular immunity in neoplastic disease is not clear, it has been demonstrated that abnormalities of DHR correlate with a bad prognosis in cancer patients[50,53-56].

The separate roles of malnutrition, advanced age, and stage of tumour growth as causes of impairment of delayed hypersensitivity response were studied in our Institution in patients with solid tumours and in 56 non-neoplastic control patients matched for age, anatomical site of disease,

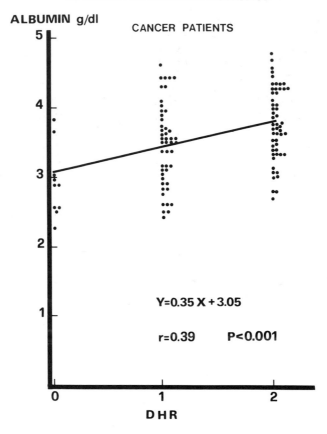

Figure 29.4 Correlation between pretreatment serum albumin level and DHR in 111 patients with solid tumours. DHR is expressed in grades ranging from 0 to 2

degree of illness, and nutritional status. Pretreatment DHR to recall antigens (tuberculin, *Candida*, streptokinase-streptodornase, trichophyton) and to dinitrochlorobenzene in cancer patients was 9% anergic, 43% hypoergic, and 48% normoergic; the distribution of DHR in controls was not significantly different.

In cancer patients, the serum albumin level showed an inverse correlation with the stage of tumour and a positive correlation with the DHR. In the controls the serum albumin level was also correlated positively with the DHR indicating that malnutrition in neoplastic or benign disease may cause depression of DHR (Figures 29.4, 29.5). In well-nourished controls, age was inversely correlated with DHR, showing that ageing itself may be another relevant cause of depression of DHR (Figure 29.6). The results of this study indicated that DHR in patients with solid tumours is similar to the DHR of non-neoplastic patients if matched for age, sex, and nutritional status. DHR impairment in

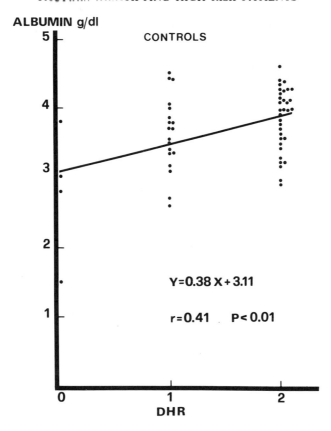

Figure 29.5 Correlation between pretreatment serum albumin level and DHR in 56 non-neoplastic patients. DHR is expressed in grades ranging from 0 to 2

cancer patients appears to be caused mainly by ageing and by malnutrition due to the advanced progression of cancer. The results of our study showed a significant correlation between DHR and serum albumin level in patients with cancer and in non-neoplastic controls and therefore support the concept that the DHR is a valuable index of nutritional status both in neoplastic and in non-neoplastic patients. Since it is known that elderly individuals present an impairment of responsiveness associated with ageing[46-48], we think that the use of DHR as an additional parameter for nutritional assessment should be restricted to subjects up to 70 y of age[46], with no immunological abnormalities.

In this study the multiple relationships between malnutrition, stage of tumour growth, age, and DHR were studied in patients with solid tumours and in carefully matched normal controls. The results provide evidence that malnutrition and ageing are leading causes of depressed cellular immunity in cancer patients. The presence of tumour *per se* does not influence the DHR significantly. This evidence is particularly convincing in patients with tumours of the digestive tract, who become severely malnourished early because of protein and/or blood loss, vomiting, obstruction, decreased food intake, and absorption and perhaps because of an increased resting metabolic expenditure[57]. These patients presented the greatest incidence of energy to skin tests. On the contrary, patients with tumours which did not cause malnutrition, such as breast cancer and melanoma, were not found to have a significant impairment of DHR when compared to age-matched controls. Bolton *et al.*[58] and Teasdale *et al.*[48] recorded significantly depressed DHR in breast cancer patients; however, the nutritional status was not adequately evaluated in their studies. The immunosuppressive factors which may[43,59,60] or may not be detected[61] in the serum of cancer patients are likely to have only a negligible *in vivo* effect on DHR and our study shows that in well-nourished patients the presence of tumour cells *per se* did not impair the DHR. It cannot be excluded, however, that the immunosuppressive factors found in the serum of some cancer patients[59], perhaps in those with advanced tumours, may add to the more relevant immunosuppressive effect of malnutrition and/or ageing.

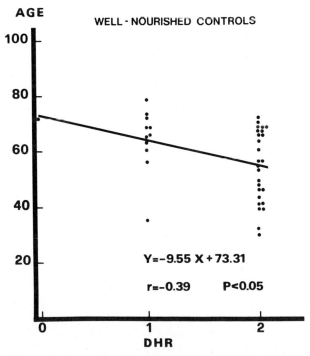

Figure 29.6 Correlation between age and pretreatment DHR in 37 well-nourished non-neoplastic patients. DHR is expressed in grades ranging from 0 to 2

In this study we evaluated only the role of malnutrition and ageing as causes of immunodepression in cancer patients. Other factors, however, such as the psychological stress, anxiety[62-64], and fear, which cannot be quantitatively studied, may also play a role. If malnutrition is a leading cause of immunological incompetence in cancer patients, as the results of this study indicate, then it is not surprising that chemical and bacterial immunostimulants failed to reverse the anergic state of patients with advanced tumours[38]. Moreover, if the immunological paralysis in such patients is determined by the lack of nutrients for immunocompetent cells, then nutritional repletion appears to be a more rational form of immunostimulation than immunostimulating drugs. Indeed, the beneficial effects of nutritional repletion on the deficient immune system of malnourished patients and animals is well-documented[65-68]. It has been shown that poor nutritional status is associated with an increased risk of post-operative infections in cancer patients[69]. It also has been observed that some malnourished anergic patients who present a higher risk of infectious complications[70] can convert to normoergic after nutritional therapy[56,71]. The reversal of anergy obtained by means of total parenteral nutrition in malnourished cancer patients was shown to correlate with a longer survival[41,56]. However, in those studies there was no evidence that the prolonged survival was actually due to improved immunological control of the malignancy. The nutritionally repleted patients might in fact have survived longer because they better resisted infection, which is a frequent cause of death in cancer patients[69].

AMINO ACIDS

Interest in plasma amino acids as prognostic indicators of human sepsis arose from observations that the plasma amino-acid profile changes with time as the septic process proceeds. Plasma aminograms correlated very closely with the physiological alterations observed during sepsis[72]. Proline seems to play a major role, since in preterminal septic patients, when the peripheral resistance and oxygen consumption fall, the proline levels rise. Determination of this amino acid has a predictive value, in fact when the plasma level reached 1200 μm/l death occurred within 3–4 days. The grouping of alanine, cystein, methionine, isoleucine, arginine, phenylalanine and tyrosine was found to successfully predict outcome in 98% of cases[73].

Freund in 1978[74] found analogous correlations between certain amino-acid levels and survival rates. Those patients who did not survive sepsis had higher levels of alanine and of branched-chain amino acids. In the same study the ratio between alanine plus methionine as numerator and tyrosine plus total branched-chain amino acids as denominator in the normal patient was 0.16, in the patient who survived it was 0.43 as opposed to 1.99 in those who died.

In a preliminary study we found that patients with minor sepsis have a significant increase in aspartic acid compared with non-septic patients.

361

Patients with major sepsis had a significant increase of glutamic acid, aspartic acid, cystein, ornithine, proline and threonine; whereas a decrease of isoleucine and lysine was observed. No significant variations of amino acid were recorded between survivors and non-survivors, even when the arginine concentration was lower (17%) in the non-survivors. These results confirm that during severe sepsis glutamic acid, aspartic acid, proline and cystein increase and isoleucine decreases; moreover, they show that ornithine and threonine increase and lysine decreases. From this study it does not appear possible to differentiate between patients who may or may not survive sepsis on the basis of a plasma amino acid pattern alone.

From all these studies it has been suggested that metabolic criteria and not standard laboratory tests or multiple physiological variables have the capacity to predict death and survival.

Urinary excretion of 3-methylhistidine (3-MH) has been suggested to be a specific marker for muscle protein breakdown. Several studies have been performed in injured subjects and in patients undergoing elective surgery of different severity, but few data are available regarding 3-MH excretion during post-operative infections. In our Institution 17 cancer patients have been studied on the day before surgery and then on the 2nd, 4th and 7th post-operative day; positive blood cultures were required to diagnose sepsis. All the patients showed a significant increase of 3-MH: on the 2nd post-operative day (52%), on the 4th post-operative day (50%) and on the 7th (49%). Nine patients developed one or more infectious episodes, whereas eight did not. Urinary excretion of 3-MH in the septic patients was constantly higher, the difference between septic and non-septic patients being statistically significant on the 4th post-operative day. These results confirm that 3-MH can be considered as a specific marker of muscle protein breakdown, which accompanies injury in man; and that during septic post-operative episodes, of moderate severity, there is increased muscle protein catabolism.

References

1 Young, U. R. and Scrinshaw, N. S. (1971). The physiology of starvation. *Sci. Am.*, **225**, 14
2 Owen, E. O., Morgan, A. P., Remph, P. *et al.* (1969). Brain metabolism during fasting. *J. Clin. Invest.*, **48**, 574
3 Jelliffe, D. B. (1966). The assessment of nutritional status of the community. *World Health Organization*, (Monograph series), **53**
4 Blackburn, G. L., Bistrian, B. R., Maini, B. S. *et al.* (1977). Nutritional and metabolic assessment of the hospitalized patient. *J. Parent. Enteral Nutr.*, **1**, 11
5 Mullen, J. L., Gertner, M. H., Buzby, G. P. *et al.* (1979). Implication of malnutrition in the surgical patient. *Arch. Surg.*, **114**, 121
6 Bistrian, B. R., Blackburn, G. L., Sherman, G. L. *et al.* (1977). Cellular immunity in adult marasmus. *Arch. Intern. Med.*, **137**, 1408
7 Bozzetti, F., Terno, G. and Longoni, C. (1973). Parenteral hyperalimentation and wound healing. *Surg. Gynecol. Obstet.*, **141**, 712
8 Daly, J. M., Vars, H. M. and Dudrick, S. J. (1972). Effect of protein depletion on strength of colonic anastomoses. *Surg. Gynecol. Obstet.*, **134**, 15
9 Law, D. K., Dudrick, S. J. and Abdou, N. J. (1974). The effect of protein calorie malnutrition on immune competence of the surgical patient. *Surg. Gynecol. Obstet.*, **139**, 357

10 Dionigi, R., Gnes, F., Bonera, A. *et al.* (1979). Nutrition and infection. *J. Parent. Enteral Nutr.*, **3**, 62

11 Selzer, M. H., Bastidas, J. A., Cooper, D. M. *et al.* (1979). Instant nutritional assessment. *J. Parent. Enteral Nutr.*, **3**, 157

12 Buzby, G. P., Mullen, J. L., Matthews, D. C. *et al.* (1980). Prognostic nutritional index in gastrointestinal surgery. *Am. J. Surg.*, **139**, 160

13 Mullen, G. L., Buzby, G. L., Waldman, T. G. *et al.* (1979). Prediction of operative morbidity and mortality by preoperative nutritional assessment. *Surg. Forum*, **30**, 80

14 Nazari, S., Dionigi, R., Comodi, I. *et al.* (1982). Preoperative prediction and quantification of septic risk caused by malnutrition. *Arch. Surg.*, **177**, 266

15 Seigel, J. H. (1977). Computers and mathematical techniques in surgery. In Sabiston, D. C. (ed.) *Davis Christopher Textbook of Surgery The Biological Basis of Modern Surgical Practice.* 11th Edn. p. 231. (Philadelphia: Saunders)

16 Goldwyn, R. M., Friedmann, H. P. and Siegel, J. H. (1977). Iteration and interaction in computer data bank analysis: a case study in the physiological classification and assessment of the critically ill. *Comput. Biomed. Res.*, **4**, 607

17 Siegel, J. H., Goldwyn, R. M. and Friedmann, H. P. (1971). Pattern and process in the evolution of human septic shock. *Surgery*, **70**, 232

18 Siegel, J. H., Goldwin, R. M., Farrel, E. J. *et al.* (1972). The surgical implications of physiologic patterns in myocardial infarction shock. *Surgery*, **72**, 126

19 Siegel, J. H., Cerra, F. B., Coleman, B. *et al.* (1979). Physiological and metabolic correlations in human sepsis. *Surgery*, **86**, 163

20 Verhagen, M., De Cock, W., De Cree, J. *et al.* (1976). Increase of serum complement levels in cancer patients with progressing tumors. *Cancer*, **38**, 1608

21 Ward, A. M. and Cooper, E. H. (1978). Acute phase proteins in the staging and monitoring of malignancy. *La Ricerca Clin Lab.* 8, Suppl., **1**, 49

22 Pisi, E., Di Feliciantonio, R., Figus, E. *et al.* (1968). Comportamento e significato prognostico della ceruloplasmina serica in relazione al quadro istopatologico nella malattia di Hodgkin. *Minerva Med.*, **59**, 944

23 Shifrine, M. and Fisher, G. L. (1976). Ceruloplasmin levels in serum from human patients with osteosarcoma. *Cancer*, **38**, 244

24 Goulian, M. and Fahey, J. L. (1961). Abnormalities in serum protein and protein-bound hexose in Hodgkin's disease. *J. Lab. Clin. Med.*, **57**, 408

25 Scanni, A., Tomirotti, M., Licciardello, L. *et al.* (1979). Variation in serum copper and ceruloplasmin levels in advanced gastrointestinal cancer treated with polichemiotherapy. *Tumori*, **65**, 331

26 Sudjivan, A. V. (1980). A hypothesis. Cancer cachexia or cachexia in cancer? *Acta Chir. Scand. Suppl.*, **498**, 155

27 Bozzetti, F., Baticci, F., Terno, G. *et al.* (1980). Impact of cancer site and treatment on the nutritional status of the patients. Abstract II Espen Congress, Newcastle, 7–10 Sept. 1980, *J. Parent. Enteral Nutr.*, **4**, 430

28 Holmberg, C. G. and Laurell, C. B. (1948). Investigations in serum copper: isolation of the copper containing protein, and a description of some of its properties. *Acta Chem. Scand.*, **2**, 550

29 Markowitz, H., Gubler, C. J., Mahoney, J. P. *et al.* (1955). Studies on copper metabolism. XIV. Copper, ceruloplasmin and oxidase activity in sera of normal human subjects, pregnant women, and patients with infections, hepatolenticular degeneration and the nephrotic syndrome. *J. Clin. Invest.*, **34**, 1498

30 Owen, C. A. and Hazerling, J. B. (1966). Metabolism of Cu-64 labelled copper by isolated rat liver. *Am. J. Physiol.*, **210**, 1059

31 Scheinberg, I. H. and Morell, A. G. (1973). Ceruloplasmin. In Scheinberg, I. H. (ed.). *Inorganic Biochemistry.* p. 306. (Amsterdam: G. L. Eichorn)

32 Searcy, R. L. (1969). Copper. In *Diagnostic Biochemistry.* p. 184. (New York: McGraw-Hill)

33 Gubler, C. J., Laney, M. E. and Cartwright, G. E. (1953). Studies on copper metabolism. The transportation of copper in blood. *J. Clin. Invest.*, **32**, 405

34 Shifrine, M. and Fisher, G. L. (1976). Ceruloplasmin levels in serum from human patients with osteosarcoma. *Cancer*, **38**, 244

35 Scanni, A., Licciardello, L. and Trovato, M. (1977). Serum copper and ceruloplasmin levels

in patients with neoplasias localized in the stomach, large intestine or lung. *Tumori*, **63**, 175

36 Crockson, R. A., Payne, C. J. and Ratcliff, A. P. (1966). Time sequence of acute phase reactive proteins following surgical trauma. *Clin. Chim. Acta*, **14**, 435

37 Dionigi, P., Dionigi, R., Pavesi, F. *et al.* (1981). Serum ceruloplasmin levels in surgical cancer patients: relationship with tumor stage and postoperative complications. *Eur. Surg. Res.*, **13**, 31

38 Mathe, G. (1976). *Cancer Active Immunotherapy*. (Berlin: Springer-Verlag)

39 Cochran, A. J., Mackie, R. M., Grant, R. M. *et al.* (1976). An examination of the immunology of cancer patients. *Int. J. Cancer*, **18**, 298

40 Hersh, E. M., Mavligit, G. M. and Guttermann, J. U. (1976). Immunodeficiency in cancer and the importance of immune evaluation of the cancer patients. *Med. Clin. Am.*, **60**, 623

41 Ota, D. M., Copeland, E. M., Corriere, J. N. *et al.* (1979). The effects of nutrition and treatment of cancer on host immunocompetence. *Surg. Gynecol. Obstet.*, **148**, 104

42 Teitell, B. C., Herson, J. and Van Eis, J. (1980). Recall antigen response in pediatric cancer patients receiving parenteral hyperalimentation. *J. Parent. Enteral Nutr.*, **4**, 9

43 Silk, M. (1967). Effect of plasma from patients with carcinoma on *in vitro* lymphocyte transformation. *Cancer*, **20**, 2088

44 Watkins, S. M. (1973). The effects of surgery on lymphocyte transformation in patients with cancer. *Clin. Exp. Immunol.*, **14**, 69

45 Hakim, A. A. (1979). Tumor-mediated immunosuppression is a challenge in cancer treatment. *Cancer Immunol. Immunother.*, **7**, 1

46 Gross, L. (1965). Immunological defect in aged population and its relationship to cancer. *Cancer*, **18**, 201

47 Waldorf, D. S., Wilkens, R. F. and Decker, J. L. (1968). Impaired delayed hypersensitivity in an aging population. *J. Am. Med. Assoc.*, **203**, 831

48 Teasdale, C., Hughes, L. E., Whitehead, R. H. *et al.* (1979). Factors affecting pretreatment immune competence in cancer patients. *Cancer Immunol. Immunother.*, **6**, 89

49 Hughes, L. E. and MacKay, W. D. (1965). Suppression of the tuberculin response in malignant disease. *Br. Med. J.*, **2**, 1346

50 Daly, J. M., Dudrick, S. J. and Copeland, E. M. (1979). Evaluation of nutritional indices as prognostic indicators in the cancer patients. *Cancer*, **43**, 925

51 Stefani, S., Kerman, R. and Abbate, J. (1976). Immune evaluation of lung cancer patients undergoing radiation therapy. *Cancer*, **37**, 2792

52 Mitchell, M. S., Wade, W. E., De Conti, R. C. *et al.* (1969). Immunosuppressive effect of cytosine arabinoside and methotrexate in man. *Ann. Int. Med.*, **70**, 535

53 Eilber, F. R. and Morton, D. L. (1970). Impaired immunologic reactivity and recurrence following cancer surgery. *Cancer*, **25**, 362

54 Dionigi, R., Gnes, F., Bonera, A. *et al.* (1979). Delayed hypersensitivity response (DHR) and infections in surgical cancer patients. *Br. J. Surg.*, **66**, 900

55 Dominioni, L., Dionigi, R., Prati, U. *et al.* (1980). Studio controllato della immunità cellulare aspecifica in pazienti neoplastici. *Minerva Med.*, **71**, 1

56 Copeland, E. M., MacFayden, B. V. Jr. and Dudrick, S. J. (1976). Effect of intravenous hyperalimentation on established delayed hypersensitivity in the cancer patient. *Ann. Surg.*, **184**, 60

57 Bozzetti, F. (1979). Correlation between resting metabolic expenditure and weight loss in cancer patients. *IRCS Med. Sci.*, **7**, 89

58 Bolton, P. M., Teasdale, C., Mander, A. M. *et al.* (1976). Immune competence in breast cancer. Relationship of pretreatment immunologic tests to diagnosis and tumor stage. *Cancer Immunol. Immunother.*, **1**, 251

59 Glasgow, A. H., Nimberg, R. B., Menzoian, J. O. *et al.* (1974). Association of anergy with an immunosuppressive peptide fraction in the serum of patients with cancer. *N. Engl. J. Med.*, **291**, 1263

60 Sample, W. F., Gertner, H. R. and Chretien, P. B. (1971). Inhibition of phytohemagglutinin-induced *in vitro* lymphocyte transformation by serum from patients with carcinoma. *J. Natl. Cancer Inst.*, **46**, 1291

61 Golob, E. K., Israsena, T., Quatrale, A. C. *et al.* (1969). Effect of serum from cancer patients on homologous lymphocyte cultures. *Cancer*, **23**, 306

62 Dominioni, L., Dionigi, R., Zonta, A. *et al.* (1978). Postoperative immunologic monitoring

of cancer patients undergoing radical surgery. *Boll. Ist. Sieroter. Milan*, **57**, 612

63 Berenbaum, M. C., Fluck, P. A. and Hurst, N. P. (1973). Depression of lymphocyte response after surgical trauma. *Br. J. Exp. Pathol.*, **54**, 597

64 Dominioni, L., Dionigi, R., Zonta, A. *et al.* (1979). Lymphocyte blastogenic response following different types of operations for cancer. Proceedings of the *14th Congress European Society for Surgical Research,* Barcelona, May 6–9, p. 136. (Basel: Karger)

65 Law, D. K., Dudrick, S. J. and Abdou, N. I. (1974). The effect of dietary protein depletion on immunocompetence: the importance of nutritional repletion prior to immunologic induction. *Ann. Surg.*, **179**, 168

66 Dionigi, R., Gnes, F., Bonera, A. *et al.* (1979). Nutrition and infection. *J. Parent. Enteral Nutr.*, **3**, 62

67 Daly, J. M., Dudrick, S. J. and Copeland, E. M. (1978). Effects of protein depletion and repletion on cell-mediated immunity in experimental animals. *Ann. Surg.*, **188**, 791

68 Dionigi, R., Zonta, A., Dominioni, L. *et al.* (1977). The effects of total parenteral nutrition on immunodepression due to malnutrition. *Ann. Surg.*, **185**, 467

69 Dionigi, R., Dominioni, L. and Campani, M. (1980). Infections in cancer patients. *Surg. Clin. N. Am.*, **60**, 145

70 Maclean, L. D., Meakins, J. L. and Taguchi, K. (1975). Host resistance in sepsis and trauma. *Ann. Surg.*, **132**, 207

71 Willcutts, H. D., Linderme, D., Chlastawa, D. *et al.* (1979). Anergy: is nutritional reversal possible/outcome significant. *J. Parent Enteral Nutr.*, **3**, 292 (Abstr)

72 Cerra, F. B. (1981). Sepsis, metabolic failure and total parenteral nutrition. *Nutritional Support Services*, **1**, 96

73 Cerra, F. B., Wiles, J. B. and Siegel, G. H. (1981). The best predictors of sepsis are metabolic. *J. Crit. Care Med.*, **8**, 230

74 Freund, H. R., Ryan, J. and Fischer, J. E. (1978). Amino acid derangement in patient in the sepsis: treatment with BCAA with infusions. *Ann. Surg.*, **188**, 423

Section 7
Home Parenteral Nutrition

1
Intestinal failure and its treatment by home parenteral nutrition

M. IRVING

INTRODUCTION

The concept of kidney failure and its treatment by dialysis is well-established. Acute renal failure is recognized as a transitory problem which, given correct treatment, will recover, as glomerular and tubular function return. Dialysis is only necessary while this reparatory process is occurring.

On the other hand chronic renal failure requires permanent dialysis until such time as kidney transplantation is provided or the patient dies. The realization that patients could learn the techniques of haemodialysis so that it was safe to allow them to administer this treatment in their own homes, was a great advance.

Although the concept of 'intestinal failure' is similar in many ways to renal failure, and although it has been propounded for many years, it is still not widely recognized. Fleming and Margot[1] define intestinal failure as 'the reduction in functioning gut mass below the minimal amount necessary for adequate digestion and absorption of nutrients'. I believe a simpler but as accurate definition is 'the inability of the body to absorb sufficient nutrients from the gut to maintain health and well-being'. Like renal failure, intestinal failure may be acute or chronic. Chronic intestinal failure may be further subdivided into partial or complete, minor degrees of the former being present in such conditions as terminal ileal Crohn's disease and chronic pancreatitis which cause selective malabsorption.

Acute intestinal failure, caused by problems such as a mid small bowel fistula, subtotal gut resection, or acute pancreatitis, produces a nutritional crisis which requires support with total parenteral nutrition but which will subside as the primary problems resolve and intestinal adaptation occurs allowing resumption of adequate enteral nutrition.

369

Chronic intestinal failure follows total or near total small bowel resection or the development of extensive small bowel disease such as Crohn's disease or motility disorders, for example that due to scleroderma of the bowel (Figure 30.1). A particular type of partial chronic intestinal failure is seen in patients with high jejunal stomas where, although nutrient absorption may be adequate, there is a considerable loss of water and electrolyte. Such patients will require prolonged parenteral nutrition if they are to stay alive for, as yet, intestinal transplantation is not a practical procedure and retrograde pacing of the bowel is at an experimental stage. An excellent account of the pathophysiology of the short bowel and the process of adaptation has been written recently by Fleming and Margot[1].

The suggestion that these patients could be treated with an 'artificial gut' was made by Scribner and his colleagues in 1970[2]. It followed from a combination of the increasing use of parenteral nutrition in European and North American hospitals over the previous 15–20 years and the success of the home dialysis programme for chronic renal failure.

In 1974 Scribner's group issued a report on Home Parenteral Nutrition (HPN) in the first 16 patients they treated[3]. The development by Broviac, Cole and Scribner[4] of the silastic rubber catheter with an integral Dacron cuff that now bears the former's name (Figure 30.2) opened the way to successful HPN. They demonstrated that the catheter could remain implanted in a major vein for a prolonged time, and that the Dacron cuff became fixed in the subcutaneous tissues. Patients were readily taught the techniques involved in the care of the catheter, and in particular how to prevent ascending infection and intraluminal thrombosis. They learned how to undertake the infusion of

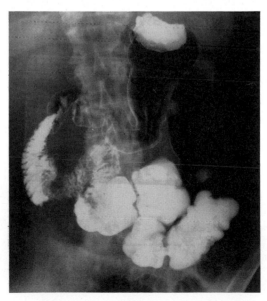

(a)

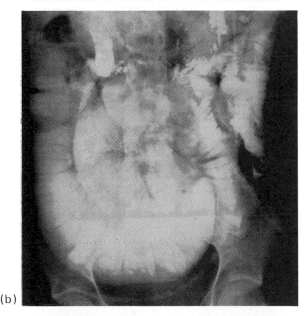

(b)

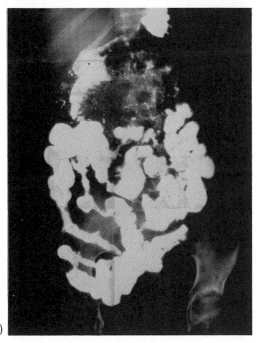

(c)

Figure 30.1 Barium radiographs of three major causes of intestinal failure. (a) Duodeno-colic anastomosis following total small bowel resection for superior mesenteric artery thrombosis, (b) extensive small bowel motility disorder − 'pseudo-obstruction', (c) extensive small bowel Crohn's disease

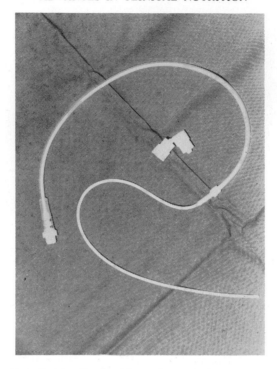

Figure 30.2   Broviac silastic rubber catheter with integral Dacron cuff

the nutrients and the recognition of the common metabolic problems that can occur. The only disadvantage was the prolonged infusion time, in the region of 12 h, that limited the mobility of the patient. It was not long before the treatment was adopted by other centres in the USA and Canada, and reports have issued from them.

With the exception of Solassol[5] and Joyeux in France the technique was slow to develop in Europe. It is only within the last 5 y that reports of HPN have appeared in the British literature[6-10].

Thus, this expensive treatment is well-established in the western world and there is now sufficient information available to allow an assessment of its value. This should reveal which patients do best and those in whom it only prolongs for a short time an inevitable death. Limited resources for health care demand that we should use this undoubtedly successful treatment carefully.

INTESTINAL FAILURE AND THE INDICATIONS FOR HOME PARENTERAL NUTRITION

Temporary intestinal failure is a common problem; for example it occurs for a matter of hours after most abdominal operations.

In such circumstances it is of little importance, for most patients can readily withstand the 2–3 days starvation before food intake is resumed. Similarly, partial intestinal failure where there is selective malabsorption of an orally ingested nutrient is both common and easily remedied by parenteral administration of the missing item. On the other hand, patients with an exacerbation of extensive Crohn's disease, a subtotal bowel resection, severe acute pancreatitis, and post-operative complications such as fistulae and abscess formation, will require parenteral nutrition if they are not to lose weight and die. Treatment should commence as soon as the problem is diagnosed and should be accompanied by vigorous efforts to treat systemic infections and to drain abscesses. In the majority of cases the problem will resolve during the course of the patients' hospital stay but in a small number it will soon be apparent that longterm treatment is required. In such circumstances the question of training the patient for HPN should be considered. Most such patients will soon revert to full oral intake as the crisis subsides but in others treatment will be needed for years or even permanently.

Cases with radiation enteritis pose a problem if the underlying malignant disease is still active. However, if it has been cured these patients are well worth treating with HPN for fistulae heal and obstructive episodes subside allowing a return to oral feeding[11].

There are, of course, many other causes of intestinal failure which can be treated by HPN. Some are obviously good indications but others such as inoperable malignant disease and post-infarction intestinal failure in elderly arteriopaths are difficult to justify. Table 30.1 shows the conditions treated by some groups who have published their results. It is interesting to note that most groups are treating similar types of patient. In all but Rault and Scribner's series Crohn's disease is the principal indication for HPN. In the other groups the short bowel syndrome accounts for only one fifth of the cases treated whereas in Rault and Scribner's group this indication accounts for 40% of the population. One is tempted to speculate that the differing emphasis may result from the fact that Rault and Scribner work from a renal unit which may receive a more varied type of patient than the other groups which have a primarily gastroenterological practice. It is similarly note-worthy that 10% of the patients in the series of Byrne et al.[20] have inoperable malignant disease as an indication for HPN. The absence of such patients from the other series probably indicates a philosophy that regards the use of HPN to sustain this type of patient as unjustifiable.

THE TECHNOLOGY OF HPN

For HPN to be successful the techniques involved must be simple, safe, effective and reliable. They must be backed up by a support service and education programme to which the patient can refer at any time.

Table 30.1 Comparative indications for home parenteral nutrition

Conditions treated	Author				
	Rault and Scribner[11]	Ladefoged and Jarnum[7]	Fleming et al.[15]	Byrne et al.[20]	UK HPN Register[21]
Age range and mean	5–69 (40)	20–68 (42)	22–67 (40)	2–77 (–)	17–76 (35)
Crohn's disease (including fistula and short bowel syndrome)	13 (25.5%)	12 (65%)	14 (66%)	43 (41%)	16 (56%)
Ulcerative colitis	–	1 (10%)	1 (5%)	–	1 (3%)
Short bowel syndrome (vascular, volvulus trauma, etc)	20 (39%)	4 (20%)	4 (19%)	16 (15%)	6 (17%)
Radiation enteritis	5 (10%)	–	–	9 (8.5%)	2 (6%)
Pseudo-obstruction	3 (9%)	–	1 (5%)	9 (8.5%)	1 (3%)
Systemic sclerosis	3 (9%)	–	–	–	3 (9%)
Intestinal fistula (not Crohn's)	1 (2%)	–	–	4 (4%)	1 (3%)
Inoperable malignant disease	–	–	–	10 (9.5%)	0
Others	6 (12%)	1 (5%)	1 (5%)	13 (13.5%)	5 (14%)
Total	51 (100%)	19 (100%)	21 (100%)	105 (100%)	(100%)

The catheter

Although originally a–v shunts and fistulae as used for home dialysis were tried they were found to be unsuitable because of the patient's normal clotting mechanisms[3]. The use of silastic rubber catheters is now virtually universal, although the need for a Dacron cuff is not generally accepted. Although most would agree that the cuff acts to prevent accidental removal, it is not agreed that it forms a barrier to ascending infection. It is acknowledged that successful management of the catheter commences with insertion under sterile conditions after meticulous skin preparation. The catheter is inserted either by puncture of the subclavian or internal jugular vein, or under direct vision by exposure of the cephalic vein in the delto-pectoral groove, or another similar superficial vein. The catheter tip is advanced until it lies in the superior vena cava at its junction with the right atrium. The distal end of the catheter is run in a subcutaneous tunnel to emerge through the skin of the chest wall near one of the nipples at a site where the patient can clearly see it. The Dacron cuff should be positioned in an intercostal space (Figure 30.3). Recently a method has been described for inserting the Broviac catheter by percutaneous puncture without the need for a cut down[12]. Occasionally a patient will be referred for management in whom all peripheral veins have thrombosed or atrophied. In these exceptional circumstances implantation of the catheter direct into the right atrium at thoracotomy may be the only way to achieve longterm access to the venous circulation. Before use the position of catheter tips must be checked by X-ray as, despite an apparently easy cannulation, they are often misplaced.

In between infusions the patient fills the catheter with 1.5 ml of heparin solution (1000 units/ml), a so-called 'heparin lock' to prevent clotting within the catheter and to allow mobility. Some workers advocate that systemic anticoagulants should be used in addition[6]; however, most do not find them necessary and thereby avoid the accompanying complications. In patients in whom lipid forms a regular part of the nutritional intake the intraluminal accumulation of lipid appears less likely if the larger version of the Broviac catheter – the Hickmann catheter – is used.

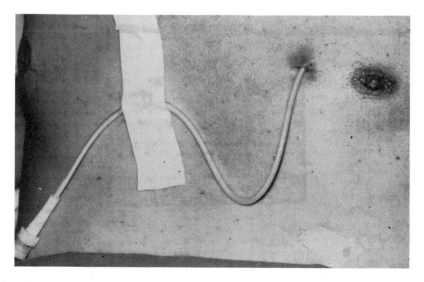

Figure 30.3 Broviac catheter emerging from its subcutaneous tunnel on the anterior wall of the chest

Administration of nutrients
Administration of the nutrients can be undertaken in one of two ways. Either multiple bags or bottles of a caloric source and amino acids can be infused individually or alternatively the day's requirements can be mixed into one 3 l bag. This latter method is the one generally preferred in the United Kingdom although lipid, as yet, is infused separately. In the UK, in contrast to some

centres in the USA, patients do not undertake their own compounding. In spite of the excellent safety record of compounding in the patient's own home it is generally considered in the UK to be a complicating factor for the patient. Thus, solutions are mixed in the hospital pharmacy, or by a pharmaceutical company and delivered to the patient at regular intervals ready for use. Most patients infuse every night but some only need 3–4 infusions a week and others, with high output stomas, only need water and electrolytes.

The composition of the infusion will obviously depend upon the patient's individual requirements. In general, caloric requirements are supplied by 20–50% dextrose solution providing 8–10 MJ a day. Nitrogen is provided as synthetic 'l' form amino acids in a dose varying between 7 and 15 g a day. The volume of normal saline administered depends upon the losses from the patient. A comprehensive range of trace elements and vitamins is also added. Some patients receive intravenous lipid daily accounting for up to half the administered energy source whereas others receive it only once a week to prevent essential fatty acid deficiency.

Infusion technique

The method by which the nutrients are infused varies. The simplest technique, that of gravity infusion, is used by some[7] but most patients prefer the security of a pump. The type of pump varies. Jeejeebhoy and his colleagues use external pneumatic compression[13] whilst Dudrick et al.[14] have pioneered the use of a portable system for selected patients. This consists of a vest with pockets into which 500 ml plastic bags containing the nutrients are fitted and a compact low pressure infusion pump maintains continuous infusion. Whilst it is less cumbersome than the system described by Solassol[5] it still does not compare with the total daytime freedom enjoyed by those who use conventional night-time infusion. However, it has its enthusiasts and some patients, particularly the elderly, seem to benefit from the continuous infusion of their requirements.

In the UK the generally preferred technique is night-time infusion with a constant volume low pressure infusion pump with inbuilt alarms (e.g. the IMED 928).

THE COMPLICATIONS OF HPN

Catheter problems

With care an implanted Broviac catheter can remain functional for years. Even if the protruding part is damaged it can be restored to active use by means of the repair kits provided by the users. Intravascular fracture is a danger[15] and in such circumstances special interventional radiological techniques have to be employed to recover the fragments if thoracotomy is to be avoided. Accidental removal is always a problem in those catheters without a Dacron cuff, and even in these it can occur in the 3–4 weeks before the cuff becomes fixed.

Blockage of the catheter lumen can result from clotting of regurgitated blood or by lipid. The likelihood of the former can be reduced by a correct heparin locking technique which ensures that heparin is still being infused when the catheter is clamped. If clotting should occur the problem can sometimes be overcome by insertion of a urokinase lock. Forcible syringing in an attempt to clear a block should be avoided for the catheters are easily disrupted in the subcutaneous tunnel.

The presence of the catheter within the lumen of a large vein such as the subclavian vein is obviously a potential stimulus to venous thrombosis. This is indeed the case with the Broviac and similar catheters as revealed by routine phlebography through the catheter. Ladefoged and Jarnum demonstrated 17 episodes in nine patients[7]. Frequently the thrombosis is asymptomatic but in some cases it can be suspected because of increasing difficulty with infusion of the nutrients and swelling of the ipsilateral arm and face. In cases of symptomatic and total occlusion the catheter will have to be removed and repositioned.

The greatest hazard is undoubtedly infection of the catheter entry site and colonization of the lumen. We believe that infection is almost totally preventable, and when it occurs is the result of breaches in the catheter care protocols. Meticulous aseptic and antiseptic technique is necessary in caring for the catheter exit site. The use of an occlusive dressing when the catheter is not in use ensures sterility between infusions. Any organisms can be involved but it is usually *Staphylococcus aureus*, *Staphylococcus epidermidis* and *Candida* that are involved. Proven infection usually demands removal of the catheter and its replacement after a few days but occasionally colonization of the catheter lumen by *Staphylococcus epidermidis* can be overcome by simultaneous systemic treatment with an appropriate antibiotic and the insertion of a urokinase lock in the lumen of the catheter.

However, the emphasis must always be upon prevention of infection. That this can be achieved is shown by results from our four-bedded surgical nutrition unit which manages patients with intestinal fistula by parenteral nutrition. In the 2 years since it was opened no patient has developed a catheter infection.

Metabolic problems
Metabolic problems arising in a patient established on an HPN programme are surprisingly infrequent but cover a wide range of disorders. Acute problems such as reactive hypoglycaemia can usually be overcome by slowing the infusion rate in the last hour before stopping for the day. Minor changes in the serum level of liver enzymes, a frequent observation, may be similarly overcome by prolonging the total infusion time[11].

Trace metal deficiencies such as those resulting from inadequate intake or excessive loss of zinc, magnesium and copper should be uncommon for the body's requirements are now known and will be included in the infusion regimen. Studies such as those of Shike *et al.*[16] have accurately demonstrated normal requirements. These workers found the normal daily copper requirements to be 0.30 mg/day increasing to 0.40–0.50 mg/day in the presence of high gastrointestinal loss. Similar studies are required for other trace metals now that selenium, chromium, manganese and antimony deficiency syndromes have been recognized.

Recently Jeejeebhoy *et al.* have described an unusual form of osteomalacia with associated hypercalcaemia and hypercalcuria which is relieved by exclusion of vitamin D from the infused regimen[17].

Psychological problems

Not surprisingly patients unable to take food or drink by mouth on a permanent or semipermanent basis, and having to rely on parenteral nutrition, can develop psychological problems. Indeed some patients find this restriction so irksome that they prefer to put up with the increased output from their stomas and the associated water and electrolyte problems that accompany eating and drinking. An excellent review of the psychiatric problems that may be encountered in patients receiving intravenous nutrition has recently been produced by Perl *et al.*[18]. On the other hand many patients coming to a HPN programme do so after years of abdominal pain and repeated prolonged hospital admission. To such patients HPN brings freedom from the tyranny of repeated hospital admissions. The ability to remain at work and at home with their families is a welcome relief. Psychological problems have not figured greatly in the data supplied to the United Kingdom HPN registry.

Most psychological problems can be prevented by adequate preparation of the patient before treatment begins and support from the base hospital and social workers in times of crisis.

TRAINING

The training of the patient in the techniques involved in HPN, and in the recognition of complications must be meticulous. This training is best provided by a nurse specializing in the management of such patients, although in the USA a pharmacist with similar interests is sometimes used in this capacity. The training programme, which should last only 10 days to 2 weeks, takes place in hospital and uses audiovisual aids such as tape–slide programmes, video cassettes and, most important of all, a comprehensive instruction manual. Introduction to another patient already on an established HPN programme can be a great encouragement.

A patient's security at home is always enhanced by the knowledge that he can contact the HPN Centre at any time he needs advice and support.

THE RESULTS OF HPN

There is no doubt that the results of putting a patient on a HPN programme can be dramatic. The literature is in universal accord that the best results occur in patients with Crohn's disease, and in patients with a short bowel syndrome other than those due to atherosclerotic mesenteric infarction. The patients who benefit most in the long run are of course those in whom the intestinal crisis subsides and residual bowel adapts.

The difficulty in assessing results arises mainly in the patients with other conditions where the outcome is variable, and in the advocated extension of HPN as a primary treatment for Crohn's disease. Although this may be justified in some cases there is also a danger that surgically remediable lesions may be inappropriately treated. Thus the statement by Rault and Scribner that in Crohn's disease 'the early institution of HPN is virtually without risk and often can prevent serious disabling illness, prolonged periods of hospitalization and extensive surgery'[19] needs subjecting to critical analysis by controlled trial. Certainly to date there is no convincing evidence that parenteral nutrition can 'switch off' an acute exacerbation of Crohn's disease.

Assessment of the value of HPN demands the collection of accurate data and the application of a modified performance status to the patients treated. In the UK we have developed a register of HPN cases, similar to the register already developed in the USA. The effectiveness of HPN is assessed by classifying the patient into 'performance status' groups as shown in Table 30.2.

Table 30.2 Performance status of HPN patients as used in UK register[21]

Group	Performance status	Number in each group at end of 1981
I	At work full-time or looking after home and family unaided	18 (I → IV) (51%)
II	At work part-time or looking after home and family with help	6 (17%)
III	Unable to work, but able to cope with HPN unaided and to go out occasionally	7 (20%)
IV	Housebound. Needs major assistance with HPN	4 (+1 from I) (12%)

Only patients in groups I and II can be considered as having had a successful response to treatment with HPN. It is my belief that eventually we should avoid commencing HPN in those patients whose underlying disease ensures they remain in groups III and IV or who soon after treatment commences move from categories I and II to categories III and IV. The results so far suggest that patients with residual malignant disease and the elderly with atherosclerotic mesenteric infarction fall into the former category whilst patients afflicted by rapidly progressive scleroderma with associated intestinal involvement fall into the latter. The scleroderma patients are particularly distressing, for after obtaining a gratifying initial response the progression of the disease in their upper limbs means they can no longer handle the catheter and pumps, thereby leaving them totally dependent upon others.

The organization of the delivery of HPN

The previous comments indicate that HPN will only rarely be required as a treatment for intestinal failure. Most patients with intestinal failure will remain in hospital during their treatment simply because the condition will be short-lived and associated with other serious complications such as sepsis, high output fistula and wound dehiscence which are not manageable at home. The actual number of new cases needing HPN arising annually is, as yet, difficult to calculate, but our initial estimate for the UK is 4 million. It is anticipated that 50% of these will only require temporary treatment leaving an annual accumulation rate of 2 million requiring permanent treatment.

The success of HPN depends upon suitable case selection, and meticulous patient training by experienced and highly skilled nursing staff. Additional support is required from the pharmaceutical services who must be used to the requirements of this type of treatment. The patient is more likely to receive this standard of service if the staff he relies upon have considerable experience in the treatment of intestinal failure.

Parenteral nutrition on a short-term basis is, and should remain, a service available in every district general hospital. However, there is a strong case for centralizing those few cases of severe intestinal failure, e.g. catastrophic short bowel, multiple high output fistulae, intestinal motility disorders and extensive Crohn's disease, that require skilled, intensive and prolonged treatment. The pattern already exists with renal failure units. The time has come to set up a limited number of 'Intestinal Failure' or 'Nutrition' units to deal with such cases. Such units would have the staff and facilities as well as the experience to ensure that longterm parenteral nutrition was administered appropriately and successfully. Additionally, in the UK such units would help solve the problem of financing HPN by being eligible for central funding.

References

1 Fleming, C. R. and Margot, R. (1981). Intestinal failure. In Hill, G. L. (ed.) *Nutrition and the Surgical Patient. Clinical Surgery International*, 2. pp. 219−235. (Edinburgh: Churchill Livingstone)
2 Scribner, B. H., Cole, J. J., Christopher, T. G., Vizzo, J. E., Atkins, R. E. and Blagg, C. R. (1970). Long term total parenteral nutrition. The concept of an artificial gut. *J. Am. Med. Assoc.*, **212**, 457

3 Broviac, J. W. and Scribner, B. H. (1974). Prolonged parenteral nutrition in the home. *Surg. Gynecol. Obstet.*, **139**, 24

4 Broviac, J. W., Cole, J. J. and Scribner, B. H. (1973). A silicone rubber atrial catheter for prolonged parenteral alimentation. *Surg. Gynecol. Obstet.*, **136**, 602

5 Solassol, Cl., Joyeux, H., Etco, L., Pujol, H. and Romieu, Cl. (1973). New techniques for long-term intravenous feeding: an artificial gut in 75 patients. *Ann. Surg.*, **179**, 529

6 Powell-Tuck, J., Nielsen, R., Farwell, J. A. and Lennard-Jones, J. E. (1978). Team approach to long-term intravenous feeding in patients with gastrointestinal disorders. *Lancet*, **2**, 825

7 Ladefoged, K. and Jarnum, S. (1978). Long term parenteral nutrition. *Br. Med. J.*, **2**, 262

8 Milewski, P. J., Gross, E., Holbrook, I., Clarke, C., Turnberg, L. A. and Irving, M. H. (1980). Parenteral nutrition at home in the management of intestinal failure. *Br. Med. J.*, **280**, 1356

9 Keighley, B. D. and MacGregor, A. R. (1980). Total parenteral nutrition at home: the implications for rural practice. *J. R. Coll. Gen. Pract.*, **215**, 354

10 Home parenteral nutrition in England and Wales – occasional review. *Br. Med. J.*, **281**, 1407

11 Rault, R. M. J. and Scribner, B. H. (1977). Parenteral nutrition in the home. In Jerzy, G. B. (ed.) *Progress in Gastroenterology*. Volume III, pp. 545–562. (New York: Glass, Grune and Stratton)

12 Linos, D. A. and Mucha, P. (1982). A simplified technique for the replacement of permanent central venous catheters. *Surg. Gynecol. Obstet.*, **154**, 248

13 Jeejeebhoy, K. N., Langer, B., Tsallas, G., Chu, R. C., Kuksis, A. and Anderson, G. B. (1976). Total parenteral nutrition at home: studies in patients surviving 4 months to 5 years. *Gastroenterology*, **6**, 954

14 Dudrick, S. J., Englert, D. M., van Buren, C. T., Rowlands, B. J. and MacFadyen, B. V. (1979). New concepts of ambulatory home hyperalimentation. *J. Parent. Enteral Nutr.*, **3**, 72

15 Fleming, C. R., Witzke, D. J. and Beart, R. W. (1980). Catheter-related complications in patients receiving home parenteral nutrition. *Ann. Surg.*, **192**, 593

16 Shike, M., Roulet, M., Kurian, R., Whitwell, J., Stewart, S. and Jeejeebhoy, K. N. (1981). Copper metabolism and requirements in total parenteral nutrition. *Gastroenterology*, **81**, 290

17 Shike, M., Sturtridge, W. E., Harrison, J. E., Tam, C. S., Bobechko, P. E., Jones, G., Murray, T. M. and Jeejeebhoy, K. N. (1980). Metabolic bone disease in patients receiving long term total parenteral nutrition. *Ann. Intern. Med.*, **92**, 343

18 Perl, M., Peterson, L. G. and Dudrick, S. J. (1981). Psychiatric problems encountered during intravenous nutrition. In Hill, G. L. (ed.) *Nutrition and the Surgical Patient. Clinical Surgery International 2.* pp. 309–318. (Edinburgh: Churchill Livingstone)

19 Rault, R. M. J. and Scribner, B. H. (1977). Treatment of Crohn's disease with home parenteral nutrition. *Gastroenterology*, **72**, 1249

20 Byrne, W. J., Ament, M. E., Burke, M. and Fonkalsrud, E. (1979). Home parenteral nutrition. *Surg. Gynecol. Obstet.*, **149**, 593

21 Irving, M. H. (1982). The United Kingdom Home Parenteral Nutrition Register. *Gut*, **23**, A438

2
Implementation of a home nutritional support programme

J. P. GRANT AND M. S. CURTAS

INTRODUCTION

Many hospitals have recognized significant clinical benefit from a nutritional support programme to assist in the metabolic management of patients unable to ingest an adequate oral diet. Such programmes have been characterized by a multidisciplinary team approach using the expertise of dietitians, pharmacists, nurses and physicians to deliver optimal care. As our ability to sustain life in the hospital improved through new methods of treatment, new surgical approaches, and aggressive nutritional support, it became apparent that the provision of home-care for the compromised but not hospital-dependent patient was necessary[1-4]. Experience has shown that as a hospital's in-patient nutritional support programme expands, the need for a home nutritional programme grows. Support by industry in meeting the needs of home nutrition programmes has been enthusiastic and effective in providing ample materials.

In spite of demonstrated need and proven effectiveness of home nutritional support programmes, little financial support is available and compensation for service is often poor. Home nutritional support programmes have therefore grown out of existing in-patient programmes with dedicated personnel giving freely of their time and expertise. Each member of the multidisciplinary team is essential for an effective home programme and if participation of one or more is lacking, it may be prudent to refer candidates to other established centres. On the other hand, with some additional effort a well-organized nutritional support programme staffed by competent and conscientious individuals can provide effective home therapy with minimal risks for the patient.

RECRUITMENT OF PERSONNEL FOR A HOME SUPPORT PROGRAMME

Although personnel to run a home programme are usually available from the in-hospital programme as the home programme grows, additional individuals may be necessary. Expertise from a core of individuals is needed including a physician, nurse, pharmacist, and dietitian, and in contrast to the in-patient programmes, the service of a psychiatrist is often quite valuable for the home patient[5]. If available, the contributions of a secretary, laboratory technician, social worker, physical therapist, and administrator can be helpful as well. Each member of this team must be dedicated to home health care delivery, maintaining up-to-date knowledge of the clinical practice of nutritional support and being willing to provide 24 h continuous patient service.

Over the past several years, with the rising number of patients receiving home nutrition, private and industrial health care agencies have been formed to assist in delivery of home-care. In addition to providing slides and other audiovisual aids for patient teaching, delivery of fluids and dressings to the patient's home, and providing and maintaining i.v. pumps, many of these agencies also provide nurses and pharmacists trained in the delivery of home nutrition. These individuals make home visits and telephone interviews and, in some cases, assist in training the patient in hospital, thereby supplementing the home support programme personnel. With a little extra effort, services of local physicians, pharmacists and visiting nurses can assist in delivery of home nutrition and their participation should be encouraged. Finally, the patient's immediate family is an invaluable resource and should be trained in all aspects of home-care to support the patient. With respect to all the numerous additional personnel available, emphasis must still be placed on maintaining control of the home nutrition programme by the hospital team to assure adherence to planned protocols and standard monitoring with every effort made to minimize complications and improve benefits.

PROGRAMME FACILITIES

Most home nutrition programmes are based within a hospital. Patient training and catheter placement usually comes during a period in the hospital although occasionally patients are referred just for home training. Minimal extra facilities are needed. If the bedside atmosphere is inappropriate for teaching, a classroom or conference room can be used. Some programmes utilize an area in the pharmacy to train patients in sterile techniques and a physical therapy facility is helpful to establish patient independence.

Out-patient follow-up can be done in an office or hospital clinic. Other than a conference and examination room and laboratory facilities little else is needed. The clinic or office should be accessible to all personnel so they may be available when the patient returns for follow-up.

Communication between the in-hospital and home nutrition teams is essential to provide optimal care. Although not necessary, shared office space as well as frequent patient conferences between both groups can help to maintain good communication.

DEVELOPMENT OF A PROTOCOL FOR HOME NUTRITIONAL SUPPORT

Of prime importance in implementing a home nutritional support programme is a detailed protocol for patient care. After careful review by the core personnel, the protocol should be submitted for review and approval by appropriate hospital committees and the hospital administration to assure their full co-operation and support. The protocol should include all of the following as well as any special areas of individual hospital concern.

Statement of objectives
An estimate of expected utilization of a home nutrition programme should be made to justify its organization. In our experience, 1 in 15 patients treated in the hospital has been sent home on intravenous or enteral support, but this ratio will vary depending on the patient profile. Objectives of establishing a programme may include standardization of treatment, optimization of patient care using a multidisciplinary approach, and assurance of continuous review to minimize complications and cost.

Home nutrition programme personnel
A complete listing should be made of all personnel, full and part-time, who will participate. These will include at a minimum a physician, nurse, dietitian and pharmacist, and, depending on the programme, a social worker, psychiatrist, administrator, secretary and laboratory technician. A full listing of the responsibilities for each individual should be given, dividing the work load according to priorities of local custom. Some centre the programme about a nurse clinician, a pharmacist[6], or the physician. However the programme is organized, a clear statement of each individual's responsibilities will give the best results. In addition, if members are drawn from other departments, the primary supervisor should be the physician in charge of the home programme rather than various department directors.

Criteria for home patient selection
Although a complete listing of criteria for patient selection cannot be given as indications will likely change with time, a general discussion should be given in the protocol[7,8]. In general, patients with significant organ failure, mental incompetency, terminal illness, or lack of strong family support are poor candidates for home nutrition. On the other hand, active individuals in good health apart from alimentary failure are good subjects. Most

385

candidates, however, fall between these extremes and a judgement as to appropriateness and feasibility of home support must be made. Emphasis on a team evaluation should be made and a psychiatrist may be helpful in decision making. Constant review of the literature is necessary to evaluate indications for treatment in specific diagnostic categories such as cancer and inflammatory bowel disease.

In addition to selecting candidates for a home nutrition programme, the best route of providing this support must also be considered. Expertise in both the enteral and parenteral systems is essential and whenever possible the least expensive, most effective method should be used. The patient's degree of anticipated compliance and ability, as well as their disease state, must be taken into consideration before selecting the appropriate method of providing home nutritional support.

Standardization of vascular and intestinal access and care
To assure maximal safety of various access techniques the protocol should carefully outline placement and maintenance procedures for the various catheters used in home support. Ongoing evaluation of new products by the team is necessary to continue improved patient care.

Development of a detailed Patient Training Manual
The home nutritional support protocol should include a Patient Training Manual basic enough for the layman to understand, yet thorough enough to cover all aspects of home support. A well-designed and clinically useful manual is available from Travenol entitled the 'Travacare Services Home Nutritional Patient Training Manual'. Other private agencies have similar manuals[9] all dealing with home parenteral nutrition principles. Similar manuals for enteral nutrition need to be developed. In general, these manuals cover aseptic technique, catheter dressing care, heparin lock and catheter access, preparation of the nutrition solution, use of infusion pumps, recognition of complications, and catheter repair. Enteral nutrition manuals should cover tube care, feeding solution preparation, infusion pump use, and recognition of complications.

Within this part of the protocol a method to assess adequate patient training should be outlined – either by written or practical testing or both.

Out-patient monitoring
The protocol should recommend procedures for routine out-patient monitoring such as frequency of visits, tests to be performed, and data to be recorded. The needs of individual patients may vary but a core of data should be standard to document effectiveness and safety.

Record keeping, review and periodic analysis
Of utmost importance is an explicit section on data recording and analysis.

Periodic review of performance with subsequent protocol revision as indicated is essential to maintain optimal patient care. These records should include, as a minimum, patient diagnosis, type of nutritional support, specific formula, pertinent laboratory data, nutritional assessment data, complications, proposed goals and therapeutic results.

PATIENT FUNDING AND PRODUCT SUPPLY IN THE USA

Funding of home nutritional support remains a significant problem in the USA. Patients possessing a Major Medical clause in a Blue Cross Blue Shield policy are now fully covered in most states for costs of home intravenous feeding. Many other insurance agencies are approving similar coverage on an individual basis upon written request and with a physician statement of necessity. In 1977 Medicare/Medicaid agreed to assume the cost of supplies and pharmaceuticals for home intravenous nutrition in patients over 65 or otherwise who qualified for Medicaid disability. Medicaid approval, however, requires a disability statement for what might otherwise be an employable patient and only becomes effective after a 2 year delay. Individual states vary in their ability to pay and most apply local adjustments before payment. During the 2 year delay, and for patients with no insurance coverage, the cost of home intravenous nutrition may prohibit support.

Funding for enteral home nutrition is even less well established. Although some insurance companies and Medicare/Medicaid will at times fund an infusion pump, most feel purchase of the enteral feeding solution is the patient's responsibility.

Some home support programmes have acquired an administrator and seek patient insurance coverage on their own. Most, however, have turned this difficult and time-consuming task over to private or industrial agencies such as the Travacare Home Nutrition Supply Management Service. In this case, a telephone call by the individual support service personnel to the agency relating the type of insurance coverage is all that is necessary. The agency will then either approve or disapprove funding based on prior experience or take 24–48 h to contact the involved insurance company to secure funding information. All aspects of patient billing and collection are assumed by the agency relieving a great burden from the patient, hospital, or home nutritional care programme.

Provision of supplies for home nutritional support in most cases is through a private or industrial home nutritional support agency. These agencies can deliver all prescribed products to the patient's home providing either pre-mixed or stock solutions as well as catheter care items, supplemental drugs, and infusion pumps and tubing. Occasionally these arrangements are made by a hospital pharmacy and rarely through a local pharmacy. Catheter care materials, feeding solutions, and pumps for enteral home feeding are also available for selected patients.

CONCLUSIONS

The implementation of a home nutritional care programme cannot be undertaken lightly. Although home support offers many benefits to the patient, improperly supervised support can lead to serious complications. Careful attention to design and operation of a home-care programme will allow safe and effective therapy with a minimum of inconvenience and optimal effectiveness.

References

1 Byrne, W. J., Ament, M. E., Burke, M. and Fonkalsrud, E. (1979). Home parenteral nutrition. *Surg. Gynecol. Obstet.*, 149, 593
2 Rault, R. M. and Scribner, B. H. (1977). Treatment of Crohn's disease with home parenteral nutrition. *Gastroenterology*, 72, 1249
3 Grundfest, S., Steiger, E., Settler, L., Washko, S. and Wateska, L. (1979). The current status of home total parenteral nutrition. *Artif. Organs*, 3, 156
4 Dudrick, S. J., Englert, D. M., Van Buren, C. T., Rowlands, B. J. and MacFayden, B. V. (1979). New concepts of ambulatory home hyperalimentation. *J. Parent. Enteral Nutr.*, 3, 72
5 Gulledge, A. D., Gipson, W. T., Steiger, E., Haoley, R. and Srp, F. (1980). Short bowel syndrome and psychological issues for home parenteral nutrition. *Gen. Hosp. Psychiatry*, 2, 271
6 Gaffron, R. E., Fleming, C. R., Berkner, S., McCallum, D., Schwartau, N. and McGill, D. (1980). Organization and operation of a home parenteral nutrition program with emphasis on the pharmacist's role. *Mayo Clin. Proc.*, 55, 94
7 Steiger, E. (ed) (1981). *Home Parenteral Nutrition.* (Washington, DC: ASPEN)
8 Fleeman, C. M. and Wright, R. A. (1981). Concepts of home parenteral nutrition. *Nutr. Support Services*, 1, 16
9 Sattler, L., Wateska, L. P., Siska, B., Washko, S., Myers, K. D., McKeown, L. and Steiger, E. (1978). *Cleveland Clinic Foundation Home TPN Manual.* (Irvine, California: McGaw Laboratories)

3
Techniques of longterm venous access

I. W. GOLDFARB

INTRODUCTION

The 1980s will be regarded as the decade during which home patient services emerged as a practical alternative to prolonged hospitalization for many patients. Early experiences with venous access for renal dialysis and hyperalimentation have provided a foundation upon which other efforts have been based. This has resulted in an increased capability for longterm venous support for a wider spectrum of problems. We have learned that longterm out-patient venous access can be accomplished safely without the many complications usually associated with this procedure. This has enabled us to eliminate the venous catheter tether which required the patient's confinement in order to achieve adequate therapy. Now, patients receiving nutritional support, chemotherapy, antibiotics and many other parenteral forms of medication can be treated safely on an out-patient basis through indwelling venous lines. Thus, therapy designed to extend life need no longer interfere with the quality of life.

Many of the early techniques of longterm venous access were designed to provide a route for nutritional support solutions. In fact, much of our knowledge about catheter placement, catheter material and catheter maintenance has come from the need for longterm access. Technical complications of insertion and the associated complications of longterm placement have provided the stimulus for the re-evaluation of techniques, materials and basic concepts. This re-evaluation has meant that there is now one catheter system that can be modified to fit many needs with a minimum risk of complications. The end result is a system of access that not only provides for safe in-house parenteral therapy but one that may also be used on an out-patient basis for longterm support.

CATHETER COMPOSITION AND TECHNIQUES OF INSERTION

The introduction of plastic catheters significantly altered intravenous therapy. These materials (polyvinyl chloride, polytetrafluroethylene) resulted in significantly increased mean infusion time. Unfortunately, these materials were also associated with an increased incidence of venous complications. Numerous publications compared the infusion times and associated complications of each of these materials[1,2]. Concepts such as fibrin sleeve formation, intimal damage and thrombus formation were formulated and associated with shortened infusion times, catheter sepsis and direct venous complications[3,4]. Ultimately, one material – silicone – was identified as 'ideal' as it had the lowest incidence of untoward complications and the longest infusion times. As the use of this material gained in popularity it was incorporated into special catheters designed for longterm out-patient venous access (e.g. Hickman, Broviac)[5,6]. These catheters, composed of silicone with various Dacron attachments or sheaths were designed for surgical implantation using a tunnelled approach. During the past 10 years thay have served as the most common means of administering nutritional support solutions and/or chemotherapeutic drugs to patients with problems of longterm access. Conceptually, the advantage of using a tunnel approach is based upon the theoretical association between catheter sepsis and migration of surface bacteria along an insertion tract. Thus, any technique that increases the anatomical distance between the insertion site and the circulation may well reduce the incidence of catheter-related sepsis. Recent evidence, however, seems to dispute this concept[7]. In addition, the

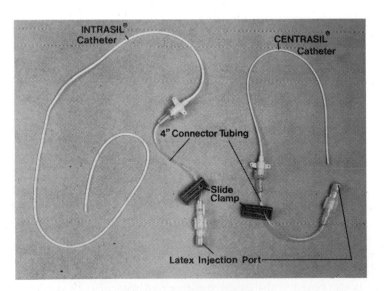

Figure 32.1 The catheter system

creation of the tunnel presents some practical shortcomings and limitations. During the past 3 years many institutions have gained experience in using silicone lines for longterm access without a tunnel approach[8,9]. The technique has proved attractive because of the 'flexibility' it offers with respect to the management of technical problems or catheter-related sepsis. Surgical implantation is not required and catheter exchanges may be performed as the need arises. This capability is particularly attractive when the question of catheter-induced sepsis occurs[10]. A trial catheter exchange is a viable alternative to the surgical removal of an 'implanted catheter' that may in fact not be the source of sepsis. The catheter system consists of a 16 gauge silicone elastomer catheter, a 16 gauge 10 cm connector tubing with Luerlock connections and slide clamp, and a Luerlock injection port (Figure 32.1). Once inserted, the site is covered with a selectively permeable, bacteriostatic dressing. The catheter is inserted into the subclavian vein using either the infraclavicular approach or by a basilic antecubital cut-down approach. The insertion is performed at the bedside[11].

(1) The necessary equipment is assembled.

(2) With the bed in the Trendelenburg position a rolled blanket is placed under the patient at the scapular level to facilitate posterior movement of the shoulder and head. The head is directed to the left.

(3) The skin is defatted with 10% acetone solution.

(4) The physician performing the insertion wears sterile gloves.

(5) The skin is prepared with Betadine solution.

(6) The area is draped with sterile towels.

(7) The skin is infiltrated using 1% lidocaine (without epinephrine) to achieve local anaesthesia. In the case of the infraclavicular approach to the subclavian vein, the skin at the site of the junction of the medial third and lateral third of the clavicle is the point of insertion. It is important to note that central venous catheterization should always be performed on the right, when possible, in order to avoid injuries to the thoracic duct.

(8) The subclavian vein is localized using a 3 cc syringe and a 22 gauge 3.8 cm needle. The needle is inserted at a 30° angle from the chest wall at approximately the junction between the medial and lateral third of the clavicle. 'Hugging' the inferior aspect of the clavicle and aiming for the sternal notch, the needle is slowly advanced, occasionally aspirating until venous blood is obtained.

(9) After localization, the needle is removed and the Centrasil® Teflon catheter introducer and needle are inserted along the same tract

(Figure 32.2). At this point, the procedure for insertion differs some-what from the typical use of this catheter.

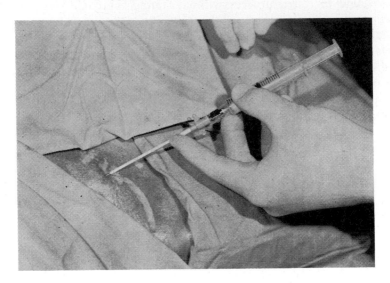

Figure 32.2   Cannulization of subclavian vein

(10)   A 0.025 cardiovascular J-wire is inserted through the introducer after having instructed the patient to inspire and hold his breath. Only about 25 cm of the J-wire is actually introduced through the Teflon sheath.

(11)   The Teflon sheath is then withdrawn over the J-wire.

(12)   The silicone elastomer catheter is then threaded over the J-wire to the site of insertion (Figure 32.3).

(13)   The silicone catheter and J-wire are 'squeezed together' and advanced together as a total unit.

(14)   When the silicone catheter has been advanced all the way to its hub, the patient is asked to hold his breath and the J-wire is removed from the catheter.

(15)   The catheter hub is then secured to the skin, using one suture placed around the reinforced point where the silicone elastomer is attached to the hub. 'Stay sutures' are placed through the eye holes on the catheter wings. It is important that these sutures be placed in a cephalad fashion in order to decrease piston motion. Ethilon is the suture material of choice because it leads to less skin reaction. The pre-flushed i.v. tubing with its 10 cm connector tubing is then attached to the catheter hub.

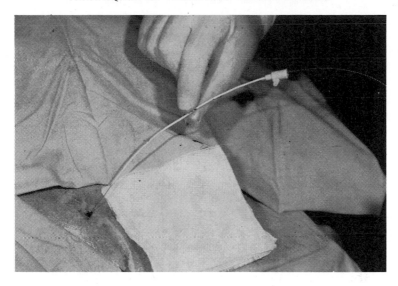

Figure 32.3 Silicone catheter is threaded over J-wire

(16) The i.v. bag is lowered below the level of the bed in order to again demonstrate adequate blood return.

(17) The insertion site is dressed.

(18) Chest auscultation is performed and an X-ray is obtained to confirm the position of the catheter (Figure 32.4).

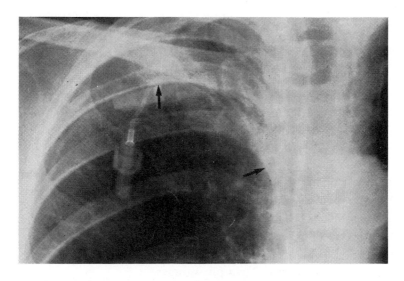

Figure 32.4 Confirmation of placement by chest X-ray

In some patients, particularly those with oncotic problems and thrombo-cytopenia, the infraclavicular percutaneous approach to the subclavian vein may be considered too hazardous. In this particular group of patients long-term silicone central venous catheterization can be obtained by the use of the basilic vein. In this group of patients we have elected to use a cut-down approach to the basilic vein approximately 5.0 cm above the elbow. Specifically, the cut-down site is along the medial aspect of the upper arm at the point where the basilic arterial pulse is palpable. When the basilic vein has been identified the catheter is delivered to the venotomy site by a separate 'stab wound'. Wound closure is routine. It is particularly important with this technique to limit the risk of phlebitis by using prewashed sterile gloves. Silicone elastomer catheters have a high affinity for particulate matter and glove talc can contribute significantly to the incidence of phlebitis. Following insertion a chest X-ray is obtained in order to confirm proper catheter place-ment. The same 10 cm Luerlock connector tubing, latex injection port and dressing are utilized.

While this particular route of catheter insertion is not as desirable as the direct infraclavicular approach to the subclavian vein, it does provide a viable alternative to catheterization in the patient with thrombocytopenia.

OUT-PATIENT CATHETER CARE

Immediately after insertion of the catheter the Nutritional and Metabolic Support Nurse begins the training programme which is designed to facilitate transition from hospital to home therapy. The nurse conducts an initial inter-view with the patient and his/her family to assess their ability to deal with the catheter system. The primary goal during this interview is to identify a 'designated family member' who will either assume the responsibility for or assist in the care of the catheter. Instruction is performed on a day-to-day basis in order to train the patient and/or the 'designated family member' in the techniques of catheter dressing changes and heparin instillation. The actual components of the out-patient dressing kit and the technique utilized in performing a dressing change are as follows[11]:

(1) Two pairs of sterile gloves.
(2) One alcohol impregnated swab.
(3) Two betadine solution impregnated swabs.
(4) One tincture of benzoin impregnated swab.
(5) One packet of betadine ointment.
(6) Two 5 × 5 cm 'all gauze' two-ply sponges.
(7) Op-site® (6 × 8.5 cm or 10 × 14 cm).

DRESSING CHANGE TECHNIQUE

(1) The dressing kit is opened and the first pair of sterile gloves is applied prior to the initiation of the dressing change. The old dressing is then removed and discarded along with the first pair of gloves.

(2) The second pair of gloves is then donned. This second pair of gloves is pre-washed because of the affinity of silicone for particulate matter and the attendant increased incidence of phlebitis. Each individual swab packet is opened and handed to the patient without touching the patient's ungloved hand. The 5 × 5 cm gauze is prepared by placing a small amount of betadine ointment in the centre of the gauze.

(3) The catheter insertion site and surrounding skin are then cleaned with the alcohol impregnated swab in order to defat skin. Alcohol is used in place of acetone to diminish skin irritation and prevent deterioration of the catheter. The cleaning procedure is started at the point of catheter insertion and then worked outward in a segmental fashion.

(4) This procedure is repeated using the betadine impregnated swab.

(5) The benzoin impregnated swab is then used to protect the surface of the skin and to enhance adhesiveness.

(6) The 5 × 5 cm gauze with its applied betadine ointment is then placed over the catheter insertion site and another 5 × 5 cm gauze is placed under the catheter hub.

(7) The Op-site dressing is then applied to the area.

(8) Heparin (10 units) is then instilled into the line (Figure 32.5).

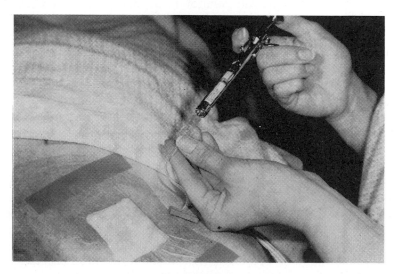

Figure 32.5 Instillation of heparin

The patient is instructed to keep accurate records with respect to temperature, skin conditions and stability of suture materials. Problems detected during dressing changes are reported directly to the Nutritional and Metabolic Support Nurse so that immediate appropriate action can be taken. The patients are also trained to change the heparin injection port every 2 weeks. The slide clamp which is located on the 10 cm connector tubing is used to prevent air embolus during this procedure. At no time is the patient or designated family member permitted to change the 10 cm connector tubing. This is done on a monthly or as-needed basis by the Nutritional and Metabolic Support Nurse in the out-patient clinic.

The designated family member and in some cases the patients themselves are trained in the technique of catheter heparinization at a fixed time each day. Maintenance of central line patency depends on daily heparinization. The actual technique involves the instillation of 10 units of heparin in 1.0 ml through the catheter injection port. It must be emphasized that the injection of heparin should not be forced because of the risk of catheter rupture. Family members are instructed that an inability to instil the heparin easily is an indication of possible thrombus formation. The Nutritional and Metabolic Support Nurse should be notified immediately. The catheter can frequently be salvaged by the use of urokinase.

CATHETER EXCHANGE

In many situations, particularly those involving a febrile response, the catheter is suspected as the source of sepsis. While the technique of documenting a catheter-related sepsis is well-known, it does not provide us with a practical and immediate solution to the management of a suspected catheter sepsis. These patients may well be febrile due to other causes. Catheter exchange is attractive in these situations as it still provides for venous access in a group of patients in whom additional cannulation may be hazardous. This catheter system readily lends itself to catheter exchange over a 0.025 J-wire without the need for repeat needle cannulization. It should be emphasized that catheter exchange is performed only when there is clinical evidence which implicates another source of the febrile response. The subsequent identification of positive blood cultures and matching organisms obtained from cultures of the tip of the exchanged catheter are an indication for catheter removal. This can usually be identified within 24 h of exchange. Catheter exchanges can also be performed if specific mechanical problems develop with an indwelling catheter. The most common problem is a catheter leak due to a disruption in catheter integrity. Patients are instructed to contact the Nutritional and Metabolic Support Nurse immediately if there is any indication of catheter leakage. Inability to instil heparin or to draw blood back through the catheter are also indications for catheter exchange. This exchange may be performed on an out-patient basis without the need for admission and subsequent surgical removal of a tunnelled catheter system. We have had no difficulty in performing catheter exchanges on catheters inserted percutaneously by the infraclavicular approach to the subclavian vein. Unfortunately, we have not been able to perform catheter exchanges on those catheters inserted via the basilic antecubital cut-down approach.

THE WESTERN PENNSYLVANIA HOSPITAL EXPERIENCE

Since the initiation of the utilization of this catheter system, 175 patients have undergone longterm venous access (>30 days), but only 79 patients have been followed on an out-patient basis. 78.8% of these patients underwent a direct infraclavicular approach to the subclavian vein while 21.2% of the lines were inserted by an antecubital cut-down approach. Mean infusion times (MIT) were 132.48 days and 151.90 days respectively (Table 32.1). In most cases lines were removed either because of the cessation of chemotherapy and nutritional support or because of death due to underlying disease. Line exchanges were required in 20.5% of the patients. The indications for line exchanges are presented in Table 32.2. Lines were exchanged or removed in any patient who developed a febrile response or manifested positive blood cultures. While eight patients underwent line exchange for this reason only four were eventually identified as having a definite catheter-related sepsis, again emphasizing the propensity for non-catheter related fever in any cancer population. Six patients underwent line exchanges for technical complications. The majority of these complications occurred during the early phases

Table 31.1   Out-patient catheter data

Total patients 79
Total lines   99

	Direct subclavian	Antecubital cut-down
Number	78	21
Days	10 334.00	3190.00
MIT	132.48	151.90
Exchanges	16	0

Table 32.2   Indications for exchange

Fever/Positive blood culture	10.3%	(8/78)*
Technical	7.7%	(6/78)
Occlusion	2.5%	(2/78) †

*4 confirmed catheter sepsis, 1 abuse (3/78 = 3.8%)
†4 catheters developed occlusion
2 were declotted by the use of urokinase

397

of our programme. They are a reflection of the learning curve for the Nursing Personnel and members of the Nutritional Support Team. The incidence of these complications has decreased significantly as experience has been gained in the utilization of this catheter system. Four patients developed line occlusion but two were 'declotted' using urokinase, thus necessitating only two line exchanges for thrombus formation.

It is interesting to note the long mean infusion times that were obtained with the antecubital cut-down approach. These results are significantly different from the infusion times reported by other investigators[8,9,12]. We can only conclude that this discrepancy may be due to the fact that all patients in our study population had insertion performed by a cut-down approach into the deep basilic system as opposed to a percutaneous superficial approach. There was a high incidence of transient 'phlebitis' (23.8%) which occurred within 72 h of insertion. 9.5% of these responded to topical therapy (warm compresses). The remainder persisted and necessitated catheter removal. The phlebitis appears to be of a non-bacterial aetiology as no positive cultures were obtained in these patients and none became febrile. Three other patients (14.3%) with this type of line ultimately developed positive catheter-related blood cultures. We do not feel that any definite conclusion should be made with respect to the antecubital approach on the basis of this data except to say that it is a viable alternative to longterm access in the thrombocytopenic patient or the patient in whom the infraclavicular approach is not feasible for other reasons (e.g. radiation, skin exfoliation). While these 'silicone long arm catheters' did have a high mean infusion time, they must be followed carefully.

FUTURE CONSIDERATIONS

The field of out-patient venous therapy remains an active area of technological advancement and clinical investigation. Newer catheter systems are continually being developed and evaluated in order to improve upon mean infusion times, biocompatibility and practical utilization[13,14]. Recently, double lumen central venous catheters have been developed and utilized in various clinical settings[15]. These newer catheter systems may well increase therapeutic capabilities without significantly hampering out-patient lifestyle. Specifically, the capability of infusing various solutions in conjunction with blood products, antibiotics or other therapeutic agents in the oncology patient is quite attractive.

New infusion devices also hold promise for providing for the maintenance of out-patient therapy without significantly interfering with day-to-day life. In the past, out-patient therapy has required pumps to accurately provide for the continuous infusion of drugs. Experience with a 24 h infusion device which is devoid of both a motor drive unit and an electrical component has been most gratifying. This device is currently being provided to our patients in a pre-filled fashion for the administration of out-patient chemotherapeutic agents. It attaches readily to the catheter system we have described and can be worn under the patient's clothing thus allowing maximum mobility. While we are presently restricting the use of this device for the administration of chemotherapeutic agents, we are interested in evaluating its value as a means of administering such other drugs as insulin, antibiotics and analgesics.

References

1 Welch, G. W., McKeel, D. W., Silverstein, P. *et al.* (1972). The role of catheter composition in the development of thrombosis. *Surg. Gynecol. Obstet.*, **138**, 421

2 Still, R. M., Saliman, F., Garcia, L. *et al.* (1977). Etiology of catheter associated sepsis: correlation with thrombogenicity. *Arch. Surg.*, **112**, 1497

3 Hoshal, V. L., Anuse, R. G. and Hoshins, P. A. (1971). Fibrin sleeve formation on indwelling subclavian central venous catheters. *Arch. Surg.*, **102**, 353

4 Hoshal, V. L. (1975). Total intravenous nutrition with peripherally inserted silastic elastomer central venous catheters. *Arch. Surg.*, **110**, 644

5 Broviac, J. W., Cole, J. J. and Scribner, B. H. (1973). A silicone rubber atrial catheter for prolonged parenteral alimentation. *Surg. Gynecol. Obstet.*, **136**, 602

6 Broviac, J. W. and Scribner, B. H. (1974). Prolonged parenteral nutrition in the home. *Surg. Gynecol. Obstet.*, **139**, 24

7 Meyenfeltd, M. M., Stapert, J., DeJong, P. C. *et al.* (1980). TPN catheter sepsis: lack of effect of subcutaneous tunnelling of PCV catheters on sepsis rate. *J. Parent. Enteral Nutr.*, **4**, 514

8 Bottino, J., McCredie, K. B., Groschel, D. H. and Lawson, M. (1979). Longterm intravenous therapy with peripherally inserted silicone elastomer central venous catheters in patients with malignant diseases. *Cancer*, **43**, 1979

9 Lawson, M., Bottino, J. C. and McCredie, K. B. (1979). Long-term IV therapy: a new approach. *Am. J. Nurs.*, **79**, 1100

10 Maher, M. M., Henderson, D. K. and Brennan, M. F. (1982). Central venous catheter exchange in cancer patients during total parenteral nutrition. *J. Nat. Intravenous Ther. Soc.*, **5**, 54

11 Breggar, F., Kim, P., DeCourcy, M. and Cavalier, A. (1981). Vascular access, catheter care and nursing implications. *Am. J. Intravenous Ther. Clin. Nutr.*, **8**, 41

12 Geis, A. C., Flanagan, S. J. and Grossman, A. (1979). Evaluation of 75 patients with the long arm silastic catheter. *J. Parent. Enteral Nutr.*, **3**, 462

13 Vandersalm, T. J. and Fitzpatrick, G. F. (1981). New technique for placement of long-term venous catheters. *J. Parent. Enteral Nutr.*, **5**, 326

14 Clift, R. A., Sanders, J. E., Stewart, P. and Thomas, E. D. (1979). A modified right atrial catheter for access to the venous system in marrow transplant recipients. *Surg. Gynecol. Obstet.*, **148**, 871

15 Sanders, J. E., Hickman, R. O., Aker, S., Hersman, J. *et al.* (1982). Experience with double lumen right atrial catheters. *J. Parent. Enteral Nutr.*, **6**, 95

4
Home parenteral nutrition

E. STEIGER

Total Parenteral Nutrition (TPN) or Intravenous Hyperalimentation (IVH) has been used in hospital patients for 15 years. It became increasingly apparent to institutions that used hyperalimentation, that there was a group of patients who required intravenous feeding at home if they were ever to leave hospital. The concept of an artificial gut system for patients became the basis for hyperalimentation at home. In the early 1970s, Joyeaux and Solassol in France, Scribner in the United States, and Jeejeebhoy in Canada pioneered the use of intravenous feeding at home for the patient with GI tract failure. Since that time, Home Parenteral Nutrition (HPN) has been used with increasing frequency in many large medical centres throughout the world.

INDICATIONS

The largest group of patients who require home hyperalimentation are those with extensive Crohn's disease who have had multiple bowel resections or severe disease of their remaining intestinal tract or both. The next largest group of patients who require home parenteral nutrition are patients who have had an intra-abdominal malignancy, usually pelvic, treated by radiation. The resultant severe radiation enteritis that occurs in a small percentage of patients requires multiple resections or bypass of significant lengths of small intestine leaving the patient with functional short bowel syndrome. The third largest group of patients are those who have had a mesenteric infarction with loss of most of their gastrointestinal tract usually from the proximal jejunum to the mid-transverse colon. The fourth category would be miscellaneous causes of short bowel syndrome such as volvulus of the small bowel with resultant gangrene and resection, pseudo-obstruction, and scleroderma. A number of these patients had been treated before home hyperalimentation was initiated by repeated hospitalizations for nutritional rehabilitation with Total Parenteral Nutrition. These patients would be restored to normal nutritional status in the hospital and then sent home on a variety of diets and drugs to control diarrhoea, and would again eventually become dehydrated and malnourished requiring re-admission within several months or weeks of their discharge.

Various types of absorption studies, motility studies, and X-ray studies are difficult to use to identify the patient in need of parenteral nutrition at home. The most consistent guideline seems to be the patient's requirement for intra-venous fluids in order to achieve normal fluid balance despite all dietary manipulations and medications.

If the patient is able to maintain positive fluid balance or fluid equilibrium, then this patient can usually be managed by diet and medication. If, however, the patient is unable to achieve positive fluid balance or fluid equilibrium, then he is almost certainly a candidate for home parenteral nutrition.

VENOUS ACCESS

A method for obtaining prolonged safe venous access is important in assuring the success of a programme of Home Hyperalimentation. Arterio-venous fistula, Scribner shunts, and standard Teflon percutaneously placed subclavian catheters have been used with limited success in the past. The silastic tunnelled Hickman or Broviac/Scribner catheters have allowed for prolonged safe venous access in most centres.

Before placing these catheters, venograms are obtained of both subclavian veins in helping to make a decision as to which side should be used. Often patients who receive TPN in the hospital using the standard infraclavicular percutaneous subclavian venipuncture have radiological evidence of subclavian vein thrombosis. This may not be apparent clinically. In the operating room under local or general anaesthesia, the Broviac/Scribner or Hickman catheter is inserted usually by a cut-down over the cephalic vein and tunnelled subcutaneously to exit at the parasternal border below the level of the nipple. The exit point of the catheter is marked pre-operatively to be sure that the exit site is at a point where the patient can readily dress it. Antibiotics are given pre-operatively and for the first 24–48 h post-operatively. A standard chest X-ray is taken, to confirm the catheter position. A non-absorbable suture is placed at the catheter exit site and tied around the catheter to firmly anchor it. This suture is not removed for 1–2 months during which time the Dacron velour cuff on the catheter in the subcutaneous tunnel is enmasked in fibrous tissue firmly fixing it in place and preventing ascending infection.

HOME PARENTERAL NUTRITION TRAINING

The training period is begun the day after catheter placement in most patients and consists of teaching the patient how to care for the catheter exit site on a daily basis as well as instruction in heparinization of the catheter. This usually takes 3–5 days of training. The following 1½–2 weeks, the patient is trained by the pharmacist on how to mix the fluids and additives in the home setting. Other techniques that have to be learned include: caring for the pump, use of the pump, metabolic self-monitoring, keeping daily weight and urine output records, trouble-shooting, and emergency repair of the catheter should it become punctured, cut, or otherwise damaged. If the patient is going to mix his/her own solutions daily at home, the average training period is approximately 21 days in the hospital. If the patient is going to use solutions prepared by regional pharmacies or distributors of PN solutions, then the in-hospital training usually takes 7–15 days. In the early days of home hyperalimentation, self-mixing in the home situation was a necessity to maintain maximal patient freedom. The patients could travel wherever they wanted to as long as their solutions were taken with them or were waiting at their destination. The patient was not tied down to any institution or any geographical location. As home parenteral nutrition has become more widespread and there are more centres involved in patient training, and as national home parenteral nutrition services have expanded, a mixing service rendered by a regional pharmacy can allow the patients to have more time for their jobs, at their homes, or at leisure. The advantages of each system should be assessed by each institution in deciding what might be best for a particular patient. Other issues that relate to this decision include cost, the patient's ability to learn home-mixing techniques and also the patient's physical capabilities of doing fine motor work with his hands.

Decisions are also made as to whether the patient will be on continuous ambulatory hyperalimentation or on an overnight schedule where the solutions are infused while the patient sleeps. In the overnight schedule, the patient slows down the rate of infusion for the last half-hour in the morning, heparinizes the catheter, caps it off, and goes about his daily activities until he has to start the infusion again at bedtime. Considerations that might favour continuous infusions include: if the patient requires large volumes of fluid (over 4 l per day) or if the patient has congestive heart failure or difficulty in tolerating rapid fluid loads. However, if the patient has an aversion to carrying the fluid with him throughout the day, or if he prefers to do his work or home activities free from the i.v. solutions and wants his life interfered with least by the programme, an overnight infusion would seem to be most appropriate. During the training programme, once the patient's requirements for fluid volume, amino acid, calorie, and additive composition is determined, this fluid volume is given in decreasing amounts of time to get the patient down to an overnight infusion schedule. For the average patient requiring 2 l of hyperalimentation fluid each night, the infusion time is 6–8 h.

TRACE ELEMENTS AND VITAMINS

Besides the usual total parenteral nutrition additives, trace elements should be given. Recommendations for trace element additives in parenteral nutrition have been developed by the American Medical Association for copper, zinc, chromium, and manganese. Increasing reports of other trace element deficiency states including selenium and molybdenum are beginning to appear in the literature. Suggested requirements for vitamins in parenteral nutrition have also been published by the AMA. Recent work regarding the use of vitamin D in patients on hyperalimentation and a possible metabolic bone disease that develops in the home parenteral nutrition patient has stimulated much work into better defining the vitamin requirements. Newer multiple vitamin preparations do provide all of the patient's vitamin requirements, including biotin, except for vitamin K which should be given separately.

PREPARING FOR HOSPITAL DISCHARGE

Prior to discharge, the patient should be proficient in all that was taught to him/her and must be able to manage possible problem situations. Arrangements should be made regarding supplying the patient in the home setting and helping the patient to set up his house for initiation of a home parenteral nutrition programme. A member of the patient's family should also be present for at least 1 day prior to discharge to see all the elements of the programme in the hospital so that he can be supportive in the home setting. It should be stressed, however, that a patient should be made primarily responsible for his own HPN programme, and in no case should a family member who lives outside the home be allowed to take responsibility. Psychological issues should also be approached during the intensive in-hospital training programme, and include fear of disease, fear of death, costs and return to a healthy lifestyle. Patients who do best on home parenteral nutrition are those who have been malnourished for many years and who for the first time in years have felt relatively well. Those patients who find it most difficult to accept the programme, at least initially, are those who have been well until a sudden catastrophic event (such as mesenteric infarction) leads to home parenteral nutrition.

HOME MONITORING

Monitoring of the home parenteral nutrition patient is divided into two separate parts. The first part of monitoring is done by the patient at home. This includes recording his daily weight, urinary dextrose measurements, intravenous fluid intake, oral fluid intake and urine output. The patient is asked to contact his physician if his urine output is less than 500 ml/day on 2 consecutive days or if he develops new or increased swelling of the feet or ankles. A weight increase or decrease in excess of 4.5 kg is also reported to the monitoring physician. He is also advised to contact his physician if the urinary dextrose measurement is persistently 2+ or greater over a 24 h period.

The other monitoring is done by the home parenteral nutritional team on a monthly follow-up visit for the patient who has started home parenteral nutrition within the preceding year, or a visit every 6–8 weeks in the patients who have become well-established in their routine and who have been on HPN for more than a year. Electrolytes, BUN, creatinine, calcium, phosphorus, magnesium, sugar, cholesterol, liver function tests, uric acid, complete blood count, prothrombin time, and serum zinc measurements are done each month. Trace elements, essential fatty acids, and vitamin levels are measured every 6 months. At each monthly follow-up visit, the patient's intake and output chart is reviewed to be certain that adequate amounts of i.v. fluid are being given. As the remaining small intestine adapts and more absorption occurs, the urine outputs will become greater, and if over 2000 ml/day, the total parenteral nutrition programme can be decreased by 500 to 1000 ml/day. This is most easily done by omitting TPN infusions one night a week to give the patient a night off. If the patient is gaining too much weight and if there are no signs of excess fluid retention, then the carbohydrate content of the fluids can be reduced, by reducing the concentrations of dextrose.

EXPENSES

Costs of home parenteral nutrition are covered in the USA by Medicare and by most major insurance carriers. The recent establishment of proprietary companies to deliver equipment, supplies, maintain inventory for the patient, directly bill the patient, and help him with insurance programmes, has provided a great service to both the physician caring for the patient as well as the patient. Concerns regarding the costs of the programme and the patient's ability to meet these costs are paramount when first going on the programme. A great deal of expertise regarding insurance coverage or disability benefits is available to allay fiscal concerns. In any event, the average yearly cost of home parenteral nutrition management is far less than keeping the patient in the hospital, and the social benefits for the patient and his family of being at home are obvious.

COMPLICATIONS

There are many complications that could occur in the home parenteral nutrition patient's course, but these are usually limited to two types. The first is catheter complications. Catheter occlusion can occur, especially with the smaller lumen Broviac/Scribner catheters. The use of urokinase has been reported to be of value and in our own limited series has saved several occluded catheters from being removed and replaced. Catheter infection is one of the biggest reasons for catheter removal and one of the most important reasons for re-admitting patients into the hospital. This is usually easily treated by correctly diagnosing the problem and then by removing the catheter. Because venous access may be limited in some of these patients, it is important to be certain one is dealing with catheter infection before pulling out the catheter. With this in mind, the patient is usually admitted to hospital, and cultures are obtained through peripheral veins as well as through the catheter itself, and the catheter is then heparinized and capped off and not used. Results of the blood cultures are awaited before making a final decision to pull the catheter. If, however, the patient remains septic and febrile despite not using the catheter and no other source of fever can be found, then the catheter is removed even if the culture reports are not available. If an infected catheter is removed, the fever usually abates and subsequent blood cultures are negative. There is no need to treat the patient with systemic antibiotics if this happens. A new catheter can be inserted if the blood cultures remain negative for 5–7 days.

The other set of complications is metabolic. These include all abnormally high or low values for all the electrolyte, vitamin, and trace element replacements. Metabolic bone disease in home parenteral nutrition patients is receiving increasing interest, and its precise treatment is uncertain. Hopefully, in the near future, some definitive studies will provide the appropriate answers.

CONCLUSIONS

Home parenteral nutrition is a life-restoring therapy that can significantly improve the life of a patient with alimentary failure. It should only be used in patients who do not respond to conventional measures. A team approach in training and monitoring these patients is essential to ensure success. Because of the prolonged periods of parenteral support, trace element and vitamin additives should be monitored closely, as well as essential fatty acid levels. Free communication between patient, members of the nutritional support team, and suppliers of HPN supplies allows for the return of an otherwise hospitalized patient to his home, family and society.

Bibliography

1 American Medical Association Department of Foods and Nutrition (1978). Multivitamin preparations for parenteral use. A statement by the Nutrition Advisory Group. *J. Parent. Enteral Nutr.*, **3**, 258

2 American Medical Association (1979). Guidelines for essential trace element preparations for parenteral use. A statement by the Nutritional Advisory Group. *J. Parent. Enteral Nutr.*, **3**, 263

3 Arakawa, T., Tamura, T., Igarashi, Y. *et al.* (1976). Zinc deficiency in two infants during total parenteral alimentation for diarrhea. *Am. J. Clin. Nutr.*, **29**, 197

4 Askanazi, J., Carpentier, Y. A., Elwyn, D. H. *et al.* (1980). Influence of total parenteral nutrition on fuel utilization in injury and sepsis. *Ann. Surg.*, **191**, 40

5 Broviac, J. W., Cole, J. J. and Scribner, B. H. (1973). A silicone rubber right atrial catheter for prolonged parenteral alimentation. *Surg. Gynecol. Obstet.*, **136**, 602

6 Burke, J. F., Wolfe, R. R., Mullany, C. J. *et al.* (1979). Glucose requirements following burn injury. *Ann. Surg.*, **190**, 274

7 Byrne, W. J., Ament, M. D., Burke, M. *et al.* (1979). Home parenteral nutrition. *Surg. Gynecol. Obstet.*, **149**, 593

8 Byrne, W. J., Halpin, T. C., Asch, M. J. *et al.* (1977). Home total parenteral nutrition: an alternative approach to the management of children with severe chronic small bowel disease. *J. Pediatr. Surg.*, **12**, 359

9 Cannon, R. A., Byrne, W. J., Ament, M. E. *et al.* (1980). Home parenteral nutrition in infants. *J. Pediatr.*, **96**, 1098

10 Collins, J. P., Oxby, C. B. and Hill, G. L. (1978). Intravenous amino acids and intravenous hyperalimentation as protein-sparing therapy after major surgery. A controlled clinical trial. *Lancet*, **1**, 788

11 Driscoll, R. H. and Rosenberg, I. H. (1978). Total parenteral nutrition, in inflammatory bowel disease. *Med. Clin. N. Am.*, **62**, 185

12 Dudrick, S. J., Englert, D. M., MacFadyen, B. V. *et al.* (1979). A rest for ambulatory patients receiving hyperalimentation. *Surg. Gynecol. Obstet.*, **148**, 587

13 Dudrick, S. J., MacFadyen, B. V., Souchon, E. A. *et al.* (1977). Parenteral nutrition techniques in cancer patients. *Cancer Res.*, **37**, 2440

14 Dudrick, S. J., Wilmore, D. W., Vars, H. M. *et al.* (1968). Long-term parenteral nutrition with growth and development and positive nitrogen balance. *Surgery*, **64**, 134

15 Dunlap, W. M., Jangs, G. W. and Hume, D. M. (1974). Anemia and netropenia caused by copper deficiency. *Ann. Intern. Med.*, **80**, 470

16 FAO/WHO Expert Consultation (1977). Dietary fats and oils in human nutrition. *FAO Food and Nutrition Paper No. 3*, Rome, p. 19

17 Faulk, D. L., Anuras, S. and Freeman, J. B. (1978). Idiopathic chronic intestinal pseudo-obstruction. Use of central venous nutrition. *J. Am. Med. Assoc.*, **240**, 2075

18 Fleming, C. R., Berkner, S., Beart, R. W., Jr. *et al.* (1980). Home parenteral nutrition for management of the severely malnourished adult patient. *Gastroenterology*, **79**, 11

19 Fleming, C. R., McGill, D. B. and Berkner, S. (1977). Home parenteral nutrition as primary therapy in patients with extensive Crohn's disease of the small bowel and malnutrition. *Gastroenterology*, **73**, 1077

20 Fleming, C. R., Witzke, D. J. and Beart, R. W., Jr. (1980). Catheter-related complications in patients receiving home parenteral nutrition. *Ann. Surg.*, **192**, 593

21 Freeman, J. B. (1980). Personal communication

22 Freeman, J. B., Stegink L. D., Fry, L. K. *et al.* (1975). Evaluation of amino acid infusions as protein-sparing agents in normal adult subjects. *Am. J. Clin. Nutr.*, **28**, 477

23 Freund, H., Atamian, S. and Fischer, J. E. (1979). Chromium deficiency during total parenteral nutrition. *J. Am. Med. Assoc.*, **241**, 496

24 Frisancho, A. R. (1974). Triceps skin fold and upper arm muscle size norms for assessment of nutritional status. *Am. J. Clin. Nutr.*, **27**, 1052

25 Goldberger, J. H., Deluca, F. G., Wesselhoeft, C. W. *et al.* (1979). A home program of long-term total parenteral nutrition in children. *J. Pediatr.*, **94**, 325

26 Graham, J. A. (1977). Conservative treatment of gastrointestinal fistulas. *Surg. Gynecol. Obstet.*, **144**, 512

27 Granberg, P. O. and Wadstrom, C. B. (1965). Kidney function during intravenous fat administration. *Acta Chir. Scand.*, **130**, 299

28 Greenberg, G. R., Marliss, E. B., Anderson, G. H. *et al.* (1976). Protein-sparing therapy in postoperative patients: effects of added hypocaloric glucose or lipid. *N. Engl. J. Med.*, **294**, 1411

29 Grischkan, D., Steiger, E. and Fazio, V. (1979). Maintenance of home hyperalimentation in patients with high output jejunostomies. *Arch. Surg.*, **144**, 838

·30 Grundfest, S., Steiger, E., Sattler, L. *et al.* (1979). The current status of home total parenteral nutrition. *Artif. Organs*, **3**, 156

31 Grundfest, S. and Steiger, E. (1980). Home parenteral nutrition. *J. Am. Med. Assoc.*, **244**, 1701

32 Gulledge, A. D., Gipson, W. T., Steiger, E. *et al.* (1980). Short bowel syndrome and psychological issues for home parenteral nutrition. *Gen. Hosp. Psychiatry*, **2**, 271

33 Hadfield, J. I. H. (1967). The indications for treatment by intravenous fat emulsions. Proceedings of the *Symposium of Parenteral Feeding. The Royal Society of Medicine*, London, October 19. p. 34. (London: Wembley Press Ltd.)

34 Hallberg, D., Holm, I., Obel, A. L. *et al.* (1967). Fat emulsions for complete intravenous nutrition. *Postgrad. Med. J.*, **43**, 307

35 Hallberg, D., Schuberth, O. and Wretland, A. (1966). Experimental and clinical studies with fat emulsion for intravenous nutrition. *Nutr. Dieta*, **8**, 245

36 Hansen, L. M., Hardie, W. R. and Hidalgo, J. (1975). Fat emulsion for intravenous administration: clinical experience with Intralipid 10%. *Ann. Surg.*, **184**, 80

37 Hartmann, G. (1976). Fat emulsions (Intralipid) in internal medicine (German). *Wein. Med. Wochenschr.*, **117**, 51

38 Hartmann, G. (1965). Parenteral fat supply: experiences with the preparation of Intralipid 'Vitrum' (German). *Schweiz. Med. Wochenschr.*, **95**, 466

39 Heimbach, D. M. and Ivey, T. D. (1976). Technique for placement of a permanent home hyperalimentation catheter. *Surg. Gynecol. Obstet.*, **143**, 634

40 Hill, G. L., Roderic, F. G. J., King, R. C. *et al.* (1979). Multi-element analysis-application to critically ill patients receiving intravenous nutrition. *Br. J. Surg.*, **66**, 868

41 Holroyde, C. P., Myers, R. N., Smink, R. D. *et al.* (1977). Metabolic response to total parenteral nutrition in cancer patients. *Cancer Res.*, **37**, 3109

42 Hooley, R. A. (1980). Nutritional assessment – a clinical perspective. *J. Am. Dietet. Assoc.*, **77**, 861

43 Hooley, R. A. (1979). *Nutritional Assessment Protocol*. Cleveland Clinic Foundation, Department of Nutritional Services

44 Jacobson, S. (1970). Complete parenteral nutrition in man for seven months. In Berg, G. (ed.) *Advances in Parenteral Nutrition*. p. 6. (Stuttgart: Georg Thieme Verlag)

45 Jacobson, S., Ericsson, J. L. E. and Obel, A. (1971). Histopathological and ultrastructural changes in the human liver during complete intravenous nutrition for seven months. *Acta Chir. Scand.*, **137**, 335

46 Jeejeebhoy, K. N., Anderson, G. H., Nakhooda, A. F. *et al.*, (1976). Metabolic studies in total parenteral nutrition with lipid in man: comparison with glucose. *J. Clin. Invest.*, **57**, 125

47 Jeejeebhoy, K. N., Anderson, G. H., Sanderson, I. and Bryan, M. H. (1974). Total parenteral nutrition: nutrient needs and technical tips. *Modern Med. Can.*, **29**, 832

48 Jeejeebhoy, K. N., Langer, B., Tsallas, G., Chu, R. C. *et al.* (1976). Total parenteral nutrition at home: studies in patients surviving 4 months to 5 years. *Gastroenterology*, **71**, 943

49 Jeejeebhoy, K. N., Zohrab, W. J., Langer, B. *et al.* (1973). Total parenteral nutrition at home for 23 months without complication, and with good rehabilitation: a study of technical and metabolic features. *Gastroenterology*, **65**, 811

50 Kapp, J. P., Duckert, F. and Hartmann, G. (1971). Platelet adhesiveness and serum lipids during and after Intralipid infusions. *Nutr. Metab.*, **13**, 92

51 Karpel, J. T. and Pedlen, V. H. (1972). Copper deficiency in long-term parenteral nutrition. *J. Pediatr.*, **80**, 32

52 Kay, R. G. and Tasman-Jones, C. (1975). Acute zinc deficiency in man during intravenous alimentation. *Aust. NZ J. Surg.*, **45**, 325

53 Kim, Y. S. and Freeman, H. J. (1977). The digestion and absorption of protein. In *Clinical Nutrition Update: Amino Acids*. p. 135. AMA

54 Kishi, H., Nishii, S., Ono, T. *et al.* (1979). Thiamin and pyridozine requirements during intravenous hyperalimentation. *Am. J. Clin. Nutr.*, **32**, 332

55 Kolanowski, J. (1977). On the mechanisms of fasting natriuresis and of carbohydrate-induced sodium retention. *Diabete et Metabolisme*, **3**, 131

56 Lavery, I. C., Steiger, E. and Fazio, V. W. (1980). Home parenteral nutrition in management of patients with severe radiation enteritis. *Dis. Colon Rectum*, **23**, 91

57 Lee, H. A. (1969). Parenteral nutrition in medicine. In Allen, C. P. and Lee, H. A. (eds.) *A Clinical Guide to Intravenous Nutrition*. p. 56. (Oxford and Edinburgh: Blackwell Scientific Publications)

58 Mackenzie, T., Blackburn, G. L. and Flatt, J. P. (1974). Clinical assessment of nutritional status using nitrogen balance. *Fed. Proc.*, **33**, 683

59 Mernagh, J. R., Harrison, J. E., McNeill, K. G. and Jeejeebhoy, K. N. (1980). Effect of total parenteral nutrition in the restitution of body nitrogen, potassium and weight. *Am. J. Clin. Nutr.*, (in press)

60 Messing, B., Latrive, J. P., Bitoun, A., Galian, A. and Bernier, J. J. (1979). La steatose hepatique au cours de la nutrition parenteral total depend-elle de l'apport calorique lipidique? *Gastroenter. Clin. Biol.*, **3**, 719

61 Messing, B., Bitoun, A. and Galian, A. (1977). La steatose hepatique au cours de la nutrition parenteral depend-elle de l'apport calorique glucidique? *Gastroenter. Clin. Biol.*, **1**, 1015

62 Moghissi, K., Hornshaw, J., Teasdale, R. R. *et al.* (1977). Parenteral nutrition in carcinoma of the oesophagus treated by surgery: nitrogen balance and clinical studies. *Br. J. Surg.*, **64**, 125

63 Oakley, J. R., Steiger, E., Lavery, I. C. *et al.* (1979). Catastrophic enterocutaneous fistulae. The role of home hyperalimentation. *Cleve. Clin. Q.*, **46**, 133

64 Pietsch, J. B., Meakins, J. L. and Maclean, L. D. (1977). The delayed hypersensitivity response: application in clinical surgery. *Surgery*, **82**, 349

65 Rault, R. J. M. and Scribner, B. H. (1977). Treatment of Crohn's disease with home parenteral nutrition. *Gastroenterology*, **72**, 1249

66 Rhoads, J. M. (1969). Approaches to an artificial gastrointestinal tract. *Am. J. Surg.*, **117**, 3

67 Rudman, D., Millikan, W. J. and Richardson, T. J. (1975). Elemental balances during intra-venous hyperalimentation of underweight adult subjects. *J. Clin. Invest.*, **55**, 94

68 Sattler, L., Wateska, L. P., Siska, B. *et al.* (1978). *Cleveland Clinic Foundation Home TPN Manual.* (Irvine, California: McGaw Laboratories)

69 Sauberlich, H. E., Dowdy, R. P. and Skala, J. H. (1973). Laboratory tests for the assessment of nutritional status. *CRC Crit. Rev. Clin. Lab. Sci.*, **4**, 215

70 Scribner, B. H., Cole, J. J. and Christopher, T. G. (1970). Long-term total parenteral nutrition. *J. Am. Med. Assoc.*, **212**, 457

71 Shenkin, A. and Wretland, A. (1977). Complete intravenous nutrition including amino acids, glucose and lipids. In *Nutritional Aspects of Care in the Critically Ill.* p. 345. (Edinburgh: Churchill Livingstone)

72 Shike, M., Harrison, J. E., Sturtridge, W. C. *et al.* (1980). Metabolic bone disease in patients receiving long-term parenteral nutrition. *Ann. Intern. Med.*, **92**, 343

73 Shike, M,. Kurian, R. and Roulet, M. (1980). Copper (Cu) requirements in total parenteral nutrition (TPN). *Gastroenterology*, **78**, 1259

74 Silvas, S. E. and Paragas, P. D., Jr. (1972). Paresthesias, weakness, seizures and hypo-phosphatemia in patients receiving hyperalimentation. *Gastroenterology*, **62**, 513

75 Solassol, C., Joyeus, H., Etco, L. *et al.* (1974). New techniques for long-term intravenous feeding and artificial gut in 75 patients. *Ann. Surg.*, **179**, 519

76 Steiger, E. and Grundfest, S. (1979). A review of home hyperalimentation. *Contemp. Surg.*, **15**, 33

77 Steiger, E. and Grundfest, S. (1980). Intravenous hyperalimentation vascular access and administration. In Deitel, M. (ed.) *Nutrition in Clinical Surgery*. pp. 53–64. (Baltimore: Williams and Wilkins)

78 Steinbereithnerk, K. and Vagecs, H. (1965). Clinical experimental examinations regarding the application of fat emulsions (German). *Wien. Klin. Wochenschr.*, **77**, 369

79 Stewart, G. R. (1979). Home parenteral nutrition for chronic short bowel syndrome. *Med. J. Aust.*, **2**, 317

80 Strobel, C. T., Byrne, W. J. and Ament, M. E. (1979). Home parenteral nutrition in children with Crohn's disease: an effective management alternative. *Gastroenterology*, **77**, 272

81 Van Rij, A. M., McKenzie, J. M., Robinson, M. F. *et al.* (1979). Selenium and total parenteral nutrition. *J. Parent. Enteral Nutr.*, **3**, 235

82 Van Rij, A. M., Thompson, C. D., McKenzie, J. M. and Robinson, M. F. (1979). Selenium

deficiency in total parenteral nutrition. *Am. J. Clin. Nutr.*, **32**, 2076
83 Viteri, F. E. and Guillermo, A. (1978). Protein-calorie malnutrition. In Goodhart, P. S. and Shils, M. E. (eds.) *Modern Nutrition in Health and Disease.* 5th Edn. (Philadelphia: Lee & Febiger)
84 Wateska, L. P., Sattler, L. L. and Steiger, E. (1980). Cost experiences with a home parenteral nutrition program. *J. Am. Med. Assoc.*, **244**, 2303
85 Wolfe, R. R., Burkot, M. J., Allsop, J. R. and Burke, J. F. (1979). Glucose metabolism in the severely burned patient. *Metabolism*, **20**, 1031
86 Wolman, S. L., Anderson, G. H., Marliss, E. B. and Jeejeebhoy, K. N. (1979). Zinc in total parenteral nutrition: requirements and metabolic effects. *Gastroenterology*, **73**, 458
87 Yeo, M. T., Gazzaniga, A. B., Bartlett, R. H. *et al.* (1973). Total intravenous nutrition: experience with fat emulsions and hypertonic glucose. *Arch. Surg.*, **106**, 792

5
Quality of life for the home-care patient

K. LADEFOGED AND S. JARNUM

The assessment of the quality of life in patients with severe chronic disorders is an important part of therapy. For patients who require permanent home parenteral nutrition (HPN) this is still more urgent for several reasons. The therapy is very expensive and makes heavy demands on social welfare systems and hospital services. It is rigorous and time-consuming and can profoundly affect the lifestyle of the patients and their families. Most patients who require permanent HPN suffer from the short bowel syndrome following extensive intestinal resection for Crohn's disease or mesenteric infarction. Their requirement for parenteral nutrition depends on the extent of any oral intake and the remaining absorptive capacity of the bowel. The infusion time required for parenteral supplementation may differ considerably from patient to patient but in most patients the infusion takes 8–12 h per day[1–4].

Few reports have been published on the psychosocial consequences of home parenteral nutrition[3,5–8]. They are based on small heterogeneous groups of patients treated for various lengths of time, which makes it difficult to segregate the influence of restitution of physical health, the presence of an ostomy, personality, financial position, support from family and hospital, duration of HPN and HPN per se. The essential question is: can patients who need permanent HPN for survival obtain an acceptable life quality?

Since the issue of self-care is fundamental for HPN, successful adaptation depends on careful training.

TRAINING FOR HPN

Training of a patient for HPN usually takes place in hospital[3,4,9,10]. The goals of the training are:

(1) To understand and accept the limitations imposed by the disease.

411

(2) To understand the principles of treatment and capability of self-care.

(3) Social adaptation.

The training programme should include (a) instruction in basic anatomy, physiology, pathophysiology, and nutrition, (b) instruction and practice in aseptic technique and use of the equipment, (c) instruction in catheter-related and metabolic complications and the proper responses to each complication, (d) assessment of the need for financial support or other supportive arrangements. The course of the training must be individually designed, based on the patient's precondition and adjusted to his emotions and reactions. The training involves a multispecialty team: physicians, nurses, surgeons, dieticians, pharmacists, social workers, and occasionally psychiatrists, with special knowledge of HPN[10-12]. The spouse or another family member, too, should learn the HPN procedure[5,8]. Contact with other well-adapted HPN patients is important[5,8]. It can be difficult to persuade patients, especially elderly ones, to take the infusions during the night-time, because of their unwillingness to allow their night's sleep to be disturbed[3]. It is, however, important to do so because it will enable them to lead a more normal social life by day. Usually the training can be completed within 2 or 3 weeks[1,3,11].

ADAPTATION TO HPN

Faced with the prospect of permanent HPN many patients go through a number of emotional reactions: disbelief, refusal, hopelessness, fear, anger, depression[5,8], which require understanding, patience and support from the staff and from the family. Little by little most patients accept their limitations, become eager to learn the principles of their new life support system, and struggle to incorporate this system into their way of life or find a new meaningful and satisfactory way of life. In patients on total parenteral nutrition major adjustment problems centre around loss of the basic function of eating[5,8]. Thus, maintenance or resumption of a normal oral intake, which is a prerequisite for optimal intestinal adaptation, may also be essential for a successful social adaptation.

Initially adjustment to HPN as well as learning of the technique seems easier for patients with Crohn's disease than for patients with mesenteric infarction[5,6,8]. Patients with Crohn's disease have often been ill for years with long hospital stays, occupational and other social resignations, and previous treatment by parenteral nutrition. They usually perceive HPN as a preferable alternative to longterm hospitalization or chronic suffering[2,8,13]. Most patients with mesenteric infarction have never been seriously ill before. For them HPN has a tremendous impact on their previous lifestyle[8]. Above all, successful adaptation depends on restitution of physical health, strong personality and stable familial and social relationships.

Table 34.1 Data from an interview study of 13 patients, 7 females and 6 males aged 24–62 y, median 53 y, on permanent home parenteral nutrition[3], occupational and social rehabilitation

	Number of patients
Job	0
Disability pension	13
Housekeeping	
Most	6
Part	6
None	1
Leisure time activities	
Active, normal physical efforts	5
Active, low physical efforts	7
Inactive	1
Feeling activity level restricted	9
Participation in social activities	
>15 times per month	4
5–15 times per month	7
<5 times per month	2
Restricted because of HPN	5

LIFE WITH HPN

Few patients on HPN are reported to manage a full-time job (Table 34.1)[1-3,8-10,13]. The lack of occupational rehabilitation may be due to several reasons. Return to a job leaves little time for other activities than work and parenteral nutrition. Many patients have been unemployed for years because of their primary disease and are adapted to this situation[3]. Frequent hospital admissions for complications, catheter exchange or metabolic studies may impede employment[3]. Furthermore, several patients feel too weak to cope with a full-time job[3,8,10]. In addition to physical distress this weakness may be due to sleep disruption because of frequent urination during the nightly infusions or because of concern about the function of the equipment[5,8]. It may be that the patients' psychological resources are exhausted by the HPN regime with nothing left to put into a job. Above all, occupational rehabilitation may depend on local welfare politics. Many patients assume household duties (Table 34.1)[1-3,8-10,13] and leave breadwinning to their partners. For many male patients this means reversal of roles which fundamentally changes the pattern of family interaction. In patients on total parenteral nutrition the family dynamics will also be affected by disturbance of the essential family event: eating together[6,8].

Most patients on HPN feel their leisure time activities restricted (Table 34.1)[3,8]. Water sports are excluded and athletics often limited. In many cases the limitations are not related to HPN but to incomplete physical restitution[3]. Participation in social activities (guests at home, visits to friends, cinema, etc.) seems rather unaffected by HPN and several patients display a very active social life (Table 34.1)[3]. Some patients, however, feel that HPN alters their togetherness with friends and relatives[3,6]. Friends may stay away or show pity for the patient or uneasiness about the treatment[3].

Table 34.2 Familial relations (13 patients)[3]

	Number of patients
Married/cohabiting	12
Divorced	1
Children at home and their reactions	(5 with 10 children)
9 children indifferent	
1 child maladjusted	
Sexual activity	
Regular	4 aged 24–34 y
Occasional	3 aged 35–55 y
None	5 aged 56–62 y
?	1 aged 53 y

Reduced sexual activity is a common event in chronically ill patients (Table 34.2)[14,15]. Although maybe not chronically ill in a strict sense, many elderly patients lose and do not regain their sex drive[3,8]. Body image distortions and fear of dislodging or damaging the catheter during intercourse may interfere with sex life. Most younger patients, however, seem to display a normal or at least fairly regular sexual activity (Table 34.2)[3,8], and, in some, sexual performance and satisfaction may even improve with correction of the state of malnutrition[5].

A major problem for some patients on HPN is drug addiction[3,10]. Treatment by narcotics often starts around the time of surgery and it can be extremely difficult to withdraw it and the patient escapes into narcomania instead of facing the problems of the HPN regime.

Table 34.3 Adaptation to HPN (13 patients)[3]

	Number of patients
Psychological symptoms	
None	7
Occasional irritability or bad temper	2
Daily irritability	3
Depression	1
Disturbed sleep	4
Feelings about HPN	
Indifferent	6
Sometimes annoyed	5
Often annoyed	2
Overall satisfaction	
Satisfied	9
Unsatisfied	4

From time to time several patients show frustration, fear, irritability, annoyance or depression (Table 34.3)[3,5], reactions which perceptibly strain the family[3]. Most patients, however, seem to comply with their conditions of life (Table 34.3)[3]. By means of structured interviews Ladefoged[3] assessed life quality in 13 patients on permanent HPN. The assessment was based on the following criteria: (1) no major physical complaints, (2) no major psychological symptoms, (3) no substantial restriction in social and leisure time activities, (4) ability to accept HPN, and (5) overall satisfaction with the conditions of life. Nine patients (69%) who fulfilled at least three of these five criteria were estimated to have a fair quality of life. Four patients (31%) who complied with less than three of the criteria were considered to have a poor life quality. The quality of life was not related to age, sex, primary disease, presence of an ostomy or duration of HPN.

Table 34.4 Data on 30 patients treated by home parenteral nutrition in the period 1970–1982. Age at start: 16–69 y, median: 45 y, 16 females, 14 males

	Number of patients
Primary disease	
Crohn's disease	17
Mesenteric infarction	7
Volvulus, intestinal gangrene	1
Radiation enteropathy	1
Intestinal pseudo-obstruction	1
Gardner's syndrome	1
Carcinoid syndrome	1
Chronic pancreatitis + enteropathy	1
Indication for HPN	
Short bowel syndrome	26
Malnutrition	2
Intestinal obstruction/pseudo-obstruction	2
Duration of HPN	
1–72 months, median 28 months	
In total: 935 patient months ~ 78 patient years	
Outcome	
Dead	9
Died from complications of the therapy	3
Alive	21
Still on HPN	16
On oral nutrition	5

PROGNOSIS

Mortality rates among different series of patients on HPN range from 1 in 1.8 to 1 in 40.4 patient years[2,4,9,10,12,16–18]. The differences in mortality rate are mainly due to different patient selection. In our case material of 30 patients, who in the period 1970–1982 had been on HPN for a total of 935 patient months (Table 34.4), nine patients died. It gives a mortality rate of 1 in 8.7 patient years. In three patients the death might be related to the therapy: one patient died from mediastinitis due to perforation of the oesophagus during endoscopy, one died from subdural haematoma possibly caused by anti-coagulant treatment, and one patient died in a state of severe malnutrition due to his reluctance to take parenteral infusions. Six patients died from their underlying disease or intercurrent diseases.

RESUMPTION OF ORAL NUTRITION

Temporary HPN has been used as a primary or supportive therapy for patients with Crohn's disease and/or gastrointestinal fistulas[1,10,13,19-21]. It may be discontinued at remission, closure of the fistulas or after operation. Some patients with subtotal enterectomy are able to maintain adequate nutrition on oral feeding after a course of HPN (Table 34.4)[1,10,22,23]. It may take ½−1 y, maybe more, with gradual reduction of the parenteral supply before patients who have had major resections can be weaned from HPN[10,19,23]. In most cases it is not possible to predict whether a patient who has had extensive resection will be able to stop HPN eventually or will need it for life.

CONCLUSION

Home parenteral nutrition (HPN) is an expensive therapy which makes great demands on social welfare systems and hospital services as well as on patient adaptation, but it offers an acceptable quality and quantity of life to previously doomed patients with severe gastrointestinal disorders, in particular extensive bowel resection. HPN is, however, still in its infancy, and life quality and quantity presumably will improve with technical advances, greater knowledge of nutritional requirements and greater experience in appropriate psychological support.

References

1 Byrne, W. J., Ament, M. E., Burke, M. and Fonkalsrud, E. (1979). Home parenteral nutrition. *Surg. Gynecol. Obstet.*, **149**, 593
2 Jeejeebhoy, K. N., Langer, B., Tsallas, G., Chu, R. C., Kuksis, A. and Anderson, G. H. (1976). Total parenteral nutrition at home: studies in patients surviving 4 months to 5 years. *Gastroenterology*, **71**, 943
3 Ladefoged, K. (1981). Quality of life in patients on permanent home parenteral nutrition. *J. Parent. Enteral Nutr.*, **5**, 132
4 Shils, M. E. (1975). A program for total parenteral nutrition at home. *Am. J. Clin. Nutr.*, **28**, 1429
5 MacRitchie, K. J. (1978). Life without eating and drinking. Total parenteral nutrition outside hospital. *Can. Psychiatr. Assoc.*, **23**, 373
6 MacRitchie, K. J. (1980). Parenteral nutrition outside hospital. Psychosocial styles of adaptation. *Can. J. Psychiatry*, **25**, 308

7 Malcolm, R., Robson, J. R. K., Vanderveen, T. W. and Mahlen, P. (1980). Psychosocial aspects of total parenteral nutrition. *Psychosomatics*, **21**, 115

8 Price, B. S. and Levine, E. L. (1979). Permanent total parenteral nutrition: psychological and social responses of the early stages. *J. Parent. Enteral Nutr.*, **3**, 48

9 Broviac, J. W. and Scribner, B. H. (1974). Prolonged parenteral nutrition in the home. *Surg. Gynecol. Obstet.*, **139**, 24

10 Fleming, C. R., Beart, R. W., Berkner, S., McGill, D. B. and Gaffron, R. (1980). Home parenteral nutrition for management of the severely malnourished adult patient. *Gastroenterology*, **79**, 11

11 Hughes, B. A., Fleming, C. R., Berkner, S. and Gaffron, R. (1980). Patient compliance with a home parenteral nutrition program. *J. Parent. Enteral Nutr.*, **4**, 12

12 Ivey, M., Riella, M., Mueller, W. and Scribner, B. (1975). Long-term parenteral nutrition in the home. *Am. J. Hosp. Pharm.*, **32**, 1032

13 Fleming, C. R., McGill, D. B. and Berkner, S. (1977). Home parenteral nutrition as primary therapy in patients with extensive Crohn's disease of the small bowel and malnutrition. *Gastroenterology*, **73**, 1077

14 Gazzard, B. G., Price, H. L., Libby, G. W. and Dawson, A. M. (1978). The social toll of Crohn's disease. *Br. Med. J.*, **2**, 1117

15 Levy, N. B. and Wynbrandt, G. D. (1975). The quality of life on maintenance haemodialysis. *Lancet*, **1**, 1328

16 Dudrick, S. J., Englert, D. M., Van Buren, C. T., Rowlands, B. J. and MacFadyen, B. V. (1979). New concepts of ambulatory home hyperalimentation. *J. Parent. Enteral Nutr.*, **3**, 72

17 Grundfest, S., Steiger, E., Sattler, L., Washko, S. and Wateska, L. (1979). The current status of home total parenteral nutrition. *Artif. Organs*, **3**, 156

18 Jeejeebhoy, K. N. (1980). Home parenteral nutrition. *Can. Med. Assoc. J.*, **122**, 143

19 Rault, R. M. J. and Scribner, B. H. (1977). Treatment of Crohn's disease with home parenteral nutrition. *Gastroenterology*, **72**, 1249

20 Byrne, W. J., Burke, M., Fonkalsrud, E. and Ament, M. E. (1979). Home parenteral nutrition: an alternative approach to the management of complicated gastrointestinal fistulas not responding to conventional medical or surgical therapy. *J. Parent. Enteral Nutr.*, **3**, 355

21 Strobel, C. T., Byrne, W. J. and Ament, M. E. (1979). Home parenteral nutrition in children with Crohn's disease: an effective management alternative. *Gastroenterology*, **77**, 272

22 Jarnum, S. and Ladefoged, K. (1981). Long-term parenteral nutrition I. Clinical experience in 70 patients from 1967 to 1980. *Scand. J. Gastroent.*, **16**, 903

23 Ladefoged, K. and Jarnum, S. (1978). Long-term parenteral nutrition. *Br. Med. J.*, **2**, 262

6
Experience with ambulatory total parenteral nutrition in Crohn's disease

J. M. MÜLLER, H. W. KELLER, J. SCHINDLER AND H. PICHLMAIER

INTRODUCTION

Preparing malnourished patients with inflammatory bowel disease for surgery by TPN Steiger[1] observed complete remissions in some cases in whom it was possible to avoid surgery. This initial report has received much attention especially by surgeons engaged in the management of Crohn's disease, because even resection with wide safety margins cannot prevent a recurrence rate up to 60% after 10 years[2,3]. In about 50% of these, who are mostly young patients, repeated operations are necessary[3]. With every operation the risk of fatal complications increases and the final outcome may be a short-bowel syndrome. Therefore any therapy that reduces the need for resection would be extremely valuable. Several authors claimed[4-9] that after a course of TPN as primary therapy for acute Crohn's disease remissions can be expected in about 50–80% of patients and some may have disease-free intervals for over 10 years[10].

The possibilities[11,12] of reversing impaired immunocompetence in malnourished patients with Crohn's disease by TPN together with reducing mucosal irritation of the inflamed bowel by elimination of food bulk to promote healing and regeneration are attractive.

But in a critical review of the published results Allan[13] pointed out that although TPN is a potentially hazardous treatment and expensive, there has been no controlled evaluation, obligatory for any therapy. In his opinion, it is impossible to isolate the value of TPN alone as patients were often receiving other medications such as corticosteroids or salicylazosulphapyridine.

We tried to evaluate prospectively:

(1)   The effect of TPN as the only therapy on the course of complicated Crohn's disease.

(2)   The metabolic situation of patients during longterm TPN.

(3)   The complications associated with TPN when carried out in a hospital which does not have a specialized nutritional support team.

To avoid the psychological stress of long periods in hospital and to reduce the costs, it was decided that TPN should be carried out by the patients at home.

PATIENTS AND METHODS

Thirty consecutive patients (Table 35.1) with clinical, radiological and patho-histological evidence of acute Crohn's disease took part in the study. They were admitted to the department of surgery of the University of Cologne for operation, because they had either repeatedly failed to respond to medical therapy or developed complications such as enterocutaneous fistulae, inter-mittent ileus or bleeding.

Table 35.1   Patients' characteristics

Number of patients	30
Sex ratio (m/f)	8/22
Age* (y)	29.6 ± 7.9
Duration of disease before TPN (months)	49 ± 23
Site of lesion	
mainly small bowel	9
mainly large bowel	6
small and large bowel	15
Medical therapy before TPN	30
duration* (months)	35 ± 22
type	
a: diet and corticosteroids	4
b: a and salicylazosulphapyridine	21
c: b and metronidazole or azathioprine	5
d: a or b or c and short-term TPN	22
Surgery before TPN	21
one resection	14
two resections	3
laparotomy/appendectomy	4

*Mean ± SD

After admission all medications were discontinued. Under strictly aseptic conditions a Broviac silastic catheter was placed in the superior vena cava[14]. The average amount of nutrients given parenterally depended on the activity of Crohn's disease. During the acute stage, patients were kept in hospital and received high caloric nutrition every 24 h. When their clinical and metabolic situation was stabilized, the infusion time was reduced gradually to 10–12 h overnight (Table 35.2). Patients were trained by a nurse in the care of the catheter, so that they were able to attach themselves to the system. The technique for home TPN was simple and inexpensive. The amino-acid, dextrose and fat solutions used contained the necessary amount of electrolytes. The patients had to add trace elements and vitamin preparations only. The delivery system consisted of an infusion-stand and a Y-shaped infusion line. A gravity flow system only was used and controlled by two roller clamps. The patients were taught to keep a constant distance between the exit of the catheter on the skin and the lower end of the infusion bottles and to adjust the flow by counting the drop-rate per min. The amino-acid solution was infused through one arm of the Y-shaped line for the whole period. The fat emulsion was infused for the first 4–5 h and then the dextrose solution was given for the remaining 7–8 h into the other arm. The infusion rates were 0.08 g/kg body weight hourly for amino acids, 0.5–0.6 g/kg h for dextrose and 0.3–0.4 g/kg h for the fat emulsion. This is within the limits of the recommendations given by the manufacturers.

Table 35.2 Average amount of nutrients given during in-hospital and home TPN

	In hospital (24 h)	At home (10–12 h)
	Daily intake/kg body weight	
Fluids	50 ml	35 ml
Energy	45 kcal	30 kcal
Amino acid	1.5 g	1 g
Dextrose	8.5 g	4.4 g
Fat	1.4 g	1.4 g
Sodium	1.8 mmol	1.2 mmol
Potassium	1.3 mmol	0.9 mmol
Chloride	2.3 mmol	1.6 mmol
Magnesium	0.1 mmol	0.07 mmol
Phosphorus	0.4 mmol	0.26 mmol
Calcium	0.2 mmol	0.2 mmol

additionally: preparations of trace elements, fat and water soluble vitamins

According to the experience of Holm[8], it was decided that TPN should be given for 12 weeks, to give the diseased bowel enough time to recover.

The clinical progress of the patients was monitored in the hospital daily and weekly at home by the same surgeons using the Crohn's disease activity index[15]. Metabolic measurements included Hb, Hct, WBC count, sodium, potassium, chloride, magnesium, calcium, phosphorus, zinc, blood urea nitrogen, creatinine, total protein, albumin, cholesterol, triglyceride, transaminases, alkaline phosphatase, bilirubin, blood and urine glucose, partial thromboplastin time, blood coagulation time, thrombin time and free plasma amino acids. The parameters were measured every third day in hospital and once a week during home TPN. When at home, patients were encouraged to work and engage in sporting activities. After the last week of

TPN, the infusion scheme was reduced gradually and oral intake started with elemental diet. 1 or 2 weeks later the central venous catheter was removed and patients ate normal meals. Again no medications were allowed. Patients were followed up every 6 months. They were asked to report immediately to the clinic, if they observed symptoms such as diarrhoea or abdominal cramps. Recurrence of Crohn's disease was defined as a relapse of symptoms or radio-logical or endoscopical signs of active Crohn's disease, which required treat-ment.

RESULTS

25 cases responded promptly to TPN. After only a few days abdominal pains ceased, stools returned to normal and patients started to gain weight. Entero-cutaneous fistulae closed spontaneously in two patients, one after a previous

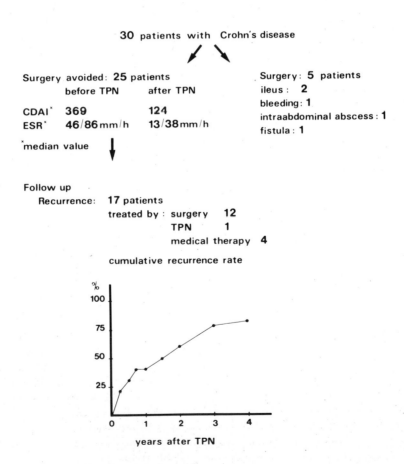

Figure 35.1 Clinical results of TPN in Crohn's disease

appendectomy, the other after a hemicolectomy. In all these patients surgery was avoided and they were discharged from the hospital within 11–23 days after the beginning of TPN. During home parenteral nutrition no relapse of active Crohn's disease was observed. When patients started to take elemental diet, their Crohn's disease activity index (CDAI) had dropped from 369 to 124 points (median value). ESR had decreased from 46/86 mm/h to 13/38 mm/h and they had gained 4.6 kg weight (median values) (Figure 35.1).

In five patients it was not possible to control the active disease by TPN. Two patients with recurrent Crohn's disease at the anastomoses after hemi-colectomies had intermittent ileus during TPN. Gastrographin studies 3 and 5 weeks after the starting of TPN showed that the stenoses had not changed since admission and it was decided to resect the affected areas. The third patient with a typical terminal ileitis developed an acute abdomen during TPN and at laparotomy an intra-abdominal abscess was found. In the fourth patient heavy bleeding occurred from the colon and a colectomy became necessary. The fifth patient had an enterocutaneous fistula system between stomach, ileum and colon.

During 3 weeks of TPN the drainage from the fistula did not decrease substantially and operative closure was performed. Despite four of these five patients being in poor nutritional status with body weight below 80% of ideal weight at admission their post-operative courses were uncomplicated. This was probably due to the intensive nutritional rehabilitation by TPN. After the resections, which were carried out with only short margins from the inflamed areas, the patients were kept on TPN for 4 weeks after surgery to give the bowel additional time for regeneration.

The 25 patients who completed the course of TPN returned to work or household duty within 4 weeks after the end of TPN. At this time they ate normal meals and have no evidence of Crohn's disease. During the first 12 months eight patients had to return to hospital because of a relapse of Crohn's disease. The common radiological findings during the first admission in six had been moderate to high grade stenoses with inflammation. Their short improvement was probably due to a reduction of the mucosal oedema at the stenoses, which facilitated the food passage and relieved their symptoms. They suffered from the same problems as before when oral feeding was resumed. In these patients and one other, in which endoscopy showed complete loss of the mucosa of the terminal ileum, resection was performed.

In the eighth case inflammation without substantial stenoses was the prominent figure in the whole colon. This patient was treated with another course of TPN and is now in remission after 3 years.

A relapse of Crohn's disease after 12–48 months occurred in nine patients. In four a resection was performed. The fifth patient was admitted to another hospital with an intra-abdominal abscess which was drained and a colostomy performed. The patient died 3 days later from sepsis. Four patients are under different forms of medication. The other 13 patients are in remission between 10 and 48 months and have no evidence of active Crohn's disease. They are not on any medication. Ten of them insist on adhering to some form of diet, avoiding milk products, sugar or fresh vegetables. The cumulative recurrence rate gives an idea of the potential of TPN in Crohn's disease. The probability of a relapse is 34.6% after 1, 59.3% after 2, 80.9% after 3 and 85.1% after 4 years.

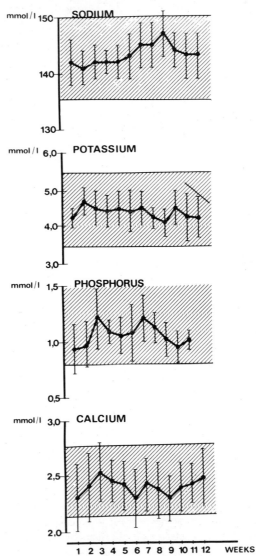

Figure 35.2 Serum concentrations of different electrolytes during TPN (hatched area = normal range)

METABOLIC SITUATION DURING HOME TPN

The patients were discharged from the hospital to home TPN only if their metabolic situation was stable and if their disease had entered a subacute phase with no more than 10 stools/day, a fistula drainage below 500 cc/day and no fever or ileus. The plasma concentrations of the electrolytes (Figure 35.2) and trace elements (Figure 35.3) stayed within our normal range during the weeks of home TPN. In five patients minor adjustments to the infusion scheme were necessary because of low levels of phosphorus or zinc.

Total protein (Figure 35.4) rose during the first 6 weeks of TPN from 6.2 g/dl to an average of 7.8 g/dl and then stayed constant. This was only partly due to the infusion regime, because in the beginning of our experience with longterm TPN, patients with a serum albumin concentration below 3.0 g/dl received 20–60 g albumin/day for 7–21 days. Despite the use of a 20% fat emulsion, which was infused in 4–5 h, the serum of the patients was never lipaemic, when it was checked 12 h later. In the amino-acid profiles (Figures 35.5 and 6) only the concentrations of methionine and ornithine were partly or constantly above our normal ranges.

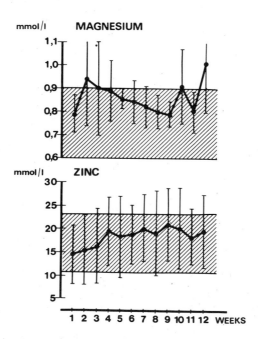

Figure 35.3 Serum concentrations of zinc and magnesium during TPN (hatched area = normal range)

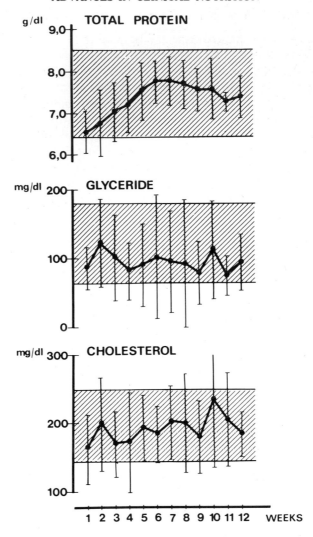

Figure 35.4 Serum concentrations of total protein, glyceride and cholesterol during TPN (hatched area = normal range)

COMPLICATIONS

Sepsis from the central venous catheter was noticed in three cases. In relation to the duration of TPN the sepsis rate was 0.9 cases per 1000 days TPN. After the removal of the catheter there was complete resolution of all symptoms. We nourished these patients for 1 or 2 days through a peripheral vein and then inserted a new central venous catheter. Two patients suffered from catheter embolism. The lost catheter tips were removed from the right

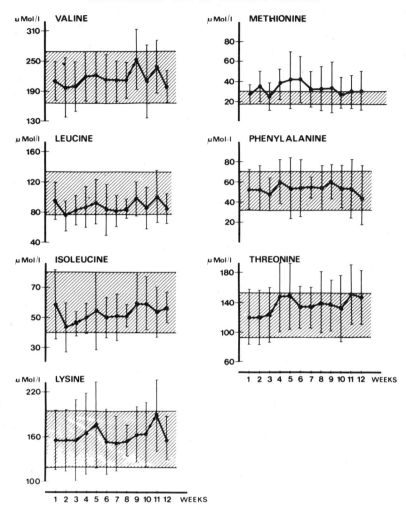

Figure 35.5 Plasma concentrations of essential amino acids during TPN (hatched area = normal range

atrium in one and the superior vena cava in the other by a transjugular technique. Thrombosis of the subclavian vein was observed in one case. The patient was treated by anticoagulation.

Metabolic disturbances observed were of minor degree and never made an interruption of TPN necessary. SGOT, SGPT and alkaline phosphatase increased slightly in all patients (Figure 35.7). Four patients showed signs of intrahepatic cholestasis with bilirubin levels over 2 g/dl. There was a rise of SGOT, SGPT and alkaline phosphatase during the sixth to tenth week of home TPN. The cholangiograms of the four patients were normal. Liver

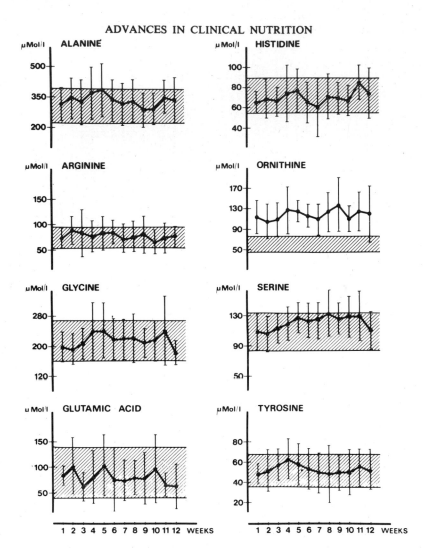

Figure 35.6 Plasma concentrations of non-essential amino acids during TPN (hatched area = normal range)

biopsies showed fatty infiltration of the peripheral type and pericholangitis. Alterations of the infusion scheme such as changing the nitrogen–energy ratio, reducing alternatively the amount of fat or dextrose, or lengthening of the infusion time led to a temporary fall of the liver enzymes but they returned to normal after the patients began to eat normally.

DISCUSSION

The results of TPN as primary therapy for Crohn's disease show some parallels to exclusion bypass or defunctioning ileostomy advocated in the

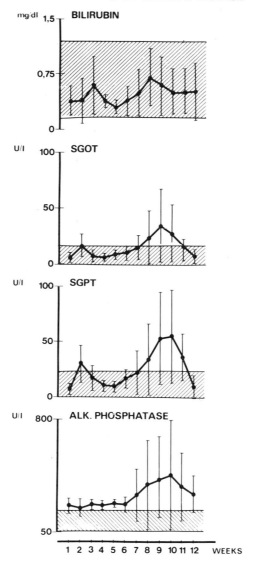

Figure 35.7 Concentration of bilirubin, SGOT, SGPT and alkaline phosphatase during TPN (hatched area = normal range)

mid-1930s by Garlock[16] and later by Oberhelman[17]. The therapeutic principle is to put the inflamed bowel to rest and postpone or avoid resection. The advantage of TPN is that operation may not be necessary, the risk of fatal complications may be less and nutritional status may improve more rapidly. On the other hand the costs of TPN are high. The early results of all three methods were promising. However, late results indicated that the optimistic expectations have not been anything like fully realized. The recurrence rate of our study is nearly the same as that reported by Oberhelman[17] for defunctioning ileostomy. In his series only in six out of 21 patients was the ileostomy closed and of these six only three lived without signs of recurrent disease.

Certainly TPN is useful in preparing malnourished patients with Crohn's disease for operation. This can be done successfully within 10 days and may lead to fewer post-operative complications as demonstrated in a controlled trial with cancer patients with reduced nutritional status[18]. TPN as primary therapy may be reserved for patients with multifocal disease, when surgery is not advisable because of the risk of the short bowel syndrome.

As long as the natural history of Crohn's disease is unpredictable, at least some patients may have the benefit of a longterm remission, and in others the extent of resection may be limited. In our few cases with remissions over several years the inflammatory component of Crohn's disease was the predominant figure.

It is therefore possible that when the first manifestations of Crohn's disease are pain, diarrhoea, anorexia and weight loss with only minor obstructive signs, parenteral feeding and intestinal rest may be sucessful in inducing remission. Against the use of TPN in these early stages stand the good results of medical therapy and the costs of TPN. The complications of home TPN can be reduced, if people interested in the problem are involved. A highly specialized team[19], preparing TPN fluids and controlling the patients' metabolism and mechanical devices for the delivering of the solution, is not necessary. The patients in this study were managed by a surgeon and a nurse. With Broviac's technique[14] for the central venous catheter, we had only two complications in the beginning, due to limited experience. Catheter-related sepsis remains the problem. A rate of 0.9 cases in 1000 days parenteral nutrition seems however to be acceptable. Using commercially available solutions in glass bottles, adding only preparations of vitamins and trace elements and using a Y-shaped infusion line proved to be a simple and reliable method to prevent infection. When the patients have been brought to a metabolically stable condition only few parameters such as electrolytes and liver enzymes need be measured once a week. The reason why four patients developed intrahepatic cholestasis, remains speculative. The histological findings in their liver biopsies are relatively common in Crohn's disease and cannot only be attributed to TPN. Others who have observed similar alterations of liver metabolism during longterm TPN suggested a deficiency of copper, essential fatty acids or folic acid, as well as hyper-ammonia or an inadequate nitrogen−energy ratio as possible causes[20−23]. We could exclude most of these factors and believe that the combination of a high amount of nutrients with already impaired liver function were responsible. The high concentration of ornithine in the plasma of all patients is to be explained by the amino-acid solution used, which contained a high level of ornithine and arginine. Because ornithine is a product of the arginine metabolism, its addition to an amino-acid preparation seems questionable.

References

1 Steiger, E., Wilmore, D. W., Dudrick, S. J. and Rhoads, J. E. (1969). Total intravenous nutrition in the management of inflammatory disease of the intestinal tract. *Fed. Proc.*, **28**, 808
2 Lennard-Jones, J. E. and Stadler, G. A. (1967). Prognosis after resection of chronic regional ileitis. *Gut*, **8**, 332
3 Stock, W., Müller, J. M. and Nohr, L. (1979). Das Rezidiv nach chirurgischer Behandlung des Morbus Crohn. *Dtsch. Med. Wschr.*, **2**, 47
4 Driscoll, R. H. and Rosenberg, I. H. (1978). Total parenteral nutrition in inflammatory bowel disease. *Med. Clin. N. Am.*, **1**, 185
5 Elson, C. O., Layden, T. J. and Nemchausky, B. A. (1980). An evaluation of total parenteral nutrition in the management of inflammatory bowel disease. *Dig. Dis. Sci.*, **25**, 42
6 Fazio, V. W., Kodner, I., Jagelman, D. G., Turnbull, R. B. and Weakley, F. L. (1976). Parenteral nutrition as primary or adjunctive treatment. *Dis. Colon Rectum*, **19**, 574
7 Greenberg, G. R., Haber, G. B. and Jeejeebhoy, K. N. (1976). Total parenteral nutrition and bowel rest in the management of Crohn's disease. *Gut*, **17**, 828
8 Holm, I. (1977). Parenteral nutrition in surgical and medical gastroenterology. *Acta Chir. Scand.*, **143**, 297
9 Reilly, J., Ryan, J. A., Strole, W. and Fischer, J. E. (1976). Hyperalimentation in inflammatory bowel disease. *Am. J. Surg.*, **131**, 192
10 Mullen, J. L., Hargrove, W. C., Dudrick, S. J., Fitts, W. T. and Rosato, E. F. (1978). Ten years' experience with intravenous hyperalimentation and inflammatory bowel disease. *Ann. Surg.*, **181**, 523
11 Law, D. U., Dudrick, S. J. and Abdou, N. J. (1974). The effect of protein calorie malnutrition on immune competence of the surgical patients. *Surg. Gynecol. Obstet.*, **139**, 258
12 Dudrick, S. J., MacFadyen, B. V. and Daly, J. M. (1976). Management of inflammatory bowel disease with parenteral hyperalimentation. In Clearfield, H. R. (ed.) *Gastrointestinal Emergencies*. (New York: Grune & Stratton) pp. 193–199
13 Allan, R. N. (1980). Invited commentary. *World J. Surg.*, **4**, 165
14 Broviac, J. M., Cole, J. J. and Scribner, B. B. (1973). A silicone rubber atrial catheter for prolonged parenteral alimentation. *Surg. Gynecol. Obstet.*, **136**, 602
15 Best, W. R., Becktel, J. M., Singleton, J. W. and Kern, F. (1976). Development of a Crohn's disease activity index. National cooperative Crohn's disease study. *Gastroenterology*, **70**, 439
16 Garlock, J. H., Crohn, B. B. and Klein, S. H. (1951). An appraisal of the longterm results of surgical treatment of regional enteritis. *Gastroenterology*, **19**, 414
17 Oberhelman, H. A. and Kohatsu, S. (1971). Diversion by ileostomy for Crohn's disease of the colon. In Krause, U. (ed.). *Skandia Symposium on Regional Enteritis*. (Stockholm: Nordiska Bokhandelns Förlag)
18 Müller, J. M., Brenner, U. and Pichlmaier, H. (1982). Preoperative parenteral feeding in patients with gastrointestinal carcinoma. *Lancet*, **1**, 68
19 Dudrick, S. J., MacFadyen, B. V., Souchon, E. A., Englert, D. M. and Copeland, E. M. (1977). Parenteral nutrition technique in cancer patients. *Cancer Res.*, **37**, 2440
20 Eckardt, K. (1978). Leberveränderungen bei morbus Crohn und colitis ulcerosa. *Dtsch. Med. Wschr.*, **103**, 1983
21 Ghadimi, H., Abaci, F., Kumar, S. and Rath, M. (1971). Biochemical aspects of intravenous alimentation. *Pediatrics*, **49**, 955
22 Johnston, J. D., Albritton, W. L. and Sunshine, P. (1972). Hyperammonemia accompanying parenteral nutrition in new born infants. *J. Pediatr.*, **81**, 154
23 Dudrick, S. J. and Long, J. M. (1976). Applications and hazards of intravenous hyper-alimentation. *Ann. Rev. Med.*, **28**, 517

7
Preliminary report on collaborative study for home parenteral nutrition patients

G. BLACKBURN, C. DI SCALA, M. M. MILLER, C. CHAMPAGNE,
M. LAHEY, A. BOTHE AND B. R. BISTRIAN

INTRODUCTION

Home-based total parenteral nutrition (HTPN) is a unique, new technology which can provide the daily nutrient requirements by central vein. For patients with gastrointestinal disability, death by starvation is prevented and self-care outside the hospital is made possible. Systematic evaluation of HTPN therapy is required to determine factors such as the incidence of new cases requiring HTPN, expenses, survival rate, and the quality of life of patients on this therapy. Such evaluation will lead to maximized patient benefit and utilization of this unique therapy.

In this report, an early assessment of HTPN by pharmacists conducted through the New England Deaconess Hospital and the Registry of Patients on Home Total Parenteral Nutrition is presented. In the future, our goal will be patient-oriented data acquisition. Information from patients on HTPN will result in a more comprehensive understanding of how this therapy can best be used.

HISTORY OF HTPN

In 1968, Dudrick showed that it is possible to administer a complete diet via a large vein and thus prevent body destruction and death from starvation[1]. Since that time, parenteral hyperalimentation has become an established life-saving tool for many patients including the severely ill[2,3], victims of multiple injury and burns[4], people with extensive weight loss prior to or recovering from major surgery[5], and patients being treated for inflammatory bowel disease[6-8] or cancer[9].

Over the past decade, refinements in nutritional assessment[10,11], innovations in solution administration[12,13], and the treatment of specific disease states[14] have led to increased sophistication in parenteral nutrition. Awareness of the complications of this therapy[15-18] has improved the safety with which parenteral nutrition can be given.

The ability to provide all the required nutrients intravenously has resulted in the survival of patients who previously would have died of starvation. The success of this therapy has created a new group of patients whose underlying disease prevents adequate nutritional absorption and thus who must be sustained with parenteral hyperalimentation. Though these patients have recovered from the acute effects of their illness, they still suffer from gastrointestinal disability because oral intake alone will not prevent starvation. Affected individuals include those with the short bowel syndrome due to congenital problems, injury to major blood vessels, or operation; those with Crohn's disease or a regional enteritis, an episodic inflammation of the bowel preventing adequate absorption of nutrients; patients with fistulae which often require prolonged nutritional support before closing; patients with radiation enteritis whose bowel is damaged severely by radiotherapy; and also the group of patients unable to eat due to side-effects of cancer treatment.

PRESENT APPROACH

Dependency on intravenous feeding by those who are otherwise healthy enough to be free of the hospital raises several complex issues. Ethically, it has been impossible to withdraw lifesaving support from alert, viable, and productive individuals. Economically, the cost of keeping such patients in hospital is prohibitive. Aware of both the economic pressures resulting from prolonged parenteral support and the improved well-being seen in home dialysis patients when returned to the family. Dr Scribner developed home parenteral nutrition in the 1970s[19].

Technique

In the present home hyperalimentation technique, solutions rich in calories, protein, vitamins and minerals are administered via a silicone catheter threaded percutaneously into the superior vena cava and tunnelled under the skin. The large vein allows mixing of concentrated solutions within the circulation; the entry site location provides protection against the spread of infection from the skin entry site into the bloodstream. Using a variable flow rate accurately controlled by a pump, patients can receive the entire volume of solutions overnight and thus remain active during the day. Many patients work as volunteers, attend school, or do full-time housework. HTPN provides therapy at home emphasizing self-care and a decrease in the need for expensive medical care in a hospital.

Unlike dialysis patients, the health and stamina of the home parenteral nutrition patient tends to be excellent. The majority of HTPN procedures are performed by the patient or a family member. Several have reported taking raft trips down river and motorcycle tours[20]. In fact, 70% of patients monitored through the New England Deaconess Hospital in 1980 (up to 6.5 years on HTPN) stated they could return to work if the present disability policies were modified to allow them to work without losing their disability benefits.

THE NEDH PROGRAMME

The New England Deaconess Hospital, Boston, Massachusetts, began home parenteral nutrition therapy in 1974. From 1974 to July, 1982 a total of 37 patients received this therapy.

An analysis of the patients receiving HTPN prior to 1980 provided our first detailed information. At this time, a total of 29 patients had received HTPN. Seventeen were still receiving home parenteral nutrition in 1980 with 6.5 years being the longest duration of therapy. Twelve patients had terminated hyperalimentation: six were alive having had favourable responses to HTPN, three died from underlying diseases, two died of systemic sepsis from intraabdominal infection and one died of heart disease.

The reasons for home parenteral nutrition prior to 1980 included inflammatory bowel disease, short bowel syndrome, radiation enteritis, malignancy, fistulae and motility disorders (Table 36.1).

Table 36.1 Diagnostic basis for home parenteral nutrition

Diagnosis	Number	(%)
Malignancy resulting in bowel dysfunction	2	7
Inflammatory bowel disease with or without resection	14	52
Bowel infarction with resection	4	15
Motility disorder (e.g. pseudo-obstruction)	1	4
Radiation enteritis	4	15
Other (multiple abdominal surgeries, multiple fistulae)	2	7
	27	100

A total of eight patients required 11 hospital admissions in 1980, and nine patients were able to avoid hospital admission entirely. The reasons for these admissions included systemic sepsis, fluid and electrolyte imbalance, mechanical catheter problems and superficial infections (Table 36.2)

The patients fell into two categories: those capable of mixing their own solutions and those requiring the regular aid of family or hospital staff. Three-quarters were in the self-mix group preparing their own solutions, managing catheter care, and ordering supplies independently of the hospital staff. An additional group involved their family in some of the tasks. Only one patient in an extended care facility participated in none of his home parenteral nutrition care (Table 36.3). Of the 82% who were not working, only 30% stated their underlying disease as their reason for not working. The remaining 70% cited their disability certification restrictions (Table 36.4).

Table 36.2 Rehospitalization during 1979 and 1980 of HPN patients directly related to home parenteral nutrition

	1979	1980
Total number of patients at home on HPN	10	17
Total number of hospital admissions	10	11
Total number of days of hospitalization for all patients combined	87	148
Number of patients with more than one admission related to HPN during this period	3	3
Causes of admissions		
Sepsis	4	5
Exit site infection	—	1
Catheter-related problems	1	2
Fluid and electrolyte problems	1	3
Other (increased fistula or ileostomy output, intractable diarrhoea)	4	1
Number of patients with no admissions related to HPN	4	9

Table 36.3 Management of care of home parenteral nutrition (patient on HPN as of December 31, 1980)

	Number	%
Patient performs entire procedure (includes preparation of solution as well as its administration)	13	76
Family member performs entire procedure	1	6
Patient performs entire procedure except solution preparation	2	12
Patient does not perform any of the procedure	1	6
	17	100

Table 36.4 Patient employment, 1980

	Number	%
Working		
(1) Full time	1	6
(2) Part time (only part time due to reimbursement requirements)	2	12
Not working		
(1) Choice	0	
(2) Physical disease, age	5	29
(3) Requirement for reimbursement	9	53
	17	100

In a 1981 census study based at the New England Deaconess Hospital, 125 questionnaires were sent to full-time pharmacists of Massachusetts hospitals. 104 (83%) of the questionnaires were answered. From this census, it was concluded that prior to 1981, 81 patients had used home TPN therapy; one-third were from the New England Deaconess.

More recently, the HTPN patients followed at the New England Deaconess Hospital between January, 1981 and July, 1982 have been described. A total of 17 HTPN patients are now on the therapy. During this period, eight new patients started HTPN therapy, three patients ended HTPN therapy with rehabilitation, and five patients died due to underlying diseases (Table 36.5).

Table 36.5   Patient diagnosis, January 1981–June 1982

New patients
   5 short bowel syndrome
   1 Crohn's disease
   1 hyperemesis gravidarum
   1 scleroderma (malabsorption)

Patients who completed HTPN
   1 hyperemesis gravidarum (pregnancy completed)
   1 short bowel syndrome (bowel rehabilitated)
   1 Crohn's disease (bowel rehabilitated)

Patients who expired
   2 short bowel syndrome
   1 scleroderma
   1 entero-gut fistula (2 ° to cancer, rad Rx)

Patient diagnoses were similar to those of the patients studied earlier, with the addition of one case of HTPN for hyperemesis gravidarum (Table 36.5).

A total of 14 readmissions occurred in the 1981–June, 1982 period. These admissions constituted a total of 288 hospital days, the vast majority of which were related to three patients. The complications are shown in Table 36.6.

Table 36.6   Patient complications related to HTPN, January 1981–January 1982

	*Number of readmissions**
Catheter sepsis	8
Catheter sepsis and thrombosis	1
Exit site infection	1
Exit site infection and thrombosis	1

*2 additional readmissions occurred during which the possibility of infection was ruled out

437

REGISTRY OF PATIENTS ON HTPN

The Board of Regents of the American College of Surgeons and the National Institutes of Health approved and funded the establishment of a Registry of patients on Total Parenteral Nutrition at Home (HTPN) in 1976 as part of the Organ Transplant Registry. Upon termination of funding of the Organ Transplant Registry, the HTPN Registry was transferred to the New York Academy of Medicine in 1977 under direction of Maurice Shils and a board of directors.*

The HTPN Registry took the opportunity to:

(1) Obtain data on human nutritional and metabolic reactions to parenteral nutrition in this unique type of patient.

(2) Serve as a source of information on the clinical and technical management of such patients.

(3) Obtain information on the medical, social, and economic problems associated with this technique and on the medical and socio-economic benefits which may result from their out-patient status.

(4) Stimulate research and teaching in this field.

The specific aims were as follows:

(1) To obtain, summarize, and distribute appropriate information concerning patients on HTPN and to analyse resulting information and issue periodic statistical reports.

(2) To serve as a documentation and reference source for interested physicians, other health personnel, governmental agencies, and insurance carriers.

HTPN Registry survey

At the outset, initial reporting forms were required for each patient from each participating physician. This became an administrative inconvenience for the respondents and approximately 1 year later, the Registry switched to a single questionnaire per institution per year. The first institution-oriented questionnaire covered the period July 1, 1977 to June 30, 1980 and collected data from 29 respondents. The second one covered the period July 1, 1978 to December 31, 1978 and 19 respondents. Since then the survey questionnaire has been covering the period January 1 to December 31, with 30 respondents in 1979 and 63 respondents in 1980 (Table 36.7).

*Advisory Board: Marvin E. Ament, George L. Blackburn and Khursheed N. Jeejeebhoy.

COLLABORATIVE STUDY FOR HPN PATIENTS

Table 36.7a Introduction to TPN Questionnaire

Dear TPN Patient,

A survey of selected Homecare TPN patients is being undertaken to ascertain if services and prescription administration can be improved for your benefit while reducing costs and ancillary problems.

Please do not place your name or identifying mark on any part of the questionnaire.

All of these questions are designed to test your opinions and estimates on approximately 15 brief questions relating to areas that we are seeking to improve for your total care.

If you have suggestions of your own relating to alternatives or areas not specifically covered, please feel free to add on your comments at the end of the form.

Table 36.7b TPN Questionnaire

---

(1) How often do you infuse your TPN solution?
daily _____   weekly _____   other _____

(2) How long does it take you to compound your product?
_____ hrs.   _____ mins.

(3) How long does it take you to set up your equipment to ready your area for compounding?
_____ hrs.   _____ mins.

(4) When do you compound your solutions?
morning _____   afternoon _____   night _____

(5) Where do you compound your solutions?
privacy of own room _____   closet _____   public area _____

(6) Have you ever experienced any of the following? (if yes, then explain)

	YES	NO
feel rushed to prepare	—	—
concern over preparation method	—	—
missing components in kits	—	—
leaking of components or preparations	—	—
contamination problems	—	—
difficulty in opening	—	—
difficulty in measuring	—	—
difficulty in handling components	—	—
lack of space to store components	—	—
interference in privacy	—	—
being interrupted during compounding	—	—
build-up of extra components not used	—	—
formulation changes	—	—
lag time in getting changes	—	—
difficulty getting to hospital for supplies	—	—
late delivery by commercial carriers	—	—
need service at night	—	—

Explanations:

_____

_____

(7) Have you ever had any episode of sepsis?   Yes _____ No _____

(8) If yes, do you feel that this episode could have been prevented by having sterile products delivered to your home ready to use?   Yes _____ No _____

(9) Are you concerned about cost reduction?   Yes _____ No _____

(10) Would you be interested in receiving a quotation from an alternate supplier?
Yes _____ No _____

439

Table 36.7a (*Continued*)

---

(11) Does your current supplier make available the following products or services?

	YES	NO
sterile supplies	—	—
injectables	—	—
prescription medications	—	—
enteral supplies	—	—
over the counter items	—	—

(12) Does current supplier handle all your insurance paperwork?

Yes _____   No _____

(13) Does current supplier provide 24 hour *local* availability of

Pharmacist	Yes _____	No _____
Nurse	Yes _____	No _____

(14) How long does it take to have unused items picked up and credited?

_____

(15) In summary, would you be interested in having your solutions precompounded and delivered to you?   Yes _____ No _____

Comments, if any:

_____

_____

_____

Summary of Registry data on patients on HTPN through 1980

An increased number of HTPN programmes participated in the 1980 survey, with 63 respondents to the questionnaire as compared to 30 in 1979, indicating that the Registry of patients on Home Total Parenteral Nutrition has become recognized as a source of information on the scope of HTPN programmes. Of the responding institutions, two are located in Canada and 61 are located in 21 states in the US.

Although it is believed that the respondents represent the great majority of hospitals sending patients home on TPN, there are a few hospitals who were unable to supply any data. As a result the number of patients discharged on HTPN, the number of children/adolescents, and some diagnoses are probably underestimated. The summary report on the 1980 survey is divided into 3 sections.

(1) Data on patients discharged on HTPN before January 1, 1980 (Table 36.8).

(2) Data on patients discharged on HTPN during 1980 (Table 36.9).

(3) For all patients on HTPN at any one time during 1980, data on hospitalization, catheter, quality of life, reimbursement (Tables 36.10, 11, 12, 13).

Table 36.8 HTPN Registry. Patients discharged on HTPN before January 1, 1980

	Number	%
Total patients discharged on HTPN before Jan. 1, 1980	552	
Patients still on HTPN as of Jan. 1, 1980	214	38.8
Patients not on HTPN as of Jan. 1, 1980	338	61.2
(a) still alive	105	19.0
(b) dead	188	34.1
(c) lost to follow-up	45	8.2
Total patients on HTPN as of December 31, 1980	384	
Patients started during 1980	171	
Patients started before Jan. 1, 1980	214	
Years on HTPN		
10–11	1	.25
9–10	1	.25
8–9	2	.5
7–8	2	.5
6–7	5	1.3
5–6	10	2.6
4–5	23	6.0
3–4	34	8.8
2–3	56	14.6
1–2	79	20.6
<1	171	44.5

Table 36.9 HTPN Registry. Patients discharged on HTPN during 1980

	Number	%
Patients discharged on HTPN during 1980	260	
Patients still on HTPN as of December 31, 1980	171	65.8
Patients not on HTPN as of December 31, 1980	89	34.2
Age of patients discharged in 1980:		
$\leq$ 20 y	41	15.8
21–30 y	26	10.0
31–40 y	53	20.3
41–50 y	45	17.3
51–60 y	47	18.1
>60 y	48	18.5
Diagnosis of patients discharged in 1980:		
Inflammatory bowel disease with/without resection		
and/or fistulae	83	31.9
Ischaemic bowel infarction with resection	38	14.6
Malignancy*	85	32.6
All others †	54	20.9

*or its treatment resulting in: obstruction, fistulae, malabsorption from radiation enteritis, malabsorption from bowel resection, tumours

†includes: motility disorder, congenital malformation, trauma, persistent bowel fistulae unrelated to above disease, short bowel syndrome unrelated to above disease, cystic fibrosis, chronic obstructive pulmonary disease, etc.

Table 36.10 HTPN Registry. All patients (*n* = 474) on HTPN during 1980: Hospitalization

	Number	%
Patients hospitalized*	312	
Admissions directly related to HTPN	210	
Total days in hospital related to HTPN †	1847	
Average number of admissions related to HTPN		67.3
Average length of hospital stay (days/admission)	12.5	
Primary cause of admission		
Sepsis	108	51.4
Catheter-related problems	53	25.2
Fluid/electrolyte problems	24	11.4
All other ‡	25	12.0

*Data available on 312 patients, since some hospitals have no data on hospitalization.
† Based on data for 239 patients with 146 admissions.
‡ Includes: bone disease work-up or treatment, retraining, hypo/hyperglycaemia, organ failure/dysfunction, psychological problems, etc.

Table 36.11 HTPN Registry. Catheters

	Number	%
Total number of patients	534	
Type of catheter		
Broviac–Scribner	235	44.0
Hickman	141	26.4
Subcutaneous CVC	76	14.2
Percutaneous	71	14.6
A–V fistula/shunt	4	0.8
Total catheter-related problems (episodes)	80	
Type of problem		
Mechanical failure	38	47.5
Clotting/occlusion	28	35.0
Dislodged catheter	13	16.2
Local infection	1	1.3

Section 3 introduces information which was not available in previous questionnaires. About 90 institutions were listed in the registry by the end of 1980. The questionnaire was mailed to the participants during the first weeks following the end of the year. The compliance was and is entirely voluntary.

Patients discharged during 1980
A total number of 260 patients were discharged on HTPN by 63 reporting institutions. As of December 31, 1980, 171 or 65.8% of these patients remained on HTPN. No further information is available yet on the other 89. Most of the patients (74.2%) were older than 30 years of age and the most

Table 36.12   HTPN Registry. Quality of Life*

	Number	%
Total patients	415	
Patients leading a normal life other than the need for HTPN	260	62.7
Patients in semi-invalid state		
(a) as a result of HTPN	14	3.4
(b) due to underlying illness	112	27.0
Patients with severely restricted activity		
(a) as a result of HTPN	4	1.0
(b) due to underlying illness	25	6.0
% Patients who are married		75.8
% Patients with children under age 18		30.0
% Patients who would be dead or institutionalized without HTPN		89.0

*Figures based on the personal estimates of those responding

Table 36.13   HTPN Registry. Reimbursement of HTPN costs for patients at home <2 years

	Number	%
Total patients	374	
Method of reimbursement		
Insurance including Major Medical	141	37.7
Medicare	62	16.6
Combination forms*	57	15.2
Medicaid	36	9.6
Provincial government funding (Canada)	29	7.8
Welfare	12	3.2
Write-off	8	2.1
Other †	29	7.8

*Medicare/insurance most frequent with 30 patients
†Includes VA, CCS, Self-pay, no insurance and non-specified

frequent diagnosis at discharge (32.6%) was malignancy or the treatment of it resulting in obstruction, fistulae, malabsorption from radiation enteritis, malabsorption from bowel resection, or tumours. Inflammatory bowel disease with or without resection was the second most frequent diagnosis covering 31.9% of the cases.

Patients discharged before January 1, 1980
The responders to the survey reported a total number of 522 patients discharged on HTPN before January 1, 1980. As of this date, 214 (38.3%) patients were still on HTPN, 105 (19.0%) alive and not on HTPN any more,

while 188 (34.1%) were dead and 45 (8.2%) were lost to follow-up. Since data on these patients go back as early as 1970, by combining the alive, off-HTPN group, possibly recovered from the underlying illness and the patients still on HTPN, it is clear that the therapy has been successful in at least 57.8% of the cases between 1 and 10 years.

All patients on HTPN during 1980

An assessment was made of all the patients who at any one time were on HTPN during 1980 to establish the cause of hospital admission, the type of catheters being used and problems, the quality of life and form of reimbursement. Not all the institutions participating in the survey were able to supply data on these questions. The number of respondents is reported in Tables 36.9, 10, 11 and 12.

A total number of 312 patients were admitted to hospital during 1980, 127 for HTPN-related problems. The total number of admissions related to HTPN was 210, with 79 patients accounting for 79 admissions, 28 patients accounting for 56 admissions (2/patient) and 20 patients accounting for 75 admissions (3 or more/patient). Hence, 48 patients (15.4%) accounted for 131 (62.4%) of the HTPN-related admissions. The average number of HTPN-related admissions was less than one per patient (67.3 per hundred patients) and the average number of days per admission was 12.5.

The most common cause of hospitalization was sepsis, not always confirmed by culture, followed by catheter-related problems and fluid or electrolyte imbalance. The frequency of catheter-related problems amounted to 80 episodes, mechanical failure and clotting being the most common with 66 episodes or 82.5% of the total. The Broviac–Scribner and Hickman catheters were used by 71.4% of the patients.

An attempt was also made by the 1980 survey to assess the quality of life led by patients on HTPN. The reporting institutions estimated that 89% of the patients would be institutionalized or dead without HTPN, that only 4.4% of the patients have limited or severely limited activity as a consequence of HTPN, while 62.7% of the patients are able to lead a normal life other than the need for HTPN. This latest figure agreed also with the reported number of patients living a common family life with 75.8% of the patients being married (an underestimated value, since not all patients are of marital age) and 30.2% of patients with children under age 18.

Almost all patients on HTPN for less than 2 years received full or partial support for the cost of medical care either by a major insurance or a government agency or a combination of them.

Five patients (Table 36.13, Category 'other') were reported to be at least partially self-paying, and one received free HTPN from the delivering centre.

However, 23 institutions or 41.8% of the responders would not accept patients unable to pay into HTPN programmes (Table 36.14). This information deserves further investigation.

Table 36.14 HTPN Registry. Availability of HTPN care

Inability of patients to pay for HTPN costs inhibits acceptance of patients in the HTPN programme.	yes: 23 institutions (41.8%) no: 32 institutions (58.2%)

Given the data available on the form of reimbursement, it is possible to speculate that financial difficulties deny a considerable number of people an effective therapy with disastrous consequences.

PROSPECTS OF THE FUTURE

The HTPN Registry is an invaluable source of information for assessing the status of this therapy. However, the willingness of nutritional services to complete and return mailed questionnaires and the shortage of data clerks at the delivering institutions, has prevented the Registry from collecting the in-depth patient information which is still needed to assess objectively the effectiveness and the problems associated with HTPN.

It would also be important in the future to be able to answer the difficult question as to how many patients need HTPN and how many do in fact receive it.

SECOND GENERATION COLLABORATIVE STUDY

A new multicentre, nationwide study is currently proposed. The supporting organization (ASPEN) and the data processing centre (NEDH) will provide data forms and core support to the participating institutions for the collection of baseline and ongoing information on the HTPN users. This will re-establish the *patient-oriented data acquisition* which the current Registry format does not provide. A battery of questionnaires will be completed for each patient by each institution at discharge, annually, and at the occurrence of major events such as hospitalization or termination of HTPN.

One questionnaire about adjustment to the therapy and quality of life will be completed periodically by each patient and/or family providing an opportunity for validation of information reported by the institutions. Information will be available on prevalence and incidence of HTPN, variables such as mortality, morbidity, number of hospital-free days, number of medication-free days, as well as changes in nutritional parameters and quality of life which are important for the characterization of disease states and the response to HTPN. In addition, collection of multicentre, nationwide data will allow a serious attempt at definition of the relationship between the aetiology and severity of disease states and the duration and outcome of HTPN therapy. Regional differences in HTPN therapy can also be assessed.

Since patients receiving HTPN represent an extremely high cost to the society, the proposed study is imperative to provide clarification of the indications for HTPN therapy.

PHYSICIAN'S PERSPECTIVE ON THE HOME TPN PATIENT

The care of patients with longterm gastrointestinal disability requires an experienced physician and the availability of social workers, physical therapists, nurses, dietitians, pharmacists, and physicians in medical specialties. A team approach is required where each member provides a perspective and analysis of patient status (physical, mental, and spiritual) and identifies patient needs which must be met to provide successful rehabilitation (Figure 36.1).

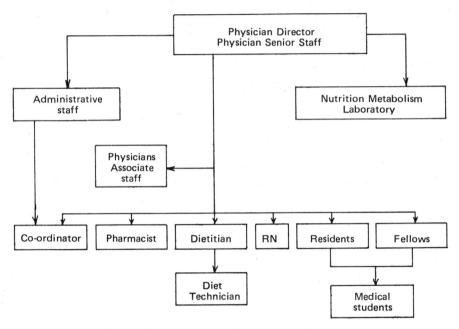

Figure 36.1 Organizational structure of nutrition support service

It is important for all members of the team, including the patient and the patient's family, to know they are dealing with a very difficult longterm illness which has rendered the patient physically weak, mentally confused, and spiritually exhausted.

Essential to the treatment are empathy, patience, and a resolve to do what is right regardless of time, effort, and ability of the patient to co-operate. This usually means avoiding multiple prescriptions, particularly of tranquillizing and addictive drugs. The approach entails repetitive, simple instruction given with encouragement and reassurance. Such attention is required for the patient to understand the rationale of TPN therapy and thus to be willing to work to get the best results. Often, this means responding to patients with words, deeds, and actions beyond what is strictly necessary.

Frequently during rehabilitation, a patient feels that his or her symptoms require medication or hospitalization. The physician and members of the team must review the treatment approach with the patient or patient's family, provide a careful integration of the patient's medical and surgical therapy and laboratory work with the ongoing treatment regimen, and give the expertise required to make the right diagnosis of the patient's symptoms and direct rehabilitation.

At all times, treatment must be consistent with the patient's needs. Treatment progresses by inches. Rarely does a giant step occur, unless remission of the disease occurs, a new treatment is implemented or the patient's environment changes. The process is a partnership among the patient, his/her family, physician, and the treatment centre and its members. It is essential that the physician provides leadership in a humane fashion.

Careful judgement, born of considerable experience, is necessary in evaluating the therapy at each clinic visit which may be on a weekly, monthly, or more longterm basis. The focus is on each day's experiences, symptoms of illness, problems, and the provision of a treatment plan that will be practical until the next visit. The treatment methods require assertiveness on the part of both the physician and the patient, but each must do his or her part to avoid aggressive, argumentative relationships. The physician must demonstrate sincere interest and give considerable reassurance for the relationship to be productive. The physician also does well to communicate to the patient the objective data upon which clinical decisions are based. Patient education and partnership are even more essential to successful HTPN therapy than they are to health-care in general.

ROLE OF HOME TPN THERAPY

Home TPN therapy is an important adjunct to the treatment of patients with longterm gastrointestinal disability. However, it is not without its problems. While it may be initiated only under a physician's direction, the knowledge and experience to do this is not very widespread. The therapy is still relatively new and educational programmes are limited. Physicians in the USA seeking help should contact: The American Society for Parenteral and Enteral Nutrition or the Home TPN Registry (2 East 103rd Street, New York City 10029, 212-876-8200 x253). For such a new therapy, it is essential to seek out regional centres experienced in providing expert care and training. The most qualified teams have treated 30–50 patients and have been in operation for at least 5 years.

HTPN therapy must be prescribed judiciously. It is an expensive form of treatment which should be used not merely to sustain life, but to enhance quality of life. While the effectiveness of HTPN has been well-established, it is important that the patient's progress be encouraged so that, as gastrointestinal function improves, parenteral therapy is reduced, and enteral feeding substituted whenever possible.

References

1 Dudrick, S. J., Wilmore, D. W., Vars, N. M. and Rhoads, J. E. (1968). Long-term total parenteral nutrition and growth, development, and positive nitrogen balance. *Surgery*, **64**, 134

2 Blackburn, G. L., Maini, B. S. and Pierce, E. C. (1977). Nutrition in the critically ill patient. *Anesthesiology*, **47**, 181

3 Steffee, W. P. (1980). Nutrition intervention in hospitalized geriatric patients. *Bull. N. Y. Acad. Med.*, **56**, 564

4 Long, J. M., Wilmore, D. W., Mason, A. D. and Pruitt, B. A. (1977). Effect of carbohydrate and fat intake on nitrogen excretion during total intravenous feeding. *Ann. Surg.*, **185**, 417

5 Mullen, J. L., Buzby, G. P., Matthews, D. C., Smale, B. F. and Rosato, E. F. (1980). Reduction of operative morbidity and mortality by combined preoperative and postoperative nutritional support. *Ann. Surg.*, **192**, 604

6 Fonkalsrud, E. W., Ament, M. E.; Fleisher, D. and Byrne, W. (1979). Surgical management of Crohn's disease in children. *Am. J. Surg.*, **138**, 15

7 Kopple, J. D. and Swendseid, M. E. (1975). Protein and amino acid metabolism in uremic patients undergoing maintenance hemodialysis. *Proc. Int. Soc. Nephrol.*, **64**, S72

8 Blackburn, G. L., Gibbons, G. W., Bothe, A., Benotti, P. N., Harkins, D. and McKinerny, T. (1977). Nutritional support in cardiac cachexia. *J. Thorac. Cardiovasc. Surg.*, **73**, 489

9 Harvey, K. B., Bothe, A. and Blackburn, G. L. (1979). Nutritional assessment and patient outcome during oncological therapy. *Cancer*, **43**, 2065

10 Mullen, J. L., Gertner, M. H. and Buzby, G. P. (1979). Implications of malnutrition in the surgical patient. *Arch. Surg.*, **114**, 121

11 Long, C. L., Schaffel, N., Geiger, J. W. *et al.* (1979). Metabolic response to injury and injury: estimation of energy and protein needs from indirect calorimetry and nitrogen balance. *J. Parent. Enteral Nutr.*, **3**, 452

12 Seltzer, M. H. (1978). Materials and devices. *J. Parent. Enteral Nutr.*, **2**, 572

13 Rutten, T., Blackburn, G. L., Flatt, J. P., Hallowel, A. B. and Chochran, G. (1975). Determination of optimal hyperalimentation infusion rate. *J. Surg. Res.*, **18**, 477

14 Abel, R. M., Beck, C. H., Abbott, W. M. *et al.* (1973). Improved survival from acute renal failure after treatment with intravenous essential L-amino acids and glucose. *N. Engl. J. Med.*, **288**, 695

15 Kaminski, M. V. (1978). A review of hyperosmolar, hyperglycemic, nonketotic dehydration: etiology, pathophysiology, and prevention during intravenous hyperalimentation. *J. Parent. Enteral Nutr.*, **2**, 690

16 Padberg, F. T., Ruggiero, J. A. and Blackburn, G. L. (1981). Central venous catheterization for parenteral nutrition. *Ann. Surg.*, **193**, 264

17 Price, B. S. and Levine, E. L. (1979). Permanent total parenteral nutrition: psychological and social responses of the early stages. *J. Parent. Enteral Nutr.*, **3**, 48

18 Sheldon, G. F. and Baker, C. (1980). Complications of nutritional support. *Crit. Care Med.*, **8**, 35

19 Scribner, E. H., Cole, J. J., Christopher, T. G. *et al.* (1970). Long-term parenteral nutrition: the concept of an artificial gut. *J. Am. Med. Assoc.*, **212**, 457

20 *Home TPN Newsletter* (1980). (Published by the University of Washington Home Parenteral Nutrition Team. Marianne F. Ivey (ed.). Seattle, Washington) Vol. VI, No. 1

Section 8
Conclusion

37
Future directions

J. M. KINNEY

I wish to make some informal comments with regard to five subjects, which I feel will undoubtedly be among the fruitful areas of investigation during the coming decade. These topics deal with metabolism, in particular the metabolic response to injury. This emphasis on metabolism in a nutritional symposium seems particularly appropriate, since many speakers have been presenting data which support the concept that the way a given individual responds to his, or her, nutrient intake depends upon the underlying metabolic status at the time.

METABOLIC BALANCE AND BODY COMPOSITION

Metabolic balance studies have tended to be down-graded in importance, as new and more sophisticated technology has been applied to the study of the acutely ill patient. Unfortunately, metabolic balance studies have seldom been performed simultaneously with serial measurements of body composition. This may be in part because balance studies are considered to be most important for looking at nitrogen balance, while body composition studies have traditionally been unable to deal with body nitrogen stores. Changes in total exchangeable potassium have been multiplied by a factor such as 8.33 to represent grams of body cell mass, or used to estimate changes in muscle protein by applying the ratio of 3 mEq/g of nitrogen. These techniques may be quite appropriate for the cross-sectional study of a population of individuals. However, we are getting increasing information from metabolic balance studies that the repletion of the depleted patient involves a restoration of body potassium at rates faster than the traditional 3 mEq per g of nitrogen, particularly in association with a high carbohydrate intake[1]. This finding may be explained in part by the association of potassium with new glycogen. Another possibility is that depletion involves a selective loss of potassium, which is out of proportion to loss of other intracellular constituents, including nitrogen. Neutron activation techniques are now being applied to the study of body composition in a way which measures body stores of nitrogen. It is of particular interest that Hill *et al.*[2] have found that the weight gain of patients receiving TPN for 2 weeks or more involves an increase in body water, body fat and potassium stores, without a corresponding increase in body nitrogen. It seems increasingly clear that we need to think of body composition in terms of the rates of loss in surgical catabolism and the rates of possible gain with surgical nutrition. The rates of restoration of a particular material may depend on the phase of convalescence and we must learn which materials take precedence in the order in which the tissue constituents are restored.

THE HYPERMETABOLIC PATIENT – ATP UTILIZATION

Resting energy metabolism has become an increasingly popular measurement in surgical patients, particularly when estimating the appropriate calorie intake. A patient is considered to be hypermetabolic if resting energy expenditure (REE) exceeds the predicted normal basal metabolic rate for a person of that age, size and sex by more than 10%. An important factor is often ignored, namely that any weight loss can be expected to produce some reduction in the resting energy expenditure of the previously normal individual. Therefore, the patient with 15% weight loss may have a resting energy expenditure of 10% above normal, which is considered insignificant, whereas an individual with this much weight loss from uncomplicated partial starvation might have an REE of minus 25%, if no catabolic stimulus such as low grade infection is present. Therefore, it is important to learn to recognize cases of such 'relative hypermetabolism' and establish the significance of this situation in regard to the optimum nutrition which such patients should receive.

When a patient is shown to have an increase in resting energy expenditure, we need to develop ways to understand how much of this increase represents an increase in production of high energy phosphate bonds versus an increase which is providing oxidative heat production, without phosphorylation. This is an important distinction for the individual without nutrient intake and additionally important when nutrient intake produces a calorigenic response which is over and above whatever hypermetabolism is already present. There used to be a common belief that the ATP content of muscle remained at normal levels at all times, except in association with major exertion or as the result of mitochondrial damage, such as following a low-flow state. This concept is generally correct because of the large reserve stores of high energy phosphate bonds in the form of creatine phosphate. However, work with needle biopsies of human muscle performed in Sweden[3] and also in the United States[4] has revealed that patients with massive injury, or serious infection, may have muscle ATP levels which are decreased as much as

25–40%, yet have a stable circulation and appear capable of full recovery. It will be important to determine whether these low levels represent impaired production or excessive utilization and what mechanisms might be involved. At least three major forms of energy utilization might be responsible for increasing the use of high energy bonds in ATP. The first is an increase in the normal work output of the cell, or tissue; the second is an increase in the synthesis, or turnover rate, of protein; and, the third, an increase in energy requirements to maintain electrolyte gradients across the cell membrane. Satisfactory techniques for measuring each of these three functions in human tissue are seriously needed. Another possible factor in the hypermetabolic patient is an increase in 'futile cycles' where energy is expended to move substrates without a net gain.

THE HYPERMETABOLIC PATIENT – HEAT METABOLISM

The presence of a fever in association with infectious disease has been recognized for many centuries. However, body temperature needs to be considered as representing the body heat content which is, of course, the balance between heat production and heat loss. We need to learn the mechanisms by which the production of fever has survival value for the patient and how to weigh the benefits versus liabilities with each level of increase in fever. Cuthbertson et al.[5] studied the influence of ambient temperature on the urinary nitrogen loss of patients with longbone fractures. It was found that the urinary nitrogen excretion could be significantly reduced during the first 12 days after convalescence if the ambient temperature was raised from 20 to 30 °C. We need to learn whether injury or infection changes the zone of thermoneutrality and, if so, what is the optimum ambient temperature for the care of such patients? Is it possible that a positive nitrogen balance can be achieved earlier and at lower levels of nitrogen intake, when an elevated ambient temperature is employed?

Controversy concerning the influence of ambient temperature on burn hypermetabolism needs to be resolved. Wilmore et al.[6] have presented evidence that the hypermetabolism of the burn patient is driven by internal factors influencing tissue metabolism, while Arturson[7] and Caldwell[8] believe that reducing heat loss to the burned surface can nearly abolish the hypermetabolism of such patients. The solution of this problem may shed light on the optimum nutritional management of such patients.

HORMONES AND MEDIATORS

During the period of classical endocrinology, which included the early years after World War II, various investigators sought to identify a 'wound hormone' which produced the catabolic response to injury. The search for a single catabolic hormone was unsuccessful. However, Sherwin *et al.*[9] introduced the concept that the hyperglycaemia associated with stress, or injury, was the result of three counter-regulatory hormones: cortisol, glucagon and one or more of the catecholamines. It may be possible that elevated levels of these three hormones, in relation to existing insulin levels, cause other features of the metabolic response to injury, but this has not been proven. Classical endocrinology has been undergoing a revolution during the past 15 years, with growing attention to the number and control of hormone receptors in various tissues. Little is known about this important new approach in regard to endocrine changes with injury. However, the list of conventional hormones is being steadily increased and new classes of chemical mediators are receiving increasing attention without necessarily being classified as hormones. Prominent in this latter group are the kinins, the prostaglandins and the materials elaborated from activated white cells. Beisel *et al.*[10] have extended their original ideas concerning a leukocyte endogenous mediator which causes an increased uptake of amino acids in the liver for acute phase protein synthesis. This laboratory has recently proposed that mediators from activated macrophages may orchestrate not only fever and an increase in acute phase protein synthesis, but a variety of other metabolic changes which we have formerly grouped together as the metabolic response to injury. The coming decade will undoubtedly witness important strides in this area and the understanding of injury may be dramatically assisted by studies of white cell function, which were first performed as a result of interest in immunology and transplantation.

THE PHYSIOLOGICAL EFFECTS OF NUTRIENTS

It is common knowledge that each of the major nutrients stimulate particular changes in metabolism, for example the thermic effect of feeding amino acids. However, there has been relatively little attention given to the fact that nutrients may influence physiological function, as well as cause metabolic changes. A recent example of this type of study[11] relates to the influence of carbohydrate intake on CO_2 production. Our group at Columbia has been using a non-invasive canopy-spirometer system for the on-line measurement of the influence of intravenous carbohydrate on the pattern of breathing. Such studies have emphasized the quantitative increases in CO_2 production, which occur with carbohydrate loads and, hence, the increase in ventilation which is required to clear the extra CO_2. It came as a surprise to us to realize that the CO_2 production in normal, post-absorptive man would be increased 40–100% above basal levels by the amount of carbohydrate which is given in conventional hyperalimentation.

In more recent studies[12], we have found that the amino acid intake appears to influence the sensitivity of the respiratory centre to CO_2 regardless of whether it is inspired CO_2 or the extra CO_2 from carbohydrate loading. Normal subjects and surgical patients tend to reduce their metabolic rate if given only 5% dextrose and water for 5 days or longer. Their minute ventilation will be decreased correspondingly. It is of interest that an infusion of isotonic amino acids by peripheral vein for only 24 h will restore the decreased metabolism and the decreased minute ventilation to normal levels. Future work is needed to determine which amino acids produce this effect and by what mechanism. It may be that the optimum nutrition for the acutely catabolic patient will have to be decided as a trade-off between the amount of calories and nitrogen which will provide the best nitrogen balance, while at the same time producing an acceptably small stress on physiological systems such as ventilation.

CONCLUSION

Many new and exciting areas in metabolism and nutrition are only becoming evident at the present time. Investigation in these areas can be expected to move from observations to the study of mechanisms and establish a new foundation for the understanding and treatment of the acutely ill and the seriously depleted patient. Much remains to be done to increase the awareness of the medical staff concerning malnutrition in the hospitalized patient. However, the length of time involved with a particular injury or illness is not limited to the period in the hospital. We must improve hospital nutrition and we must sensitize ourselves to the level of malnutrition which is still present when the patient goes home. In these days of hospital cost containment, fewer and fewer patients can be kept in the hospital long enough to make a substantial start on rebuilding tissues that were lost during the catabolic phase of their convalescence. The last thing to return in convalescence after serious injury, or illness, is not ambulation, or eating solid foods, or the healing of a wound so that sutures are not required, but the ability to work a full day without fatigue. Undoubtedly the entire period after injury or illness could be shortened for many patients by minimizing tissue loss as a result of proper nutrition in the hospital and accelerating convalescence after hospital discharge by the combination of graduated physical activity with nutritional intake.

References

1 Becker, M. A., Askanazi, J., Carpentier, Y. A., Elwyn, D. H. and Kinney, J. M. (1981). Correlations between nitrogen and potassium balance (Abstract). *J. Parent. Enteral Nutr.*, **5**, 259
2 Hill, G. L., King, R. F. G. J., Smith, R. C., Smith, A. H., Oxby, C. B., Sharafi, A. and Burkinshaw, L. (1979). Multi-element analysis of the living body by neutron activation analysis – application to critically ill patients receiving intravenous nutrition. *Br. J. Surg.*, **66**, 868

3 Bergstrom, J., Bostrom, H., Furst, P. *et al.* (1976). Preliminary studies of energy rich phosphagens in muscle from severely ill patients. *Crit. Care Med.*, **4**, 197

4 Ltaw, K. Y., Askanazi, J., Michelson, C. B., Kantrowitz, L. R., Furst, P. and Kinney, J. M. (1980). Effect of injury and sepsis on high-energy phosphates in muscle and red cells. *J. Trauma.*, **20**, 755

5 Tilstone, W. J. and Cuthbertson, D. P. (1970). The protein component of the disturbance of energy metabolism in trauma. In Porter, R. and Knight, J. (eds.) *Energy Metabolism in Trauma.* p. 43. (London: J. & A. Churchill)

6 Wilmore, D. W., Long, J. M., Mason, A. D., Jr., Skreen, R. W. and Pruitt, B. A. (1974). Catecholamines: mediator of the hypermetabolic response to thermal injury. *Ann. Surg.*, **180**, 653

7 Arturson, M. G. S. (1977). Transport and demand of oxygen in severe burns. *J. Trauma*, **17**, 179

8 Caldwell, F. T. (1976). Energy metabolism following thermal burns. *Arch. Surg.*, **111**, 181

9 Shamoon, H., Hendler, R. and Sherwin, R. S. (1981). Synergistic interactions among anti-insulin hormones in the pathogenesis of stress hyperglycemia in humans. *J. Clin. Endocr. Metab.*, **52**, 1235

10 Powanda, M. C. and Beisel, W. R. (1982). Hypothesis: leukocyte endogenous mediator/ endogenous pyrogen/lymphocyte-activating factor modulates the development of nonspecific and specific immunity and affects nutritional status. *Am. J. Clin. Nutr.*, **35**, 762

11 Askanazi, J., Carpentier, Y. A., Elwyn, D. H., Nordenstrom, J., Jeevanandam, M., Rosenbaum, S. H., Gump, F. E. and Kinney, J. M. (1980). Influence of total parenteral nutrition on fuel utilization in injury and sepsis. *Ann. Surg.*, **191**, 40

12 Weissman, C., Askanazi, J. and Rosenbaum, S. H. (1982). Amino acids and respiration. *Ann. Int. Med.*, (In press)

38
Thoughts for the future of clinical nutrition

A. W. GOODE

The last two decades have seen the availability of both the nutrient solutions and safe techniques for the provision of appropriate nutritional support. Significant clinical expertise is required to successfully manage patients needing intravenous nutritional support and many problems remain if the ideal is to be achieved: maximum benefit in minimum time with minimal complications.

NUTRITIONAL ASSESSMENT

The assessment of intracellular deficiencies in malnutrition, or following trauma or sepsis is essential if the nutrients and energy supplied are to be the most appropriate for the needs of the patient. The non-invasive study of metabolites in living cells may now be possible using the technique of nuclear magnetic resonance (NMR).

Like most spectroscopic techniques it involves the perturbation of molecules by an external energy source and measurement of the resultant signal. Nuclei when placed in a magnetic field and subjected to radio-frequency electromagnetic radiation will absorb energy. NMR is unusual in that the signal is produced by perturbing the nuclei rather than the electron shells of atoms. There are marked differences in the frequencies at which different nuclei absorb energy and much smaller but easily distinguishable differences caused by the local environment of the nuclei.

Studies of tumour cells[1], liver cells[2] and muscle[3] in isolated cell preparations and heart[4], kidneys[5] and liver[6] in perfused organs have been made. The technique enables clinicians to now study intracellular metabolite concentrations and pH states in human subjects. It may allow us to study energy consumption in the cell, metabolite deficiencies and accumulations in disease states and results in a more specific therapy for the correction of intracellular metabolite abnormalities in disease and a more accurate identification of conditions facilitating cellular anabolism.

LEAN TISSUE PRESERVATION

The concept that prevention is the epitome of medical practice has not been evident in clinical nutrition where the basis of practice is the restoration of lost lean tissue. Because of this constant clinical demand, attention has been focussed on perfecting methods of nutritional repletion and less on mechanisms which may influence protein metabolism, energy consumption and effects on peripheral tissues[7]. Surgery is associated with changes in the blood levels of the circulating thyroid hormones, thyroxine (T_4), triiodothyronine (T_3) and reverse triiodothyronine (rT_3) in the post-operative period. Active thyroid hormone has a multiplicity of metabolic effects in the cell influencing nuclear transcription, mitochondrial activation and the sodium pump mechanism. The active hormone available at the receptor site may be reduced by the formation from T_4 of the relatively inert compound rT_3, by the removal of an atom of iodine from the inner ring of T_4 rather than from the outer ring, as in T_3. This change has been described after surgery[8], illness[9], starvation[10] and hepatic cirrhosis[11]. However, after surgery rT_3 production cannot be abolished by carbohydrate infusion[12] unlike in the

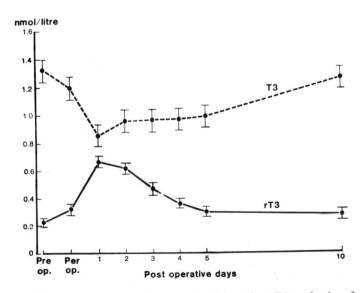

Figure 38.1 Triiodothyronine (T_3) and reverse triiodothyronine (rT_3) production after surgery of moderate severity with a carbohydrate intake of 3000 calories per day. (Reference 12)

other conditions, suggesting it has a major role in peripheral autoregulatory mechanisms (Figure 38.1).

Recent work[13] has shown that in the post-operative period there is a correlation between T_3 to rT_3 ratio and protein synthesis and breakdown (Figures 38.2, 3). This could arise from several factors and carbohydrate and

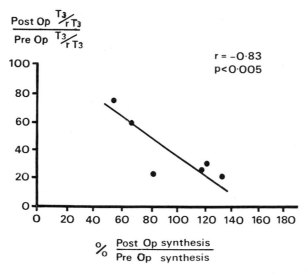

Figure 38.2 Post-operative changes in T_3/rT_3 ratio and whole body protein synthesis. (Reference 13)

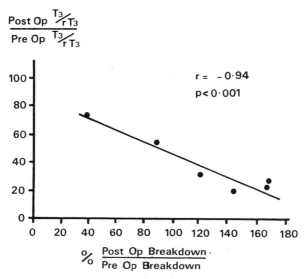

Figure 38.3 Post-operative changes in T_3/rT_3 ratio and whole body protein breakdown. (Reference 13)

fat metabolism may be involved in a potentially complex mechanism. Thus the availability of inactive thyroid hormone at peripheral cell receptor sites may modify the energy consumption by the cell, mediated both by thyroid and adrenergic stimulation. Further elucidation of the mechanism of thyroxine monodeiodination to reverse T_3 may allow changes in the T_3/rT_3 ratio, which may both regulate energy demand by the cell and influence protein turnover.

LEAN TISSUE REPLETION

Repletion of the lean body mass is a fundamental aim in patients enterally or parenterally fed. Alterations in the amino-acid content of solutions may influence the efficacy as simply measured by daily nitrogen balance[14]. Formation of protein in cells needs all the individual amino acids and Harper and Rogers[15] have shown that the relative amount of the amino acids with each other is important for growth in rats. By implication the concepts of amino-acid imbalance toxicity and antagonism may be of some importance. The improvement in methods of measuring protein turnover[16] has given a tool of greater sensitivity to determine appropriate and optimal amino-acid mixtures. This may determine more precisely synergism or toxicity of various amino-acid mixtures or, indeed, if it is of any clinical consequence.

The post-operative patient has been well studied following various surgical procedures such as herniorrhaphy and cholecystectomy. These have provided models following uncomplicated surgery of varying magnitude allowing studies in similar groups of patients and valid conclusions to be drawn. However, the clinical discipline to produce similar models in other situations has been lax. Too often heterogeneous groups of patients have been studied under a common title such as 'cancer'. Similar clinical groups only must be studied; certainly it may provide further insight into metabolic changes with the progression of disease. As an example Glass and Goode[17] studying amino-acid profiles in the presence of a histologically staged large bowel tumour have prescribed a plasma amino-acid profile similar to that found in starvation with benign disease (Figure 38.4). However, weight loss in the presence of a tumour without distant metastases resulted in a significant change in plasma arginine levels compared with controls. This rise did not correspond to changes in blood urea and the cause is uncertain.

Enteral nutrition has grown markedly in the last 5 years, with the advent of whole protein and chemically defined or 'elemental' diets. Knowledge of their effect upon bowel motor activity and their absorption in man is important. Such information will help to determine the efficiency of enteral regimens and the specific indications for the use of different formulations, as well as influencing the clinical decision to use an enteral or parenteral feeding

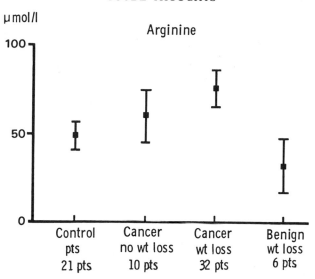

Figure 38.4 Plasma free arginine concentration μmol/l (mean±SEM) in control patients, patients with staged carcinoma of the colon, with and without weight loss, and patients with weight loss and a benign lesion of the gastrointestinal tract. (Reference 17)

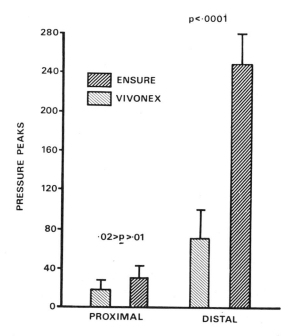

Figure 38.5 Number of pressure peaks recorded over a standard time of study as a measure of the motility response, proximal and distal, to the infusion site of a whole protein or amino-acid based enteral diet. (Reference 18)

regimen. Wilson and Goode[18] have studied the effect of different compositions of enteral feeding on gastric and small bowel motility. Two distinct patterns of motility predominate (Figure 38.5). The first occurs during fasting and is a recurrent cyclical pattern characterized by brief bursts of powerful regular contractions separated by quiescent periods which migrate along the bowel from proximal stomach to terminal ileum – the migrating motor complex. On feeding, this is replaced by a pattern of irregular contractions induced by the presence of nutrient in the bowel. A whole protein diet induced this feeding pattern, but an amino-acid preparation did not. In dogs the development of a feeding pattern is associated with a 30% increase in nutrient absorption per unit length of bowel[19]. Should, therefore, a whole protein feed be always the first choice, or would a prolonged mucosal contact time with an amino-acid preparation be of benefit in the short bowel patient? Certainly the documentation of efficiency in enteral feeding will help clarify the interface between the use of enteral or parenteral feeding.

CONCLUSION

Clinical nutrition is now at the forefront of good clinical practice. Undoubtedly the best results are obtained by the expertise of a multi-disciplinary nutritional care team. Such teams should document accurately the indications for and possible advantages of nutritional support from a solid scientific basis. This means the patient groups studied must be comparable and anecdotal conjecture avoided. Clinicians and scientists, with a major clinical or research commitment to clinical nutrition, also have a particular responsibility to more widely disseminate their knowledge and expertise to those who have responsibility for clinical care. Too often patients in hospital are starving by default in the midst of plenty – perhaps this is the greatest challenge.

References

1 Navon, G., Navon, R., Shulman, R. G. and Yamane, T. (1978). Phosphorus metabolites in lymphoid, friend erythroleukaemia and Hela cells observed by high resolution ^{31}P nuclear magnetic resonance. *Proc. Nat. Acad. Sci. USA*, **75**, 891

2 Cohen, S. M., Ogawa, S., Rottenberg, H., Glynn, P., Yamane, T., Brown, T. R., Shulman, R. G. and Williamson, J. R. (1978). ^{31}P nuclear magnetic resonance studies of isolated rat liver cells. *Nature (London)*, **273**, 554

3 Hoult, D. J., Busby, S. J. W., Gadian, D. G., Radda, G. K., Richards, R. E. and Seeley, P. J. (1974). Observation of tissue metabolites using ^{31}P nuclear magnetic resonance. *Nature London*, **252**, 285

4 Garlick, P. B., Radda, G. K., Seeley, P. J. and Chance, B. (1977). Phosphorus NMR studies on perfused heart. *Biochem. Biophys. Res. Commun.*, **74**, 1256

5 Sehr, P. A., Radda, G. K., Bore, P. J. and Sells, R. A. (1977). A model kidney transplant studied by phosphorus nuclear magnetic resonance. *Biochem. Biophys. Res. Commun.*, **77**, 195

6 Iles, R. A., Baron, P. G. and Cohen, R. D. (1979). The effect of reduction of perfusion rate on lactate and oxygen uptake glucose output and energy supply in the isolated perfused liver of starved rats. *Biochem. J.*, **184**, 635

7 Jung, R. T., Shetty, P. S. and James, W. P. T. (1980). Nutritional effects of thyroid and catecholamine metabolism. *Clin. Sci.*, **58**, 183

8 Prescot, R. W. G., Yeo, P. P. B., Watson, M. J., Johnston, I. D. A., Ratcliffe, J. G. and Evered, D. C. (1979). Total and free thyroid hormone concentrations following elective surgery. *J. Clin. Pathol.*, **32**, 321

9 Carter, J. N., Eastman, C. J., Corcoran, J. M. and Lazarus, L. (1974). Effect of severe chronic illness on thyroid function. *Lancet*, **2**, 971

10 Vagenakis, A. G., Burger, A., Portnay, G. I. *et al.* (1975). Diversion of peripheral thyroxine metabolism from active to inactive pathways during complete fasting. *J. Clin. Endocrinol. Metab.*, **41**, 191

11 Normura, S., Pitman, C. S., Chambers, J. B., Buck, M. W. and Shimizu, T. (1975). Reduced peripheral conversion of thyroxine to triiodothyronine in patients with hepatic cirrhosis. *J. Clin. Invest.*, **36**, 643

12 Goode, A. W., Herring, A. N., Orr, J. S., Ratcliffe, W. A. and Dudley, H. A. F. (1981). The effect of surgery with carbohydrate infusion on circulating thiiodothyronine and reverse triiodothyronine. *Ann. R. Coll. Surg. Engl.*, **63**, 168

13 Goode, A. W., Orr, J. S., Ratcliffe, W. A., Clague, M. B. and Johnston, I. D. A. (1982). Protein dynamics and thyroid hormones in the post-operative period. Unpublished observations

14 Tweedle, D. E. F., Spivey, J. and Johnston, I. D. A. (1973). Choice of intravenous amino acid solutions for use after surgical operation. *Metabolism*, **22**, 173

15 Harper, A. E. and Rogers, O. R. (1965). Amino acid imbalance. *Proc. Nutr. Soc.*, **24**, 173

16 Waterlow, J. C., Golden, M. H. N. and Garlick, P. J. (1978). Protein turnover in man measured with ^{15}N. Comparison of end products and dose regimes. *Am. J. Physiol.*, **235**, E165

17 Glass, R. E. and Goode, A. W. (1981). Amino acid profiles and weight loss in the presence of a gastrointestinal tumour. *J. Parent. Enteral Nutr.*, **5**, 566

18 Wilson, I. A. I. and Goode, A. W. (1982). Patterns of bowel motility with various enteral feeding regimens. Unpublished observations

19 Pearce, E. A. and Wingate, D. L. (1980). Myolectric and absorptive activity in the transected canine small bowel. *J. Physiol.*, **302**, 11

39
Concluding remarks to the symposium on clinical nutrition

G. F. CAHILL

I am honoured by the task of being invited to summarize this Second Bermuda Symposium on Clinical Nutrition. It has been 5 years since the first symposium was held and many advances have been made to this newly emerging medical discipline since that time. Nutritional support is now considered an important component of every aspect of patient therapy and in certain disease states it is the critical factor which decides whether there will or will not be survival. The technical advances in both enteral and parenteral alimentation have been so marked that these are now components of routine medical practice whereas 5 years ago they were isolated events except in a few centres. The challenge is no longer directed to whether it should be done or how it can be done. Instead the challenge is now directed to the specific patterning of the nutriments as well as the major clinical decision, or I should say commitment, based on morals and ethics as to whether it should be done. In other words, clinical nutrition has now achieved a degree of technical sophistication and has unfortunately brought with it a number of problems, and these will be discussed subsequently.

Three and four decades ago the flame photometer for electrolyte analysis was developed for clinical use. This resulted in a great number of clinical and theoretical studies in the control of electrolyte levels, but on the clinical side, now that these could be determined with accuracy, there was made the somewhat facetious statement that no patient died without normalization of their electrolyte levels. In an analogous fashion we are now concerned with what we consider to be an optimal state of nutrition; namely, an appropriate amount of fat and muscle in the body as determined by anthropometric indices and, more related to this conference, a normal serum or plasma aminogram, normal meaning the pattern as occurs in a healthy individual after an overnight fast. In fact, the overnight or post-absorptive aminogram has almost become a kind of pinnacle of appropriate nutrition in the eyes of some individuals without realizing that the post-absorptive state amino-acid pattern may not be the most appropriate setting other than for the post-absorptive state.

Cells experience only their immediate environment. A given muscle cell located, let's say, in the right biceps of a patient does not know whether the patient has had multiple fractures of the femur or bacterial peritonitis, or a 60% third degree burn, it only appreciates that it is bathed in fluids bearing nutrients and hormonal signals which tell it to behave accordingly. The other thing it knows is its stimulation by motor nerves and these likewise dictate its metabolic response. As has been shown by a number of investigators, if the stimuli are infrequent and result in twitches, the muscle increases its content of enzymes and actin and myosin toward the direction of white muscle and conversely if it receives sustained repetitive stimuli, the biochemical changes will be more in the direction of red muscle with increased oxidative enzymes and lesser amounts of glycogen and glycogenolytic enzymes. In fact, the neuromechanical events dominate the nutritional events as was shown by Goldberg[1] over 10 years ago when he demonstrated that a muscle in a fasting animal could be made to hypertrophy by chronic stimulation. Conversely, a muscle not used will undergo atrophy regardless of the richness of its nutritional and hormonal environment. Thus the dominant effect on a given muscle cell is its mechanical use but what the nutrient and hormonal environment serves to do is to integrate it and the remaining muscles in the body into the nutritional state of the organism, so that if there be a demand for amino acids for gluconeogenesis, then the muscle mass in concert will respond by proteolysis and amino-acid release. However, the relative contributions of any single muscle will be mainly a function of its mechanical use. This can even be interpreted to say that physical therapy is equally as important, if not more so, for any given muscle group than is overall nutrition. I mention this primarily since this has received so little recognition at this symposium.

Now let us consider some other tissues directly. One of the major advances that has occurred in the past 5 years, as discussed in a number of papers in this volume, is the fact that nutrition can be specifically patterned for individual tissues. There is now clear evidence that increases in the levels of the branched-chain amino acids can, in both the normal and stressed individual, cause a greater degree of nitrogen retention than a similar amount of nitrogen given as a mixture of a number of amino acids. This appears to be extremely logical since what peripheral tissues see after a meal is mainly only an increase in the branched-chain amino acids as well as methionine and one or two other amino acids. In other words, as I mentioned above, since a muscle only sees its immediate environment, if one wants to keep muscle in an anabolic or fed state, it appears extremely logical to keep muscles in a milieu which they would see under the usual physiological circumstances after a meal. This means higher levels of insulin and higher levels of the branched-chain amino acids. Thus for muscle 'well-being' the administration of the branched-chain amino acids appears extremely logical in addition to their now being shown by a number of investigators to be effective in maintaining muscle nitrogen. However, when one looks at liver, again as discussed by a number of papers in this symposium, the pattern of amino acids as seen in the anabolic state is essentially what comes up the portal vein from the gut. As shown by a number of workers this includes both the branched-chains and, more importantly, the entire spectrum of non-essential amino acids, particularly glutamine, alanine, glycine, and the other predominant non-essentials. Dr Glen Mortimore at Hershey, Pennsylvania, in his elegant studies[2] on the control of protein synthesis and breakdown in rat liver has shown that the primary effectors in promoting synthesis and decreasing breakdown are a number of non-essential amino acids and thus if one is considering specific feeding of the liver, one should consider in therapy high levels of the non-essential amino acids and again with much emphasis on glutamine. The focus to date on the therapy of hepatic disease has been more on the brain transport of amino acids than on the specific problem of which amino acids might be most appropriate for the well-being of the liver cell itself. Dr Joyeux has presented some interesting data on hepatic regeneration in dog and man and has found that the branched-chain amino acids, if enriched in the nutritional mix, provoked a greater increase in mitotic index, and this certainly bears more examination since it is somewhat in contrast with the experience of Mortimore looking at protein turnover in the rat model.

What was not addressed directly but indirectly in the discussion relative to using delayed hypersensitivity and other immunological parameters as an index of nutritional well-being is the substrate requirements for the various tissues and cells involved in all levels of immune responsivity. It is known that leukocytes, whether they be mononuclear cells or lymphocytes, or polymorphonuclear leukocytes, use glucose as their main fuel, but which amino acids are the primary stimuli or even which are essential for these cells for their protein metabolism has, to my knowledge, not been studied. In other words, should one not only consider patterning of amino acids for the benefit of muscle as well as for kidney or for liver without considering which is the appropriate milieu for those cells involved in host defence.

After saying all of the above, and as has been reviewed in the many papers in this volume, we are now at the level where we are sharpening the pencil to a fine point in trying to pattern nutrition towards helping specific situations. It is my personal feeling that the versatility of the body is so great that, unless one provokes in a major fashion a gross imbalance by being too enthusiastic with one kind of nutrient or another, the body will be able to adapt itself moderately well even in the presence of severe trauma or infection. In other words, the slight alterations in amino-acid contents in various enteral and parenteral therapies may help in redirecting the aminogram back towards the post-absorptive paragon, but whether this does or does not do any good to the patient as evidenced by survival as the best index will be extremely hard to assess. One can only think of the hundreds of studies directed to whether anti-coagulation of the patient with the acute myocardial infarction is or is not beneficial as an example of having to judge therapeutic efficacy by survival. If any of these manipulations in altering amino-acid profiles would result in a major beneficial outcome, I think we might have seen it by now and I therefore wonder how rewarding will be the minor modifications in patterns currently under study by a number of investigators. Since most of the technical problems in getting the amino acids into solution, their sterilization and combining these with glucose and with various formulations of lipid have been mainly solved, we are now presented with the opportunity of too many variables with which to deal. It is my personal opinion that the most prudent pattern of nutrition would mimic what occurs under normal non-diseased nutrition with 15% of the calories as amino acids and half of the balance as carbohydrate and the other half as lipid, but this is a very unscientific opinion based on speculation.

Finally, throughout this symposium has been mentioned the fact that clinical nutrition is now under certain circumstances 'life-saving', under other circumstances it is 'life-extending' and we have also heard the term 'life-prolonging' being used. To me these three are all synonymous with the exception that they represent an emotional bias on the person using the term. There is no question but that a severely traumatized young individual unable to eat and rapidly wasting away is saved by his or her alimentation, be it enteral or parenteral. There is also no question but that the patient with renal failure or hepatic failure has his viability extended by appropriate nutritional therapy, and finally no one will argue that the elderly patient in the nursing home with an oesophageal stricture or inoperable cancer will have his life prolonged by appropriate nutritional therapy. The same would be said of prolonging the life of the newborn with major congenital deformities about which much has recently appeared relating to their nutritional therapy in the press. There is no doubt in this latter case that life can be prolonged almost indefinitely should the therapeutist be sufficiently enthusiastic to want to do this. Thus, we now have a tool which is more than we can ethically and morally handle and we in the medical care industry should give priority to the use of clinical nutrition before its use becomes even more political than it already is. Now that it is law that life must be prolonged at the taxpayers' expense in renal failure, there may be similar laws for the prolongation of the life of the individual with pulmonary failure or with alimentary failure, or perhaps if not directly in nutrition, in one in which nutrition can play a central life-prolonging role. Thus, as emphasized by the various papers in this symposium, the field of clinical nutrition has very rapidly matured into a complex medical and ethical specialty and one cannot develop science and ethics independent of each other. As an endocrinologist, I can best liken the field to an adolescent who has become physically mature but is uncoordinated, unsophisticated and undisciplined and requires intellectual and emotional growth to catch up with his or her physical assets. Hopefully by the next symposium, not only will enteral and parenteral alimentation be physically more mature, but we will also have the ethical and moral wisdom to be able to apply it optimally for both the individual as well as for the society.

References

1 Goldberg, A. L. (1971). In Albert, N. (ed.) *Cardiac Hypertrophy* p. 39. (New York: Academic Press)
2 Mortimore, G. (1982). Mechanisms of cellular protein catabolism. *Nutr. Rev.*, **40**, 1

Index

acetoacetyl CoA thiolase in lactation *xx*
adipose tissue metabolism 294
ageing and immunosuppression 359–61
alanine
  BCAA nitrogen sparing 42–4
  BCAA oxidation 6
  levels and surgical stress 7
  output and stress with BCAA 55
  protein sparing 47
  and supplement pH 86
albumin
  acute renal failure 117
  bilirubin binding 193
  binding and fat emulsions 186, 187
  synthesis and fatty liver 263
  synthesis and TPN in rats 267
alkaline phosphatase and TPN 60, 61
allosteric enhancement 200
ambulatory total parenteral nutrition
  anatomical changes 422, 423
  Crohn's disease 419–31
American Society for Parenteral and Enteral
  Nutrition 448
amino acids *see also* branched chain (BCAA)
  analysis 38, 245
  antagonism in uraemia and nutrition 33
  aromatic in catabolic states 29
  body pool 26–8
  branched chain see BCAA, isoleucine,
    leucine and valine
  concentration and TPN-fat stability
    245–8
  daily requirements 3
  dextrose solution pH 249
  dietary manipulations 11
  enteral formula 156, 157
  enzymes initiating metabolism *xvi*
  fat emulsion mixing 231, 243, 244
  flux control 32
  formulation for catabolic chronic dialysis
    143
  imbalance and food intake 140
  input and clinical trauma 54
  intracellular in injury 31
  keto and low protein diet in chronic renal
    failure 125–7
  liver disease 148, 149
  muscle and BCAA supplement in injury
    41
  new formula 34
  pH buffering capacity 258
  pH at isoelectric points 250
  plasma levels 12
  plasma levels and BCAA in stress 60–2
  plasma levels and BCAA supplement in
    injury 41
  pool normalization 34
  rate limiting in chronic dialysis 142
  regulation of kinetics and BCAA 67
  respiratory effects 454, 455
  selective solution and liver regeneration 99
  sepsis indicators 361, 362
  serum levels in acute renal failure 117, 118
  serum levels and HPN 428
  toxicity 261
  urinary free and stress with BCAA 59
  wound healing 346
amino acids, essential 3
  acute renal failure 119
  and extra-, intracellular water 26, 27
  serum and HPN 427
aminograms 4, 65, 466
  uraemia 141
Aminosyn, modified and nitrogen balance
  45, 46
animal studies *see also* injured rat
  BCAA supplementation 15
  hepatic insufficiency 84
  monoacetoacetin infusion 332, 333
anorexia 154
antagonism by BCAA in uraemia 30
appetite suppression and leucine 140
arginine
  bowel turnover 460, 461
  chronic dialysis 143
artificial gut 370
ascitis
  cause and management 153

ATP, muscle in injury   452, 453

bacterial contamination   242
basilic vein catheterization   394
BCAA see branched chain amino acids
   enriched solutions see supplement
α-keto analogues   47
BCKA see branched chain keto acids
bed rest, nitrogen balance   28
bile salt deficiency and malabsorption in
   cirrhosis   152, 153
bilirubin absorption spectrum and albumin
   binding   194, 195
bilirubin binding
   borate, salicylate buffer   198, 199
   fatty acids, in vitro   198–200
   intravenous fat emulsions   193–244
   lipid emulsions in vitro   196, 197
   linoleic acid : albumin ratio   198, 199
   sites   195
bilirubin concentrations and fat emulsion
   186, 187
bilirubin encephalopathy   193–204
biopsy
   muscle, measurements   26
   muscle, technique   25, 452
blood–brain barrier
   amino acid transport constants   12
   biphasic transport constants   12
   deranged permeability   13
   permeability and drugs   13
blood flow and substrate utilization   xiv
branched chain amino acids (BCAA) see also
   isoleucine, leucine and valine
   anabolism   8, 9
   antagonism in uraemia   30
   clinical studies   15, 51–62, 54–6, 77–81,
      93–8, 104
   daily requirements   3
   dehydrogenases
      factors affecting   4
      liver and dietary protein   5
      tissue distribution   4, 5
   effects of central neurotransmitters   9
   injection, metabolic effects   8, 9
   liver disease   150, 151, 157
   metabolic roles   3
   metabolism, muscle   3, 32
   metabolism regulation   52
   neurotransmitter metabolism   9–15
   nitrogen retention   466
   plasma levels
      and amino acids in liver disease   12
      amino acids and stress   60–2
      relevence in disease   25
      surgical stress and supplementation   79,
         80
   protein sparing effect   6, 47, 66

protein synthesis   9, 32, 47, 49
   roles   3, 32, 33, 47
      anti-catabolic   37
   sepsis   28, 29, 32
   starvation, plasma levels   3
   stress   6, 7, 147
   surgical stress   77–81
   transaminases
      factors affecting   4
      tissue distribution   4, 5, 147
   trauma   51–63, 147
branched-chain amino acid–alanine cycle in
   fasting   5
branched-chain amino acids/aromatic amino
   acid ratio (BCAA/AAA)
   blood–brain barrier competition   11, 12
   hepatic encephalopathy   12
   hepatic insufficiency   85, 90, 91
   hepatic resection with BCAA   97
   liver disease   10, 11
branched-chain keto acids
   decarboxylation   31
   generation   6
   insulin effects in uraemia   32
breast feeding and fertility   174, 175
breast milk
   amino acids   171
   composition   171, 172
   and growth   172
   immunoglobulins   172, 173
   neonatal survival   171
   prematurity   173, 174
   short bowel syndrome   173
   and stool pH   173
burns
   calorie requirement   164, 165, 169
   degree and calorie requirement   169, 170
   fat emulsions   168, 169
   hypermetabolism   163, 164, 453
   management   163
   nutritional assessment   164–6
   preparations   168, 169
   supplementation   163–70
   tube feeding and extent   167
   wound healing   166
γ-butyrobetaine hydroxylase in neonates
   317

calcium, serum in TPN   424
calcium ions and emulsion aggregation   236,
   237
Candida spp.   377
carbohydrate
   enzymes initiating metabolism   xvi
   fat emulsion mixture   230
   fat metabolism in TPN   286, 287
   intake and free fatty acid metabolism   294,
      295

parenteral nutrition 283–90
    sources 283
    supplement level and amino acids 39–42
carbon dioxide production and nutrition 454
    455
carcinoid syndrome 415
cardiac disease, carnitine role 319
carnitine
    blood stream delivery 321
    cardiac disease 319
    dose 321, 322
    excretion 321
    enzymes 317, 318
    fatty acid transport 315
    formulation for supplementation 320
    functions 315
    gastrointestinal uptake 320, 321
    haemodialysis patients 318
    independence 331
    infant diets 317, 318
    liver disease 318
    muscle disease 318, 319
    patients benefiting 316
    role in clinical nutrition 315–22
    route of administration 320
    stress 319
    tissue concentration parameters 316, 317
    transport facilitation 322
catabolic conditions 28, 37
catabolic index 68, 69
    calculation 68, 72
catabolism, BCAA 51, 52
    interorgan participation 6
    intracellular pattern 26
    muscle, liver and kidney 4, 5, 52
    organ distribution of enzymes 5, 7
    rates of protein 37
catheters for home parenteral nutrition
    blockage 377, 406
    Broviac 374, 375, 390, 402
    composition 390–4
    dressing change 395, 396
    exchange 396, 397
    exit point 402
    Hickmann 355, 390, 402
    infection 377, 391, 396, 397, 406, 426
    insertion 374, 375, 384, 385, 391–4, 420
    introducer 391
    leak 396
    lifetime 376
    long-term access 389
    long-term assessment 397, 398
    new 398
    out-patient care 394
    plastic materials 390
    problems 376, 377, 397, 398, 406, 426,
        427, 430
    questionnaire findings 442

silastic rubber 370, 372
silicone 390
silicone long-arm 398
suture 390, 391
system 390, 391
tip recovery 426, 427
ceruloplasmin 353
    prognosis and significance 356, 357
    properties 356
cholesterol in HPN 426
chronic dialysis
    amino acid kinetics 141
    biochemical changes and hyperalimentation
        135, 136
    carnitine 318
    catabolism 131
    hyperparathyroidism 135, 137
    intradialysis hyperalimentation 131–44
cirrhosis
    fat absorption 152, 153
    small bowel dysfunction 152
citrulline in uraemia 30
clinical studies
    BCAA infusion 15
    hepatic resection 93–8
    liver disease 104
    output data and BCAA 54–6
    parameters monitored 54, 60
    patient characteristics and stress 69
    surgical stress 77–81
    trauma patients 51–62
clotting factors
    dog hepatectomy 89, 90
    hepatic resection 93, 94
cluster analysis 352–6
    and prognosis 362
coma, causes 10
complications, HPN 405, 413, 434 see also
        metabolic, psychological
    catheter 376, 377, 397, 398, 406, 426, 427
        430, 437, 442
    exit-site 437
    metabolism 377, 378, 406, 427, 428
    post-operative 430
copper, daily requirement 378
Coulter counter 47, 49, 225, 227
critically ill
    fat metabolism 293–300
    glucose infusion 278, 279
    monoglyceride use 335
    protein sparing 279
Crohn's disease 369, 373, 419–31
    HPN 373, 374, 401, 402, 412, 415
    manifestations 430
    patient characteristics 420
    recurrence rate 419
    relapse rates 423, 424
    response to TPN 422–4

results of HPN   379
surgery preparation and recovery   419
TPN assessment   419–31
TPN and remission   419
Crohn's disease Activity Index   422, 423
cuff
   fixing   377
   need for   374, 376
   positioning   375
   Scribner Dacron   370, 372
Curreri's formula   164
cystine deficiency   158

data acquisition, patient-orientated   445
deficiencies, trace metal   377, 378
delayed hypersensitivity
   cancer patients   357, 358
   factors affecting   359, 360
   serum albumin and hard tumours   358
   therapy effects   357
diabetes and muscle amino acids   26, 27
dialysis see chronic dialysis
diet
   amino acid imbalance   140, 141
   low protein   113, 123–8
difference spectroscopy, bilirubin   194, 195
disease and plasma amino acids   25
   substrate partitioning   xxi
diurnal changes in metabolite levels
  · glycogen synthetase deficiency   xiv, xv
dogs, mixed system metabolism   249–58
drug addiction   414
duodenal–colic anastamosis   370, 371

egg lecithin
   composition   213
   fatty acid residues and composition   214
egg phosphatides, bilirubin binding   197
electrolyte
   critical aggregation concentrations   236
   emulsion stability   218, 219, 241, 243, 248
   fat emulsion mixture   231, 232
   Intralipid zeta potential   223
elimination and mixed system nutrition   251,
   255, 256
emulsifiers
   ionized   217, 218
   mode of action   215–217
   zwitterionic   218
emulsion
   accelerated stability testing   228, 229, 236
   aggregation   228, 236, 237
   breaking   224
   creaming   224
   DVLO theory   217, 219
   electrolyte effects   218, 219
   electrostatic forces   217, 218
   flocculation   224

laser light scattering   225–7
microscopy   227
mixed systems, formulation   231
particle size analysis   225–7, 243, 244, 248
particle size limits   227, 228
stability assessment   224–9
stabilization   215–17
surface potential   221
zeta potential   228
encephalopathy see also hepatic
   associated factors   9
   causes   10
   oral intake   154
energy sources
   alternatives   325–35
   body regions   329
   dual   191
   fat   185, 189
   ketone bodies   329–31
enteral feeding formulae   156, 157
enzymes
   sites of lipid initiation   xvii
   substrate initiation   xvi
epidermal growth factor   172
essential fatty acid
   deficiency and correction   180–2, 189
   deficiency symptoms   181
   placental transfer   181
ethical considerations   434
ethilon sutures   392
ethylacetoacetate properties and parenteral
   feeding   332
exit site complications   437

F080 composition   14
   and nitrogen balance   39, 42
fasting metabolism   326
fat
   caloric substitute   182–5
   endogenous, metabolism   293, 294
   exogenous, metabolism   296–8
   liver accumulation, sources   264–6
   metabolism in critically ill   293–300
   oxidation inhibition   294
   oxidation in injury   296–8
   serum and HPN   425, 426
fat emulsions see also Intralipid, TPN-fat
   advantages   190, 208
   amino acid interaction and pH   249
   amino acid mixing   231
   bilirubin binding in vitro   196, 197
   bilirubin competition   186, 187
   burn patients   168, 169
   caloric source   185, 189
   carbohydrate mixing and pH   230
   carbon dioxide generation   191
   clearance of preparations   237
   component properties and stability   219–24

electrolyte effects 223
intravenous introduction 189
intravenous in neonates 179–87
hypermetabolic patients 208
liver function 299, 300
mixed systems 229, 230
mixed systems zeta potential 231, 232
nitrogen sparing 287
pH 222, 230
respiratory effects 298, 299
safflower 181
soybean emulsion 181
stability 207, 213–38
trade names and composition 219
uses 179, 180, 191
fat emulsions, 20%
clinical applications 189–92
commercial preparations 191, 192
cost 190
isotonic 190
fatty acid
chain length and bilirubin binding 200
free, metabolism and carbohydrate uptake
294, 295
medium chain 268–70
metabolic enzymes 286
metabolism and mixed systems in dogs
253, 254
fetus weight and development 179
fluid intake
fat emulsion, 20% 190
lipid emulsion in neonates 185
fluid overload
critically ill 191
premature neonates 183
Freamine formulation and nitrogen balance
39, 42
futile circles 453

Gardner's syndrome 415
gastrointestinal function in liver disease 152,
143
gastrointestinal tract cancer
cluster analysis, nutritional indicators
354, 356
gloves, home use of sterile 394, 395
glucagon in stress 63
gluconeogenesis, BCAA nitrogen sparing
42–4
glucose
alternatives 325–35
blood levels in parenteral nutrition
308–10
clamp 277
concentration and TPN-fat stability
245–8
disposal rate and clamp 277–9
hepatomegaly 265–7

intake and metabolism in injury 285, 293
intake and ventilation 288, 289
lipid substitute 311
liver cell permeability 310
liver function 306, 307
metabolism in depleted patient 284, 285
metabolism in hypermetabolic patient
284–6
physiological range 330
plasma and stress 63
protein sparing 279, 280
and TPN-fat particle size distribution
243
use in lactation xviii
glutamine
excretion and stress with BCAA 55
muscle in catabolic states 26, 27
nitrogen flow 4
glyceride, serum and HPN 426
glycogen synthetase deficiency, metabolic
levels xiv, xv

Harris–Benedict formula 299
energy expenditure 326
HBABA dye 193
heparin lock 375, 395
port changing 396
hepatectomy see also hepatic resection
amino acid solutions 88
anaesthesia 86
BCAA/AAA ratio 85, 90, 91
clotting factors 90
dogs and treatments 87, 88
experimental model 86
mitotic index and nutrition 91, 92
nitrogen balance 89, 90
nutrient solution 88
parameters evaluated 89
technique 87
Hepatic Aid
formula comparison 156
pudding 158
hepatic encephalopathy 262
amino acid correction 13
BCAA hypothesis 10–14
BCAA infusion data 14
biological correlations 11
complex factors 14
mechanisms 10, 11
toxins 10
hepatic insufficiency see also hepatectomy,
hepatic resection, liver
amino acid responses 84
amino acid solution 85, 86
solution formulation 84
surgical in dogs 84
hepatic regeneration 467
BCAA role 83–99

factors governing   83
hepatectomy   83
   mitotic index   91, 92
   procedure   83
hepatic resection
   clotting factors and BCCA   93, 94
   parenteral nutrition support   93
   patient characteristics   93
   plasma amino acids and BCAA   94–8
hepatic steatosis   261
   glucose induced   265, 266
   medium chain fatty acids   268, 269
   nutritional depletion   264
hepatomegaly, causes in TPN   264
high risk patient, identification   351–62
home-care agencies   384
home parenteral nutrition (HPN)   369–448
   adaptation   412, 413
   attitudes   412
   cases needing   380, 415, 433
   changing needs   447
   collaborative study, second generation
      445, 446
   continuous ambulatory   405
   costs   387, 405, 429, 434, 448
   costs and cover   387, 405
   costs and reimbursement methods   443
   delivery   380, 387
   doctors' view   446, 447
   duration   415
   history   433
   indications in intestinal failure   372–4,
      401, 406, 415, 435, 437
   infection   377, 397
   infusion technique and schedules   376,
      403, 420, 434
   insurance cover   387, 405
   leisure restrictions   413
   management   436
   metabolism   425, 427
   monitoring   404, 405, 424–9
   mortality rates   415, 444
   nutrient administration   375, 376, 404
   nutrient mixing   403
   outcome   415, 416
   overnight   403
   patient selection   385, 386
   patient training   394, 411, 412
   prognosis   415, 416
   protein levels   425, 425
   quality of life   411–16
   psychological problems   378
   readmission   436
   rehabilitation   413, 446
   results and assessment   379, 380
   return to oral   416
   role   447
   schedule choice   403

serum proteins   425, 426
support programme   383–8, 446
technology and techniques   373–6
use in Europe   370, 372
venous access   389–99, 402
Home TPN Registry   448
hormones and nutrition   454
hospital, discharge preparations   404
HPN see home parenteral nutrition
hyaline membrane disease   183
3-hydroxybutyrate dehydrogenase   xx
   site and product   xvi
hyperalimentation, intradialysis   131–44
   administration   131
   BCAA   140, 141
   composition   133
   pre-dialysis chemistry   135, 136
   recommended formula   143
   responses   134
   weight profiles   132, 133, 137–9
hyperglycaemia   210
hyperketonaemia and protein breakdown in
   stress   68
hypermetabolism
   aetiology in burn patients   164
   ATP utilization   452, 453
   definition   452
   and degree of burning   163, 164
   fat oxidation inhibition   294
   heat metabolism   453
hyperosmolar dehydration   210
hyperparathyroidism in chronic dialysis
   patients   135–7
hypoglycaemia and infusion rate   377

immunosuppression in cancer patients   360
infection
   catheter organisms   377
   long term, incidence   397
   prevention   377
   topical nutrition   346
   wound cleaning   340
infusion of BCAA
   clinical and animal studies   15
   clinical investigations   14
   effects and protein synthesis   33
   F080   14
   redistribution   8
infusion, HPN
   new devices   398, 399
   overnight   403, 434
   pump   376
   schedule choice   403
   technique   376
   times   420
injured rat   37–48
   effective BCAA composition   46
   injuries   37, 38, 47

modified Aminosyn 45, 46
nitrogen balance 42–6
plasma and muscle amino acids 40, 41
protein breakdown 44
injury
  amino acid patterns 31
  BCAA effects 28, 29, 65
  fat inhibition 65
  fat oxidation 297, 298
  glucose metabolism 285
  metabolic response 46, 275
insulin
  BCAA metabolism 32
  and ketone bodies in mammary tissue xx
  leucine agonism 31
  plasma and glucose clearance 280
  resistance causes 275, 276
  resistance in stress 275
  resistance in uraemia 31
insulin clamp technique 276
intestinal failure 369–80
  actue causes 369
  chronic, causes 370
  definitions 369
  HPN 372–4 380
  major causes 370–2
  temporary 372, 373
intestinal fistula, HPN use 374
Intralipid 207
  clearance 237
  composition 219
  electrolyte effects 223
  fatty acid accumulation 264, 265
  injury and plasma release 297
  liver accumulation 266, 267
  particle composition 290
  surface potential and pH 221
ionic lipids 219–22
isoleucine 3
  blood–brain barrier transport constants 12
  brain uptake and blood valine 12, 13
  membrane permeability 31
  plasma and hepatic resection 97
  plasma levels 12
  uraemia 30, 31

j-wire 392, 393

kernicterus 193 see also bilirubin encephalopathy
ketone bodies
  and infection 331
  parenteral nutrient 329
  physiological effects 330, 331
ketosis, beneficial effects 329
!Kung nomads, breast feeding and fertility 174, 175

lactate, plasma levels in stress 63
lactation see also breast feeding
  glucose use xviii, xix
  ketone bodies xx
  substrate partitioning xviii, xxi
  substrate requirement xviii
  triglyceride partitioning xviii, xix
laser light scattering 225–7
lean tissue preservation 458–60
lean tissue repletion 460–2
lecithin
  definition 213, 214
  egg yolk and soybean composition 213
leucine
  antagonism in uraemia 33
  blood brain barrier transport constants 12
  body metabolism and plasma levels 106
  catabolism and lactation xx
  continuous infusion ^{14}C and amino acid flux 71–3, 103
  excretion and BCAA in stress 57
  kinetics evaluation 72, 73
  metabolic role 3, 4
  metabolism kinetics 328
  metabolism model 8
  muscle effects 107
  obesity 37
  oxidation control 9
  plasma and hepatic resection 96
  plasma levels 12
  plasma in stress 63
  protein synthesis 8, 10, 47
  turnover equation 105
  uraemia 30, 31
linoleic acid
  : albumin ratio and bilirubin binding 198, 199
  daily requirement 189
  liver depletion and fat-free TPN 263
lipid see fat
lipids, enzymes initiating metabolism xvi, xvii
lipid metabolism in parenteral nutrition 261–70, 286, 287
Lipofundin
  clearance 237
  composition 219
lipogenesis rate and lactation xix
lipoprotein lipase
  activity 296
  gram negative sepsis xxi
  induction 256
  site and product xvi
  starvation xvii
Liposyn 219
liver see also hepatic
  fatty and glucose intake 286
  nutritional support 99

protein metabolism   147
liver disease
  body potassium   153, 154
  carnitine   318
  chronic BCAA effects   103
  clinical details and intravenous nutrition
    104
  fluid/electrolyte balance   153, 154
  intestinal function   152, 153
  malabsorption   152, 153
  plasma amino acid levels   12, 103
  protein metabolism   147, 148
  vitamin deficiencies   151
liver failure
  BCAA   150, 151
  causes   150
  enteral feeding   154–60
  oral feeding   154, 155
  protein sources   154
liver function
  fat emulsion   299, 300
  values and parenteral nutrition   306
liver malfunction
  blood sugar   306, 307
  glucose levels and TPN   308–10
  nutrition, patient characteristics   304, 305
  parenteral nutrition   303–11
liver regeneration *see also* hepatic
  assessment   89
  nutritional factors   98, 99
  l-system   29 *see also* transport, leucine
    preferring
  hepatic insufficiency   85
  rate limiting in catabolism   31
lysine, muscle in catabolic states   26, 27
lysophosphatidylcholine, ionization
  characteristics   222

magnesium and HPN   425–7
malignant disease *see also* radiation enteritis
  maintenance of inoperable   373, 374
malnutrition   415 *see also* nutritional
  assessment
  classification   351
  features in hospital   351
  immune function   119
mammary gland, substrate sink   xx, xxi
Medicare/Medicaid, home nutritional support
  in US   387, 405
medium chain fatty acids in TPN   268–70
  advantages   268
  metabolism   268, 269
  metabolism in rats   269
mesenteric infarction   412, 415, 435
  gastrointestinal tract loss   401
metabolic balance studies   451, 452
metabolic complications of HPN   377, 378,
    406, 427, 428

metabolism
  initiating enzymes   xv, xvi
  yield of substrates   xv, xvi
methionine
  blood levels and liver failure   12, 13
  catabolic states   29
  emulsion stability   247
  plasma levels in hepatic resection   98
  plasma levels in stress · 63
  toxicity   141
3-methylhistidine
  excretion and protein breakdown   362
  excretion in surgical stress and BCAA   80,
    81
  stress   63
mitotic index, hepatectomy   91, 92
mixed fuel substrates   268
mixed systems   229–31
  accelerated testing   235
  divalent cation effect   232–4
  electrolyte content   232, 233, 235
  elimination kinetics in dogs   255, 256
  infusion profiles   258
  metabolism in dogs   249–58
  particle size analysis   232, 233
  stable composition   257
  storage   234
  zeta potential of particles   232
model
  injured rat   37–48
  leucine metabolism   8
monitoring, HPN
  metabolic parameters   421, 424–9
  patient   404, 405
  team   405
monkey, septic and monoglycerides   335
monoacetoacetin
  dose–response in animals   335
  maximal intravenous infusion rates   332,
    333
  physical properties   332
  urinary nitrogen   334
monobutyrin
  physical properties   332
  urinary nitrogen   334
motility, bowel and enteral nutrition   460,
    461
muscle
  amino acids in catabolic states   26–8, 33
  BCAA oxidation   46
  BCAA uptake   3
  hyperstimulation   466
  nitrogen flow   5, 6, 37
  safe biopsy and measurements   26
muscle disease, carnitine role   318, 319

neonate
  bilirubin levels   201

breast milk 171–5
carnitine biosynthesis 317
caogulopathy 182
energy expenditure in premature 182
lung, fat globules 187
premature and breast milk 173, 174
response to fat emulsions 180–2
safflower oil and bilirubin binding 200–3
survival of premature 179
vitamin E 185, 186
neutron activation techniques 452
New England Deaconess Hospital programme
435–7 *see also* questionnaire
assessment and discharge time 443–5
care availability 445
diagnosis 437
HPN indications 435
patient registry survey 438–45
reasons for not working 436
rehospitalization 436, 442
nitrogen
excretion and stress with BCAA 58, 59, 63
nitrogen balance
acute renal failure 116
hepatectomy 89, 90
nutritional status in burns 165
stress 70
surgical stress 79
nitrogen conservation 37–9, 45, 287
alanine role 42–4
BCAA effects in rats 38, 39, 45
carbohydrate levels 39–42
gluconeogenesis role 42–4
radioactive studies 44, 45
nitrogen shuttle 5, 6
nuclear magnetic resonance 457
nurse, nutritional and metabolic support 394
Nutrafundin composition 230
nutrient administration, HPN 375, 376
composition 376
frequency 376
home and hospital TPN 421
infusion technique 376
mixing 403
volume and time 404
nutritional assessment
cluster analysis 352–6
reference groups 353
sepsis susceptibility 353, 354
studies 457, 458
nutritional support *see* supplementation
nutrients, requirements 403
nutrition, return to oral 416

occupation
disability benefits 435, 436
fulltime and HPN 413, 435, 436
octanoic acid and bilirubin binding 198

orangutans, young rearing practices 174
ornithine, blood levels 425, 428, 438, 431
3-oxoacid CoA transferase
lactation *xx*
site and product *xvi*

pancreatitis 369
and enteropathy 415
parathyroidectomy and weight gain in dailysis
patients 137, 138
parenteral nutrition
carbohydrate sources 293–90
partitioning of substrates *xiii, xxi*
disease *xxi*
lactation *xviii–xxi*
signals in mammary gland *xx, xxi*
tissue *xvii, xviii*
patient training, HPN 378, 403, 404
catheter care 394
composition 412
exit site care 403
hospital 411
nurse 394
objectives 411, 412
relatives 412
patient training manual 386
peripheral metabolism 4–8
peripheral parenteral nutrition (PPN)
adverse effects 191
fat emulsion use 191
permeability, substrate and utilization *xiv,
xv*
pH
buffering of amino acids 258
emulsion stability 221, 249, 250
stool and breast milk 173
phenylalanine
breast milk 171
excretion and BCAA in stress 58
levels and surgical stress 7
plasma and hepatic resection 94, 95
plasma and liver failure 12
plasma in stress 63
protein synthesis 141
phlebitis 394
transient 398
phosphatidic acid, ionic species and acidity
222
phosphatidylcholine, ionization characteristics
222
phosphatidylethanolamine, ionization
characteristics 222
phosphatidylinositol, ionization characteristics
222
phosphatidylserine, ionization characteristics
222
phospholipids
film, mechanical strength 221

ionization characteristics 222
liquid crystalline, surface potential and pH 222
metabolism and mixed systems in dogs 254, 255
minor and coalescence behaviour 220, 221
oil-water absorption 215
properties and emulsion stability 219–21
structure and functional groups 214
phosphorus, serum and TPN 424
photon correlation spectroscopy 226
particle size analysis 225, 226
polycystic kidney disease, low protein diet and amino acid supplement 126
polyunsaturated fatty acids (PUFA) 186
liver accumulation and TPN composition 264, 265
oil emulsions 186
vitamin E intake 186
potassium, serum and TPN 424
potential energy–separation curves 216, 217
prealbumin in surgical stress 79
pregnancy, triglyceride partitioning xviii, xix
prognosis, metabolic criteria 362
prolactin xx
proline, plasma levels in stress 63
protein
breakdown in liver disease 103
breakdown rate in injured rat 44
metabolism and BCAA 32
metabolism calculation and liver disease 106, 107
mixed tissue, amino acid distribution 27
turnover [^{14}C] leucine use 105
protein catabolic rate (PCR) 166
protein catabolism
estimation 328
rates 66
uraemic rats 120
urea nitrogen 165
protein metabolism in liver failure 147, 148
protein sparing 6, 47, 66
carbohydrate levels 39–42
glucose in critically ill 279, 280
ketones 333
protein synthesis
BCAA role 9, 32, 47, 49
dextrose effects 107, 108
factor affecting 33
free amino acid pools 26–8
liver disease 103, 467
muscle and valine 45
rate and amino acids 28
whole body and TPN 265, 266
proteolysin, muscle and redox state 9

pseudo-obstruction, small bowel 370, 371, 415
HPN use 374, 435
psychological effects of HPN 378, 404, 413–15
adjustments 412, 413
quality of life
assessment 415, 443
home care patient 411–16
questionnaire, TPN at home 438–45
catheter findings 442
content 439, 440
cost reimbursemnt 443
data 440–2
introduction 439
quality of life 443, 445
response 437

radiation enteritis 373, 415, 435
HPN use 374, 401
record keeping 386, 387, 404, 405
HPN patient 396, 403
rehabilitation, HPN
elements of successful 446
social and occupational 413
renal failure, acute 369
amino acid formulations 114
amino acid pattern 30
catabolism 121
characteristics 113
essential amino acid role 119, 120
mortality 121
nitrogen balance 116
nutritional intake 115
nutritional support 113–21
patient characteristics 116
protein restriction 113
rats and essential amino acids 120
renal function recovery 115, 116, 118
serum protein and TPN 117
serum total amino acids 117, 118
renal failure, chronic 369
amino acid supplements 124, 127
low protein diet 123–8
renal mass and outcome 128
serum creatinine and low protein diet 125, 126
respiration
amino acid intake 455
fat emulsion effects 298, 299
glucose intake 288, 289
respiratory distress syndrome and fat in neonates 187
respiratory quotient
fat emulsion 298, 299
glucose 288, 289
resting energy expenditure (REE) 452
fever 304

trauma 326, 327
retinol binding protein 353
Rose formula 140

safflower oil
    composition 186
    concentration and bilirubin binding 196, 197
    use in neonates 181, 183, 184
salicylate and bilirubin binding 195, 196
Schutze Hardy rule, colloid stability 219
scleroderma of bowel 370, 371
sepsis 190, 210
    amino acid indicators 361, 362
    BCAA 28, 29, 32
    catheter-related 210, 211
sexual activity 414
short bowel syndrome 370, 434, 435
    breast milk 173
    HPN use 374, 401, 415
skin, preparation for catheter 391
social activity and HPN 413, 414
sodium, serum and TPN 424
soybean emulsion
    composition 186, 213
    concentration and bilirubin binding 196, 197
    use in neonates 181, 183, 184
stability see also emulsions, fat emulsion
    accelerated testing 228, 229
    amino acid : glucose additives 245-7
    fat emulsions 213-38
    freeze : thaw 228, 229, 232
    lipid containing TPN solution 208, 241-58
    measurements 225-9
    mixed systems 232-6
    physical and chemical TPN-fat 247
Staphylococcus aureus 377
Staphylococcus epidermidis 377
starvation, BCAA plasma levels 3
stay sutures 392
stereophotogrammetry 341-5
    camera positions 344
    small objects 344
    wound assessment 345
storage times, lipid-TPN mixtures 243, 244
    and additives 244, 245
stress 15 see also surgical stress, trauma, injury
    assessment of extent 67
    BCAA source 6
    BCAA utilization 7
    carnitine role 319
    insulin resistance 276, 277
    intensive care 70
    metabolic response 51-63, 65
    metabolic response data 63, 72-4

metabolism of fuel alteration 275-81
modulation of metabolic response 61, 63
muscle conversion, reversal 7
types, metabolic parameters 63
stress diabetes 275
study design
    catabolic index 68, 69
    crossover 70, 72
    duration 71
    efficacy of BCAA 67
    enriched solution use 65-74
subclavian vein
    approaches 394
    cannulization 392
    localization and catheter insertion 391
substrates
    availability xiv
    partitioning in circulation xiii-xxi
    physiological ranges 330
    regulation of utilization xiii-xvii
    sinks xx, xxi
    tissue partitioning xvii, xviii
    utilization, factors affecting xiv
supplementation 14-16 see also, nutrition, amino acids, fat, HPN, TPN
    acute renal failure 113-21
    burn patients 163-70
    carnitine 311-22
    conditions benefiting 66, 67
    effective composition 46
    formula and leucine kinetics 73
    hepatic insufficiency, solution 84-6
    injured rat 37-48
    input-output data 68
    liver failure 147-60
    nitrogen balance in stress and BCAA 70
    nutritional data 52, 53
    radiolabelled leucine 71-3
    solution and nitrogen balance in rats 38, 39
    study design and efficacy 65-74
    surgical stress 79, 80
    trauma 51-63
support programme, HPN 383-8
    access standardization 386
    enteral and parenteral 386
    facilities 384, 385
    financial 383
    financing and insurance 387
    objectives 385
    organizational structure 446
    out-patient monitoring 386
    patient selection 385, 386
    personnel recruitment 384, 385
    responsibilities 385
surgical stress 77-81
    BCAA use 77-81
    input-output 77, 80, 81

nitrogen balance 79
patient characteristics 77, 78
serum prealbumin 79
TPN efficacy 78
urinary BCAA and supplementation 81
surgery
  septic and BCAA 7
  thyroid hormone effects 458–60
Synthamin, *see* Intralipid
systemic sclerosis, HPN use 374

threonine, muscle in catabolic states 26, 27
tidal volume and glucose TPN 288, 289
topical nutrition 339–48
  cell nutrition 346, 347
  infection control 346
  vascularization 346, 347
total parenteral nutrition (TPN) 401, 402
  see also ambulatory, home
  fatty liver 262, 263
  liver malfunction 303–11
  metabolic effects 425–7
  multibottle 242
  objectives 290, 460
  remission in Crohn's disease 419
  response to surgery 423
  three litre bag 242
total parenteral nutrition-fat containing
  see also Intralipid, fat, mixed systems
  advantages 208
  clinical use 207–11
  complications 210, 211
  composition of stable 257
  container and compartments 241
  contraindications 210
  delivery 209
  electrolyte effects 243
  European use 242, 243
  formulations 209
  indications 209
  metabolic complications 210
  mixed systems 229, 230
  Montpelier system 242
  particle size distribution
    amino acid effects 243, 244
    and electrolytes 248
    and glucose 243
    and time 244
  physical and chemical changes 247
  safety 107
  stability 208, 241–58
  stability and additive concentration
    245–8
TPN *see* total parenteral nutrition
toxins and hepatic failure 10
trace element
  addition to TPN 245
  particle distribution and TPN-fat mix 248

trace metals 421
  daily requirements 378
  recommendations 404
training *see* patient training
transferrin in acute renal failure 117
transport, leucine-preferring (l-system) 29
trauma
  alanine output 55
  BCAA supplementation 51–63
  bowel and urinary output 54, 56
  clinical studies and data 52–4
  glucose–insulin interactions 276, 277
  glutamine excretion 55
  leucine output 57
  metabolic stress response 61
  nitrogen balance 59, 70
  resting energy expenditure 326, 327
trauma-septic state 6, 7
  cellular energy levels 32, 33
Travasorb hepatic
  amino acid normalization 158, 159
  formula comparison 155, 156
Travamulsion composition 219
triacylglycerol metabolism *xv*
triene/tetraene ratio, lipid emulsion correction
  180, 181
triglyceride, metabolism and mixed system in
  dogs 252, 253
triiodithyronine in surgical stress 458, 459
triiodothyronine/reverse triiodothyronine
  ratio 459, 460
trisegmentectomy *see* hepatic resection
Trive 1000 mixed system 230
tryptophan
  emulsion stability 247
  neurotransmitter production 11
tryptophan/BCAA ratio 11
tube feeding
  burns 167
  liver disease 159
tumours, gastrointestinal and TPN 262
tyrosine
  blood levels and liver failure 12
  deficiency 158
  emulsion stability 247
  excretion and BCAA in stress 57
  hepatic resection and BCAA 94, 95

ulcerative colitis, HPN 374
uraemia
  amino acid abnormalities 140
  BCAA antagonism 30
  insulin resistance 31
  intracellular amino acid pattern 31, 33
  muscle amino acids 26, 27, 33
  parathyroid hormone toxicity 137
urea generation in burn patients 166

urine, BCAA and surgical stress 81
urokinase lock 377, 396

valine
hyperalimentation in dialysis 140, 141
muscle fractional synthesis rate 44
muscle levels and catabolism 26, 27
muscle protein synthesis 45, 48
plasma level and brain isoleucine uptake 12, 13
plasma levels 12
plasma levels and hepatic resection 96
transport constants, blood brain barrier 12
trauma 26
uraemia 30, 33
venous access techniques 402
assessment of long term 397, 398
change, reasons for 397
long term 389–99
safety 389
vitamin
deficiencies in liver disease 151, 152
multiple preparations 404
vitamin D
metabolic bone disease 378, 404
vitamin E
deficiency in neonates 185
free-radical scavenger 186
oil emulsions 186

: PUFA rátios 186
volvulus of small bowel 402, 415

water, intracellular and amino acid distribution 27, 28
weight gain, intravenous hyperalimentation 132–5, 137–9
weight loss
BCAA and alanine prevention 43, 44
intravenous hyperalimentation 134
wound healing 339, 340
amino acid mechanisms 346
applications 339
burns 166
factors delaying 340
nutrition solution 339, 340
pressure sore 342
quantification 341
ulcer 343
wound hormone 453, 454

X-ray and catheter placement 393, 394

zeta potential 222–3
electrolyte effects 248–9
measurement 228
particles in mixed systems 232
zinc, role in wound healing 347
serum and HPN 425